ELEMENTS OF MATHEMATICS

NICOLAS BOURBAKI

ELEMENTS OF MATHEMATICS

Theory of Sets

Springer

Originally published as
Éléments de Mathématique,
Théorie des Ensembles
Paris, 1970

Mathematics Subject Classification (2000): 03-02, 03Exx

Library of Congress Control Number: 2004110815

Softcover edition of the 1st printing 1968

ISBN 3-540-22525-0 Springer Berlin Heidelberg New York

Springer is a part of Springer Science+Business Media
springeronline.com

Printed in Germany

Typesetting by the Translator
Cover design: design+production GmbH, Heidelberg
Printed on acid-free paper 41/3142/XT – 5 4 3 2 1 0

TO THE READER

1. This series of volumes, a list of which is given on pages VII and VIII, takes up mathematics at the beginning, and gives complete proofs. In principle, it requires no particular knowledge of mathematics on the reader's part, but only a certain familiarity with mathematical reasoning and a certain capacity for abstract thought. Nevertheless, it is directed especially to those who have a good knowledge of at least the content of the first year or two of a university mathematics course.

2. The method of exposition we have chosen is axiomatic and abstract, and normally proceeds from the general to the particular. This choice has been dictated by the main purpose of the treatise, which is to provide a solid foundation for the whole body of modern mathematics. For this it is indispensable to become familiar with a rather large number of very general ideas and principles. Moreover, the demands of proof impose a rigorously fixed order on the subject matter. It follows that the utility of certain considerations will not be immediately apparent to the reader unless he has already a fairly extended knowledge of mathematics; otherwise he must have the patience to suspend judgment until the occasion arises.

3. In order to mitigate this disadvantage we have frequently inserted examples in the text which refer to facts the reader may already know but which have not yet been discussed in the series. Such examples are always placed between two asterisks : * ... *. Most readers will undoubtedly find that these examples will help them to understand the text, and will prefer not to leave them out, even at a first reading. Their omission would of course have no disadvantage, from a purely logical point of view.

4. This series is divided into volumes (here called " Books "). The first six Books are numbered and, in general, every statement in the text

assumes as known only those results which have already been discussed in the preceding volumes. This rule holds good within each Book, but for convenience of exposition these Books are no longer arranged in a consecutive order. At the beginning of each of these Books (or of these chapters), the reader will find a precise indication of its logical relationship to the other Books and he will thus be able to satisfy himself of the absence of any vicious circle.

5. The logical framework of each chapter consists of the *definitions*, the *axioms*, and the *theorems* of the chapter. These are the parts that have mainly to be borne in mind for subsequent use. Less important results and those which can easily be deduced from the theorems are labelled as "propositions", "lemmas", "corollaries", "remarks", etc. Those which may be omitted at a first reading are printed in small type. A commentary on a particularly important theorem appears occasionally under the name of "scholium".

To avoid tedious repetitions it is sometimes convenient to introduce notations or abbreviations which are in force only within a certain chapter or a certain section of a chapter (for example, in a chapter which is concerned only with commutative rings, the word "ring" would always signify "commutative ring"). Such conventions are always explicitly mentioned, generally at the beginning of the chapter in which they occur.

6. Some passages in the text are designed to forewarn the reader against serious errors. These passages are signposted in the margin with the sign ☡ ("dangerous bend").

7. The Exercises are designed both to enable the reader to satisfy himself that he has digested the text and to bring to his notice results which have no place in the text but which are nonetheless of interest. The most difficult exercises bear the sign ¶.

8. In general, we have adhered to the commonly accepted terminology, except where there appeared to be good reasons for deviating from it.

9. We have made a particular effort always to use rigorously correct language, without sacrificing simplicity. As far as possible we have drawn attention in the text to *abuses of language*, without which any mathematical text runs the risk of pedantry, not to say unreadability.

10. Since in principle the text consists of the dogmatic exposition of a theory, it contains in general no references to the literature. Bibliographical references are gathered together in *Historical Notes*, usually at the end of each chapter. These notes also contain indications, where appropriate, of the unsolved problems of the theory.

The bibliography which follows each historical note contains in general only those books and original memoirs which have been of the greatest importance in the evolution of the theory under discussion. It makes no sort of pretence to completeness; in particular, references which serve only to determine questions of priority are almost always omitted.

As to the exercices, we have not thought it worthwhile in general to indicate their origins, since they have been taken from many different sources (original papers, textbooks, collections of exercises).

11. References to a part of this series are given as follows :

a) If reference is made to theorems, axioms, or definitions presented *in the same section*, they are quoted by their number.

b) If they occur *in another section of the same chapter*, this section is also quoted in the reference.

c) If they occur *in another chapter in the same Book*, the chapter and section are quoted.

d) If they occur *in another Book*, this Book is first quoted by its title.

The *Summaries of Results* are quoted by the letter R : thus *Set Theory*, R signifies "*Summary of Results of the Theory of Sets*".

CONTENTS OF THE ELEMENTS OF MATHEMATICS SERIES

I. Theory of sets

II. Algebra

CONTENTS OF THE ELEMENTS OF MATHEMATICS SERIES

III. General topology

1. Topological structures. 2. Uniform structures. 3. Topological groups. 4. Real numbers. 5. One-parameter groups. 6. Real number spaces, affine and projective spaces. 7. The additive groups $\mathbf{R}^n$. 8. Complex numbers. 9. Use of real numbers in general topology. 10. Function spaces.

IV. Functions of a real variable

1. Derivatives. 2. Primitives and integrals. 3. Elementary functions. 4. Differential equations. 5. Local study of functions. 6. Generalized Taylor expansions. The Euler-Maclaurin summation formula. 7. The gamma function. Dictonary.

V. Topological vector spaces

1. Topological vector spaces over a valued field. 2. Convex sets and locally convex spaces. 3. Spaces of continuous linear mappings. 4. Duality in topological vector spaces. 5. Hilbert spaces: elementary theroy. Dictionary.

VI. Integration

1. Convexity inequalities. 2. Riesz spaces. 3. Measures on locally compact spaces. 4. Extension of a measure. L^p spaces. 5. Integration of measures. 6. Vectorial integration. 7. Haar measure. 8. Convolution and representation.

Lie groups and Lie algebras

1. Lie algebras.

Commutative algebra

1. Flat modules. 2. Localization. 3. Graduations, filtrations and topologies. 4. Associated prime ideals and primary decomposition. 5. Integers. 6. Valuations. 7. Divisors.

Spectral theories

1. Normed algebras. 2. Locally compact groups.

CONTENTS

CONTENTS

INTRODUCTION

Ever since the time of the Greeks, mathematics has involved proof; and it is even doubted by some whether proof, in the precise and rigorous sense which the Greeks gave to this word, is to be found outside mathematics. We may fairly say that this sense has not changed, because what constituted a proof for Euclid is still a proof for us; and in times when the concept has been in danger of oblivion, and consequently mathematics itself has been threatened, it is to the Greeks that men have turned again for models of proof. But this venerable bequest has been enlarged during the past hundred years by important acquisitions.

By analysis of the mechanism of proofs in suitably chosen mathematical texts, it has been possible to discern the structure underlying both vocabulary and syntax. This analysis has led to the conclusion that a sufficiently explicit mathematical text could be expressed in a conventional language containing only a small number of fixed "words", assembled according to a syntax consisting of a small number of unbreakable rules : such a text is said to be *formalized*. The description of a game of chess, in the usual notation, and a table of logarithms, are examples of formalized texts. The formulae of ordinary algebraic calculation would be another example, if the rules governing the use of brackets were to be completely codified and strictly adhered to; in practice, some of these rules are never made explicit, and certain derogations of them are allowed.

The verification of a formalized text is a more or less mechanical process, the only possible causes of error being the length or intricacy of the text : this is why a mathematician will normally accept with confidence the result of an algebraic calculation performed by another, provided he knows that the calculation is not unduly long and has been done with care. On the other hand, in an unformalized text, one is exposed to the dangers of faulty reasoning arising from, for example, incorrect use of intuition or argument by analogy. In practice, the mathematician who wishes to satisfy himself of the perfect correctness or "rigour" of a proof or a theory hardly ever has recourse to one or another of the complete formalizations available nowadays, nor even usually to the incomplete and partial formalizations provided by algebraic and other calculi. In general he is content to bring the exposition to a point where his experience and mathematical flair tell him that translation into formal language would be no more than an exercise of patience (though doubtless a very tedious one). If, as happens again and again, doubts arise as to the correctness of the text under consideration, they concern ultimately the possibility of translating it unambiguously into such a formalized language: either because the same word has been used in different senses according to the context, or because the rules of syntax have been violated by the unconscious use of modes of argument which they do not specifically authorize, or again because a material error has been committed. Apart from this last possibility, the process of rectification, sooner or later, invariably consists in the construction of texts which come closer and closer to a formalized text until, in the general opinion of mathematicians, it would be superfluous to go any further in this direction. In other words, the correctness of a mathematical text is verified by comparing it, more or less explicitly, with the rules of a formalized language.

The *axiomatic method* is, strictly speaking, nothing but this art of drawing up texts whose formalization is straightforward in principle. As such it is not a new invention; but its systematic use as an instrument of discovery is one of the original features of contemporary mathematics. As far as reading or writing a formalized text is concerned, it matters little whether this or that meaning is attached to the words, or signs in the text, or indeed whether any meaning at all is attached to them; the only important point is the correct observance of the rules of syntax. Thus, as everyone knows, the same algebraic calculation can be used to solve problems about pounds weight or pounds sterling, about parabolas or motion under gravity. The same advantage attaches to every text written according to the axiomatic method, and for the same reasons: once the theorems of general topology have been established, they may be applied at will to ordinary space, Hilbert space, and many others. This faculty of being able to give different meanings to the words or prime concepts of a theory is indeed an important source of enrichment of the mathematician's intuition, which is not necessarily spatial or sensory, as is sometimes believed, but is far more a certain feeling for the behaviour of mathematical objects, aided often by images from very varied sources, but founded above all on everyday experience. Thus one is often led to study with profit those parts of a theory which traditionally have been neglected in this theory but studied systematically in a general axiomatic context, of which the given theory is a special case (for example, properties which have their historical origin in another specialization of the general theory). Moreover — and this is what concerns us most particularly in this series of volumes — the axiomatic method allows us, when we are concerned with complex mathematical objects, to separate their properties and regroup them around a small number of concepts : that is to say, using a word which will receive a precise definition later, to classify them according to the *structures* to which they belong. (Of course, the same structure can arise in connection with various different mathematical objects.) For example, some of the properties of the sphere are topological, others are algebraic, others again can be considered as belonging to differential geometry or the theory of Lie groups. Although this principle may occasionally become somewhat artificial when applied to closely interwoven structures, it is the basis of the division into Books of the subject-matter of this series.

¶ Just as the art of speaking a language correctly precedes the invention of grammar, so the axiomatic method had been practised long before the invention of formalized languages; but its conscious practice can rest only on the knowledge of the general principles governing such languages and their relationship with current mathematical texts. In this Book our first object is to describe such a language, together with an exposition of general principles which could be applied to many other similar languages; however, one of these languages will always be sufficient for our purposes. For whereas in the past it was thought that every branch of mathematics depended on its own particular intuitions which provided its concepts and prime truths, nowadays it is known to be possible, logically speaking, to derive practically the whole of known mathematics from a single source, the Theory of Sets. Thus it is sufficient for our purposes to describe the principles of a single formalized language, to indicate how the Theory of Sets could be written in this language, and then to show how the various branches of mathematics, to the extent that we are concerned with them in this series, fit into this framework. By so doing we do not claim to legislate for all time. It may happen at some future date that mathematicians will agree to use modes of reasoning which cannot be formalized in the language described here : it would then be necessary, if not to change the language completely, at least to enlarge its rules of syntax. But that is for the future to decide.

It goes without saying that the description of the formalized language is made in ordinary language, just as the rules of chess are. We do not propose to enter into a discussion of the psychological and metaphysical problems which underlie the use of ordinary language in such circumstances (for example, the possibility of recognizing that a letter of the alphabet is "the same" in two different places on the page, etc.). Moreover, it is scarcely possible to undertake such a description without making use of numeration. It is objected by some that the use of numbers in such a context is suspect, even tantamount to *petitio principii*. It is clear, however, that in fact we are using numbers merely as marks (and that we could for that matter replace them by other signs, such as colours or letters), and that we are not making use of any mathematical reasoning when we number the signs which occur in an explicitly written formula. We shall not enter into the question of teaching the principles of a formalized language to beings whose intellectual development has not reached the stage of being able to read, write, and count.

If formalized mathematics were as simple as the game of chess, then once our chosen formalized language had been described there would remain only the task of writing out our proofs in this language, just as the author of a chess manual writes down in his notation the games he proposes to teach, accompanied by commentaries as necessary. But the matter is far from being as simple as that, and no great experience is necessary to perceive that such a project is absolutely unrealizable : the tiniest proof at the beginning of the Theory of Sets would already require several hundreds of signs for its complete formalization. Hence, from Book I of this series onwards, it is imperative to condense the formalized text by the introduction of a fairly large number of new words (called *abbreviating symbols*) and additional rules of syntax (called *deductive criteria*). By doing this we obtain languages which are much more manageable than the formalized language in its strict sense. Any mathematician will agree that these condensed languages can be considered as merely shorthand transcriptions of the original formalized language. But we no longer have the certainty that the passage from one of these languages to another can be made in a purely mechanical fashion : for to achieve this certainty it would be necessary to complicate the rules of syntax which govern the use of the new rules to such a point that their usefulness became illusory; just as in algebraic calculation and in almost all forms of notation commonly used by mathematicians, a workable instrument is preferable to one which is theoretically more perfect but in practice far more cumbersome.

As the reader will see, the introduction of this condensed language is accompanied by "arguments" of a particular type, which belong to what is called *metamathematics*. This discipline ignores entirely any meaning which may originally have been attributed to the words or phrases of formalized mathematical texts, and considers these texts as particularly simple objects, namely as assemblies of previously given objects in which only the assigned order is of importance. Moreover, just as for example a chemistry textbook announces in advance the result of an experiment performed under given conditions, so metamathematical "arguments" usually assert that when a succession of operations has been performed on a text of a given type, then the final text will be of another given type. In the simplest cases, the assertions are indeed the purest truisms (comparable, for example, to the following: "if a bag of counters contains black counters and white counters, and if we replace all the black counters by white ones, then there will be only white counters in the bag"). But very soon we meet examples where the argument takes a typically mathematical turn, in which the use of arbitrary integers and mathematical induction predominates. Although we have countered the objection made earlier against the use of numeration in the description of a formalized language, it is no longer possible at this point to deny the danger of begging the question, because right from the beginning it seems that we are making use of all the resources of arithmetic, of which (among other things) we propose later to lay down the foundations. To this objection the reply is sometimes made that arguments of this type merely describe operations which can be actually carried out and verified, and for this reason carry conviction of another order from that which can be accorded to mathematics proper. But it seems simpler to say that we *could* avoid all these metamathematical arguments if the formalized text were explicitly written out : instead of using the "deductive criteria", we should recommence each time the sequences of operations which the purpose of the criteria is to abbreviate by predicting their result. But formalized mathematics cannot in practice be written down in full, and therefore we must have confidence in what might be called the common sense of the mathematician : a confidence analogous to that accorded by a calculator or an engineer to a formula or a numerical table without any awareness of the existence of Peano's axioms, and which ultimately is based on the knowledge that it has never been contradicted by facts.

We shall therefore very quickly abandon formalized mathematics, but not before we have carefully traced the path which leads back to it. The first "abuses of language" thus introduced will allow us to write the rest of this series (and in particular the Summary of Results of Book I) in the same way as all mathematical texts are written in practice, that is to say partly in ordinary language and partly in formulae which constitute partial, particular, and incomplete formalizations, the best-known examples of which are the formulae of algebraic calculation. Sometimes we shall use ordinary language more loosely, by voluntary abuses of language, by the pure and simple omission of passages which the reader can safely be assumed to be able to restore easily for himself, and by indications which cannot be translated into formalized language and which are designed to help the reader to reconstruct the complete text. Other passages, equally untranslatable into formalized language, are introduced in order to clarify the ideas involved, if necessary by appealing to the reader's intuition; this use of the resources of rhetoric is perfectly legitimate, provided only that the possibility of formalizing the text remains unaltered. The first examples of this will appear in Chapter III of this Book, which describes the theory of integers and cardinal numbers.

Thus, written in accordance with the axiomatic method and keeping always in view, as it were on the horizon, the possibility of a complete formalization, our series lays claim to perfect rigour : a claim which is not in the least contradicted by the preceding considerations, nor by the need to correct errors which slip into the text from time to time.

¶ We take the same realistic attitude towards the question of consistency (or non-contradiction). This question has been one of the main preoccupations of modern logicians, and is partly responsible for the creation of formalized languages (cf. Historical Note). A mathematical theory is said to be contradictory if a theorem and its negation have both been proved in the theory; from the usual rules of reasoning, which are the basis of the rules of syntax in formalized languages, it follows then that every theorem is both true and false in the theory, and the theory is consequently of no interest. If therefore we are involuntarily led to a contradiction, we cannot allow it to remain without rendering futile the theory in which it occurs.

Can we be certain that this will never happen? Without entering into a discussion — which is outside our competence — of the very notion of certainty, we may observe that metamathematics can set itself the task of examining the problems of consistency by its own methods. To say that a theory is contradictory means in effect that it contains a correct formalized proof which leads to the conclusion $0 \neq 0$. Now metamathematics can attempt, using methods of reasoning borrowed from mathematics, to investigate the structure of such a formalized text, in the hope of "proving" that such a text cannot exist. In fact such "proofs" have been given for certain partial formalized languages, which are less rich than that which we propose to introduce, but rich enough to express a good part of classical mathematics. We may of course reasonably ask what in fact has been "proved" in this way : for, if mathematics were contradictory, then some of its applications to material objects, and in particular to formalized texts, would stand in danger of being illusory. To escape this dilemma, the consistency of a formalized language would have to be "proved" by arguments which could be formalized in a language less rich and consequently more worthy of confidence; but a famous theorem of metamathematics, due to Gödel, asserts that this is impossible for a language of the type we shall describe, which is rich enough in axioms to allow the formulation of the results of classical arithmetic.

On the other hand, in the proofs of "relative" consistency (that is to say, those which establish the consistency of a theory on the supposition that some other theory, for example set theory, is non-contradictory), the metamathematical part of the argument (cf. Chapter I, § 2, no. 4) is so simple that it seems hardly possible to deny it without renouncing all rational use of our intellectual faculties. Since the various mathematical theories are now logically attached to the Theory of Sets, it follows that any contradiction encountered in one or another of these theories must give rise to a contradiction in the Theory of Sets itself. Of course, this is not an argument from which we can infer the consistency of the Theory of Sets. Nevertheless, during the half-century since the axioms of this theory were first precisely formulated, these axioms have been applied to draw conclusions in the most diverse branches of mathematics without leading to a contradiction, so that we have grounds for hope that no contradiction will ever arise.

If it should turn out otherwise, it would be that the observed contradiction was inherent in the fundamental principles of the Theory of Sets; these principles would therefore require to be modified, if possible without compromising those parts of mathematics we wish most to retain, and it is clear that the task of modification would be made easier by the use of the axiomatic method and formalized language, which allow us to formulate these principles more precisely and to separate out their consequences more clearly. Indeed, this is more or less what has happened in recent times, when the "paradoxes" of the Theory of Sets were eliminated by adopting a formalized language essentially equivalent to that which we shall describe here; and a similar revision would have to be undertaken if this language in its turn should prove to be contradictory.

To sum up, we believe that mathematics is destined to survive, and that the essential parts of this majestic edifice will never collapse as a result of the sudden appearance of a contradiction; but we cannot pretend that this opinion rests on anything more than experience. Some will say that this is small comfort; but already for two thousand five hundred years mathematicians have been correcting their errors to the consequent enrichment and not impoverishment of their science; and this gives them the right to face the future with serenity.

CHAPTER I

Description of Formal Mathematics

1. TERMS AND RELATIONS

1. SIGNS AND ASSEMBLIES

The *signs* of a mathematical theory $\mathcal{T}$ (*) are the following :

(1) The *logical signs* (†) : $\square$, τ, $\vee$, $\neg$.

(2) The *letters*.

By letters we mean upper and lower case Roman letters, with or without accents. Thus A, A′, A″, A‴, ... are letters. At any place in the text it is possible to introduce letters other than those which have appeared in previous arguments.

(3) The *specific signs*, which depend on the theory under consideration.

In the theory of sets we shall use only the following three specific signs : $=$, $\in$, $\supset$.

An *assembly* in $\mathcal{T}$ is a succession of signs of $\mathcal{T}$ written next to one another; certain signs, other than letters, may be joined in pairs by bars above the line, which are called *links*. *For example, in the Theory of Sets, in which $\in$ is a specific sign,

$$\tau \vee \neg \in \square A' \in \square A''$$

is an assembly.*

(*) The meaning of this expression will become clear as the chapter progresses.

(†) For the intuitive meanings of these signs, see no. 3, Remark.

The exclusive use of assemblies would lead to insuperable difficulties both for the printer and for the reader. For this reason current texts use abbreviating symbols (notably words of ordinary speech) which do not belong to formal mathematics. The introduction of such symbols is the object of *definitions*. Their use *is not indispensable to the theory*, and can often lead to confusion which only a certain familiarity with the subject will enable the reader to avoid.

Examples

(1) The assembly $\vee\neg$ is represented by $\Rightarrow$.
(2) The following symbols represent assemblies (and very long ones at that):

"3 and 4"

$\varnothing$

N

Z

"the real line"

"the Γ function"

$f \circ g$

$\pi = \sqrt{2} + \sqrt{3}$

$1 \in 2$

"Every finite division ring is a field"

"The zeros of $\zeta(s)$ other than $-2, -4, -6, \ldots$ lie on the line

$$\mathrm{R}(s) = 1/2".$$

In general, the symbol used to represent an assembly contains all the letters which appear in the assembly. Nevertheless, this principle can sometimes be infringed without risk of confusion. *For example, "the completion of X" represents an assembly which contains the letter X, but which also contains the letter which represents the set of entourages of the uniform structure of X. On the other hand,

$$\int_0^1 f(x)\, dx$$

represents an assembly in which the letter x (and the letter d) do not appear; and the assemblies represented by **N**, **Z**, "the Γ function" do not contain any letters.*

A *mathematical theory* (or simply a *theory*) contains rules which allow us to assert that certain assemblies of signs are *terms* or *relations* of the theory, and other rules which allow us to assert that certain assemblies are *theorems* of the theory.

The description of these rules, which will appear in this chapter, *does not belong* to formal mathematics; the rules involve assemblies which are more or less undetermined, for example undetermined letters. To simplify the exposition it is convenient to denote such assemblies by less cumbersome symbols. We shall use, especially, combinations of signs (of a mathematical theory), bold-face italic letters (with or without indices or accents), and particular symbols, of which some examples will be given. *Since our object is only to avoid circumlocutions* (cf. note (*), § 3, no. 1, p. 28) we shall not enunciate strict general rules for the use of these symbols; the reader will be able to reconstruct without trouble the assembly in question, in each particular case. By abuse of language we shall often say that the symbols *are* assemblies, rather than that they *denote* assemblies : expressions such as "the assembly $\boldsymbol{A}$" or "the letter $\boldsymbol{x}$", in the statements of the following rules, should therefore be replaced by "the assembly denoted by $\boldsymbol{A}$" or "the letter denoted by $\boldsymbol{x}$".

Let $\boldsymbol{A}$ and $\boldsymbol{B}$ be assemblies. We shall denote by $\boldsymbol{AB}$ the assembly obtained by writing the assembly $\boldsymbol{B}$ on the right of the assembly $\boldsymbol{A}$. We shall denote by $\vee \boldsymbol{A} \urcorner \boldsymbol{B}$ the assembly obtained by writing, from left to right, the sign $\vee$, the assembly $\boldsymbol{A}$, the sign $\urcorner$, the assembly $\boldsymbol{B}$. And so on.

Let $\boldsymbol{A}$ be an assembly and let $\boldsymbol{x}$ be a letter. We shall denote by $\tau_{\boldsymbol{x}}(\boldsymbol{A})$ the assembly constructed as follows : form the assembly $\tau\boldsymbol{A}$, link each occurrence of $\boldsymbol{x}$ in $\boldsymbol{A}$ to the τ written on the left of $\boldsymbol{A}$, and then replace $\boldsymbol{x}$ everywhere it occurs by the sign $\square$. The assembly denoted by $\tau_{\boldsymbol{x}}(\boldsymbol{A})$ therefore *does not contain* $\boldsymbol{x}$.

Example. The symbol $\tau_x(\in xy)$ represents the assembly

$$\tau \in \square y.$$

Let $\boldsymbol{A}$ and $\boldsymbol{B}$ be assemblies and let $\boldsymbol{x}$ be a letter. The assembly obtained by replacing $\boldsymbol{x}$, wherever it occurs in $\boldsymbol{A}$, by the assembly $\boldsymbol{B}$ is denoted by $(\boldsymbol{B}|\boldsymbol{x})\,\boldsymbol{A}$ (read : $\boldsymbol{B}$ replaces $\boldsymbol{x}$ in $\boldsymbol{A}$). If $\boldsymbol{x}$ does not appear in $\boldsymbol{A}$, then $(\boldsymbol{B}|\boldsymbol{x})\,\boldsymbol{A}$ is identical with $\boldsymbol{A}$; in particular,

$$(\boldsymbol{B}|\boldsymbol{x})\ \tau_{\boldsymbol{x}}(\boldsymbol{A})$$

is identical with $\tau_{\boldsymbol{x}}(\boldsymbol{A})$.

Example. If we replace x by $\square$ wherever x occurs in the assembly $\vee \in xy = xx$, we obtain the assembly $\vee \in \square y = \square\square$.

If $\boldsymbol{A}$ is an assembly and we are interested particularly in a letter $\boldsymbol{x}$, or two distinct letters $\boldsymbol{x}$ and $\boldsymbol{y}$ (which may or may not appear in $\boldsymbol{A}$), we shall often write $\boldsymbol{A}\{\boldsymbol{x}\}$ or $\boldsymbol{A}\{\boldsymbol{x}, \boldsymbol{y}\}$. In this case we write $\boldsymbol{A}\{\boldsymbol{B}\}$ instead of $(\boldsymbol{B}|\boldsymbol{x})\,\boldsymbol{A}$. We denote by $\boldsymbol{A}\{\boldsymbol{B}, \boldsymbol{C}\}$ the assembly obtained by *simultaneously* replacing $\boldsymbol{x}$ by $\boldsymbol{B}$ and $\boldsymbol{y}$ by $\boldsymbol{C}$ wherever they occur in $\boldsymbol{A}$ (note that $\boldsymbol{x}$ and $\boldsymbol{y}$ may appear in $\boldsymbol{B}$ and in $\boldsymbol{C}$); if $\boldsymbol{x}'$ and $\boldsymbol{y}'$ are distinct letters, other than $\boldsymbol{x}$ and $\boldsymbol{y}$, which appear in neither $\boldsymbol{A}$, $\boldsymbol{B}$, nor $\boldsymbol{C}$, then $\boldsymbol{A}\{\boldsymbol{B}, \boldsymbol{C}\}$ is the same as $(\boldsymbol{B}|\boldsymbol{x}')\,(\boldsymbol{C}|\boldsymbol{y}')\,(\boldsymbol{x}'|\boldsymbol{x})\,(\boldsymbol{y}'|\boldsymbol{y})\,\boldsymbol{A}$.

Remark. When an abbreviating symbol Σ is introduced, by means of a definition, to represent a certain assembly, the (usually tacit) convention is made of representing the assembly obtained by substituting an assembly B for a letter x in the original assembly, by the symbol obtained by the replacing the letter x in Σ by the assembly B (or, more often, by an abbreviating symbol representing the assembly B).

¶ *For example, having defined what assembly is represented by the symbol $E \otimes F$, where E and F are letters — an assembly which, incidentally, contains other letters besides E and F — the symbol $Z \otimes F$ can be used without further explanation.*

¶ This rule can lead to confusions, which are avoided by the use of various typographical devices; the most common consists in replacing x by (B) in place of B.

¶ *For example, $M \cap N$ denotes an assembly containing the letter N. If we substitute for N the assembly represented by $P \cup Q$, we get an assembly denoted by $M \cap (P \cup Q)$.*

2. CRITERIA OF SUBSTITUTION

Formal mathematics contains only explicitly written assemblies. Nevertheless, even with the use of abbreviating symbols, the development of mathematics strictly in accordance with this principle would lead to extremely long chains of reasoning. For this reason we shall establish *criteria* relating to indeterminate assemblies; each of these criteria will describe once for all the final result of a definite sequence of manipulations on these assemblies. These criteria are therefore not indispensable to the theory; their justification belongs to *metamathematics*.

The development of metamathematics itself requires, in practice, the use of abbreviating symbols, some of which have already been indicated. Most of these symbols are also used in mathematics.

We shall make use of the following criteria, called the *criteria of substitution*:

CS1. *Let A and B be assemblies and let x and x' be letters. If x' does not appear in A, then $(B|x)\,A$ is identical with $(B|x')\,(x'|x)\,A$.*

CS2. *Let A, B, and C be assemblies and let x and y be distinct letters* (*). *If y does not appear in B, then $(B|x)\,(C|y)\,A$ is identical with*

$$(C'|y)\,(B|x)\,A,$$

where *C' is the assembly $(B|x)\,C$.*

(*) In accordance with what was said in no. 1, the phrase " x and y are distinct letters " is an abuse of language : it means that x and y *denote* distinct letters in the assemblies under consideration.

CS3. *Let* A *be an assembly and let* x *and* x' *be letters. If* x' *does not appear in* A, *then* $\tau_x(A)$ *is identical with* $\tau_{x'}(A')$, *where* A' *is the assembly* $(x'|x)\,A$.

CS4. *Let* A *and* B *be assemblies and let* x *and* y *be distinct letters. If* x *does not appear in* B, *then* $(B|y)\tau_x A$ *is identical with* $\tau_x(A')$, *where* A' *is the assembly* $(B|y)\,A$.

CS5. *Let* A, B, C *be assemblies and let* x *be a letter. The assemblies* $(C|x)\,(\neg A)$, $(C|x)\,(\vee AB)$, $(C|x)\,(\Longrightarrow AB)$, $(C|x)\,(sAB)$ *(where* s *is a specific sign) are respectively identical with* $\neg A'$, $\vee A'B'$, $\Longrightarrow A'B'$, $sA'B'$, *where* A', B' *are respectively* $(C|x)\,A$, $(C|x)\,B$.

As an example, let us indicate the principle of the verification of CS2. Compare the operation which takes us from A to $(B|x)\,(C|y)\,A$ with the operation which takes us from A to $(C'|y)\,(B|x)\,A$. In each operation, no sign which appears in A and is distinct from x and y is altered. At every place where x appears in A, we have to substitute B for x in the first and in the second operation; this is clear in regard to the first operation, and in regard to the second it follows from the fact that y does not appear in B. Finally, at every place where y appears in A, the first operation consists in replacing C for y, then B for x at every place where x appears in C; but it is clear that this comes to the same thing as substituting for y, wherever it appears in A, the assembly $(B|x)\,C$.

3. FORMATIVE CONSTRUCTIONS

Some of the specific signs of a theory are called *relational*, and the others are called *substantific*. With every specific sign is associated a natural number called its *weight* (which is practically always the number 2).

¶ An assembly is said to be of the *first species* if it begins with a τ, or with a substantific sign, or if it consists of a single letter; otherwise it is of the *second species*.

¶ A *formative construction* in a theory $\mathcal{T}$ is a sequence of assemblies which has the following property: for each assembly A of the sequence, one of the following conditions is satisfied:

(a) A is a letter.

(b) There is in the sequence an assembly B of the second species, preceding A, such that A is $\neg B$.

(c) There are two assemblies B and C of the second species (distinct or not), preceding A, such that A is $\vee BC$.

(d) There is an assembly B of the second species, preceding A, and a letter x such that A is $\tau_x(B)$.

(e) There is a specific sign s of weight n (*) in $\mathcal{T}$, and n assemblies $A_1, A_2, \ldots, A_n$ of the first species, preceding A, such that A is $sA_1A_2\ldots A_n$.

¶ The assemblies of the first species (resp. of the second species) *which appear in the formative constructions of* $\mathcal{T}$ are called *terms* (resp. *relations*) in $\mathcal{T}$.

Example. *In the theory of sets, in which $\in$ is a relational sign of weight 2, the following sequence of assemblies is a formative construction :

$$A$$
$$A'$$
$$A''$$
$$\in AA'$$
$$\in AA''$$
$$\neg \in AA'$$
$$\vee \neg \in AA' \in AA''$$

Hence the assembly given as an example in no. 1 is a term in the theory of sets.*

Remark. Intuitively, terms are assemblies which represent *objects*, and relations are assemblies which represent *assertions* which can be made about these objects. Condition (a) means that the letters represent objects. Condition (b) means that if B is an assertion, then $\neg B$, called the *negation* of B, is an assertion (which is read : not B). Condition (c) means that if B and C are assertions, $\vee BC$, which is called the *disjunction* of B and C, is an assertion (which is read : either B or C); thus $\Longrightarrow BC$ is an assertion (in words : "either not B, or C", or "B implies C"). Condition (d) means that if B is an assertion and x a letter, then $\tau_x(B)$ is an object. Let us consider the assertion B as expressing a property of the object x; then, if there exists an object which has the property in question, $\tau_x(B)$ represents a distinguished object which has this property; if not, $\tau_x(B)$ represents an object about which nothing can be said. Finally, condition (e) means that if $A_1, A_2, \ldots, A_n$ are objects, and if s is a relational (resp. substantific) sign of weight n, then $sA_1A_2\ldots A_n$ is an assertion about the objects $A_1, \ldots, A_n$ (resp. an object depending on $A_1, \ldots, A_n$).

Examples. The symbols $\varnothing$, N, "the real line", "the Γ function", $f \circ g$ represent terms. The symbols $\pi = \sqrt{2} + \sqrt{3}$, $1 \in 2$, "every finite

---

(*) As was said above, it would be possible, for the development of present-day mathematical theories, to limit our consideration to specific signs of weight 2, and consequently to avoid using the expression "natural number n" in the definition of a formative construction.

division ring is a field", "the zeros of $\zeta(s)$ other than $-2, -4, -6, \ldots$ lie on the line $\mathcal{R}(s) = 1/2$" represent relations. The symbol "3 and 4" represents neither a term nor a relation.

The initial sign of a relation is $\vee$, $\neg$, or a relational sign. The initial sign of a term is either τ or a substantific sign, provided that the term does not consist of a single letter. The latter assertion follows from the fact that a term is an assembly of the first species. If A is a relation, then A features in a formative construction, is not a letter, and does not begin with τ, so that three cases are possible: (1) A is preceded by an assembly B such that A is $\neg B$; (2) A is preceded by two assemblies B and C such that A is $\vee BC$; (3) A is preceded by assemblies A_1, A_2, ..., A_n such that A is $sA_1A_2 \ldots A_n$, s being a relational sign.

4. FORMATIVE CRITERIA

CF1. *If A and B are relations in a theory $\mathcal{T}$, then $\vee AB$ is a relction in $\mathcal{T}$.*

Consider two formative constructions (in $\mathcal{T}$), one of which contains A and the other B. Consider the sequence of assemblies obtained by writing first the assemblies of the first construction, then the assemblies of the second construction, and finally $\vee AB$. Since A and B are of the second species, it is immediately verified that this sequence is a formative construction of $\mathcal{T}$. The assembly $\vee AB$ is of the second species, hence it is a relation in $\mathcal{T}$.

The three following criteria are established similarly:

CF2. *If A is a relation in a theory $\mathcal{T}$, then $\neg A$ is a relation in $\mathcal{T}$.*

CF3. *If A is a relation in a theory $\mathcal{T}$, and if x is a letter, then $\tau_x(A)$ is a term in $\mathcal{T}$.*

CF4. *If $A_1, A_2, \ldots, A_n$ are terms in a theory $\mathcal{T}$, and if s is a relational* (resp. *substantific*) *sign of weight n in $\mathcal{T}$, then $sA_1A_2\ldots A_n$ is a relation* (resp. *a term*) *in $\mathcal{T}$.*

These criteria immediately imply the following:

CF5. *If A and B are relations in a theory $\mathcal{T}$, then $\Rightarrow AB$ is a relation in $\mathcal{T}$.*

CF6. *Let $A_1, A_2, \ldots, A_n$ be a formative construction in a theory $\mathcal{T}$, and let x and y be letters. Suppose that y does not appear in any A_i. Then $(y|x)\, A_1$, $(y|x)\, A_2$, ..., $(y|x)\, A_n$ is a formative construction in $\mathcal{T}$.*

To prove CF6, let A'_i be the assembly $(y|x)\,A_i$. If A_i is a letter, then A'_i is a letter. If A_i is of the form $\neg A_j$, where A_j is an assembly of the second species which precedes A_i in the construction, then A'_i is identical with $\neg A'_j$ by CS5, and A'_j is an assembly of the second species. The reasoning is similar if A_i is of the form $\vee A_j A_k$ or $sA_{j_1}A_{j_2}\ldots A_{j_m}$, s being a specific sign of $\mathcal{T}$. If finally A_i is of the form $\tau_z(A_j)$, where A_j is an assembly of the second species which precedes A_i in the construction, there are various cases to consider:

(a) z is a letter distinct from x and y. Then A'_i is identical with $\tau_z(A'_j)$ by CS4, and A'_j is an assembly of the second species.

(b) z is identical with x. Then A_i does not contain x, hence A'_i is identical with A_i, that is to say with $\tau_x(A_j)$; since y does not appear in A_j, $\tau_x(A_j)$ is identical with $\tau_y(A_j)$ by CS3.

(c) z is identical with y. Then A_i is the assembly τA_j, because y does not appear in A_j; therefore A'_i is the assembly $\tau A'_j$, that is $\tau_u(A'_j)$, where u is a letter which does not appear in A'_j.

CF7. *Let A be a relation* (resp. *a term*) *in a theory $\mathcal{T}$, and let x and y be letters. Then $(y|x)\,A$ is a relation* (resp. *a term*) *in $\mathcal{T}$.*

Let $A_1, A_2, \ldots, A_n$ be a formative construction in which A appears. We shall show step by step that, if A_t is a relation (resp. a term) then $(y|x)A_t$, which we shall denote by A'_t, is also a relation (resp. a term). Suppose that this point has been established for $A_1, A_2, \ldots, A_{i-1}$; let us prove it for A_i. If A_i is a letter, then A'_i is a letter. If A_i is preceded in the construction by a relation A_j such that A_i is $\neg A_j$, then A'_i is identical with $\neg A'_j$ by CS5, and $\neg A'_j$ is a relation by CF2. The argument is similar if A_i is preceded by relations A_j, A_k such that A_i is $\vee A_j A_k$, or if A_i is preceded by terms $A_{j_1}, \ldots, A_{j_m}$ such that A_i is $sA_{j_1}\ldots A_{j_m}$, where s is a specific sign of $\mathcal{T}$ of weight m. Finally, if A_i is preceded by a relation A_j such that A_i is $\tau_z(A_j)$, there are various cases to consider:

(a) z is distinct from both x and y. Then A'_i is identical with $\tau_z(A'_j)$ by CS4, and we know already that A'_j is a relation; hence A_i is a term, by CF3.

(b) z is identical with x. Then A_i does not contain x, therefore A'_i is identical with A_i, and consequently is a term.

(c) z is identical with y. Then let u be a letter, distinct from both x and y, which does not appear in $A_1, A_2, \ldots, A_j$. By CF6, the sequence of assemblies $(u|y)\,A_1, \ldots, (u|y)\,A_j$, which we shall denote by $A''_1, \ldots, A''_j$, constitutes a formative construction in $\mathcal{T}$. Since y no longer appears in this new construction, $(y|x)\,A''_1, \ldots, (y|x)\,A''_j$ is a formative construction by CF6, so that $(y|x)\,A''_j$ is a relation in $\mathcal{T}$; consequently $\tau_u((y|x)A''_j)$

is a term of $\mathcal{T}$. But this term is identical with $(y|x\tau)_u(A''_j)$ by CS4, hence with $(y|x)\tau_y(A_j)$ by CS3, hence is identical with A_i.

CF8. *Let* A *be a relation* (resp. *a term*) *in a theory* $\mathcal{T}$, *let* x *be a letter and* T *a term in* $\mathcal{T}$. *Then* $(T|x)\,A$ *is a relation* (resp. *a term*) *in* $\mathcal{T}$.
Let $A_1, A_2, \ldots, A_n$ be a formative construction in which A appears. Let $x_1, x_2, \ldots, x_p$ be the distinct letters which appear in T. Let us associate with each letter x_i a letter x'_i, distinct from each of the letters $x_1, \ldots, x_p$ and the letters which appear in $A_1, \ldots, A_n$, in such a way that the letters $x'_1, \ldots, x'_p$ are all distinct. The assembly

$$(x'_1|x_1)(x'_2|x_2)\ldots(x'_p|x_p)T$$

is a term T' by CF7, and $(T|x)A$ is identical with

$$(x_1|x'_1)(x_2|x'_2)\ldots(x_p|x'_p)(T'|x)A$$

by application of CS1. It is therefore enough to show that $(T'|x)\,A$ is a relation (resp. a term); in other words, we may suppose from now on that the letters which appear in T do not appear in $A_1, \ldots, A_n$.
¶ We shall show step by step that, if A_t is a relation (resp. a term), then $(T|x)\,A_t$, which we shall denote by A'_t, is a relation (resp. a term). Suppose this point has been established for $A_1, A_2, \ldots, A_{i-1}$, and let us prove it for A_i. If A_i is a letter, then A'_i is either the same letter or T, and therefore a term. If A_i is of the form $\urcorner A_j$, where A_j is a relation which precedes A_i in construction, then A_i is identical with $\urcorner A'_j$ by CS5, and we know already that A'_j is a relation, hence A'_i is a relation by CF2. The proof is analogous if A_i is of the form $\vee A_jA_k$, or $sA_{j_1}\ldots A_{j_m}$. Finally, if A_i is of the form $\tau_z(A_j)$, where A_j is a relation which precedes A_i in the construction, there are various cases to be considered:

(a) z is distinct from x and from the letters which appear in T. Then A'_i is identical with $\tau_z(A'_j)$ by CS4, and we know already that A'_j is a relation; hence A'_i is a term by CF3.

(b) z is identical with x. Then A_i does not contain x, hence A'_i is identical with A_i and consequently is a term.

(c) z appears in T. Then z does not appear in A_j, so that $A^!$ is identical with τA_j; hence A'_i is identical with $\tau A'_j$. Now, we know already that A'_j is a relation, and $\tau A'_j$ is identical with $\tau_u(A'_j)$, where u is a letter which does not appear in A'_j; it follows by CF3 that A'_i is a term.

Intuitively, if A is a relation in $\mathcal{T}$, which we may regard as expressing a property of an object x, the assertion $(B|x)\,A$ amounts to saying that the object B has this property. If A is a term in $\mathcal{T}$, it represents an object which depends in some way on the object denoted by x; the term $(B|x)\,A$ represents what the object A becomes when we take x to be the object B.

2. THEOREMS

From now on, if $\boldsymbol{A}$ *is a relation, we shall write* not($\boldsymbol{A}$) *instead of* $\urcorner \boldsymbol{A}$. *If* $\boldsymbol{A}$ *and* $\boldsymbol{B}$ *are relations, we shall write* "($\boldsymbol{A}$) or ($\boldsymbol{B}$)" *instead of* $\vee \boldsymbol{AB}$, *and* $(\boldsymbol{A}) \Longrightarrow (\boldsymbol{B})$ *instead of* $\Longrightarrow \boldsymbol{AB}$. *Sometimes we shall leave out the brackets. In each case the reader will be able to determine without difficulty the assembly under consideration.*

1. AXIOMS

We have already seen that the specific signs determine the terms and the relations in a theory $\mathcal{T}$. To construct $\mathcal{T}$, we proceed as follows :

(1) First we write down a certain number of relations in $\mathcal{T}$; these are called the *explicit axioms* of $\mathcal{T}$. The letters which appear in the explicit axioms are called the *constants* of $\mathcal{T}$.

(2) We lay down one or more rules (*), called the *schemes* of $\mathcal{T}$, which must have the following properties : (a) the application of such a rule $\mathcal{R}$ furnishes a relation in $\mathcal{T}$; (b) if $\boldsymbol{T}$ is a term in $\mathcal{T}$, if $\boldsymbol{x}$ is a letter, and if $\boldsymbol{R}$ is a relation in $\mathcal{T}$ constructed by applying the scheme $\mathcal{R}$, then the relation $(\boldsymbol{T}|\boldsymbol{x})\boldsymbol{R}$ can also be constructed by applying $\mathcal{R}$.

In all the cases we envisage, the verification of these conditions is always easy.

Every relation contructed by applying a scheme of $\mathcal{T}$ is called an *implicit axiom* of $\mathcal{T}$.

Intuitively, the axioms represent either self-evident assertions or else hypotheses from which one wishes to draw consequences. The constants represent well-defined objects for which the properties expressed by the explicit axioms are supposed to be true. On the other hand, if the letter $\boldsymbol{x}$ is not a constant, it represents a completely undetermined object; if a property of the object $\boldsymbol{x}$ is assumed to be true by means of an axiom, then this axiom is necessarily implicit, so that the property remains true for any object $\boldsymbol{T}$.

(*) For the sake of brevity, these rules are expressed by using the symbols mentioned in § 1, no. 1 (and especially bold-faced italic letters); but it would be easy to avoid the use of these symbols completely in the formulation of the rules (see § 3, no. 1, note (*) on p. 28).

2. PROOFS

A *demonstrative text* in a theory $\mathcal{T}$ comprises:

(1) An auxiliary formative construction of relations and terms in $\mathcal{T}$.

(2) A *proof in* $\mathcal{T}$, that is to say a sequence of relations in $\mathcal{T}$ which appear in the auxiliary formative construction, such that for every relation $\boldsymbol{R}$ in the sequence at least one of the following conditions is satisfied:

(a_1) $\boldsymbol{R}$ is an explicit axiom of $\mathcal{T}$.

(a_2) $\boldsymbol{R}$ results from the application of a scheme of $\mathcal{T}$ to terms or relations which appear in the auxiliary formative construction.

(b) there are two relations $\boldsymbol{S}$, $\boldsymbol{T}$ in the sequence which precede $\boldsymbol{R}$, such that $\boldsymbol{T}$ is $\boldsymbol{S} \Longrightarrow \boldsymbol{R}$.

¶ A *theorem* in $\mathcal{T}$ is a relation *which appears in a proof in* $\mathcal{T}$.

This notion is therefore essentially dependent on the state of the theory under consideration, at the time when it is being described. A relation in a theory $\mathcal{T}$ *becomes* a theorem in $\mathcal{T}$ when one succeeds in inserting it into a proof in $\mathcal{T}$. To say that a relation in $\mathcal{T}$ "is not a theorem in $\mathcal{T}$" cannot have any meaning without reference to the stage of development of the theory $\mathcal{T}$.

A theorem in $\mathcal{T}$ is also called a "*true* relation in $\mathcal{T}$" (or "proposition", "lemma", "corollary", etc.). Let $\boldsymbol{R}$ be a relation in $\mathcal{T}$, let $\boldsymbol{x}$ be a letter and $\boldsymbol{T}$ a term in $\mathcal{T}$; if $(\boldsymbol{T}|\boldsymbol{x})\boldsymbol{R}$ is a theorem in $\mathcal{T}$, $\boldsymbol{T}$ is said to *satisfy the relation* $\boldsymbol{R}$ in $\mathcal{T}$ (or to be a *solution of* $\boldsymbol{R}$), when $\boldsymbol{R}$ is considered as a relation in $\boldsymbol{x}$.

¶ A relation is said to be *false* in $\mathcal{T}$ if its negation is a theorem in $\mathcal{T}$. A theory $\mathcal{T}$ is said to be *contradictory* when one has written a relation which is both true and false in $\mathcal{T}$.

Here again, we are dealing with a notion that depends on the particular state of development of a theory. The reader should beware of the confusion (unfortunately suggested by the intuitive meaning of the word "false") which consists in believing that, once one has proved that a relation $\boldsymbol{R}$ is false in $\mathcal{T}$, one has thereby established that $\boldsymbol{R}$ "is not true" in $\mathcal{T}$ (strictly speaking, this phrase has no precise meaning *in mathematics*, as we have remarked above).

¶ In what follows we shall give metamathematical criteria, called *deductive criteria*, which will allow us to shorten proofs. These criteria will be denoted by the letter C followed by a number.

C1 (*Syllogism*). *Let* $\boldsymbol{A}$ *and* $\boldsymbol{B}$ *be relations in a theory* $\mathcal{T}$. *If* $\boldsymbol{A}$ *and* $\boldsymbol{A} \Longrightarrow \boldsymbol{B}$ *are theorems in* $\mathcal{T}$, *then* $\boldsymbol{B}$ *is a theorem in* $\mathcal{T}$.

Let $R_1, R_2, \ldots, R_n$ be a proof in $\mathcal{T}$ in which A appears, and let $S_1, S_2, \ldots, S_p$ be a proof in $\mathcal{T}$ in which $A \Longrightarrow B$ appears. Clearly $R_1, R_2, \ldots, R_n, S_1, S_2, \ldots, S_p$ is a proof in $\mathcal{T}$ in which both A and $A \Longrightarrow B$ appear. Hence

$$R_1, R_2, \ldots, R_n, S_1, S_2, \ldots, S_p, B$$

is a proof in $\mathcal{T}$, and therefore B is a theorem in $\mathcal{T}$.

3. SUBSTITUTIONS IN A THEORY

Let $\mathcal{T}$ be a theory, let $A_1, A_2, \ldots, A_n$ be its explicit axioms, let x be a letter and T a term of $\mathcal{T}$. Let $(T|x)\mathcal{T}$ be the theory whose signs and schemes are the same as those of $\mathcal{T}$, and whose explicit axioms are $(T|x)A_1, (T|x)A_2, \ldots, (T|x)A_n$.

C2. *Let A be a theorem in a theory $\mathcal{T}$, let T be a term of $\mathcal{T}$, and let x be a letter. Then $(T|x)A$ is a theorem in the theory $(T|x)\mathcal{T}$.*

Let $R_1, R_2, \ldots, R_n$ be a proof in $\mathcal{T}$ in which A appears. Consider the sequence $(T|x)R_1, (T|x)R_2, \ldots, (T|x)R_n$, which is a sequence of relations in $\mathcal{T}$ by reason of CF8 (§ 1, no. 4). We shall show that this sequence is a proof in the theory $(T|x)\mathcal{T}$; this will establish the criterion. If R_k is an implicit axiom of $\mathcal{T}$, then $(T|x)R_k$ is again an implicit axiom of $\mathcal{T}$ (no. 1) and therefore of $(T/x)\mathcal{T}$. If R_k is an explicit axiom of $\mathcal{T}$, then $(T|x)R_k$ is an explicit axiom of $(T|x)\mathcal{T}$. Finally, if R_k is preceded by relations R_i and R_j, where R_j is $R_i \Longrightarrow R_k$, then $(T|x)R_k$ is preceded by $(T|x)R_i$ and $(T|x)R_j$, and the latter is identical with $(T|x)R_i \Longrightarrow (T|x)R_k$ (criterion CS5).

C3. *Let A be a theorem of a theory $\mathcal{T}$, let T be a term of $\mathcal{T}$, and let x be a letter which is not a constant of $\mathcal{T}$. Then $(T|x)A$ is a theorem of $\mathcal{T}$.*

This follows immediately from C2, because x does not feature in the explicit axioms of $\mathcal{T}$. More particularly, if $\mathcal{T}$ contains no explicit axioms, or if the explicit axioms contain no letters, then the criterion C3 applies without restriction on the letter x.

4. COMPARISON OF THEORIES

A theory $\mathcal{T}'$ is said to be *stronger* than a theory $\mathcal{T}$ if all the signs of $\mathcal{T}$ are signs of $\mathcal{T}'$, all the explicit axioms of $\mathcal{T}$ are theorems in $\mathcal{T}'$, and the schemes of $\mathcal{T}$ are schemes of $\mathcal{T}'$.

C4. *If a theory $\mathcal{T}'$ is stronger than a theory $\mathcal{T}$, then all the theorems of $\mathcal{T}$ are theorems of $\mathcal{T}'$.*

Let $R_1, R_2, \ldots, R_n$ be a proof in $\mathcal{T}$. We shall show, step by step, that each R_i is a theorem in $\mathcal{T}'$. Suppose that this is true for the relations preceding R_k. If R_k is an axiom of $\mathcal{T}$, it is a theorem in $\mathcal{T}'$ by hypothesis. If R_k is preceded by relations R_i and $R_i \Longrightarrow R_k$, we know already that R_i and $R_i \Longrightarrow R_k$ are theorems in $\mathcal{T}'$, and therefore R_k is a theorem in $\mathcal{T}'$ by virtue of C1. Hence, in every case, R_k is a theorem in $\mathcal{T}'$, and the proof is complete.

If each of two theories $\mathcal{T}$ and $\mathcal{T}'$ is stronger than the other, $\mathcal{T}$ and $\mathcal{T}'$ are said to be *equivalent*. Then every theorem of $\mathcal{T}$ is a theorem of $\mathcal{T}'$, and vice versa.

C5. *Let $\mathcal{T}$ be a theory, let $A_1, A_2, \ldots, A_n$ be its explicit axioms, $a_1, a_2, \ldots, a_h$ its constants, and let $T_1, T_2, \ldots, T_h$ be terms in $\mathcal{T}$. Suppose that*

$$(T_1|a_1)\,(T_2|a_2)\ldots(T_h|a_h)A_i \qquad (\text{for } i = 1, 2, \ldots, n)$$

are theorems in a theory $\mathcal{T}'$, that the signs of $\mathcal{T}$ are signs of $\mathcal{T}'$, and that the schemes of $\mathcal{T}$ are schemes of $\mathcal{T}'$. Then, if A is a theorem in $\mathcal{T}$,

$$(T_1|a_1)\ldots(T_h|a_h)A$$

is a theorem in $\mathcal{T}'$.

For $\mathcal{T}'$ is stronger than the theory $(T_1|a_1)\ldots(T_n|a_n)\,\mathcal{T}$, and we can apply C2 and C4.

When we use this procedure to deduce a theorem in $\mathcal{T}'$ from a theorem in $\mathcal{T}$, we say that *we are applying in $\mathcal{T}'$ the results of $\mathcal{T}$*. Intuitively, the axioms of $\mathcal{T}$ express properties of $a_1, a_2, \ldots, a_h$, and A expresses a property which is a consequence of these axioms. If the objects $T_1, T_2, \ldots, T_h$ in $\mathcal{T}'$ have the properties expressed by the axioms of $\mathcal{T}$, then they also have the property A.

*For example, in the theory of groups $\mathcal{T}$, the explicit axioms contain two constants G and μ (the group and the law of composition). In the theory of sets $\mathcal{T}'$, we define two terms : the real line and addition of real numbers. If we substitute these terms for G and μ respectively in the explicit axioms of $\mathcal{T}$, we obtain theorems in $\mathcal{T}'$. Moreover, the schemes and signs of $\mathcal{T}$ and $\mathcal{T}'$ are the same. We may therefore "apply the results of group theory to the additive group of real numbers". We say that we have constructed a *model* for group theory in the theory of sets. (Note that since the theory of groups is stronger than the theory of sets, we can also apply the results of the theory of sets to the theory of groups.)*

Remark. Under the hypotheses of C5, if the theory $\mathfrak{T}$ turns out to be contradictory, the same will be true of $\mathfrak{T}'$. For if A and "not A" are theorems in $\mathfrak{T}$, then $(T_1|a_1)\dots(T_h|a_h)A$ and $\text{not}(T_1|a_1)\dots(T_h|a_h)A$ are theorems in $\mathfrak{T}'$. * For example, if the theory of groups were contradictory, the theory of sets would also be contradictory. *

3. LOGICAL THEORIES

1. THE AXIOMS

A *logical theory* is any theory $\mathfrak{T}$ in which the schemes S1 to S4 below provide implicit axioms.

S1. *If* A *is a relation in* $\mathfrak{T}$, *the relation* $(A \text{ or } A) \Longrightarrow A$ *is an axiom of* $\mathfrak{T}$ (*).

S2. *If* A *and* B *are relations in* $\mathfrak{T}$, *the relation* $A \Longrightarrow (A \text{ or } B)$ *is an axiom of* $\mathfrak{T}$.

S3. *If* A *and* B *are relations in* $\mathfrak{T}$, *the relation* $(A \text{ or } B) \Longrightarrow (B \text{ or } A)$ *is an axiom of* $\mathfrak{T}$.

S4. *If* A, B, *and* C *are relations in* $\mathfrak{T}$, *the relation*

$$(A \Longrightarrow B) \Longrightarrow ((C \text{ or } A) \Longrightarrow (C \text{ or } B))$$

is an axiom of $\mathfrak{T}$.

These rules are in fact schemes; let us verify this, for example for S2. Let R be a relation obtained by applying S2; then there are relations A and B in $\mathfrak{T}$ such that R is the relation $A \Longrightarrow (A \text{ or } B)$. Let T be a term in $\mathfrak{T}$, let x be a letter, and let A' and B' be the relations $(T|x)A$ and $(T|x)B$; then $(T|x)R$ is the same as $A' \Longrightarrow (A' \text{ or } B')$, and can therefore be obtained by applying S2.

Intuitively, the rules S1 through S4 merely express the meaning which is attached to the words "or" and "implies" in the usual language of mathematics (†).

(*) This scheme may be expressed without using the letter A or the abbreviating symbol $\Longrightarrow$ as follows : *whenever we have a relation, we obtain a theorem by writing, from left to right,* $\vee$, $\neg$, $\vee$, *and then the given relation three times.* The reader may, as an exercise, translate in a similar way the expressions of the other schemes.

(†) In everyday speech, the word "or" has two different meanings, according to the context : when we link two statements by the word "or" we may mean to assert at least one of the two (and possibly both together), or we may mean to assert one to the exclusion of the other.

If a logical theory $\mathcal{T}$ is contradictory, *every relation in* $\mathcal{T}$ *is a theorem in* $\mathcal{T}$. For let A be a relation in $\mathcal{T}$ such that A and "not A" are theorems in $\mathcal{T}$, and let B be any relation in $\mathcal{T}$. By S2, (not A) $\Longrightarrow$ ((not A) or B) is a theorem in $\mathcal{T}$; therefore, by C1 (§ 2, no. 2), "(not A) or B", that is to say $A \Longrightarrow B$, is a theorem in $\mathcal{T}$. A second application of C1 shows that B is a theorem in $\mathcal{T}$.

¶ *From now on* $\mathcal{T}$ *will denote a logical theory.*

2. FIRST CONSEQUENCES

C6. *Let* A, B, C *be relations in* $\mathcal{T}$. *If* $A \Longrightarrow B$ *and* $B \Longrightarrow C$ *are theorems in* $\mathcal{T}$, *then* $A \Longrightarrow C$ *is a theorem in* $\mathcal{T}$.

For $(B \Longrightarrow C) \Longrightarrow ((A \Longrightarrow B) \Longrightarrow (A \Longrightarrow C))$ is an axiom of $\mathcal{T}$, by replacing A by B, B by C, and C by "not A" in S4. By C1 (§ 2, no. 2), $(A \Longrightarrow B) \Longrightarrow (A \Longrightarrow C)$ is a theorem in $\mathcal{T}$. A further application of C1 completes the proof.

C7. *If* A *and* B *are relations in* $\mathcal{T}$, *then* $B \Longrightarrow (A$ or $B)$ *is a theorem in* $\mathcal{T}$.

For $B \Longrightarrow (B$ or $A)$ and $(B$ or $A) \Longrightarrow (A$ or $B)$ are axioms of $\mathcal{T}$ by virtue of S2 and S3. Now use C6.

C8. *If* A *is a relation in* $\mathcal{T}$, $A \Longrightarrow A$ *is a theorem in* $\mathcal{T}$.

For $A \Longrightarrow (A$ or $A)$ and $(A$ or $A) \Longrightarrow A$ are axioms, by S2 and S1. Now use C6.

C9. *If* A *is a relation and* B *a theorem in* $\mathcal{T}$, *then* $A \Longrightarrow B$ *is a theorem in* $\mathcal{T}$.

For $B \Longrightarrow$ ((not A) or B) is a theorem by C7, and therefore "(not A) or B", that is to say $A \Longrightarrow B$, is a theorem by C1.

C10. *If* A *is a relation in* $\mathcal{T}$, *then* "A or (not A)" *is a theorem in* $\mathcal{T}$.

For "(not A) or A" is a theorem by C8; now use S3 and C1.

C11. *If* A *is a relation in* $\mathcal{T}$, "$A \Longrightarrow$ (not not A)" *is a theorem in* $\mathcal{T}$.

For this relation is "(not A) or (not not A)", and the result follows from C10.

C12. *Let* A *and* B *be two relations in* $\mathcal{T}$. *Then the relation*

$$(A \Longrightarrow B) \Longrightarrow ((\text{not } B) \Longrightarrow (\text{not } A))$$

is a theorem in $\mathcal{T}$.

For

$$((\text{not } A) \text{ or } B) \Longrightarrow ((\text{not } A) \text{ or } (\text{not not } B))$$

is a theorem, by C11, S4 and C1. On the other hand,

$$((\text{not } A) \text{ or } (\text{not not } B)) \Longrightarrow ((\text{not not } B) \text{ or } (\text{not } A))$$

is an axiom, by S3. Therefore

$$((\text{not } A) \text{ or } B) \Longrightarrow ((\text{not not } B) \text{ or } (\text{not } A))$$

is a theorem by C6. Hence the result.

C13. *Let A, B, C be relations in $\mathcal{T}$. If $A \Longrightarrow B$ is a theorem in $\mathcal{T}$, then $(B \Longrightarrow C) \Longrightarrow (A \Longrightarrow C)$ is a theorem in $\mathcal{T}$.*

For $(\text{not } B) \Longrightarrow (\text{not } A)$ is a theorem, by C12 and C1. Therefore $(C \text{ or } (\text{not } B)) \Longrightarrow (C \text{ or } (\text{not } A))$ is a theorem, by S4 and C1. By a double application of S3 and C6 we infer that

$$((\text{not } B) \text{ or } C) \Longrightarrow ((\text{not } A) \text{ or } C)$$

is a theorem; but this is the given relation.

¶ *From now on, we shall generally use* C1 *and* C6 *without quoting them explicitly.*

3. METHODS OF PROOF

I. *Method of the auxiliary hypothesis.* This rests on the following rule:

C14 (*Criterion of deduction*). *Let A be a relation in $\mathcal{T}$, and let $\mathcal{T}'$ be the theory obtained by adjoining A to the axioms of $\mathcal{T}$. If B is a theorem in $\mathcal{T}'$, then $A \Longrightarrow B$ is a theorem in $\mathcal{T}$.*

Let $B_1, B_2, \ldots, B_n$ be a proof in $\mathcal{T}'$ in which B appears. We shall show, step by step, that the relations $A \Longrightarrow B_k$ are theorems in $\mathcal{T}$. Suppose that this has been established for the relations which precede B_i, and let us show that $A \Longrightarrow B_i$ is a theorem in $\mathcal{T}$. If B_i is an axiom of $\mathcal{T}'$, then B_i is either an axiom of $\mathcal{T}$ or is A. In both cases, $A \Longrightarrow B_i$ is a theorem in $\mathcal{T}$ by applying C9 or C8. If B_i is preceded by relations B_j and $B_j \Longrightarrow B_i$, we know that $A \Longrightarrow B_j$ and $A \Longrightarrow (B_j \Longrightarrow B_i)$ are theorems in $\mathcal{T}$. Hence $(B_j \Longrightarrow B_i) \Longrightarrow (A \Longrightarrow B_i)$ is a theorem in $\mathcal{T}$ by C13. Hence, by C6, $A \Longrightarrow (A \Longrightarrow B_i)$, that is to say "$(\text{not } A) \text{ or } (A \Longrightarrow B_i)$", is a theorem in $\mathcal{T}$, and therefore so is "$(A \Longrightarrow B_i) \text{ or } (\text{not } A)$" by S3. Now, $(\text{not } A) \Longrightarrow ((\text{not } A) \text{ or } B_i)$, that is to say $(\text{not } A) \Longrightarrow (A \Longrightarrow B_i)$, is a theorem in $\mathcal{T}$, by S2. By application of S4 we see then that

$$((A \Longrightarrow B_i) \text{ or } (\text{not } A)) \Longrightarrow ((A \Longrightarrow B_i) \text{ or } (A \Longrightarrow B_i))$$

is a theorem in $\mathcal{T}$, and hence that "$(A \Longrightarrow B_i) \text{ or } (A \Longrightarrow B_i)$" is a theorem in $\mathcal{T}$. By S1 we conclude that $A \Longrightarrow B_i$ is a theorem in $\mathcal{T}$.

In practice we indicate that we are going to use this criterion by a phrase such as "suppose that A is true". This phrase means that for the time being the reasoning will be performed in the theory $\mathcal{T}'$, until the relation B has been proved. When this has been achieved it has been established that $A \Longrightarrow B$ is a theorem in $\mathcal{T}$, and one continues thereafter to reason in $\mathcal{T}$ without in general indicating that one has abandoned the theory $\mathcal{T}'$. The relation A introduced as a new axiom is called the *auxiliary hypothesis.* * For example, when we say "let x be a real number", we are constructing a theory in which the relation "x is a real number" is an auxiliary hypothesis. *

II. *Method of reductio ad absurdum.* This is founded on the following rule :

C15. *Let A be a relation in $\mathcal{T}$, and let $\mathcal{T}'$ be the theory obtained by adjoining the axiom* "not A" *to the axioms of $\mathcal{T}$. If $\mathcal{T}'$ is contradictory, then A is a theorem in $\mathcal{T}$.*

For A is a theorem in $\mathcal{T}'$; consequently (method of the auxiliary hypothesis) "(not A) $\Longrightarrow A$" is a theorem in $\mathcal{T}$. By S4,

$$(A \text{ or } (\text{not } A)) \Longrightarrow (A \text{ or } A)$$

is a theorem in $\mathcal{T}$; by C10, "A or A" is a theorem in $\mathcal{T}$. Now use S1.

In practice, we indicate that we are going to use this criterion by a phrase such as "suppose that A is false". This phrase means that for the time being the reasoning will be performed in the theory $\mathcal{T}'$, until two theorems of the form B and "not B" have been proved. When this has been achieved it is then established that A is a theorem in $\mathcal{T}$, which is generally indicated by a phrase such as "Now this (i.e., in the preceding notation, B and "not B") is absurd; hence A is true". One then continues in the original theory $\mathcal{T}$.

¶ As first applications of these methods, let us establish the following criteria:

C16. *If A is a relation in $\mathcal{T}$, then* (not not A) $\Longrightarrow A$ *is a theorem in $\mathcal{T}$.*

For suppose that "not not A" is true; then we have to prove A. Suppose A is false. In the theory so defined, "not not A" and "not A" are theorems, which is absurd; therefore A is true.

C17. *If A and B are relations in $\mathcal{T}$, then*

$$((\text{not } B) \Longrightarrow (\text{not } A)) \Longrightarrow (A \Longrightarrow B)$$

is a theorem in $\mathcal{T}$.

For suppose that (not B) $\Longrightarrow$ (not A) is true. We have to show that $A \Longrightarrow B$ is true. Suppose that A is true, and let us show that B is true. Suppose "not B" is true. Then "not A" is true, which is absurd.

III. *Method of disjunction of cases.* This rests on the following rule:

C18. *Let A, B, C be relations in $\mathcal{T}$. If "A or B" $A \Longrightarrow C$, $B \Longrightarrow C$ are theorems in $\mathcal{T}$, then C is a theorem in $\mathcal{T}$.*

For, by S4, "(A or B) $\Longrightarrow$ (A or C)" and "(C or A) $\Longrightarrow$ (C or C)" are theorems in $\mathcal{T}$. By S3 and S1, it follows that (A or B) $\Longrightarrow C$ is a theorem in $\mathcal{T}$; hence the result.

To prove C it is therefore enough, when we have at our disposal a theorem "A or B", first to prove C by adjoining A to the axioms of $\mathcal{T}$, and then to prove C by adjoining B to the axioms of $\mathcal{T}$. The interesting feature of this method lies in the fact that if "A or B" is true, we cannot in general assert either that A is true or that B is true.

In particular, by C10, if "$A \Longrightarrow C$" and "(not A) $\Longrightarrow C$" are both theorems in $\mathcal{T}$, then C is a theorem in $\mathcal{T}$.

IV. *Method of the auxiliary constant.* This is founded on the following rule:

C19. *Let x be a letter and let A and B be relations in $\mathcal{T}$ such that:*

(1) *the letter x is not a constant of $\mathcal{T}$ and does not appear in B;*
(2) *there is a term T in $\mathcal{T}$ such that $(T|x)A$ is a theorem in $\mathcal{T}$.*

Let $\mathcal{T}'$ be the theory obtained by adjoining A to the axioms of $\mathcal{T}$. If B is a theorem in $\mathcal{T}'$, then B is a theorem in $\mathcal{T}$.

Indeed, $A \Longrightarrow B$ is a theorem in $\mathcal{T}$ (criterion of deduction). Since x is not a constant of $\mathcal{T}$, $(T|x)(A \Longrightarrow B)$ is a theorem in $\mathcal{T}$ by virtue of C3. Since x does not appear in B, $(T|x)(A \Longrightarrow B)$ is identical with $((T|X)A) \Longrightarrow B$, by CS5 (§ 1, no. 2). Finally, $(T|x)A$ is a theorem in $\mathcal{T}$, and therefore so is B.

Intuitively, the method consists in using, in order to prove B, an arbitrary object x (the *auxiliary constant*) which is supposed to be endowed with certain properties, denoted by A. * For example, in a proof in geometry which involves, among other things, a line D, we may "take" a point x on this line; the relation A is then $x \in \mathrm{D}$. * In order that one should be able to use an object endowed with certain properties during the course of a proof, it is clearly necessary that such objects should exist. The theorem $(T|x)A$, called the *theorem of legitimation*, guarantees this existence.

In practice we indicate that we are going to use this method by a phrase such as "let x be an object such that A". By contrast with the method of the auxiliary hypothesis, the conclusion of the argument does not involve x.

4. CONJUNCTION

Let A, B be assemblies. The assembly

$$\text{not } ((\text{not } A) \text{ or } (\text{not } B))$$

will be denoted by "A and B".

CS6. *Let A, B, T be assemblies and x a letter. Then the assembly*

$$(T|x)(A \text{ and } B)$$

is identical with "$(T|x)A$ and $(T|x)B$".

This is an immediate consequence of CS5 (§ 1, no. 2).

CF9. *If A, B are relations in $\mathcal{T}$, then* "A and B" *is a relation in* $\mathcal{T}$ (called the *conjunction* of A and B).

This follows immediately from CF1 and CF2 (§ 1, no. 4).

C20. *If A, B are theorems in $\mathcal{T}$, then* "A and B" *is a theorem in* $\mathcal{T}$.

Suppose that "A and B" is false, that is to say,

$$\text{not not } ((\text{not } A) \text{ or } (\text{not } B))$$

is true. By C16, "(not A) or (not B)", that is to say, $A \Longrightarrow (\text{not } B)$ is true, hence "not B" is true; but this is absurd. Hence "A and B" is true.

C21. *If A, B are relations in $\mathcal{T}$, then*

$$(A \text{ and } B) \Longrightarrow A, \qquad (A \text{ and } B) \Longrightarrow B$$

are theorems in $\mathcal{T}$.

The relations $(\text{not } A) \Longrightarrow ((\text{not } A) \text{ or } (\text{not } B))$, $(\text{not } B) \Longrightarrow ((\text{not } A) \text{ or } (\text{not } B))$ are theorems in $\mathcal{T}$, by S2 (no. 1) and C7 (no. 2). Now $((\text{not } A) \text{ or } (\text{not } B)) \Longrightarrow (\text{not } (A \text{ and } B))$ is a theorem in $\mathcal{T}$ by C11. Hence $(\text{not } A) \Longrightarrow (\text{not } (A \text{ and } B))$, $(\text{not } B)) \Longrightarrow (\text{not } (A \text{ and } B))$ are theorems in $\mathcal{T}$. The result follows by applying C17.

¶ We shall denote by "A and B and C" (resp. "A or B or C") the relation "A and (B and C)" (resp. "A or (B or C)"). More generally, if

$$A_1, A_2, \ldots, A_n$$

are relations, we denote by "A_1 and A_2, and ... and A_p" a relation which is constructed step by step by means of the convention that "A_1 and A_2 and ... and A_h" denotes the same relation as "A_1 and (A_2 and ... and A_h)". The relation "A_1 or A_2 or ... or A_h" is defined similarly. The relation "A_1 and A_2 and ... and A_h" is a theorem in $\mathcal{T}$ if and only if each of the relations $A_1, A_2, \ldots, A_h$ is a theorem in $\mathcal{T}$.

It follows that every logical theory $\mathcal{T}$ is equivalent to a logical theory $\mathcal{T}'$ which has at most one explicit axiom. This is clear if $\mathcal{T}$ has no explicit axiom. If $\mathcal{T}$ has explicit axioms $A_1, A_2, \ldots, A_h$, let $\mathcal{T}'$ be the theory which has the same signs and schemes as $\mathcal{T}$, and the explicit axiom "A_1 and A_2 and ... and A_h". It is immediately seen that every axiom of $\mathcal{T}$ (resp. $\mathcal{T}'$) is a theorem of $\mathcal{T}'$ (resp. $\mathcal{T}$).

Let $\mathcal{T}_0$ be the theory with no explicit axioms which has the same signs as $\mathcal{T}$ and S1, S2, S3, S4 as its only schemes. Then the study of $\mathcal{T}$ reduces, in principle, to the study of $\mathcal{T}_0$: for the relation A to be a theorem in $\mathcal{T}$ it is necessary and sufficient that there exist axioms $A_1, A_2, \ldots, A_h$ of $\mathcal{T}$ such that (A_1 and A_2 and ... and A_h) $\Longrightarrow A$ is a theorem in $\mathcal{T}_0$. This condition is evidently sufficient. Suppose conversely that A is a theorem in $\mathcal{T}$, and let $A_1, A_2, \ldots, A_h$ be the axioms of $\mathcal{T}$ appearing in a proof in $\mathcal{T}$ which contains A. Let $\mathcal{T}'$ (resp. $\mathcal{T}''$) be the theory constructed from $\mathcal{T}_0$ by adjoining the axioms $A_1, A_2, \ldots, A_h$ (resp. the axiom "A_1 and A_2 and ... and A_h"). The proof of A in $\mathcal{T}$ is a proof of A in $\mathcal{T}'$, therefore A is a theorem in $\mathcal{T}'$ and consequently in $\mathcal{T}''$, because (as we noted above) $\mathcal{T}'$ and $\mathcal{T}''$ are equivalent. By the criterion of deduction, (A_1 and A_2 and ... and A_h) $\Longrightarrow A$ is a theorem in $\mathcal{T}_0$.

If $\mathcal{T}$ is contradictory, then it follows from what has been said that there exists a conjunction A of axioms of $\mathcal{T}$ and a relation R in $\mathcal{T}$ such that $A \Longrightarrow$ (R and (not R)) is a theorem in $\mathcal{T}_0$. Therefore

$$((\text{not } R) \text{ or } (\text{not not } R)) \Longrightarrow (\text{not } A)$$

is a theorem in $\mathcal{T}_0$, and since "(not R) or (not not R)" is a theorem in $\mathcal{T}_0$, "not A" is a theorem in $\mathcal{T}_0$. Conversely, if there exists a conjunction A of axioms of $\mathcal{T}$ such that "not A" is a theorem in $\mathcal{T}_0$, then A and "not A" are theorems in $\mathcal{T}$, so that $\mathcal{T}$ is contradictory.

5. EQUIVALENCE

Let A and B be assemblies. The assembly

$$(A \Longrightarrow B) \text{ and } (B \Longrightarrow A)$$

will be denoted by $A \Longleftrightarrow B$.

CS7. *Let A, B, T be assemblies, and let x be a letter. Then the assembly $(T|x)(A \Longleftrightarrow B)$ is identical with $(T|x)A \Longleftrightarrow (T|x)B$.*

This follows immediately from CS5 (§ 1, no. 2) and CS6 (no. 4).

CF10. *If A and B are relations in $\mathcal{T}$, then $A \Longleftrightarrow B$ is a relation in* T.

This follows immediately from CF5 (§ 1, no. 4) and CF9 (no. 4).

¶ If $A \Longleftrightarrow B$ is a theorem in $\mathcal{T}$, we shall say that A and B are *equivalent* in $\mathcal{T}$; if x is a letter which is not a constant of $\mathcal{T}$, and if A and B are considered as relations in x, then every term in $\mathcal{T}$ which satisfies one also satisfies the other.

¶ It follows from the criteria C20, C21 (no. 4) that in order to prove a theorem in $\mathcal{T}$ of the form $A \Longleftrightarrow B$, it is necessary and sufficient to be able to prove $A \Longrightarrow B$ and $B \Longrightarrow A$ in $\mathcal{T}$. This is often done by proving B in the theory deduced from $\mathcal{T}$ by adjoining the axiom A, and then by proving A in the theory deduced from $\mathcal{T}$ by adjoining the axiom B. These remarks lead immediately to the following criteria, whose proofs we leave to the reader :

C22. *Let A, B, C be relations in $\mathcal{T}$. If $A \Longleftrightarrow B$ is a theorem in $\mathcal{T}$, then $B \Longleftrightarrow A$ is a theorem in $\mathcal{T}$. If $A \Longleftrightarrow B$ and $B \Longleftrightarrow C$ are theorems in $\mathcal{T}$, then $A \Longleftrightarrow C$ is a theorem in $\mathcal{T}$.*

C23. *Let A, B be equivalent relations in $\mathcal{T}$, and let C be a relation in $\mathcal{T}$. Then the following are theorems in $\mathcal{T}$:*

$$(\text{not } A) \Longleftrightarrow (\text{not } B); \qquad (A \Longrightarrow C) \Longleftrightarrow (B \Longrightarrow C);$$
$$(C \Longrightarrow A) \Longleftrightarrow (C \Longrightarrow B);$$
$$(A \text{ and } C) \Longleftrightarrow (B \text{ and } C); \quad (A \text{ or } C) \Longleftrightarrow (B \text{ or } C).$$

C24. *Let A, B, C be relations in $\mathcal{T}$. Then the following are theorems in $\mathcal{T}$:*

$$(\text{not not } A) \Longleftrightarrow A; \qquad (A \Longrightarrow B) \Longleftrightarrow ((\text{not } B) \Longrightarrow (\text{not } A));$$
$$(A \text{ and } A) \Longleftrightarrow A; \qquad (A \text{ and } B) \Longleftrightarrow (B \text{ and } A);$$
$$(A \text{ and } (B \text{ and } C)) \Longleftrightarrow ((A \text{ and } B) \text{ and } C);$$
$$(A \text{ or } B) \Longleftrightarrow \text{not } ((\text{not } A) \text{ and } (\text{not } B));$$
$$(A \text{ or } A) \Longleftrightarrow A; \qquad (A \text{ or } B) \Longleftrightarrow (B \text{ or } A);$$
$$(A \text{ or } (B \text{ or } C)) \Longleftrightarrow ((A \text{ or } B) \text{ or } C);$$
$$(A \text{ and } (B \text{ or } C)) \Longleftrightarrow ((A \text{ and } B) \text{ or } (A \text{ and } C));$$
$$(A \text{ or } (B \text{ and } C)) \Longleftrightarrow ((A \text{ or } B) \text{ and } (A \text{ or } C));$$
$$(A \text{ and } (\text{not } B)) \Longleftrightarrow \text{not } (A \Longrightarrow B);$$
$$(A \text{ or } B) \Longleftrightarrow ((\text{not } A) \Longrightarrow B).$$

C25. *If A is a theorem in $\mathcal{T}$ and B is a relation in $\mathcal{T}$, then*

$$(A \text{ and } B) \Longleftrightarrow B$$

is a theorem in $\mathcal{T}$. *If* "not A" *is a theorem in* $\mathcal{T}$, *then* (A or B) $\Longleftrightarrow B$ *is a theorem in* $\mathcal{T}$.

¶ *In principle, from now on throughout the rest of this series, criteria* C1 *through* C25 *will be used without reference.*

4. QUANTIFIED THEORIES

1. DEFINITION OF QUANTIFIERS

The only logical signs which played a role in § 3 are $\neg$ and $\vee$. The rules we shall now state are essentially concerned with the use of the logical signs τ and $\square$.

¶ If R is an assembly and x a letter, the assembly $(\tau_x(R)|x)R$ is denoted by "there exists x such that R", or by $(\exists x)R$. The assembly

"not $((\exists x)$ not $R)$"

is denoted by "for all x, R", or by "given any x, R", or by $(\forall x)R$. The abbreviating symbols $\exists$ and $\forall$ are called respectively the *existential quantifier* and the *universal quantifier*. The letter x does not appear in the assembly denoted by $\tau_x(R)$ *and therefore does not appear* in the assemblies denoted by $(\exists x)R$ and $(\forall x)R$.

CS8. *Let* R *be an assembly and let* x *and* x' *be letters. If* x' *does not appear in* R, *then* $(\exists x)R$ *and* $(\forall x)R$ *are respectively identical with* $(\exists x')R'$ *and* $(\forall x')R'$, *where* R' *is* $(x'|x)R$.

For $(\tau_x(R)|x)R$ is identical with $(\tau_x(R)|x')R'$ by CS1 (§1, no. 2), and $\tau_x(R)$ is identical with $\tau_{x'}(R')$ by CS3 (§1, no. 2). Hence $(\exists x)R$ is identical with $(\exists x')R'$. It follows that $(\forall x)R$ is identical with $(\forall x')R'$.

CS9. *Let* R *and* U *be assemblies and let* x *and* y *be distinct letters. If* X *does not appear in* U, *then* $(U|y)(\exists x)R$ *and* $(U|y)(\forall x)R$ *are respectively identical with* $(\exists x)R'$ *and* $(\forall x)R'$, *where* R' *is* $(U|y)R$.

For, by CS2 (§1, no. 2), $(U|y)(\tau x(R)|x)R$ is identical with

$$(T|x)(U|y)R,$$

where T is $(U|y)\tau_x(R)$, that is $\tau_x(R')$, by CS4 (§1, no. 2). Hence $(U|y)(\exists x)R$ is identical with $(\exists x)\mathrm{R}'$, and consequently $(U|Y)(\forall x)R$ is identical with $(\forall x)R'$.

CF11. *If **R** is a relation in a theory 𝔗 and if **x** is a letter, then* $(\exists x)R$ *and* $(\forall x)R$ *are relations in 𝔗.*

This follows immediately from CF3, CF8, and CF2 (§ 1, no. 4).

> Intuitively, let us consider ***R*** as expressing a property of the object denoted by ***x***. From the intuitive meaning of the term $\tau_x(R)$, the assertion $(\exists x)R$ means that there is an object which has the property ***R***. The assertion "not $(\exists x)$(not ***R***)" means that there is no object which has the property "not ***R***", and therefore that every object has the property ***R***.

In a logical theory 𝔗, if we have a theorem of the form $(\exists x)R$, in which the letter ***x*** is not a constant of 𝔗, this theorem can serve as a theorem of legitimation in the method of the auxiliary constant (§3, no. 3), because it is identical with $(\tau_x(R)|x)R$. Let 𝔗′ be the theory obtained by adjoining ***R*** to the axioms of 𝔗. If we can prove a relation ***S***, in which ***x*** does not appear, in the theory 𝔗′, then ***S*** is a theorem in 𝔗.

C26. *Let 𝔗 be a logical theory, let **R** be a relation in 𝔗, and let **x** be a letter. The relations* $(\forall x)R$ *and* $(\tau_x(\text{not } R)|x)R$ *are then equivalent in 𝔗.*

For $(\forall X)R$ is identical with "not $(\tau_x(\text{not } R)|x)(\text{not } R)$" and therefore with "not not $(\tau_x(\text{not } R)|x)R$".

C27. *If **R** is a theorem in a logical theory 𝔗 in which the letter **x** is not a constant, then* $(\forall x)R$ *is a theorem in 𝔗.*

For $(\tau_x(\text{not } R)|X)R$ is a theorem in 𝔗, by C3 (§2, no. 3).

> On the other hand, if ***x*** is a constant of 𝔗, the truth of ***R*** does not imply that of $(\forall x)R$. Intuitively, the fact that ***R*** is a true property of ***x***, which is a definite object of 𝔗, clearly does not imply that ***R*** is a true property of every object.

C28. *Let 𝔗 be a logical theory, let **R** be a relation in 𝔗, and let **x** be a letter. Then the relations* "not $(\forall x)R$" *and* $(\exists x)(\text{not } R)$ *are equivalent in 𝔗.*

For "not $(\forall x)R$" is identical with "not not $(\exists x)(\text{not } R)$".

2. AXIOMS OF QUANTIFIED THEORIES

A *quantified theory* is any theory 𝔗 in which the schemes S1 to S4 (§3, no. 1) and the scheme S5 below provide implicit axioms.

S5. *If **R** is a relation in 𝔗, if **T** is a term in 𝔗, and if **x** is a letter, then the relation* $(T|x)R \Longrightarrow (\exists x)R$ *is an axiom.*

This rule is indeed a scheme. For let A be an axiom of $\mathcal{T}$ obtained by applying S5; there is then a relation R in $\mathcal{T}$, a term T in $\mathcal{T}$, and a letter x such that A is $(T|x)R \Longrightarrow (\exists x)R$. Let U be a term in $\mathcal{T}$ and let y be a letter. We shall show that $(U|y)A$ can also be obtained by applying S5. Using CS1 (§1, no. 2) and CS8 (no. 1), we can confine ourselves to the case in which x is distinct from y and does not appear in U. Let R' be the relation $(U|y)R$ and T' the term $(U|y)T$. The criteria CS2 (§1, no. 2) and CS9 (no. 1) show that $(U|y)A$ is identical with $(T'|x)R' \Longrightarrow (\exists x)R'$.

The scheme S5 says that if there is an object T for which the relation R, considered as expressing a property of x, is true, then R is true for the object $\tau_x(R)$; this accords with the intuitive meaning we have attributed to $\tau_x(R)$ (§1, no. 3, Remark).

3. PROPERTIES OF QUANTIFIERS

From now on we shall have to consider only quantified theories. For the rest of this section, $\mathcal{T}$ will denote such a theory, and $\mathcal{T}_0$ will denote the theory without explicit axioms which has the same signs as $\mathcal{T}$ and whose only schemes are S1 through S5. $\mathcal{T}$ is stronger than $\mathcal{T}_0$.

C29. *Let* R *be a relation in* $\mathcal{T}$ *and let* x *be a letter. Then the relations* "not $(\exists x)R$" *and* $(\forall x)(\text{not } R)$ *are equivalent in* $\mathcal{T}$.

It is sufficient to prove the criterion in the theory $\mathcal{T}_0$ in which x is not a constant. The theorem $R \Longleftrightarrow (\text{not not } R)$ gives us, by C3 (§2, no. 3), the theorems

$$(\exists x)R \Longrightarrow (\tau_x(R)|x)(\text{not not } R)$$

and

$$(\exists x)\,(\text{not not } R) \Longrightarrow (\tau_x(\text{not not } R)|X)R.$$

Applying S5, we deduce the theorems in $\mathcal{T}_0$

$$(\exists x)R \Longrightarrow (\exists x)\,(\text{not not } R), \quad (\exists x)\,(\text{not not } R) \Longrightarrow (\exists x)R,$$

whence the theorem $(\exists x)R \Longleftrightarrow (\exists x)\,(\text{not not } R)$. Now

$$(\exists x)\,(\text{not not } R)$$

is equivalent in $\mathcal{T}_0$ to "not not $(\exists x)\,(\text{not not } R)$", that is to "not $(\forall x)\,(\text{not } R)$". Hence the result.

The criteria C28 and C29 enable us to deduce properties of one of the quantifiers from properties of the other.

C30. *Let **R** be a relation in $\mathcal{T}$, let **T** be a term in $\mathcal{T}$, and let **x** be a letter. Then the relation* $(\forall x)R \Longrightarrow (T|x)R$ *is a theorem in* $\mathcal{T}$.

By S5, $(T|x)(\text{not } R) \Longrightarrow (\tau_x(\text{not } R)|x)(\text{not } R)$ is an axiom. This relation is identical with

$$(\text{not } (T|x)R) \Longrightarrow \text{not}(\tau_x(\text{not } R)|X)R).$$

Hence $(\tau_x(\text{not } R)|X)R \Longrightarrow (T|x)R$ is a theorem in $\mathcal{T}$. Now use C26 (no. 1).

> Let **R** be a relation in $\mathcal{T}$. By C26, C27, and C30 it is the same (provided the letter **x** is not a constant of $\mathcal{T}$) whether we state the theorem **R** in $\mathcal{T}$, or the theorem $(\forall x)R$, or the metamathematical rule : if **T** is any term in $\mathcal{T}$, then $(T|x)R$ is a theorem in $\mathcal{T}$.

C31. *Let **R** and **S** be relations in $\mathcal{T}$, and let **x** be a letter which is not a constant of $\mathcal{T}$. If* $R \Longrightarrow S$ (resp. $R \Longleftrightarrow S$) *is a theorem in* $\mathcal{T}$, *then*

$$(\forall x)R \Longrightarrow (\forall x)S, \qquad (\exists x)R \Longrightarrow (\exists x)S$$
$$[\text{resp. } (\forall x)R \Longleftrightarrow (\forall x)S, \qquad (\exists x)R \Longleftrightarrow (\exists x)S]$$

are theorems in $\mathcal{T}$.

Suppose that $R \Longrightarrow S$ is a theorem in $\mathcal{T}$. Let us adjoin the hypothesis $(\forall x)R$ (in which **x** does not appear). Then **R**, hence **S**, and therefore also $(\forall x)S$, are true. Consequently $(\forall x)R \Longrightarrow (\forall x)S$ is a theorem in $\mathcal{T}$. It follows that if $R \Longleftrightarrow S$ is a theorem in $\mathcal{T}$, then so is

$$(\forall x)R \Longleftrightarrow (\forall x)S.$$

The rules relating to $\exists$ can now be deduced by using C29.

C32. *Let **R** and **S** be relations in $\mathcal{T}$, and let **x** be a letter. Then the relations*

$$(\forall x)(R \text{ and } S) \Longleftrightarrow ((\forall x)R \text{ and } (\forall x)S),$$
$$(\exists x)(R \text{ or } S) \Longleftrightarrow ((\exists x)R \text{ or } (\exists x)S)$$

are theorems in $\mathcal{T}$.

It is sufficient to prove these criteria in $\mathcal{T}_0$, in which **x** is not a constant. If $(\forall x)(R \text{ and } S)$ is true, then "**R** and **S**" is true, and therefore each of the relations **R**, **S** is true. Consequently each of the relations $(\forall x)R$, $(\forall x)S$ is true, and hence "$(\forall x)R$ and $(\forall x)S$" is true. Similarly one shows that if "$(\forall x)R$ and $(\forall x)S$" is true, then $(\forall x)(R \text{ and } S)$ is true. Hence the first theorem. The second follows by applying C29.

It should be noted that if $(\forall x)(R \text{ or } S)$ is a theorem in $\mathcal{T}$, we may not conclude that $((\forall x)R \text{ or } (\forall x)S)$ is a theorem in $\mathcal{T}$. Intuitively, to say that the relation $(\forall x)(R \text{ or } S)$ is true means that for each object x, at least one of the relations R, S is true; but in general only one of the two will be true, and whether it is R or S will depend on the choice of x. Likewise, if $((\forall x)R \text{ and } (\exists x)S)$ is a theorem in $\mathcal{T}$, we may not conclude that $(\exists x)(R \text{ and } S)$ is a theorem in $\mathcal{T}$. However, there is the following criterion :

C33. *Let R and S be relations in $\mathcal{T}$, and let x be a letter which does not appear in R. Then the relations*

$$(\forall x)(R \text{ or } S) \iff (R \text{ or } (\forall x)S),$$
$$(\exists x)(R \text{ and } S) \iff (R \text{ and } (\exists x)S)$$

are theorems in $\mathcal{T}$.

It is sufficient to establish the criterion in $\mathcal{T}_0$, in which x is not a constant. Let $\mathcal{T}'$ be the theory obtained by adjoining $(\forall x)(R \text{ or } S)$ to the axioms of $\mathcal{T}_0$. In $\mathcal{T}'$, "R or S", and therefore $(\text{not } R) \implies S$, are theorems. If "not R" is true (a hypothesis in which x does not feature), then S and therefore also $(\forall x)S$ are true. Consequently

$$(\text{not } R) \implies (\forall x)S$$

is a theorem in $\mathcal{T}'$, and hence $(\forall x)(R \text{ or } S) \implies (R \text{ or } (\forall x)S)$ is a theorem in $\mathcal{T}_0$. Likewise, if "R or $(\forall x)S$" is true, then "R or S" and therefore $(\forall x)(R \text{ or } S)$ are true. Consequently

$$(R \text{ or } (\forall x)S) \implies (\forall x)(R \text{ or } S)$$

is a theorem in $\mathcal{T}_0$. The rule relating to $\exists$ follows by applying C29.

C34. *Let R be a relation and let x and y be letters. Then the relations*

$$(\forall x)(\forall y)R \iff (\forall y)(\forall x)R,$$
$$(\exists x)(\exists y)R \iff (\exists y)(\exists x)R,$$
$$(\exists x)(\forall y)R \implies (\forall y)(\exists x)R$$

are theorems in $\mathcal{T}$.

It is sufficient to prove these theorems in $\mathcal{T}_0$, in which x and y are not constants. If $(\forall x)(\forall y)R$ is true, then $(\forall y)R$, and therefore R, hence $(\forall x)R$, hence $(\forall y)(\forall x)R$, are true. Likewise, if $(\forall y)(\forall x)R$ is true, then $(\forall x)(\forall y)R$ is true; and the first theorem follows. The second now follows by use of C29. Finally, since $(\forall y)R \implies R$ is a theorem in $\mathcal{T}_0$, so is $(\exists x)(\forall y)R \Longrightarrow (\exists x)R$ by C31; if $(\exists x)(\forall y)R$ is true, then $(\exists x)R$ is true, and therefore so is $(\forall y)(\exists x)R$. Hence the third theorem.

On the other hand, if $(\forall y)(\exists x)R$ is a theorem in $\mathcal{T}$, we may not conclude that $(\exists x)(\forall y)R$ is a theorem in $\mathcal{T}$. Intuitively, to say that the relation $(\forall y)(\exists x)R$ is true means that, given any object y, there is an object x such that R is a true relation between the objects x and y. But in general the object x will depend on the choice of the object y, whereas to say that $(\exists x)(\forall y)R$ is true means that there is a *fixed* object x such that R is a true relation between this object and *any* object y.

4. TYPICAL QUANTIFIERS

Let A and R be assemblies and let x be a letter. We denote the assembly $(\exists x)(A \text{ and } R)$ by $(\exists_A x)R$, and the assembly

$$\text{"not } (\exists_A x) \text{ (not } R)\text{"}$$

by $(\forall_A x)R$. The abbreviating symbols $\exists_A$ and $\forall_A$ are called *typical quantifiers*. Observe that the letter x does not appear in the assemblies denoted by $(\exists_A x)R$, $(\forall_A x)R$.

CS10. *Let A and R be assemblies and let x and x' be letters. If x appears neither in R nor in A, then $(\exists_A x)R$ and $(\forall_A x)R$ are respectively identical with $(\exists_{A'} x')R'$ and $(\forall_{A'} x')R'$, where R' is $(x'|x)R$ and A' is $(x'|x)A$.*

CS11. *Let A, R, U be assemblies, and let x, y be distinct letters. If x does not appear in U, the assemblies $(U|y)(\exists_A x)R$ and $(U|y)(\forall_A x)R$ are respectively identical with $(\exists_{A'} x')R'$ and $(\forall_{A'} x')R'$, where R' is $(U|y)R$ and A' is $(U|y)A$.*

These rules are immediate consequences of criteria CS8, CS9 (no. 1), CS5 (§1, no. 2), and CS6 (§3, no. 4).

CF12. *Let A and R be relations in $\mathcal{T}$, and let x be a letter. Then*

$$(\exists_A x)R \quad \text{and} \quad (\forall_A x)R$$

are relations in $\mathcal{T}$.

This follows directly from CF11 (no. 1), CF9 (§3, no. 4), and CF2 (§1, no. 4).

Intuitively, consider A and R as expressing properties of x. It may happen that in a series of proofs, we are concerned only with objects satisfying A. To say that there exists an object satisfying A such that R means that there exists an object such that "A and R"; whence the definition of $\exists_{A}$. To say that all objects which satisfy A have the property R means that there is no object satisfying A such that "not R"; whence the definition of $\forall_{A}$. In practice, these signs are replaced by various phrases, depending on the nature of the relation A. * For example: "for all integers x, R"; "there exists an element x of the set E such that R"; and so on. *

C35. *Let A and R be relations in $\mathfrak{T}$, and let x be a letter. Then the relations $(\forall_{A}x)R$ and $(\forall x)(A \Longrightarrow R)$ are equivalent in $\mathfrak{T}$.*

For the relation $(\forall_{A}x)R$ is identical with

$$\text{"not}(\exists x)(A \text{ and } (\text{not } R))\text{"}.$$

Now, "A and (not R)" is equivalent in $\mathfrak{T}_0$ to "not $(A \Longrightarrow R)$"; therefore "not $(\exists x)(A$ and (not R))" is equivalent in $\mathfrak{T}_0$ to

$$\text{"not }(\exists x)(\text{not }(A \Longrightarrow R))\text{"},$$

by C31 (no. 3), and the latter relation is identical with $(\forall x)(A \Longrightarrow R)$. The criterion is therefore established in $\mathfrak{T}_0$, and consequently in $\mathfrak{T}$.

We shall often have to prove relations of the form $(\forall_{A}x)R$; generally we shall use one of the following two criteria:

C36. *Let A and R be relations in $\mathfrak{T}$, and let x be a letter. Let $\mathfrak{T}'$ be the theory obtained by adjoining A to the axioms of $\mathfrak{T}$. If x is not a constant of $\mathfrak{T}$, and if R is a theorem in $\mathfrak{T}'$, then $(\forall_{A}x)R$ is a theorem in $\mathfrak{T}$.*

For $A \Longrightarrow R$ is a theorem in $\mathfrak{T}$, by the criterion of deduction, therefore $(\forall_{A}x)R$ is a theorem in $\mathfrak{T}$ by C27 (no. 1) and C35.

In practice, we indicate that we are going to use this rule by a phrase such as "Let x be any element such that A". In the theory $\mathfrak{T}'$ so defined, we seek to prove R. Of course, we cannot assert that R itself is a theorem in $\mathfrak{T}$.

C37. *Let A and R be relations in $\mathfrak{T}$ and let x be a letter. Let $\mathfrak{T}'$ be the theory obtained by adjoining the relations A and* "not R" *to the axioms of $\mathfrak{T}$. If x is not a constant of $\mathfrak{T}$, and if $\mathfrak{T}'$ is contradictory, then $(\forall_{A}x)R$ is a theorem in $\mathfrak{T}$.*

For the theory $\mathfrak{T}'$ is equivalent to the theory obtained by adjoining "not $(A \Longrightarrow R)$" to the axioms of $\mathfrak{T}$. By the method of *reductio ad absurdum*, $A \Longrightarrow R$ is a theorem in $\mathfrak{T}$, and therefore so is $(\forall_{A}x)R$ by C27 (no. 1) and C35.

In practice we say : "Suppose that there exists an object x satisfying A for which R is false", and seek to establish a contradiction.

The properties of typical quantifiers are analogous to those of quantifiers :

C38. *Let A and R be relations in $\mathfrak{T}$, and let x be a letter. Then the relations*

$$\text{not } (\forall_A x)R \Longleftrightarrow (\exists_A x)(\text{not } R),$$
$$\text{not } (\exists_A x)R \Longleftrightarrow (\forall_A x)(\text{not } R)$$

are theorems in $\mathfrak{T}$.

C39. *Let A, R, and S be relations in $\mathfrak{T}$, and let x be a letter which is not a constant of $\mathfrak{T}$. If the relation $A \Longrightarrow (R \Longrightarrow S)$ [resp. $A \Longrightarrow (R \Longleftrightarrow S)$] is a theorem in $\mathfrak{T}$, then the relations*

$$(\exists_A x)R \Longrightarrow (\exists_A x)S, \qquad (\forall_A x)R \Longrightarrow (\forall_A x)S$$
$$[\text{resp. } (\exists_A x)R \Longleftrightarrow (\exists_A x)S, \qquad (\forall_A x)R \Longleftrightarrow (\forall_A x)S]$$

are theorems in $\mathfrak{T}$.

C40. *Let A, R, and S be relations in $\mathfrak{T}$ and let x be a letter. Then the relations*

$$(\forall_A x)(R \text{ and } S) \Longleftrightarrow ((\forall_A x)R \text{ and } (\forall_A x)S),$$
$$(\exists_A x)(R \text{ or } S) \Longleftrightarrow ((\exists_A x)R \text{ or } (\exists_A x)S)$$

are theorems in $\mathfrak{T}$.

C41. *Let A, R, and S be relations in $\mathfrak{T}$, and let x be a letter which does not appear in R. Then the relations*

$$(\forall_A x)(R \text{ or } S) \Longleftrightarrow (R \text{ or } (\forall_A x)S),$$
$$(\exists_A x)(R \text{ and } S) \Longleftrightarrow (R \text{ and } (\exists_A x)S)$$

are theorems in $\mathfrak{T}$.

C42. *Let A, B, R be relations in $\mathfrak{T}$ and let x and y be letters. If x does not appear in B, and if y does not appear in A, then the relations*

$$(\forall_A x)(\forall_B y)R \Longleftrightarrow (\forall_B y)(\forall_A x)R,$$
$$(\exists_A)x(\exists_B y)R \Longleftrightarrow (\exists_B y)(\exists_A x)R,$$
$$(\exists_A x)(\forall_B y)R \Longrightarrow (\forall_B y)(\exists_A x)R$$

are theorems in $\mathfrak{T}$.

By way of example, let us prove part of C42. The relation

$$(\exists_A x)(\exists_B y)R$$

is identical with $(\exists x)(A$ and $(\exists y)(B$ and $R))$, and therefore (because y does not appear in A) is equivalent in $\mathcal{T}_0$ to

$$(\exists x)(\exists y)(A \text{ and } (B \text{ and } R))$$

by C33 and C31. Likewise, $(\exists_B x)(\exists_A y)R$ is equivalent to

$$(\exists y)(\exists x)(B \text{ and } (A \text{ and } R)).$$

Now apply C31 and C34 (no. 3).

* As an example of the application of these criteria, consider the following relation : "the sequence of real-valued functions (f_n) converges uniformly to 0 in [0, 1]". This means "for each $\varepsilon > 0$ there exists an integer n such that for each $x \in [0, 1]$ and each integer $m \geqslant n$ we have $|f_m(x)| \leqslant \varepsilon$". Suppose we wish to take the negation of this relation (for example, to obtain a proof by contradiction); the criterion C38 shows that this negation is equivalent to the following relation : " there exists an $\varepsilon > 0$ such that for each integer n there exists an $x \in [0, 1]$ and an $m \geqslant n$ for which $|f_m(x)| > \varepsilon$".

5. EQUALITARIAN THEORIES

1. THE AXIOMS

An *equalitarian theory* is a theory $\mathcal{T}$ which has a relational sign of weight 2, written $=$ (read "equals"), and in which the schemes S1 through S5 (§§3 and 4), together with the schemes S6 and S7 below, provide implicit axioms. If T and U are terms in $\mathcal{T}$, the assembly $= TU$ is a relation in $\mathcal{T}$ (called the *relation of equality*) by virtue of CF4; in practice it is denoted by $T = U$ or $(T) = (U)$.

S6. *Let x be a letter, let T and U be terms in $\mathcal{T}$, and let $R\{x\}$ be a relation in $\mathcal{T}$; then the relation $(T = U) \Rightarrow (R\{T\} \Leftrightarrow R\{U\})$ is an axiom.*

S7. *If R and S are relations in $\mathcal{T}$ and if x is a letter, then the relation $((\forall x)(R \Leftrightarrow S)) \Rightarrow (\tau_x(R) = \tau_x(S))$ is an axiom.*

To show that the rule S6 is a scheme, let A be an axiom of $\mathcal{T}$, obtained by applying S6; then there is a relation R in $\mathcal{T}$, terms T and U in $\mathcal{T}$, and a letter x, such that A is $(T=U) \Longrightarrow ((T|x)R \Longleftrightarrow (U|x)R)$. We shall show that if y is a letter and V a term in $\mathfrak{T}$, the relation $(V|x)A$ can be obtained by applying S6. By means of CS1 (§1, no. 2) we may assume that x is distinct from y and does not appear in V. Let T', U', R' denote the assemblies $(V|y)T$, $(V|y)U$, $(V|y)R$ respectively. By CS2 and CS5 (§1, no. 2), $(V|y)A$ is identical with

$$(T' = U') \Longrightarrow ((T'|x')R' \Longleftrightarrow (U'|x')R')$$

and the proof is complete. The verification that S7 is a scheme is similar.

Intuitively, the scheme S6 means that if two objects are equal, they have the same properties. Scheme S7 is more remote from everyday intuition; it means that if two properties R and S of an object x are equivalent, then the distinguished objects $\tau_x(R)$ and $\tau_x(S)$ (chosen respectively from the objects which satisfy R, and those which satisfy S, if such objects exist) are equal. The reader will note that the presence in S7 of the quantifier $\forall x$ is essential (cf. Exercise 7).

The negation of the relation $= TU$ is denoted by $T \neq U$ or $(T) \neq (U)$ (where the sign $\neq$ is read "is different from").

¶ From S6 we deduce the following criterion :

C43. *Let x be a letter, let T and U be terms in $\mathfrak{T}$, and let $R\{x\}$ be a relation in $\mathfrak{T}$. Then the relations*

$$(T = U \text{ and } R\{T\}), \quad (T = U \text{ and } R\{U\})$$

are equivalent.

For if we adjoin the hypotheses $T = U$ and $R\{T\}$, then $R\{U\}$ is true by S6; therefore $(T = U$ and $R\{U\})$ is true.

When a relation of the form $T = U$ has been proved in a theory $\mathfrak{T}$, it is often said (by abuse of language) that T and U "are the same" or are "identical". Likewise, when $T \neq U$ is true in $\mathfrak{T}$, we say that T and U "are distinct" in place of saying that T is different from U.

2. PROPERTIES OF EQUALITY

From now on we shall consider only equalitarian theories. Let $\mathfrak{T}$ be such a theory, and let $\mathfrak{T}_0$ be the theory whose signs are those of $\mathfrak{T}$ and whose only axioms are those provided by schemes S1 through S7. The theory $\mathfrak{T}_0$ is weaker than $\mathfrak{T}$ (§2, no. 4) and has no constants. The following three theorems are theorems in $\mathfrak{T}_0$.

Theorem 1. $x = x$.

Let S denote the relation $x = x$ in $\mathcal{T}_0$. By C27 (§4, no. 1), for every relation R in $\mathcal{T}_0$, $(\forall x)(R \Longleftrightarrow R)$ is a theorem in $\mathcal{T}_0$, and therefore, by S7, $\tau_x(R) = \tau_x(R)$, that is to say $(\tau_x(R)|x)S$, is a theorem in $\mathcal{T}_0$. Taking R to be the relation "not S" and considering C26 (§4, no. 1), we see that $(\forall x)S$ is a theorem in $\mathcal{T}_0$. By C30 (§4, no. 3), S is therefore a theorem in $\mathcal{T}_0$.

The relation $(\forall x)(x = x)$ is also a theorem in $\mathcal{T}_0$; and if T is a term in $\mathcal{T}_0$, then $T = T$ is a theorem in $\mathcal{T}_0$ (cf. §4, no. 3). It is possible to transform later theorems in the same way into theorems in which no letter appears or into metamathematical criteria. From now on we shall not explicitly perform these transformations, but we shall often implicitly make use of them.

Theorem 2. $(x = y) \Longleftrightarrow (y = x)$.

Suppose that the relation $x = y$ is true. By S6, the relation

$$(x = y) \Longrightarrow ((x|y)(y = x) \Longleftrightarrow (y|y)(y = x)),$$

that is

$$(x = y) \Longrightarrow ((x = x) \Longleftrightarrow (y = x)),$$

is true. Therefore $(x = x) \Longleftrightarrow (y = x)$ is true. By Theorem 1 it follows that $y = x$ is true, and the theorem is proved.

Theorem 3. $((x = y)$ and $(y = z)) \Longrightarrow (x = z)$.

Let us adjoin the hypotheses $x = y$, $y = z$ to the axioms of $\mathcal{T}_0$. By S6 the relation $(x = y) \Longrightarrow ((x = z) \Longleftrightarrow (y = z))$ is true. Hence

$$(x = z) \Longleftrightarrow (y = z),$$

and consequently $x = z$, are true.

C44. *Let x be a letter and let T, U, $V\{x\}$ be terms in $\mathcal{T}_0$. Then the relation $(T = U) \Longrightarrow (V\{T\} = V\{U\})$ is a theorem in $\mathcal{T}_0$.*

For let y and z be two distinct letters which are distinct from x and from the letters which appear in T, U, V. Adjoin the hypothesis $y = z$. Then, by S6,

$$((Y|z)(V\{y\} = V\{z\})) \Longleftrightarrow (V\{y\} = V\{z\}),$$

that is to say $(V\{y\} = V\{y\}) \Longleftrightarrow (V\{y\} = V\{z\})$, is true. Now, $V\{y\} = V\{y\}$ is true by Theorem 1; hence $V\{y\} = V\{z\}$ is true.

From all this it follows that $(y = z) \Longrightarrow (V\{y\} = V\{z\})$ is a theorem in $\mathcal{T}_0$, say A. But $(T|y)(U|z)A$ is precisely

$$(T = U) \Longrightarrow (V\{T\} = V\{U\}).$$

¶ A relation of the form $T = U$, where T and U are terms in $\mathcal{T}$, is called an *equation*; a *solution* (in $\mathcal{T}$) of the relation $T = U$, considered as an equation in a letter x, is therefore (§2, no. 2) a term V in $\mathcal{T}$ such that $T\{V\} = U\{V\}$ is a theorem in $\mathcal{T}$.

¶ Let T and U be two terms in $\mathcal{T}$, and let $x_1, x_2, \ldots, x_n$ be the letters which appear in T but not in U. If the relation

$$(\exists x_1)\ldots(\exists x_n)(T = U)$$

is a theorem in $\mathcal{T}$, we say that *U can be put in the form T* (in $\mathcal{T}$). Let R be a relation in $\mathcal{T}$ and let y be a letter. Let V be a solution (in $\mathcal{T}$) of R, considered as a relation in y. If every solution (in $\mathcal{T}$) of R, considered as a relation in y, can be put in the form V, then V is said to be the *complete solution* (or *general solution*) of R (in $\mathcal{T}$).

3. FUNCTIONAL RELATIONS

Let R be an assembly and x a letter. Let y and z be distinct letters which are distinct from x and which do not appear in R. Let y', z' be two other letters with the same properties. By CS8, CS9 (§4, no. 1), CS2, CS5 (§1, no. 2), and CS6 (§3, no. 4), the assemblies

$$(\forall y)(\forall z)(((y|x)R \text{ and } (z|x)R) \Longrightarrow (y = z))$$

and

$$(\forall y')(\forall z')(((y'|x)R \text{ and } (z'|x)R) \Longrightarrow (y' = z'))$$

are identical. If R is a relation in $\mathcal{T}$, the assembly thus defined is a relation in $\mathcal{T}$ which is denoted by "there exists at most one x such that R"; the letter x does not appear in this relation. When this relation is a theorem in $\mathcal{T}$, R is said to be *single-valued* in X in $\mathcal{T}$. To prove that R is single-valued in the theory $\mathcal{T}$, it is enough to prove $y = z$ in the theory obtained by adjoining to $\mathcal{T}$ the axioms $(y|x)R$ and $(z|x)R$, where y and z are distinct letters which are distinct from x and appear neither in R nor in the explicit axioms of $\mathcal{T}$.

C45. *Let R be a relation in $\mathcal{T}$, and let x be a letter which is not a constant of $\mathcal{T}$. If R is single-valued in x in $\mathcal{T}$, then $R \Longrightarrow (x = \tau_x(R))$ is a theorem in $\mathcal{T}$. Conversely if, for some term T in $\mathcal{T}$ which does not contain x, $R \Longrightarrow (x = T)$ is a theorem in $\mathcal{T}$, then R is single-valued in x in $\mathcal{T}$.*

Suppose R is single-valued in x in $\mathcal{T}$, and let us show that

$$R \Longrightarrow (x = \tau_x(R))$$

is a theorem in $\mathcal{T}$. Adjoin the hypothesis R. Then $(\tau_x(R)|x)R$ is true by S5, and hence "R and $(\tau_x(R)|x)R$" is true. Now, since R is single-valued in x,

$$(R \text{ and } (\tau_x(R)|x)R) \Longrightarrow (x = \tau_x(R))$$

is a theorem in $\mathcal{T}$ by C30 (§4, no. 3). Therefore $x = \tau_x(R)$ is true.

¶ Conversely, suppose that $R \Longrightarrow (x = T)$ is a theorem in $\mathcal{T}$. Let y, z be distinct letters which are distinct from x and which appear neither in R nor in the explicit axioms of $\mathcal{T}$. Since x is not a constant of $\mathcal{T}$ and does not appear in T, the relations

$$(y|x)R \Longrightarrow (y = T), \qquad (z|x)R \Longrightarrow (z = T)$$

are theorems in $\mathcal{T}$. Adjoin the hypotheses $(y|x)R$ and $(z|x)R$. Then $y = T$ and $z = T$ are true, hence $y = z$ is true.

¶ Let R be a relation in $\mathcal{T}$. The relation

"$(\exists x)R$ and there exists at most one x such that R"

is denoted by "there exists exactly one x such that R". If this relation is a theorem in $\mathcal{T}$, R is said to be a *functional relation in* x in the theory $\mathcal{T}$.

C46. *Let R be a relation in $\mathcal{T}$, and let x be a letter which is not a constant of $\mathcal{T}$. If R is functional in x in $\mathcal{T}$, then $R \Longleftrightarrow (x = \tau_x(R))$ is a theorem in $\mathcal{T}$. Conversely, if for some term T in $\mathcal{T}$ which does not contain x,*

$$R \Longleftrightarrow (X = T)$$

is a theorem in $\mathcal{T}$, then R is functional in x in $\mathcal{T}$.

Suppose R is functional in x in $\mathcal{T}$. Then $R \Longrightarrow (x = \tau_x(R))$ is a theorem in $\mathcal{T}$ by C45. On the other hand, $(\exists x)R$ is a theorem in T. By S6 the relation

$$(x = \tau_x(R)) \Longrightarrow (R \Longleftrightarrow (\exists x)R)$$

is a theorem in $\mathcal{T}$. If we adjoin the hypothesis $x = \tau_x(R)$, it follows that R is true. Therefore $(x = \tau_x(R)) \Longrightarrow R$ is a theorem in $\mathcal{T}$.

¶ Conversely, if $R \Longleftrightarrow (x = T)$ is a theorem in $\mathcal{T}$, then R is single-valued in x in $\mathcal{T}$, by C45. Moreover, $(T|x)R \Longleftrightarrow (T = T)$ is a theorem in $\mathcal{T}$; hence $(T|x)R$ and therefore $(\exists x)R$ are theorems in $\mathcal{T}$.

¶ If a relation R is functional in x in T, then R is equivalent to the relation $x = \tau_x(R)$, which is often more manageable. Generally an abbreviating symbol Σ is introduced to represent the term $\tau_x(R)$. Such a symbol is called a *functional symbol* in $\mathcal{T}$.

Intuitively, Σ represents the unique object which has the property defined by R. * For example, in a theory where "y is a real number $\geqslant 0$" is a theorem, the relation "x is a real number $\geqslant 0$ and $y = x^2$" is functional in x. The corresponding functional symbol is taken to be either $\sqrt{y}$ or $y^{1/2}$. *

C47. *Let x be a letter which is not a constant of $\mathcal{T}$, and let $R\{x\}$ and $S\{x\}$ be two relations in $\mathcal{T}$. If $R\{x\}$ is functional in x in $\mathcal{T}$, then the relation $S\{\tau_x(R)\}$ is equivalent to* $(\exists x)(R\{x\}$ and $S\{x\})$.

For it follows from C46 and C43 that ($R\{x\}$ and $S\{x\}$) is equivalent to ($R\{x\}$ and $S\{\tau_x(R)\}$); since $S\{\tau_x(R)\}$ does not contain x,

$$(\exists x)(R\{x\} \text{ and } S\{\tau_x(R)\})$$

is equivalent to ($S\{\tau_x(R)\}$ and $(\exists x)R$) by C33 (§ 4, no. 3); and the result follows from the fact that $(\exists x)R$ is true, because R is functional in x.

APPENDIX

CHARACTERIZATION OF TERMS AND RELATIONS

Metamathematics, when it goes beyond the very elementary level of the present chapter, makes considerable use of the results of mathematics, as we remarked in the Introduction. The purpose of this Appendix is to give a simple example of this type of reasoning (*). We shall begin by establishing certain results which belong to the mathematical theory of *free semigroups*; we shall then use these results in a metamathematical " application " to obtain a characterization of the terms and relations in a theory.

1. SIGNS AND WORDS

* Let S be a non-empty set, whose elements will be called *signs* (this terminology being appropriate to the metamathematical application we have in mind). Let $L_0(S)$ be the free semigroup generated by S; the elements of $L_0(S)$ are called *words* and are identified with finite sequences $A = (s_i)_{0 \leqslant i \leqslant n}$ of elements of S. The law of composition in $L_0(S)$ will be written multiplicatively, so that AB is the sequence obtained by juxtaposition of A and B. The empty word $\varnothing$ is the identity element of $L_0(S)$. We recall that the *length* $l(A)$ of a word $A \in L_0(S)$ is the number of elements in the sequence A; thus $l(AB) = l(A) + l(B)$, and the words of length 1 are the signs. Let L(S) denote the set of non-empty words in $L_0(S)$.

¶ Suppose, moreover, that we are given a mapping $s \rightarrow n(s)$ of S into

(*) The results established in this Appendix will not be used anywhere else in this series.

the set **N** of integers $\geqslant 0$. For each non-empty word $A = (s_i)_{0 \leqslant i \leqslant k}$ of L(S), we put

$$n(A) = \sum_{i=0}^{k} n(s_i)$$

and $n(\emptyset)=0$. $n(A)$ is called the *weight* of A. Clearly $n(AB)=n(A)+n(B)$.
¶ If $A=A'BA''$, the word B is said to be a *segment* of A (a *proper* segment if also $B \neq A$). If A' (resp. A'') is empty, B is said to be an *initial* (resp. *final*) segment of A. If $l(A') = k$, then B is said to *begin at the* $(k+1)$th *place*.

¶ If $A = BCDEF$ (where the words B, C, D, E, F may be empty) the segments C and E of A are said to be *disjoint*.

2. SIGNIFICANT WORDS

A *significant sequence* is any sequence $(A_j)_{1 \leqslant j \leqslant n}$ of words of $L_0(S)$ with the following property : for each word A_i of the sequence, one of the following two conditions is satisfied :

(1) A_i is a sign of weight 0.

(2) There exist p words $A_{i_1}, A_{i_2}, \ldots, A_{i_p}$ in the sequence, with indices less than i, and a sign f of weight p such that

$$A_i = f A_{i_1} A_{i_2} \ldots A_{i_p}.$$

Words which appear in significant sequences are called *significant words.* Then we have :

Proposition 1. *If $A_1, A_2, \ldots, A_p$ are p significant words and if f is a sign of weight p, then the word $f A_1 A_2 \ldots A_p$ is significant.*

3. CHARACTERIZATION OF SIGNIFICANT WORDS

A word $A \in L_0(S)$ is said to be *balanced* if it has the following two properties :

(1) $l(A) = n(A) + 1$ (which implies that A is not empty).

(2) For every proper initial segment B of A, $l(B) \leqslant n(B)$.

Proposition 2. *A word is significant if and only if it is balanced.*

Let A be a significant word belonging to a significant sequence

$$A_1, A_2, \ldots, A_n.$$

We shall show by induction on k that each A_k is balanced. Suppose that this has been established for the A_j with index $j < k$, and let us show that it is true for A_k. If A_k is a sign of weight 0 (which is the only possibility when $k = 1$), A_k is balanced because

$$l(A_k) = 1 \quad \text{and} \quad n(A_k) = 0.$$

If A_k is not a sign of weight 0, then $A_k = fB_1B_2 \ldots B_p$, where f is a sign of weight p, and the B_j are of the form A_{i_j}, where $i_j < k$, and are therefore balanced words by the inductive hypothesis. We have

$$\begin{aligned} l(A_k) &= 1 + l(B_1) + l(B_2) + \cdots + l(B_p) \\ &= 1 + (n(B_1) + 1) + (n(B_2) + 1) + \cdots + (n(B_p) + 1) \\ &= 1 + p + n(B_1) + n(B_2) + \cdots + n(B_p) \\ &= 1 + n(A_k). \end{aligned}$$

Let C be a proper initial segment of A, and let q be the largest of the integers $m < p$ such that B_m is a segment of C, so that

$$C = f B_1B_2 \ldots B_qD,$$

where D is a proper initial segment of B_{q+1}. Then

$$\begin{aligned} l(C) &= 1 + l(B_1) + l(B_2) + \cdots + l(B_q) + l(D) \\ &\leqslant 1 + (n(B_1) + 1) + (n(B_2) + 1) + \cdots + (n(B_q) + 1) + n(D) \\ &\leqslant p + n(B_1) + \cdots + n(B_q) + n(D) = n(C). \end{aligned}$$

Hence A_k is balanced.

¶ To prove that, conversely, every balanced word is significant, we need the following two lemmas :

Lemma 1. *Let A be a balanced word. Then for each integer k such that $0 \leqslant k < l(A)$ there exists exactly one balanced segment S of A which begins at the $(k + 1)$th place.*

The uniqueness of S is an immediate consequence of the following remark : if T is a balanced word, then by definition no proper initial segment of T is balanced. Let us prove the existence of S. Write $A = BC$ where $l(B) = k$. For each i such that $0 \leqslant i \leqslant q = l(C)$, let C_i be the initial segment of C of length i. Since B is a proper initial segment of A, we have

$$l(C_q) = l(A) - l(B) \geqslant n(A) + 1 - n(B) = n(C_q) + 1.$$

On the other hand, we have $0 = l(C_0) \leqslant n(C_0) = 0$. Let i be the largest of the integers $j < q$ such that $l(C_h) \leqslant n(C_h)$ for $0 \leqslant h \leqslant j$; then we have $l(C_i) \leqslant n(C_i)$ and $l(C_{i+1}) \geqslant n(C_{i+1}) + 1$. We show that C_{i+1} is balanced : the condition relating to proper initial segments is satisfied by reason of the definition of i; on the other hand, we have

$$n(C_{i+1}) + 1 \leqslant l(C_{i+1}) = l(C_i) + 1 \leqslant n(C_i) + 1 \leqslant n(C_{i+1}) + 1,$$

so that $l(C_{i+1}) = n(C_{i+1}) + 1$, and the proof of Lemma 1 is complete.

LEMMA 2. *Every balanced word A can be put in the form*

$$A = f A_1 A_2 \ldots A_p,$$

where the A_i are balanced and $n(f) = p$.

Let f be the initial sign of A. By Lemma 1, A can be written as

$$f A_1 A_2 \ldots A_p,$$

where the A_i are balanced : we need only define A_i inductively as the balanced segment of A which begins at the $k(i)$th place, where

$$k(i) = 2 + \sum_{j<i} l(A_j).$$

Moreover, we have

$$\begin{aligned} 1 + l(A_1) + \cdots + l(A_p) = l(A) &= n(A) + 1 \\ &= n(f) + n(A_1) + \cdots + n(A_p) + 1 \\ &= n(f) + (l(A_1) - 1) + \cdots + (l(A_p) - 1) + 1, \end{aligned}$$

from which it follows that $n(f) = p$.

¶ Now that the two lemmas have been proved, it is obvious by induction on the length of A that every balanced word A is significant, by reason of Lemma 2 and Proposition 1.

COROLLARY 1. *Let A be a significant word. For each integer k such that $0 \leqslant k < l(A)$ there is exactly one significant segment of A which begins at the $(k+1)$th place.*

COROLLARY 2. *Every significant word may be written in exactly one way in the form $f A_1 A_2 \ldots A_p$, where the A_i are significant and $n(f) = p$.**

4. APPLICATION TO ASSEMBLIES IN A MATHEMATICAL THEORY

Suppose that the set S is the set of signs of a mathematical theory $\mathcal{T}$. We put $n(\square) = 0$, $n(\tau) = n(\urcorner) = 1$, $n(\vee) = 2$, $n(x) = 0$ for every letter x;

and finally, for every specific sign s of $\mathcal{T}$, $n(s)$ is the weight of s, which is fixed when $\mathcal{T}$ is given.

¶ Let A be an assembly in $\mathcal{T}$. We denote by A^* the word obtained by deleting the links in A, and we shall say that A is *balanced* if A^* is balanced [in $L_0(S)$]. A *segment* of A is any assembly obtained by replacing, in a segment S of A^*, the links which, in A, join pairs of signs in S.

CRITERION 1. *If A is a term or a relation in $\mathcal{T}$, then A is balanced.*

Let $A_1, A_2, \ldots, A_n$ be a formative construction in $\mathcal{T}$, in which A appears. Let us argue by induction and suppose that we have proved that the A_j of indices $j < i$ are balanced; then we have to prove that A_i is balanced. The proof goes just as in the first part of the proof of Proposition 2, except when A_i is of the form $\tau_x(B)$ with $B = A_j$ and $j < i$. In this case, let C be the assembly obtained by replacing x, wherever it occurs in B, by $\square$. The word A_i^* is then identical with τC^*; now B^* is balanced, and therefore C^* is balanced (because $n(\square) = n(x) = 0$). Consequently A_i^* is balanced.

¶ We have thus obtained a necessary condition for an assembly in $\mathcal{T}$ to be a term or a relation. But this condition is not sufficient, as we shall see.
¶ Let A be a balanced assembly in $\mathcal{T}$. If A begins with a letter or a $\square$, then A must consist of this sign alone (Proposition 2, Corollary 2). For all other cases, we shall now define the assembly or assemblies *antecedent* to A.

(1) If A begins with a $\neg$, or a $\vee$, or a specific sign, then A^* can be put in exactly one way in the form $fB_1B_2 \ldots B_p$, where f is a sign of weight $p \geqslant 1$ and the B_i are balanced (Proposition 2, Corollary 2). The segments $A_1, A_2, \ldots, A_p$ of A which correspond to the segments $B_1, B_2, \ldots, B_p$ of A^* are called the assemblies *antecedent* to A. Moreover, we shall say that A is *perfectly balanced* if A is identical with

$$fA_1A_2 \ldots A_p,$$

in other words if no link in A joins f to one of the B_i, or joins two distinct B_i.

(2) If A begins with a τ, then A^* is of the form τB, where B is balanced (Proposition 2, Corollary 2). In this case an assembly antecedent to A is any of the assemblies A_1 defined as follows : replace the signs $\square$ in B which are linked in A to the initial τ by a letter x distinct from the other letters which appear in B, and replace the links which join two signs of B in A. (If, instead of x, we substitute a letter y which also does not appear in B, we obtain an assembly which is just $(y|x)A_1$.) Moreover, we shall say that A is *perfectly balanced* if A is identical with

$\tau_{x}(A_1)$, in other words if no link joins the initial τ to any sign of B other than a $\square$.

¶ We can now state the following criterion :

CRITERION 2. *Let A be a balanced assembly in* $\mathcal{T}$.

¶ *For A to be a term it is necessary and sufficient that one of the following conditions be satisfied :* (1) *A consists of a single letter;* (2) *A begins with a* τ, *is perfectly balanced, and its antecedent assemblies are relations* (by CF8, it is enough to verify that *one* antecedent assembly is a relation); (3) *A begins with a substantific sign, is perfectly balanced, and its antecedent assemblies are terms.*
¶ *For A to be a relation it is necessary and sufficient that one of the following conditions be satisfied :* (1) *A begins with a* $\vee$ *or a* $\neg$, *is perfectly balanced, and its antecedent assemblies are relations;* (2) *A begins with a relational sign, is perfectly balanced, and its antecedent assemblies are terms.*

The criteria CF1 to CF4 (§ 1, no. 4) show that the conditions are sufficient. Let us show that they are necessary. We have already seen (§ 1, no. 3) that if A is a relation, then A begins with a $\vee$, or a $\neg$, or a relational sign. The reasoning is similar in each of the three cases. If, for example, A begins with a $\vee$, then A is of the form $\vee BC$, where B and C are relations, so that B and C are the assemblies antecedent to A; hence A is perfectly balanced. If A is a term, then there are two cases : it consists of a single letter, or it begins with either a substantific sign or a τ. In the second case, we argue as above. If A begins with a τ, the definition of a formative construction shows that A is of the form $\tau_{x}(B)$, where B is a relation and x a letter, so that we may take B as assembly antecedent to A, and A is perfectly balanced.

When we wish to test whether a given assembly A (not consisting of a single letter) is a relation (resp. a term) in $\mathcal{T}$, we first verify that A is balanced and that it begins with a $\vee$, a $\neg$, or a relational sign (resp. with a τ or a substantific sign). Then we form the antecedent assembly or assemblies, and verify (if appropriate) that A is perfectly balanced. Having done this, we are left with an analogous problem relating to shorter assemblies. Thus, step by step, we come down to assemblies each of which consists of a single sign, for which the solution is immediate.

Remark. Except in certain mathematical theories which are particularly weak in axioms (cf. Exercise 7) we have no general procedure of this type to enable us to test whether or not a given relation R in a theory $\mathcal{T}$ is a theorem in $\mathcal{T}$.

EXERCISES

§ 1

1. Let $\mathfrak{T}$ be a theory with no specific signs. Show that no assembly in $\mathfrak{T}$ is a relation, and that the only assemblies in $\mathfrak{T}$ which are terms are assemblies consisting of a single letter.

2. Let $\boldsymbol{A}$ be a term or a relation in a theory $\mathfrak{T}$. Show that every sign $\square$ in $\boldsymbol{A}$ (if there is any) is linked to a single sign τ situated to its left. Show that every sign τ in $\boldsymbol{A}$ (if there is any) is either not linked at all or else is linked to certain signs $\square$ situated to its right. Show also that no other sign is linked.

3. Let $\boldsymbol{A}$ be a term or a relation in a theory $\mathfrak{T}$. Show that every specific sign, if there is any, is followed by $\square$, or τ, or a letter, or a substantific sign.

¶ 4. Let $\boldsymbol{A}$ be a term or a relation in a theory $\mathfrak{T}$, and let $\boldsymbol{B}$ be an assembly in $\mathfrak{T}$. Show that $\boldsymbol{AB}$ is neither a term nor a relation in $\mathfrak{T}$. (Use induction on the number of signs in $\boldsymbol{A}$.)

5. Let $\boldsymbol{A}$ be an assembly in a theory $\mathfrak{T}$ and let x be a letter. Show that if $\tau_x(\boldsymbol{A})$ is a term in $\mathfrak{T}$, then $\boldsymbol{A}$ is a relation in $\mathfrak{T}$.

6. Let $\boldsymbol{A}$ and $\boldsymbol{B}$ be assemblies in a theory $\mathfrak{T}$. If $\boldsymbol{A}$ and $\Longrightarrow \boldsymbol{AB}$ are relations in $\mathfrak{T}$, prove that $\boldsymbol{B}$ is a relation in $\mathfrak{T}$ (use Exercise 4).

§ 2

1. Let $\mathfrak{T}$ be a theory, let $\boldsymbol{A}_1, \boldsymbol{A}_2, \ldots, \boldsymbol{A}_n$ be its explicit axioms and $a_1, a_2, \ldots, a_h$ its constants.

(a) Let $\mathfrak{T}'$ be the theory whose signs and schemes are those of $\mathfrak{T}$, and whose explicit axioms are $A_1, A_2, \ldots, A_{n-1}$. The axiom A_n is said to be *independent* of the other axioms of $\mathcal{T}$ if $\mathcal{T}'$ is not equivalent to $\mathcal{T}$. This will be so if and only if A_n is not a theorem of $\mathcal{T}'$.

(b) Let $\mathcal{T}''$ be a theory whose signs and schemes are the same as those of $\mathcal{T}$. Let $T_1, T_2, \ldots, T_h$ be terms of $\mathcal{T}$ such that

$$(T_1|a_1)(T_2|a_2) \ldots (T_h|a_h)A_i$$

is a theorem of $\mathcal{T}''$ for $i = 1, 2, \ldots, n-1$, and such that "not $(T_1|a_1)(T_2|a_2) \ldots (T_h|a_h)A_n$" is a theorem of $\mathcal{T}''$. Then either A_n is independent of the other axioms of $\mathcal{T}$, or else $\mathcal{T}''$ is contradictory.

§ 3

1. Let A, B, C be relations in a logical theory $\mathcal{T}$. Show that the following relations are theorems in $\mathcal{T}$:

$$A \Rightarrow (B \Rightarrow A),$$
$$(A \Rightarrow B) \Rightarrow ((B \Rightarrow C) \Rightarrow (A \Rightarrow C)),$$
$$A \Rightarrow ((\text{not } A) \Rightarrow B),$$
$$(A \text{ or } B) \Leftrightarrow ((A \Rightarrow B) \Rightarrow B),$$
$$(A \Leftrightarrow B) \Leftrightarrow ((A \text{ and } B) \text{ or } ((\text{not } A) \text{ and } (\text{not } B))),$$
$$(A \Leftrightarrow B) \Leftrightarrow \text{not } ((\text{not } A) \Leftrightarrow B),$$
$$(A \Rightarrow (B \text{ or } (\text{not } C))) \Leftrightarrow ((C \text{ and } A) \Rightarrow B),$$
$$(A \Rightarrow (B \text{ or } C)) \Leftrightarrow (B \text{ or } (A \Rightarrow C)),$$
$$(A \Rightarrow B) \Rightarrow ((A \Rightarrow C) \Rightarrow (A \Rightarrow (B \text{ and } C))),$$
$$(A \Rightarrow C) \Rightarrow ((B \Rightarrow C) \Rightarrow ((A \text{ or } B) \Rightarrow C)),$$
$$(A \Rightarrow B) \Rightarrow ((A \text{ and } C) \Rightarrow (B \text{ and } C)),$$
$$(A \Rightarrow B) \Rightarrow ((A \text{ or } C) \Rightarrow (B \text{ or } C)).$$

2. Let A be a relation in a logical theory $\mathcal{T}$. If $A \Leftrightarrow (\text{not } A)$ is a theorem of $\mathcal{T}$, then $\mathcal{T}$ is contradictory.

3. Let $A_1, A_2, \ldots, A_n$ be relations in a logical theory $\mathcal{T}$.

(a) To prove the relation "A_1 or A_2 or ... or A_n" in $\mathcal{T}$, it is enough to prove A_n in the theory obtained by adjoining to $\mathcal{T}$ the axioms not A_1, not A_2, ..., not A_{n-1}.

(b) If "A_1 or A_2 or ... or A_n" is a theorem in $\mathcal{T}$, then to prove a theorem A in $\mathcal{T}$ it is sufficient to prove the theorems

$$A_1 \Rightarrow A, \quad A_2 \Rightarrow A, \quad \ldots, \quad A_n \Rightarrow A.$$

4. Let A and B be relations in a logical theory $\mathcal{T}$. Let $A|B$ denote the relation "(not A) or (not B)". Prove the following theorems in $\mathcal{T}$:

$$\begin{aligned}(\text{not } A) &\Longleftrightarrow (A|A),\\ (A \text{ or } B) &\Longleftrightarrow ((A|A)|(B|B)),\\ (A \text{ and } B) &\Longleftrightarrow ((A|B)|(A|B)),\\ (A \Longrightarrow B) &\Longleftrightarrow (A|(B|B)).\end{aligned}$$

5. Let $\mathcal{T}$ be a logical theory with explicit axioms $A_1, A_2, \ldots, A_n$. Then A_n is independent of the other axioms of $\mathcal{T}$ (§ 2, Exercise) if and only if the theory whose signs and schemes are the same as those of $\mathcal{T}$, and whose explicit axioms are $A_1, A_2, \ldots, A_{n-1}$, "not A_n", is not contradictory.

§ 4

In all the Exercises for § 4, $\mathcal{T}$ *denotes a quantified theory.*

1. Let A and B be relations in $\mathcal{T}$, and let x be a letter which does not appear in A. Then $(\forall x)(A \Longrightarrow B) \Longleftrightarrow (A \Longrightarrow (\forall x)B)$ is a theorem in $\mathcal{T}$.

2. Let A and B be relations in $\mathcal{T}$, and let x be a letter, distinct from the constants of $\mathcal{T}$, which does not appear in A. If $B \Longrightarrow A$ is a theorem in $\mathcal{T}$, then $((\exists x)B) \Longrightarrow A$ is a theorem in $\mathcal{T}$.

3. Let A be a relation in $\mathcal{T}$ and let x and y be letters. The relations

$$(\forall x)(\forall y)A \Longrightarrow (\forall x)((x|y)A), \qquad (\exists x)((x|y)A) \Longrightarrow (\exists x)(\exists y)A$$

are then theorems in $\mathcal{T}$.

4. Let A and B be relations in $\mathcal{T}$ and let x be a letter. Show that the relations

$$\begin{aligned}&(\forall x)(A \text{ or } B) \Longrightarrow ((\forall x)A \text{ or } (\exists x)B),\\ &(\exists x)A \text{ and } (\forall x)B \Longrightarrow (\exists x)(A \text{ and } B)\end{aligned}$$

are theorems in $\mathcal{T}$.

5. Let A and B be relations in $\mathcal{T}$, and let x and y be letters. If x does not appear in B, nor y in A, then

$$(\forall x)(\forall y)(A \text{ and } B) \Longleftrightarrow ((\forall x)A \text{ and } (\forall y)B)$$

is a theorem in $\mathcal{T}$.

6. Let A and R be relations in $\mathcal{T}$, and let x be a letter. Then the relations $(\exists_A x)R \Longrightarrow (\exists x)R$, $(\forall x)R \Longrightarrow (\forall_A x)R$ are theorems in $\mathcal{T}$.

7. Let A and R be relations in $\mathfrak{T}$, and let x be a letter distinct from the constants of $\mathfrak{T}$. If $R \Longrightarrow A$ is a theorem in $\mathfrak{T}$, then

$$(\exists x)R \Longleftrightarrow (\exists_A x)R$$

is a theorem in $\mathfrak{T}$. If (not R) $\Longrightarrow A$ is a theorem in $\mathfrak{T}$, then

$$(\forall x)R \Longleftrightarrow (\forall_A x)R$$

is a theorem in $\mathfrak{T}$. In particular, if A is a theorem in $\mathfrak{T}$,

$$(\exists x)R \Longleftrightarrow (\exists_A x)R \quad \text{and} \quad (\forall x)R \Longleftrightarrow (\forall_A x)R$$

are theorems in $\mathfrak{T}$.

8. Let A and R be relations in $\mathfrak{T}$, let T be a term in $\mathfrak{T}$ and let x be a letter. If $(Tx|)A$ is a theorem in $\mathfrak{T}$, then $(T|x)R \Longrightarrow (\exists_A x)R$ and $(\forall_A x)R \Longrightarrow (T|x)R$ are theorems in $\mathfrak{T}$.

§ 5

In all the Exercises for § 5, $\mathfrak{T}$ *denotes an equalitarian theory.*

1. The relation $x = y$ is functional in x in $\mathfrak{T}$.

2. Let R be a relation in $\mathfrak{T}$ and let x, y be distinct letters. Then the relations $(\exists x)(x = y$ and $R)$, $(y|x)R$ are equivalent in $\mathfrak{T}$.

3. Let R and S be relations in $\mathfrak{T}$, let T be a term in $\mathfrak{T}$, and let x and y be distinct letters. Suppose that y is not a constant of T and that x does not appear in T. Let $\mathfrak{T}'$ be the theory obtained by adjoining S to the axioms of $\mathfrak{T}$. If R is functional in x in $\mathfrak{T}'$, and if $(T|y)S$ is a theorem in $\mathfrak{T}$, then the relation $(T|y)R$ is functional in X in $\mathfrak{T}$.

4. Let R and S be relations in $\mathfrak{T}$, and let x be a letter which is not a constant of $\mathfrak{T}$. If R is functional in x in $\mathfrak{T}$, and if $R \Longleftrightarrow S$ is a theorem in $\mathfrak{T}$, then S is functional in x in $\mathfrak{T}$.

5. Let R, S, T be relations in $\mathfrak{T}$ and let x be a letter. If R is functional in x in $\mathfrak{T}$, show that the following relations are theorems in $\mathfrak{T}$:

$$(\text{not } (\exists x)(R \text{ and } S)) \Longleftrightarrow (\exists x)(R \text{ and } (\text{not } S)),$$

$$(\exists x)(R \text{ and } (S \text{ and } T)) \Longleftrightarrow (\exists x)(R \text{ and } S) \text{ and } (\exists x)(R \text{ and } T)),$$
$$(\exists x)(R \text{ and } (S \text{ or } T)) \Longleftrightarrow ((\exists x)(R \text{ and } S) \text{ or } (\exists x)(R \text{ and } T)).$$

6. Show that if the scheme $(\exists x)R \Longrightarrow R$ provides implicit axioms in $\mathfrak{T}$, then $x = y$ is a theorem in $\mathfrak{T}$ (cf. Exercise 1).

7. Show that if the scheme $(\boldsymbol{R} \iff \boldsymbol{S}) \implies (\tau_x(\boldsymbol{R}) = \tau_x(\boldsymbol{S}))$ provides implicit axioms in $\mathcal{T}$, then $x = y$ is a theorem in $\mathcal{T}$. (Take $\boldsymbol{R}$ to be the relation $x = x$, $\boldsymbol{S}$ the relation $x = y$, and then substitute x for y in the axiom thus obtained.) (*)

Appendix

1. Let S be a set of signs, let A be a word in $L_0(S)$, and B, C be two significant segments of A. Then either B is a segment of C, or C is a segment of B, or else B and C are disjoint.

2. Let S be a set of signs, let A be a significant word in $L_0(S)$, and suppose that $A = A'BA''$, where B is significant. Show that, if C is a significant word, then the word $A'CA''$ is significant. (Use Exercise 1.)

3. Let E be a set and let f be a mapping of $E \times E$ into E ("internal law of composition"). Let S be a set of signs which is the disjoint union of E and the set consisting of the single element f. Put $n(f) = 1$, and $n(x) = 0$ for all $x \in E$.

(a) Let M be the set of significant words of $L_0(S)$. Show that there is a unique mapping v of M into E satisfying the following conditions: (1) $v(x) = x$ for all $x \in E$; (2) if A and B are two significant words, then $v(fAB) = f(v(A), v(B))$.

(b) For each word $A = (s_i)_{0 \leqslant i \leqslant n}$ in $L_0(S)$ let A^* be the word (s_{i_k}), where the i_k are the indices i such that $s_i \neq f$, arranged in ascending order. Two words A, B of $L_0(S)$ are said to be similar if $A^* = B^*$. Show that if the law of composition f is associative, i.e., if

$$f(f(x, y), z) = f(x, f(y, z)),$$

then $v(A) = v(B)$ whenever A and B are similar significant words ("general theorem of associativity"). (A significant word

$$A = (s_i)_{0 \leqslant i \leqslant 2n}$$

is said to be normal if $s_i = f$ for $i = 0, 2, 4, \ldots, 2n - 2$, and $s_i \neq f$ for the other indices. Show that every significant word is similar to a unique normal word A', and prove that $v(A) = v(A')$ by induction on the length of A.)

¶ 4. Let $\boldsymbol{A}$ be a term or a relation in a theory $\mathcal{T}$. Consider the sequence of assemblies defined as follows. First of all, write $\boldsymbol{A}$. If $\boldsymbol{A}$

(*) We shall see that in the theory of sets $(\exists x)(\exists v)(x \neq y)$ is a theorem (Chapter II, § 1, Exercise 2).

consists of a single letter, stop there. If not, write next the assembly or assemblies antecedent to A (if A begins with a τ, choose *one* of the antecedent assemblies arbitrarily). Then write the assembly or assemblies antecedent to those of the preceding assemblies which do not consist of a single letter; and so on.

(a) If we reverse the order of this sequence of assemblies, we get a formative construction.

(b) Let B be a balanced segment of A such that no sign of B is linked *in* A to a sign outside B. Show that B is either a term or a relation (use (a) and Exercise 1).

(c) Replace B in A by a term (resp. a relation) if B is a term (resp. a relation). Show that the assembly thus obtained is a term if A is a term, a relation if A is a relation.

5. Let A be an assembly in a theory $\mathfrak{T}$, let T be a term in $\mathfrak{T}$, and let x be a letter. If $(T|x)A$ is a term (resp. a relation), then A is a term (resp. a relation). (Use Exercise 4.)

6. A relation in a theory $\mathfrak{T}$ is said to be *logically irreducible* if it begins with a relational sign. Let $R_1, R_2, \ldots, R_n$ be distinct logically irreducible relations in $\mathfrak{T}$. A *logical construction* with base $R_1, R_2, \ldots, R_n$ is any sequence $A_1, A_2, \ldots, A_p$ of assemblies in $\mathfrak{T}$ such that each A_i satisfies one of the following conditions : (1) A_i is one of the relations $R_1, R_2, \ldots, R_n$; (2) there exists an assembly A_j preceding A_i such that A_i is $\neg A_j$; (3) there exist two assemblies A_j and A_k preceding A_i such that $\vee A_i$ is $A_j A_k$.

(a) Show that the assemblies in a logical construction with base

$$R_1, R_2, \ldots, R_n$$

are relations in $\mathfrak{T}$. A relation which appears in such a logical construction is said to be *logically constructed* from $R_1, R_2, \ldots, R_n$.

(b) If R and S are logically constructed from $R_1, R_2, \ldots, R_n$, then so are $\neg R$, $\vee RS$, $\Rightarrow RS$, "R and S", $R \Leftrightarrow S$.

(c) Let R be a relation in $\mathfrak{T}$. Consider the sequence of relations defined as follows. First of all, write R. If R is logically irreducible, stop there. If not, write next the assembly or assemblies antecedent to R (which are well determined *relations*). Then write the assembly or assemblies antecedent to those of the preceding assemblies which are not logically irreducible; and so on. Let $R_1, R_2, \ldots, R_n$ be the distinct logically irreducible relations which arise from this construction; they are called the *logical components* of R. Show that R is logically constructed from its logical components, but that if one of the relations of the sequence

$\boldsymbol{R}_1, \boldsymbol{R}_2, \ldots, \boldsymbol{R}_n$ is omitted, then $\boldsymbol{R}$ is not logically constructed from the remaining $(n-1)$ relations.

(d) Let $\boldsymbol{R}$ be a relation and let $\boldsymbol{R}_1, \boldsymbol{R}_2, \ldots, \boldsymbol{R}_n$ be distinct logically irreducible relations such that (1) $\boldsymbol{R}$ is logically constructed from $\boldsymbol{R}_1, \boldsymbol{R}_2, \ldots, \boldsymbol{R}_n$, (2) if we omit one relation from the sequence $\boldsymbol{R}_1, \boldsymbol{R}_2, \ldots, \boldsymbol{R}_n$, then $\boldsymbol{R}$ is not logically constructed from the remaining relations. Show that $\boldsymbol{R}_1, \boldsymbol{R}_2, \ldots, \boldsymbol{R}_n$ are the logical components of $\boldsymbol{R}$.

¶ 7. Let $\boldsymbol{R}_1, \boldsymbol{R}_2, \ldots, \boldsymbol{R}_n$ be distinct logically irreducible relations (Exercise 6) in a theory $\mathcal{T}$. Let $\boldsymbol{A}_1, \boldsymbol{A}_2, \ldots, \boldsymbol{A}_m$ be a logical construction with base $\boldsymbol{R}_1, \boldsymbol{R}_2, \ldots, \boldsymbol{R}_n$. Suppose that each $\boldsymbol{R}_j$ is endowed with one of the signs 0, 1. Then we give each $\boldsymbol{A}_i$ one of the signs 0, 1 according to the following rules: (1) if $\boldsymbol{A}_i$ is identical with $\boldsymbol{R}_j$, give $\boldsymbol{A}_i$ the same sign as $\boldsymbol{R}_j$; (2) if $\boldsymbol{A}_i$ is identical with $\urcorner\boldsymbol{A}_j$, where $\boldsymbol{A}_j$ precedes $\boldsymbol{A}_i$, then give $\boldsymbol{A}_i$ the sign 1 (resp. 0) if $\boldsymbol{A}_j$ has the sign 0 (resp. 1); (3) if $\boldsymbol{A}_i$ is identical with $\vee\boldsymbol{A}_j\boldsymbol{A}_k$, give $\boldsymbol{A}_i$ the sign 1 if $\boldsymbol{A}_j$ and $\boldsymbol{A}_k$ both have the sign 1; otherwise give $\boldsymbol{A}_i$ the sign 0. (We are applying the following "symbolical rule":

$$\urcorner 0 = 1, \qquad \urcorner 1 = 0, \qquad \vee 11 = 1, \qquad \vee 10 = \vee 01 = \vee 00 = 0.)$$

(a) Show that there is only one way of attributing a sign to each $\boldsymbol{A}_i$ in accordance with these rules.

(b) If $\boldsymbol{R}$ is logically constructed from $\boldsymbol{R}_1, \boldsymbol{R}_2, \ldots, \boldsymbol{R}_n$, then the sign carried by $\boldsymbol{R}$ is independent of the logical construction, with base

$$\boldsymbol{R}_1, \boldsymbol{R}_2, \ldots, \boldsymbol{R}_n,$$

in which $\boldsymbol{R}$ appears.

(c) If $\boldsymbol{R}$ and $\boldsymbol{S}$ are logically constructed from $\boldsymbol{R}_1, \boldsymbol{R}_2, \ldots, \boldsymbol{R}_n$, and if the signs carried by $\boldsymbol{R}$ and $\Longrightarrow \boldsymbol{RS}$ are 0, then the sign carried by $\boldsymbol{S}$ is 0.

(d) Suppose from now on that the *only* axioms of $\mathcal{T}$ are those provided from the schemes S1 through S4. Let $\boldsymbol{R}$ be a theorem in $\mathcal{T}$, and let $\boldsymbol{R}_1, \boldsymbol{R}_2, \ldots, \boldsymbol{R}_n$ be its logical components. Show, however, that the signs 0 and 1 are given to $\boldsymbol{R}_1, \boldsymbol{R}_2, \ldots, \boldsymbol{R}_n$, and the sign carried by $\boldsymbol{R}$ is 0. (Prove this first for the case when $\boldsymbol{R}$ is an axiom of $\mathcal{T}$; in the general case, consider a proof of $\boldsymbol{R}$, and apply (c)) $\mathcal{T}$.

(e) Let $\boldsymbol{R}$ be a logically irreducible relation in $\mathcal{T}$. Show that neither $\boldsymbol{R}$ nor "not $\boldsymbol{R}$" is a theorem in $\mathcal{T}$. In particular, $\mathcal{T}$ is non-contradictory (use (d)).

(f) Let $\boldsymbol{R}_1, \boldsymbol{R}_2, \ldots, \boldsymbol{R}_n$ be distinct, logically irreducible relations in $\mathcal{T}$. Consider all relations of the form "$\boldsymbol{R}'_1$ or $\boldsymbol{R}'_2$ or ... or $\boldsymbol{R}'_n$" where, for each i, $\boldsymbol{R}'_i$ is one of the two relations $\boldsymbol{R}_i$, "not $\boldsymbol{R}_i$". Let $\boldsymbol{S}_1, \boldsymbol{S}_2, \ldots, \boldsymbol{S}_p$ be these relations. Let $\boldsymbol{T}_1, \boldsymbol{T}_2, \ldots, \boldsymbol{T}_q$ be the relations of the form

"S_{i_1} and S_{i_2} and ... and S_{i_r}", where $i_1, i_2, \ldots, i_r$ is any strictly increasing sequence of indices. Finally, let T_0 be the relation "R_1 or (not R_1)", which is a theorem in $\mathcal{T}$. Show that every relation logically constructed from $R_1, R_2, \ldots, R_n$ is equivalent in $\mathcal{T}$ to exactly one of the relations $T_0, T_1, \ldots, T_q$. (Show first that every relation R_i is equivalent in $\mathcal{T}$ to one of the relations $T_0, T_1, \ldots, T_q$. If R is logically constructed from $R_1, R_2, \ldots, R_n$, argue step by step on a logical construction with base $R_1, R_2, \ldots, R_n$ which contains R. Finally, use (d) to prove uniqueness.)

(g) Let R be a relation in $\mathcal{T}$, and let $R_1, R_2, \ldots, R_n$ be its logical components. Then R is a theorem in $\mathcal{T}$ if and only if, however the signs 0 and 1 are assigned to $R_1, R_2, \ldots, R_n$, the corresponding sign carried by R is 0.

¶ 8. Let $R_1, R_2, \ldots, R_n$ be the logically irreducible relations in a theory $\mathcal{T}$ (Exercise 6). Assign to each of the relations R_i one of the signs 0, 1, 2. To each relation in a logical construction with base $R_1, R_2, \ldots, R_n$ we assign one of the signs 0, 1, 2 in accordance with the following symbolical rules (Exercise 7) :

$$\neg 0 = 1, \quad \neg 1 = 0, \quad \neg 2 = 2;$$
$$\vee 00 = \vee 01 = \vee 02 = \vee 10 = \vee 20 = \vee 22 = 0;$$
$$\vee 11 = 1, \quad \vee 12 = \vee 21 = 2.$$

(a) If R is logically constructed from $R_1, R_2, \ldots, R_n$, the sign carried by R is independent of the logical construction, with base $R_1, R_2, \ldots, R_n$, in which R appears.

(b) Suppose that the only axioms of $\mathcal{T}$ are those provided by the schemes S2, S3, S4. Let R be a theorem in $\mathcal{T}$. Show that however the signs 0, 1, 2 are assigned to the logical components of R, the corresponding sign carried by R is 0. On the other hand, if S is logically irreducible and has sign 2, then the sign carried by $(S$ or $S) \Longrightarrow S$ is 2. Deduce that $\mathcal{T}$ is not equivalent to the theory which has the same signs as $\mathcal{T}$ and whose axioms are provided by the schemes S1, S2, S3, S4.

(c) Establish an analogous result for theories based only on the schemes S1, S3, S4, or on the schemes S1, S2, S4. (Use respectively the following rules :

$$\neg 0 = 1, \quad \neg 1 = 0, \quad \neg 2 = 2, \quad \vee 00 = \vee 01 = \vee 10 = \vee 02 = \vee 20 = 0,$$
$$\vee 11 = 1, \quad \vee 12 = \vee 21 = 1, \quad \vee 22 = 1;$$
$$\neg 0 = 1, \quad \neg 1 = 2, \quad \neg 2 = 0,$$
$$\vee 00 = \vee 01 = \vee 10 = \vee 02 = \vee 20 = \vee 21 = 0,$$
$$\vee 11 = \vee 12 = 1, \quad \vee 22 = 2.)$$

(d) Establish an analogous result for a theory based only on the schemes S1, S2, S3. (Use four signs 0, 1, 2, 3 and the following rules :

$$\neg 0 = 1, \quad \neg 1 = 0, \quad \neg 2 = 3, \quad \neg 3 = 0,$$

$$\vee 00 = \vee 01 = \vee 10 = \vee 02 = \vee 20 = \vee 03 = \vee 30 = \vee 23 = \vee 32 = 0,$$

$$\vee 11 = 1, \quad \vee 12 = \vee 21 = \vee 22 = 2, \quad \vee 13 = \vee 31 = \vee 33 = 3.)$$

CHAPTER II

Theory of Sets

1. COLLECTIVIZING RELATIONS

1. THE THEORY OF SETS

The *theory of sets* is a theory which contains the relational signs $=$, $\in$ and the substantific sign $\supset$ (all these signs being of weight 2); in addition to the schemes S1 to S7 given in Chapter I, it contains the scheme S8, which will be introduced in no. 6, and the explicit axioms A1 (no. 3.) A2 (no. 5), A3 (§ 2, no. 1), A4 (§ 5, no. 1), and A5 (Chapter III, § 6, no. 1), These explicit axioms contain no letters; in other words, the theory of sets is a theory *without constants.*

Since the theory of sets is an equalitarian theory, the results of Chapter I are applicable.

¶ From now on, unless the contrary is expressly stated, we shall always argue in a theory which is stronger (Chapter I, § 2, no. 4) than the theory of sets; if the theory is not mentioned, it is to be assumed that the theory of sets is implied. It will be evident in many cases that this hypothesis is superfluous, and the reader should have no difficulty in determining in what theory weaker than the theory of sets the results stated are valid.

If T and U are terms, the assembly $\in TU$ is a relation (called the *relation of membership*) which in practice we write in one of the following ways: $T \in U$, $(T) \in (U)$, "T belongs to U", "T is an element of U". The relation "not $(T \in U)$" is written $T \notin U$.

From a "naive" point of view, many mathematical entities can be considered as collections or "sets" of objects. We do not seek to formalize this notion, and in the formalistic interpretation of what follows, the word "set" is to be considered as strictly synonymous with "term". In particular, phrases such as "let X be a set" are, in principle, quite superfluous, since every letter is a term. Such phrases are introduced only to assist the intuitive interpretation of the text.

2. INCLUSION

DEFINITION 1. *The relation denoted by* $(\forall z)((z \in x) \Longrightarrow (z \in y))$, *in which only the letters* x *and* y *appear, is written in one of the following ways :* $x \subset y$, $y \supset x$, "x *is contained in* y", "y *contains* x", "x *is a subset of* y". *The relation* "not $(x \subset y)$" *is written* $x \not\subset y$ *or* $y \not\supset x$.

In accordance with the conventions mentioned in Chapter I, § 1, no. 1, this definition entails the following metamathematical convention. Let $\boldsymbol{T}$ and $\boldsymbol{U}$ be assemblies; if we substitute $\boldsymbol{T}$ for x and $\boldsymbol{U}$ for y in the assembly $x \subset y$, we obtain an assembly which is denoted by $\boldsymbol{T} \subset \boldsymbol{U}$; if we denote by $\mathbf{x}$, $\mathbf{y}$ letters which are distinct from x, y and distinct from each other, and which appear neither in $\boldsymbol{T}$ nor in $\boldsymbol{U}$, the assembly $\boldsymbol{T} \subset \boldsymbol{U}$ is then identical with $(\boldsymbol{T}|\mathbf{x})(\boldsymbol{U}|\mathbf{y})(\mathbf{x}|x)(\mathbf{y}|y)(x \subset y,)$ and hence by CS8, CS9 (Chapter I, § 4, no. 1), and CS5 (Chapter I, § 1, no. 2) with $(\forall z)((z \in \boldsymbol{T}) \Longrightarrow (z \in \boldsymbol{U}))$, provided . that z is a letter which appears neither in $\boldsymbol{T}$ nor in $\boldsymbol{U}$.

From now on, whenever we state a mathematical definition, we shall not mention the metamathematical convention which it entails.

CS12. *Let* $\boldsymbol{T}$, $\boldsymbol{U}$, $\boldsymbol{V}$ *be assemblies and let* $\mathbf{x}$ *be a letter. Then the assembly* $(\boldsymbol{V}|\mathbf{x})(\boldsymbol{T} \subset \boldsymbol{U})$ *is identical with* $(\boldsymbol{V}|\mathbf{x})\boldsymbol{T} \subset (\boldsymbol{V}|\mathbf{x})\boldsymbol{U}$.

This follows immediately from CS9 (Chapter I, § 4, no. 1) and CS5 (Chapter I, § 1, no. 2).

CF13. *If* $\boldsymbol{T}$ *and* $\boldsymbol{U}$ *are terms,* $\boldsymbol{T} \subset \boldsymbol{U}$ *is a relation.*

This follows immediately from CF8 (Chapter I, § 1, no. 4).

¶ Every relation of the form $\boldsymbol{T} \subset \boldsymbol{U}$ (where $\boldsymbol{T}$ and $\boldsymbol{U}$ are terms) is called an *inclusion relation.*

From now on we shall no longer write down explicitly the criteria of substitution and the formative criteria which should follow the definitions. It should be noted, however, that these criteria will often be used implicitly in proofs.

To prove the relation $x \subset y$ in a theory $\mathcal{T}$, it is enough, by C27 (Chapter I, § 4, no. 1), to prove that $z \in y$ in the theory obtained by adjoining $z \in x$ to the axioms of $\mathcal{T}$, where z is a letter distinct from x, y and the constants of the theory. In pratice we say "let z be an element of x", and we attempt to prove $z \in y$.

PROPOSITION 1. $x \subset x$.

Obvious.

PROPOSITION 2. $(x \subset y$ and $y \subset z) \Longrightarrow (x \subset z)$.

Adjoin the hypotheses $x \subset y$, $y \subset z$, and $u \in x$. Then the relations

$$(u \in x) \Longrightarrow (u \in y), \quad (u \in y) \Longrightarrow (u \in z),$$

are true, and therefore the relation $u \in z$ is true.

3. THE AXIOM OF EXTENT

The *axiom of extent* is the following axiom :

A1. $$(\forall x)(\forall y)((x \subset y \text{ and } y \subset x) \Longrightarrow (x = y)).$$

Intuitively, this axiom expresses the fact that two sets which have the same elements are equal.

To prove that $x = y$ it is therefore enough to prove $z \in y$ in the theory obtained by adjoining the hypothesis $z \in x$, and $z \in x$ in the theory obtained by adjoining the hypothesis $z \in y$, where z is a letter distinct from x, y, and the constants.

C48. *Let $\mathbf{R}$ be a relation, x a letter, and y a letter distinct from x which does not appear in $\mathbf{R}$. Then the relation $(\forall x)((x \in y) \Longleftrightarrow \mathbf{R})$ is single-valued in y.*

Let z be a letter distinct from x which does not appear in $\mathbf{R}$. Adjoin the hypotheses

$$(\forall x)((x \in y) \Longleftrightarrow \mathbf{R}), \quad (\forall x)((x \in z) \Longleftrightarrow \mathbf{R}).$$

Then we have successively the theorems

$$(\forall x)(((x \in y) \Longleftrightarrow \mathbf{R}) \quad \text{and} \quad ((x \in z) \Longleftrightarrow \mathbf{R})),$$
$$(\forall x)((x \in y) \Longleftrightarrow (x \in z)),$$
$$y \subset z, \qquad z \subset y.$$

By A1 we have then $y = z$. This proves C48.

4. COLLECTIVIZING RELATIONS

Let $\mathbf{R}$ be a relation and let x be a letter. If y and y' denote letters distinct from x *which do not appear in* $\mathbf{R}$, then the relations

$$(\exists y)(\forall x)((x \in y) \Longleftrightarrow \mathbf{R}), \quad (\exists y')(\forall x)((x \in y') \Longleftrightarrow \mathbf{R})$$

are identical by CS8 (Chapter I, §4, no. 1). The relation thus defined (which does not contain x) is denoted by $\text{Coll}_x R$.

¶ If $\text{Coll}_x R$ is a theorem in a theory $\mathcal{T}$, R is said to be *collectivizing in* x in $\mathcal{T}$. When this is so, we may introduce an auxiliary constant a, distinct from x and the constants of $\mathcal{T}$, and which does not appear in R, with the introductory axiom $(\forall x)((x \in a) \Longleftrightarrow R)$, or equivalently (if x is not a constant of $\mathcal{T}$) $(x \in a) \Longleftrightarrow R$.

Intuitively, to say that R is collectiving in x is to say that there exists a set a such that the objects x which possess the property R are precisely the elements of a.

Examples

(1) The relation $x \in y$ is clearly collectivizing in x.

(2) The relation $x \notin x$ *is not collectivizing in* x; in other words, (not $\text{Coll}_x (x \notin x)$) is a theorem. Let us argue by contradiction and suppose that $x \notin x$ is collectivizing. Let a be an auxiliary constant, distinct from x and the constants of the theory, with the introductory axiom

$$(\forall x)((x \notin x) \Longleftrightarrow (x \in a)).$$

Then the relation

$$(a \notin a) \Longleftrightarrow (a \in a)$$

is true by C30 (Chapter I, §4, no. 3). The method of disjunction of cases (Chapter I, §3, no. 3) shows first that the relation $a \notin a$ is true, and then that the relation $a \in a$ is true, which is absurd.

C49. *Let R be a relation and x a letter. If R is collectivizing in x, the relation $(\forall x)((x \in y) \Longleftrightarrow R)$, where y is a letter distinct from x which does not appear in R, is functional in y.*

This follows immediately from C48.

Very often in what follows we shall have at our disposal a theorem of the form $\text{Coll}_x R$. To represent the term

$$\tau_y (\forall x)((x \in y) \Longleftrightarrow R),$$

which does not depend on the choice of the letter y (distinct from x and not appearing in R), we shall introduce a functional symbol $\mathcal{E}_x(R)$; the corresponding term does not contain X. This term is denoted by "*the set of all x such that R*". By definition (Chapter I, §4, no. 1) the relation

$$(\forall x)((x \in \mathcal{E}_x(R)) \Longleftrightarrow R)$$

is *identical* with $\text{Coll}_{\mathbf{x}}\boldsymbol{R}$; consequently the relation $\boldsymbol{R}$ is *equivalent* to

$$\mathbf{x} \in \mathcal{E}_{\mathbf{x}}(\boldsymbol{R}).$$

C50. *Let* $\boldsymbol{R}$, $\boldsymbol{S}$ *be two relations and let* $\mathbf{x}$ *be a letter. If* $\boldsymbol{R}$ *and* $\boldsymbol{S}$ *are collectivizing in* $\mathbf{x}$, *the relation* $(\forall \mathbf{x})(\boldsymbol{R} \Rightarrow \boldsymbol{S})$ *is equivalent to*

$$\mathcal{E}_{\mathbf{x}}(\boldsymbol{R}) \subset \mathcal{E}_{\mathbf{x}}(\boldsymbol{S}),$$

and the relation $(\forall \mathbf{x})(\boldsymbol{R} \Leftrightarrow \boldsymbol{S})$ *is equivalent to* $\mathcal{E}_{\mathbf{x}}(\boldsymbol{R}) = \mathcal{E}_{\mathbf{x}}(\boldsymbol{S})$.

This follows immediately from the preceding remark and from Definition 1 and axiom A1.

5. THE AXIOM OF THE SET OF TWO ELEMENTS

A2. $$(\forall x)(\forall y) \ \text{Coll}_z(z = x \text{ or } z = y).$$

> This axiom says that if x and y are objects, then there is a set whose only elements are x and y.

DEFINITION 2. *The set* $\mathcal{E}_z\,(z = x \text{ or } z = y)$, *whose only elements are* x *and* y, *is denoted by* $\{x, y\}$.

The relation $z \in \{x, y\}$ is therefore equivalent to "$z = x$ or $z = y$"; it follows from C50 that $\{y, x\} = \{x, y\}$.

> Let $\boldsymbol{R}\{z\}$ be a relation and let x, y be letters distinct from z. From the criteria C32, C33 (Chapter I, §4, no. 3), and C43 (Chapter I, §5, no. 1) it follows easily that the relation $(\exists z)((z \in \{x, y\}) \text{ and } \boldsymbol{R}\{z\})$ is equivalent to "$\boldsymbol{R}\{x\}$ or $\boldsymbol{R}\{y\}$"; consequently the relation $(\forall z)((z \in \{x, y\}) \Rightarrow \boldsymbol{R}\{z\})$ is equivalent to "$\boldsymbol{R}\{x\}$ and $\boldsymbol{R}\{y\}$".

The set $\{x, x\}$, which is denoted simply by $\{x\}$, is called *the set whose only element is* x; the relation $z \in \{x\}$ is equivalent to $z = x$, and the relation $x \in X$ is equivalent to $\{x\} \subset X$.

6. THE SCHEME OF SELECTION AND UNION

The *scheme of selection and union* is the following:

S8. *Let* $\boldsymbol{R}$ *be a relation, let* $\mathbf{x}$ *and* $\mathbf{y}$ *be distinct letters, and let* $\boldsymbol{X}$ *and* $\boldsymbol{Y}$ *be letters distinct from* $\mathbf{x}$ *and* $\mathbf{y}$ *which do not appear in* $\boldsymbol{R}$. *Then the relation*

(1) $$(\forall \mathbf{y})(\exists \boldsymbol{X})(\forall \mathbf{x})(\boldsymbol{R} \Rightarrow (\mathbf{x} \in \boldsymbol{X})) \Rightarrow (\forall \boldsymbol{Y})\ \text{Coll}_{\mathbf{x}}((\exists \mathbf{y})((\mathbf{y} \in \boldsymbol{Y}) \text{ and } \boldsymbol{R}))$$

is an axiom.

Let us first show that this rule is indeed a scheme. Let S denote the relation (1), and let us substitute a term T for a letter z in S; by CS8 (Chapter I, §4, no. 1) we may suppose that x, y, X, Y are distinct from z and do not appear in T. Then $(T|z)S$ is identical with

$$(\forall y)(\exists X)(\forall x)(R' \Longrightarrow (x \in X)) \Longrightarrow (\forall Y)\ \mathrm{Coll}_X((\exists y)((y \in Y) \text{ and } R')),$$

where R' is $(T|z)R$.

Intuitively, the relation $(\forall y)(\exists X)(\forall x)(R \Longrightarrow (x \in X))$ means that for every object y there exists a set X (which may depend on y) such that the objects x which are in the relation R with the given object y are elements of X (but not necessarily the whole of X). The scheme of selection and union asserts that if this is the case and if Y is any set, then there exists a set whose elements are precisely the objects x which are in the relation R with at least one object y of the set Y.

C51. *Let P be a relation, let A be a set, and let x be a letter which does not appear in A. Then the relation "P and $x \in A$" is collectivizing in x.*

Let R denote the relation "P and $x = y$", where y is a letter distinct from x which appears neither in P nor in A. The relation

$$(\forall x)(R \Longrightarrow (x \in \{y\}))$$

is true by C27 (Chapter I, §4, no. 1). Let X be a letter distinct from x and y which does not appear in P. The preceding relation is identical with $(\{y\}|X)((\forall x)(R \Longrightarrow (x \in X)))$ (because x is distinct from y), and therefore the relation $(\forall y)(\exists X)(\forall x)(R \Longrightarrow (x \in X))$ is true by virtue of S5 and C27 (Chapter I, §4, nos. 1 and 2). If follows from S8 and C30 (Chapter I, §4, no. 3) that the relation

$$(A|Y)\ \mathrm{Coll}_x(\exists y)(y \in Y \text{ and } R)$$

(where Y is a letter which does not appear in R) is true, and this relation is identical with $\mathrm{Coll}_x(\exists y)(y \in A$ and $R)$ (because neither x nor y appears in A). Finally, the relation "$y \in A$ and R" is equivalent to "$x = y$ and $x \in A$ and P" by C43 (Chapter I, §5, no. 1); since x appears neither in P nor in A, the relation

$$(\exists y)(x = y \text{ and } x \in A \text{ and } P)$$

is equivalent to "$((\exists y)(x = y))$ and $x \in A$ and P" by C33 (Chapter I, §4, no. 3) and therefore to "P and $x \in A$" because $(\exists y)(x = y)$ is true.

¶ The set $\mathcal{E}_x(P \text{ and } x \in A)$ is called *the set of all $x \in A$ such that P* (* thus we may speak of the set of all real numbers such that P *).

C52. *Let R be a relation, A a set, and x a letter which does not appear in A. If the relation $R \Longrightarrow (x \in A)$ is a theorem, then R is collectivizing in x.*

For R is then equivalent to "R and $x \in A$".

Remark. Let R be a relation which is collectivizing in x, and let S be a relation such that $(\forall x)(S \Longrightarrow R)$ is a theorem. Then S is collectivizing in x; for R is equivalent to $x \in \mathcal{E}_x(R)$, so that

$$S \Longrightarrow (x \in \mathcal{E}_x(R))$$

is a theorem, and the assertion follows from C52. Notice also that in this case we have $\mathcal{E}_x(S) \subset \mathcal{E}_x(R)$ by C50.

C53. *Let T be a term, A a set, x and y distinct letters. Suppose that x does not appear in A and that y does not appear in A nor in T. Then the relation $(\exists x)(y = T \text{ and } x \in A)$ is collectivizing in y.*

Let R be the relation $y = T$. The relation $(\forall y)(R \Longrightarrow (y \in \{T\}))$ is true, hence so is $(\forall x)(\exists X)(\forall y)(R \Longrightarrow (y \in X))$, where X is a letter, distinct from y, which does not appear in R. By virtue of S8, the relation $(\exists x)(x \in A \text{ and } R)$ is collectivizing in y, and C53 is proved.

¶ The relation $(\exists x)(y = T \text{ and } x \in A)$ is often read as follows : "y can be put in the form T for some x belonging to A". The set

$$\mathcal{E}_y((\exists x)(y = T \text{ and } x \in A))$$

is generally called *the set of objects of the form T for $x \in A$.* The assembly so denoted contains neither x nor y, and does not depend on the choice of the letter y satisfying the conditions of C53.

7. COMPLEMENT OF A SET. THE EMPTY SET

The relation $(x \notin \mathrm{A} \text{ and } x \in X)$ is collectivizing in x by C51.

DEFINITION 3. *Let A be a subset of a set X. The set of elements of X which do not belong to A, that is to say the set $\mathcal{E}x\,(x \notin \mathrm{A} \text{ and } x \in \mathrm{X})$, is called the complement of A in X, and is denoted by* $\complement_{\mathrm{X}}\mathrm{A}$ *or* $\mathrm{X} - \mathrm{A}$ (or by $\complement\mathrm{A}$ if there is no risk of confusion).

Let A be a subset of a set X; the relations "$x \in \mathrm{X}$ and $x \notin \mathrm{A}$" and $x \in \complement_{\mathrm{X}}\mathrm{A}$ are then equivalent. Consequently the relation "$x \in \mathrm{X}$ and $x \notin \complement_{\mathrm{X}}\mathrm{A}$" is equivalent to "$x \in \mathrm{X}$ and ($x \notin \mathrm{X}$ or $x \in \mathrm{A}$)", hence to

$x \in A$. In other words, $A = \complement_X(\complement_X A)$ is a true relation. Similarly, one shows that if B is a subset of X, the relations $A \subset B$ and $\complement_X B \subset \complement_X A$ are equivalent.

THEOREM 1. *The relation* $(\forall x)(x \notin X)$ *is functional in* X.

For the relation $(\forall x)(x \notin X)$ implies $(\forall Y)(X \subset Y)$; by virtue of the axiom of extent, the relation $(\forall x)(x \notin X)$ is therefore single-valued in X. On the other hand, the relation $(\forall x)(x \notin \complement_Y Y)$ is true, which proves that $(\exists X)(\forall x)(x \notin X)$ is true.

¶ The term $\tau_x((\forall x)(x \notin X))$ corresponding to this functional relation is represented by the functional symbol $\varnothing$, and is called *the empty set* (*); the relation $(\forall x)(x \notin X)$, which is equivalent to $X = \varnothing$, is read as follows: "*the set* X *is empty*". We have the theorems $x \notin \varnothing$, $\varnothing \subset X$, $\complement_X X = \varnothing$, $\complement_X \varnothing = X$. The relation $X \subset \varnothing$ is equivalent to $X = \varnothing$. If $\boldsymbol{R}\{x\}$ is a relation, the relation $(\forall x)((x \in \varnothing) \Longrightarrow \boldsymbol{R}\{x\})$ is true.

Remark. There exists no set of which every object is an element; in other words, "not $(\exists X)(\forall x)(x \in X)$" is a theorem. For if there were such a set, then by C52 every relation would be collectivizing. But we have seen (no. 4) that the relation $x \notin x$ is not collectivizing.

2. ORDERED PAIRS

1. THE AXIOM OF THE ORDERED PAIR

As we have said in §1, no. 1, the sign $\supset$ is a substantific sign of weight 2 in the theory of sets. If T, U are terms, then $\supset$TU is a term, which in practice is denoted by (T, U). In this notation, the *axiom of the ordered pair* is the following:

A3. $(\forall x)(\forall x')(\forall y)(\forall y')(((x, y) = (x', y')) \Longrightarrow (x = x' \text{ and } y = y'))$.

Since the relation "$x = x'$ and $y = y'$" implies $(x, y) = (x', y')$ by C44 (Chapter I, § 5, no. 2), the relation $(x, y) = (x', y')$ is *equivalent* to "$x = x'$ and $y = y'$".

(*) The term denoted by $\varnothing$ is therefore $\tau \neg\neg\neg \in \tau \neg\neg \in \square\square\square$.

¶ The relation $(\exists x)(\exists y)(z = (x, y))$ is written "*z is an ordered pair*". If z is an ordered pair, the relations

$$(\exists y)(z = (x, y)), \quad (\exists x)(z = (x, y))$$

are functional with respect to x and y respectively; this is an immediate consequence of A3.

The terms

$$\tau_x((\exists y)(z = (x, y))) \qquad \text{and} \qquad \tau_y((\exists x)(z = (x, y)))$$

are denoted by $\text{pr}_1 z$ and $\text{pr}_2 z$, respectively, and are called the ***first coordinate*** (or ***first projection***) and the ***second coordinate*** (or ***second projection***) of z. If z is an ordered pair, the relation $(\exists y)(z = (x, y))$ is equivalent to $x = \text{pr}_1 z$, and the relation $(\exists x)(z = (x, y))$ to $y = \text{pr}_2 z$ (Chapter I, §5, no. 3).

¶ The relation $z = (x, y)$ is equivalent to "z is an ordered pair, and $x = \text{pr}_1 z$ and $y = \text{pr}_2 z$"; for the latter relation is equivalent to

$$(\exists x')(\exists y')(\exists x'')(\exists y'')(z = (x', y') \text{ and } z = (x, y'') \text{ and } z = (x'', y));$$

by A3, "$z = (x', y')$ and $z = (x, y'')$ and $z = (x'', y)$" is equivalent to "$z = (x, y)$ and $x = x'$ and $x = x''$ and $y = y'$ and $y = y''$"; hence by C33 (Chapter I, §4, no. 3), "z is an ordered pair, and $x = \text{pr}_1 z$ and $y = \text{pr}_2 z$" is equivalent to

$$z=(x, y) \text{ and } (\exists x')(\exists y')(\exists x'')(\exists y'')(x=x' \text{ and } x=x'' \text{ and } y=y' \text{ and } y=y''),$$

which proves our assertion. Evidently we have

$$\text{pr}_1(x, y) = x, \qquad \text{pr}_2(x, y) = y,$$

and the relation $z = (\text{pr}_1(z), \text{pr}_2(z))$ is equivalent to "z is an ordered pair".

¶ Let $\boldsymbol{R}\{\boldsymbol{x}, \boldsymbol{y}\}$ be a relation, where the letters $\boldsymbol{x}$ and $\boldsymbol{y}$ are distinct and appear in $\boldsymbol{R}$. Let $\boldsymbol{z}$ be a letter, distinct from $\boldsymbol{x}$ and $\boldsymbol{y}$, which does not appear in $\boldsymbol{R}$. Let $\boldsymbol{S}\{\boldsymbol{z}\}$ denote the relation

$$(\exists \boldsymbol{x})(\exists \boldsymbol{y})(\boldsymbol{z} = (\boldsymbol{x}, \boldsymbol{y}) \text{ and } \boldsymbol{R}\{\boldsymbol{x}, \boldsymbol{y}\});$$

this is a relation which contains one letter fewer than $\boldsymbol{R}$, and which is ***equivalent*** to "$\boldsymbol{z}$ is an ordered pair and $\boldsymbol{R}\{\text{pr}_1 \boldsymbol{z}, \text{pr}_2 \boldsymbol{z}\}$", this follows from the fact that $\boldsymbol{z} = (\boldsymbol{x}, \boldsymbol{y})$ is equivalent to "$\boldsymbol{z}$ is an ordered pair and $\boldsymbol{x} = \text{pr}_1 \boldsymbol{x}$ and $\boldsymbol{y} = \text{pr}_2 \boldsymbol{z}$", and from criteria C33 (Chapter I, §4, no. 3) and C47 (Chapter I, §5, no. 3). It follows immediately that $\boldsymbol{R}\{x, y\}$ is equivalent to $\boldsymbol{S}\{(x, y)\}$, and also to

$$(\exists z)(z = (x, y) \text{ and } \boldsymbol{S}\{z\})$$

by C47.

This means that we may interpret a relation between the objects x and y as a property of the ordered pair formed by these objects.

2. PRODUCT OF TWO SETS

THEOREM 1. *The relation*

$$(\forall X)(\forall Y)(\exists Z)(\forall z)((z \in Z) \Longleftrightarrow (\exists x)(\exists y)(z = (x, y) \text{ and } x \in X \text{ and } y \in Y)$$

is true. In other words, for all X *and all* Y *the relation "z is an ordered pair and* $\mathrm{pr}_1 z \in X$ *and* $\mathrm{pr}_2 z \in Y$*" is collectivizing in* z.

Let A_y denote the set of all objects of the form (x, y) for $x \in X$ (cf. §1, no. 6, criterion C53). Let R be the relation $z \in A_y$, which is equivalent to $(\exists x)(z = (x, y)$ and $x \in X)$. It is clear that the relation

$$(\forall y)(\exists A)(\forall z)(R \Longrightarrow (z \in A))$$

is true, by virtue of S5 (Chapter I, §4, no. 2). It then follows from S8 that the relation $(\exists y)(y \in Y$ and R) is collectivizing in z. But this relation is equivalent to $(\exists x)(\exists y)(y \in Y$ and $x \in X$ and $z = (x, y))$; hence the result.

DEFINITION 1. *Given two sets* X *and* Y, *the set*

$$\mathcal{E}_z((\exists x)(\exists y)(z = (x, y) \text{ and } x \in X \text{ and } y \in Y))$$

is called the product of X *and* Y *and is denoted by* $X \times Y$.

The relation $z \in X \times Y$ is thus equivalent to "z is an ordered pair and $\mathrm{pr}_1 z \in X$ and $\mathrm{pr}_2 z \in Y$". The sets X and Y are called the *first* and *second factors* of $X \times Y$.

PROPOSITION 1. *If* A′, B′ *are non-empty sets, the relation* $A' \times B' \subset A \times B$ *is equivalent to* "$A' \subset A$ and $B' \subset B$".

In the first place, the relation $z \in A' \times B'$ is equivalent to "z is an ordered pair and $\mathrm{pr}_1 z \in A'$ and $\mathrm{pr}_2 z \in B'$"; therefore, without any restriction on A′ and B′, the relation "$A' \subset A$ and $B' \subset B$" implies

$$A' \times B' \subset A \times B.$$

Conversely, let us first show that if $B' \neq \emptyset$ (without restriction on A'), the relation $A' \times B' \subset A \times B$ implies $A' \subset A$. Let x be an element of A'; since $B' \neq \emptyset$, there is an object y which is an element of B'; we have $(x, y) \in A' \times B'$, hence $(x, y) \in A \times B$, and consequently $x \in A$; thus $A' \subset A$. Similarly, if $A' \neq \emptyset$, the relation $A' \times B' \subset A \times B$ implies $B' \subset B$. Hence the result.

PROPOSITION 2. *Let* A *and* B *be two sets. The relation* $A \times B = \emptyset$ *is equivalent to* "$A = \emptyset$ or $B = \emptyset$".

For the relation $z \in A \times B$ implies $\mathrm{pr}_1 z \in A$ and $\mathrm{pr}_2 z \in B$; hence $A \neq \emptyset$ and $B \neq \emptyset$. Conversely, the relation "$x \in A$ and $y \in B$" implies $(x, y) \in A \times B$ and hence $A \times B \neq \emptyset$. In other words, the relation $A \times B \neq \emptyset$ is equivalent to "$A \neq \emptyset$ and $B \neq \emptyset$"; hence the result.

If A, B, C are sets, we write

$$(A \times B) \times C = A \times B \times C;$$

an element $((x, y), z)$ of $A \times B \times C$ is written (x, y, z) and is called a *triple*. Again, if A, B, C, D are sets, we write

$$(A \times B \times C) \times D = A \times B \times C \times D;$$

and so on.

3. CORRESPONDENCES

1. GRAPHS AND CORRESPONDENCES

DEFINITION 1. G *is said to be a graph if every element of* G *is an ordered pair, i.e., if the relation*

$$(\forall z)(z \in G \Longrightarrow (z \text{ is an ordered pair}))$$

is true.

If G is a graph, the relation $(x, y) \in G$ is expressed by saying that "y corresponds to x under G".

Let $\boldsymbol{R}\{\boldsymbol{x}, \boldsymbol{y}\}$ be a relation, where $\boldsymbol{x}$ and $\boldsymbol{y}$ are distinct letters. Let $\boldsymbol{G}$ be a letter, distinct from $\boldsymbol{x}$ and $\boldsymbol{y}$, which does not appear in $\boldsymbol{R}$. If the relation

$$(\exists \boldsymbol{G})(\boldsymbol{G} \text{ is a graph and } (\forall \boldsymbol{x})(\forall \boldsymbol{y})(\boldsymbol{R} \Longleftrightarrow ((\boldsymbol{x}, \boldsymbol{y}) \in \boldsymbol{G})))$$

is true, the relation $\boldsymbol{R}$ is said to *have a graph* (with respect to the letters $\boldsymbol{x}$ and $\boldsymbol{y}$). The graph $\boldsymbol{G}$ is then unique by virtue of the axiom of extent, and is called the *graph* of $\boldsymbol{R}$ with respect to $\boldsymbol{x}$ and $\boldsymbol{y}$.

¶ Let Z be a letter, distinct from x and y, which does not appear in R. If the relation

$$(\exists Z)(\forall x)(\forall y)(R \Longrightarrow ((x, y) \in Z))$$

is true, then R has a graph; we may take this graph to be the set of ordered pairs z such that $z \in Z$ and $R\{\mathrm{pr}_1 z, \mathrm{pr}_2 z\}$ (z being a letter distinct from x, y, Z which does not appear in R). This condition is satisfied if we know a term T, which does not contain either x or y, such that $R \Longrightarrow ((x, y) \in T)$ is true.

PROPOSITION 1. *Let* G *be a graph. There exists exactly one set* A *and exactly one set* B *with the following properties*:
(1) *the relation* $(\exists y)((x, y) \in \mathrm{G})$ *is equivalent to* $x \in \mathrm{A}$;
(2) *the relation* $(\exists x)((x, y) \in \mathrm{G})$ *is equivalent to* $y \in \mathrm{B}$.

For it is sufficient to take A (resp. B) to be the set of all objects of the form $\mathrm{pr}_1 z$ (resp. $\mathrm{pr}_2 z$), where $z \in \mathrm{G}$ (§ 1, no. 6). Precisely,

$$\mathrm{A} = \mathcal{E}_x((\exists y)((x, y) \in \mathrm{G})) \quad \text{and} \quad \mathrm{B} = \mathcal{E}_y((\exists x)((x, y) \in \mathrm{G}));$$

these sets are called respectively *the first and second projections* of the graph G, or the *domain* and *range* of G; they are denoted by $\mathrm{pr}_1\langle \mathrm{G}\rangle$ and $\mathrm{pr}_2\langle \mathrm{G}\rangle$ (or by $\mathrm{pr}_1\mathrm{G}$ and $\mathrm{pr}_2\mathrm{G}$ if there is no risk of confusion). It is immediately verified that $\mathrm{G} \subset (\mathrm{pr}_1\mathrm{G}) \times (\mathrm{pr}_2\mathrm{G})$; every set of ordered pairs is therefore a subset of a product, and conversely. If one of the two sets $\mathrm{pr}_1\mathrm{G}$, $\mathrm{pr}_2\mathrm{G}$ is empty, we have $\mathrm{G} = \varnothing$ (§ 2, Proposition 2).

Remark. The relation $x = y$ has no graph, for the first projection of the graph, if it existed, would be the set of all objects (cf. § 1, no. 7, Remark).

DEFINITION 2. *A correspondence between a set* A *and a set* B *is a triple*

$$\Gamma = (\mathrm{G}, \mathrm{A}, \mathrm{B}),$$

where G *is a graph such that* $\mathrm{pr}_1\mathrm{G} \subset \mathrm{A}$ *and* $\mathrm{pr}_2\mathrm{G} \subset \mathrm{B}$. G *is said to be the graph of* Γ, A *is the source, and* B *the target of* Γ.

If $(x, y) \in \mathrm{G}$, we say that "y corresponds to x in the correspondence Γ". For each $x \in \mathrm{pr}_1\mathrm{G}$ the correspondence Γ is said to be *defined at* x, and $\mathrm{pr}_1\mathrm{G}$ is called the *domain of* Γ; each $y \in \mathrm{pr}_2\mathrm{G}$ is said to be a *value taken by* Γ, and $\mathrm{pr}_2\mathrm{G}$ is called the *range of* Γ.

If $R\{x, y\}$ is a relation which has a graph G (with respect to the letters x, y), and if A, B are two sets such that $\mathrm{pr}_1 G \subset A$ and $\mathrm{pr}_2 G \subset B$, we say that R is a *relation between an element of A and an element of B* (with respect to the letters x, y). The correspondence $\Gamma = (G, A, B)$ is said to be the correspondence between A and B *defined by the* relation R (with respect to x and y).

Let G be a graph and let X be a set. The relation

$$x \in \mathrm{X} \text{ and } (x, y) \in \mathrm{G}$$

implies $(x, y) \in \mathrm{G}$ and therefore has a graph G′. The second projection of G′ consists of all the objects which correspond under G to objects of X.

Definition 3. *Let G be a graph and X a set. The set of all objects which correspond under G to elements of X is called the image of X under G and is denoted by* $\mathrm{G}\langle \mathrm{X} \rangle$ *or* $\mathrm{G}(\mathrm{X})$.

¶ *Let* $\Gamma = (\mathrm{G}, \mathrm{A}, \mathrm{B})$ *be a correspondence and let X be a subset of A. The set* $\mathrm{G}\langle \mathrm{X} \rangle$ *is also denoted by* $\Gamma\langle \mathrm{X} \rangle$ *or* $\Gamma(\mathrm{X})$ *and is called the image of X under* Γ.

Remarks

(1) Precisely, $\mathrm{G}\langle \mathrm{X} \rangle$ denotes the set $\mathcal{E}_y((\exists x)(x \in \mathrm{X} \text{ and } (x,y) \in \mathrm{G}))$. From now on we shall not usually translate our definitions into formal language.

(2) The notations $\mathrm{G}(\mathrm{X})$ and $\Gamma(\mathrm{X})$ can occasionally lead to confusion with the notation introduced later (cf. no. 4, Remark following Definition 9).

Let G be a graph. Since the relation $(x, y) \in \mathrm{G}$ implies $y \in \mathrm{pr}_2\mathrm{G}$, we have $\mathrm{G}\langle \mathrm{X} \rangle \subset \mathrm{pr}_2\mathrm{G}$ for every set X. Since $(x, y) \in \mathrm{G}$ implies $x \in \mathrm{pr}_1\mathrm{G}$, we have $\mathrm{G}\langle \mathrm{pr}_1\mathrm{G} \rangle = \mathrm{pr}_2\mathrm{G}$. We have $\mathrm{G}\langle \varnothing \rangle = \varnothing$, since $x \notin \varnothing$ is a theorem. If $\mathrm{X} \subset \mathrm{pr}_1\mathrm{G}$ and $\mathrm{X} \neq \varnothing$, we have $\mathrm{G}\langle \mathrm{X} \rangle \neq \varnothing$.

Proposition 2. *Let G be a graph and let X, Y be two sets; then the relation* $\mathrm{X} \subset \mathrm{Y}$ *implies* $\mathrm{G}\langle \mathrm{X} \rangle \subset \mathrm{G}\langle \mathrm{Y} \rangle$.

This is an immediate consequence of the definitions and of C50 (§ 1, no. 4).

Corollary. *If* $\mathrm{A} \supset \mathrm{pr}_1\mathrm{G}$, *we have* $\mathrm{G}\langle \mathrm{A} \rangle = \mathrm{pr}_2\mathrm{G}$.

Definition 4. *Let G be a graph and x an object. The set* $\mathrm{G}\langle \{x\} \rangle$ (which is sometimes denoted by $\mathrm{G}(x)$, by abuse of language) *is called the section of G at x.*

It follows immediately from C43 (Chapter I, § 5, no. 1) that the relation $y \in \mathrm{G}\langle \{x\} \rangle$ is equivalent to $(x, y) \in \mathrm{G}$. If G and G′ are two graphs, the relation $\mathrm{G} \subset \mathrm{G}'$ is thus equivalent to

$$(\forall x)(\mathrm{G}\langle \{x\} \rangle \subset \mathrm{G}'\langle \{x\} \rangle).$$

If $\Gamma = (G, A, B)$ is a correspondence between A and B, then for every $x \in A$ the section of G at x is also called the *section of* Γ *at* x and is denoted by $\Gamma\langle\{x\}\rangle$ (or $\Gamma(x)$).

2. INVERSE OF A CORRESPONDENCE

Let G be a graph and $A = \mathrm{pr}_1 G$, $B = \mathrm{pr}_2 G$ its projections. The relation $(y, x) \in G$ implies $(x, y) \in B \times A$; this relation therefore has a graph which consists of all ordered pairs (x, y) such that $(y, x) \in G$.

DEFINITION 5. *Let* G *be a graph. The graph whose elements are the ordered pairs* (x, y) *such that* $(y, x) \in G$ *is called the inverse of* G *and is denoted by* $\overset{-1}{G}$.

For every set X, $\overset{-1}{G}\langle X\rangle$ is called the *inverse image of* X *under* G.

¶ It is clear that the inverse of $\overset{-1}{G}$ is G, and that

$$\mathrm{pr}_1 \overset{-1}{G} = \mathrm{pr}_2 G, \qquad \mathrm{pr}_2 \overset{-1}{G} = \mathrm{pr}_1 G.$$

In particular, if X, Y are two sets, we have

$$\overset{-1}{\overbrace{X \times Y}} = Y \times X.$$

A graph G is said to be *symmetric* if $\overset{-1}{G} = G$.

¶ Let $\Gamma = (G, A, B)$ be a correspondence between A and B. Since $\mathrm{pr}_1 \overset{-1}{G} \subset B$ and $\mathrm{pr}_2 \overset{-1}{G} \subset A$, the triple $(\overset{-1}{G}, B, A)$ is a *correspondence between* B and A, called the *inverse* of the correspondence Γ, and denoted by $\overset{-1}{\Gamma}$. For every subset Y of B, the image $\overset{-1}{\Gamma}\langle Y\rangle$ of Y under $\overset{-1}{\Gamma}$ is also called the *inverse image* of Y under Γ. Clearly the inverse of $\overset{-1}{\Gamma}$ is Γ.

3. COMPOSITION OF TWO CORRESPONDENCES

Let G, G′ be two graphs. Let A denote the set $\mathrm{pr}_1 G$ and let C denote the set $\mathrm{pr}_2 G'$. The relation $(\exists y)((x, y) \in G$ and $(y, z) \in G')$ implies that $(x, z) \in A \times C$, and therefore has a graph with respect to x and z.

DEFINITION 6. *Let* G, G′ *be two graphs. The graph (with respect to* x *and* z*) of the relation* $(\exists y)((x, y) \in G$ and $(y, z) \in G')$ *is called the composition of* G′ *and* G, *and is denoted by* $G' \circ G$ (or sometimes by $G'G$).

PROPOSITION 3. *Let* G, G′ *be two graphs. The inverse of* $G' \circ G$ *is then* $\overset{-1}{G} \circ \overset{-1}{G'}$.

For the relation "$(x, y) \in G$ and $(y, z) \in G'$" is equivalent to

$$(z, y) \in \overset{-1}{G'} \text{ and } (y, x) \in \overset{-1}{G}.$$

PROPOSITION 4. *Let* G_1, G_2, G_3 *be graphs. Then*

$$(G_3 \circ G_2) \circ G_1 = G_3 \circ (G_2 \circ G_1).$$

The relation $(x, t) \in (G_3 \circ G_2) \circ G_1$ is equivalent to the relation

$$(\exists y)((x, y) \in G_1 \text{ and } \exists z((y, z) \in G_2 \text{ and } (z, t) \in G_3))$$

and therefore (by C33 (Chapter I, § 4, no. 3)) to the relation

(1) $$(\exists y)(\exists z)((x, y) \in G_1 \text{ and } (y, z) \in G_2 \text{ and } (z, t) \in G_3).$$

Similarly, the relation $(x, t) \in G_3 \circ (G_2 \circ G_1)$ is equivalent to

(2) $$(\exists z)(\exists y)((x, y) \in G_1 \text{ and } (y, z) \in G_2 \text{ and } (z, t) \in G_3).$$

But the relations (1) and (2) are equivalent; hence the result.

¶ The graph $G_3 \circ (G_2 \circ G_1)$ is denoted by $G_3 \circ G_2 \circ G_1$. Similarly, if G_1, G_2, G_3, G_4 are graphs, we put

$$G_4 \circ (G_3 \circ G_2 \circ G_1) = G_4 \circ G_3 \circ G_2 \circ G_1;$$

and so on.

PROPOSITION 5. *Let* G, G′ *be graphs and let* A *be a set. Then*

$$(G' \circ G)\langle A \rangle = G'\langle G\langle A \rangle\rangle.$$

For by virtue of C33 (Chapter I, § 4, no. 3) the relation $z \in (G' \circ G)\langle A \rangle$ is equivalent to

$$(\exists y)((\exists x)(x \in A \text{ and } (x, y) \in G) \text{ and } (y, z) \in G')$$

and is therefore equivalent to $(\exists y)(y \in G\langle A \rangle$ and $(y, z) \in G')$; hence the result.

¶ If G and G′ are two graphs, we have $\mathrm{pr}_1(G' \circ G) = \overset{-1}{G}\langle \mathrm{pr}_1 G' \rangle$, and $\mathrm{pr}_2(G' \circ G) = G'\langle \mathrm{pr}_2 G \rangle$. For example, to prove the second of these relations it is enough to note that the relation $z \in \mathrm{pr}_2(G' \circ G)$ is equivalent to $(\exists x)((x, z) \in G' \circ G)$ and therefore to

$$(\exists y)((\exists x)((x, y) \in G) \text{ and } (y, z) \in G');$$

but this is equivalent to $z \in G'\langle \mathrm{pr}_2 G \rangle$.

¶ If G is a graph and X a set such that $X \subset \mathrm{pr}_1 G$, we have

$$X \subset \overset{-1}{G}\langle G\langle X \rangle\rangle.$$

For the relation $x \in X$ implies by hypothesis that $(\exists y)((x, y) \in G)$; but $(x, y) \in G$ is equivalent to $(y, x) \in \overset{-1}{G}$, and on the other hand $(x, y) \in G$ implies $(\exists z)(z \in X$ and $(z, y) \in G)$; hence $x \in X$ implies

$$(\exists y)((\exists z)(z \in X \text{ and } (z, y) \in G) \text{ and } (y, x) \in \overset{-1}{G}),$$

that is to say, $x \in \overset{-1}{G}\langle G\langle X \rangle\rangle$.

¶ It is clear that if G_1, G_2, G'_1, G'_2 are graphs, the relations $G_1 \subset G_2$ and $G'_1 \subset G'_2$ imply $G'_1 \circ G_1 \subset G'_2 \circ G_2$.

¶ Let $\Gamma = (G, A, B)$ and $\Gamma' = (G', B, C)$ be two correspondences such that the target of Γ is the same as the source of Γ'. From the above discussion we have $\mathrm{pr}_1(G' \circ G) \subset \mathrm{pr}_1 G \subset A$ and $\mathrm{pr}_2(G' \circ G) \subset \mathrm{pr}_2 G' \subset C$; hence we may state the following definition :

Definition 7. *Let* $\Gamma = (G, A, B)$ *and* $\Gamma' = (G', B, C)$ *be two correspondences such that the target of* Γ *is the source of* Γ'. *Then the correspondence* $(G' \circ G, A, C)$ *is called the composition of* Γ' *and* Γ, *and is denoted by* $\Gamma' \circ \Gamma$ (or sometimes $\Gamma'\Gamma$).

It follows immediately from Proposition 5 that if X is a subset of A we have $(\Gamma' \circ \Gamma)\langle X \rangle = \Gamma'\langle \Gamma\langle X \rangle\rangle$. Furthermore, since the target of $\overset{-1}{\Gamma'}$ is the same as the source of $\overset{-1}{\Gamma}$, the inverse of $\Gamma' \circ \Gamma$ is $\overset{-1}{\Gamma} \circ \overset{-1}{\Gamma'}$, by Proposition 3.

Definition 8. *If* A *is a set, the set* Δ_A *of all objects of the form* (x, x), *where* $x \in A$, *is called the diagonal of* $A \times A$.

Clearly we have $\mathrm{pr}_1 \Delta_A = \mathrm{pr}_2 \Delta_A = A$. The correspondence

$$I_A = (\Delta_A, A, A)$$

is called the *identity correspondence* on A; it is its own inverse.

¶ If Γ is a correspondence between A and B and if I_A, I_B are the identity correspondences on A, B, respectively, then $\Gamma \circ I_A = I_B \circ \Gamma = \Gamma$.

4. FUNCTIONS

DEFINITION 9. *A graph* F *is said to be a functional graph if for each* x *there is at most one object which corresponds to* x *under* F (Chapter I, § 5, no. 3). *A correspondence* $f = (F, A, B)$ *is said to be a function if its graph* F *is a functional graph and if its source* A *is equal to its domain* pr_1F. *In other words, a correspondence* $f = (F, A, B)$ *is a function if for every* x *belonging to the source* A *of* f *the relation* $(x, y) \in F$ *is functional in* y (Chapter I, § 5, no. 3); *the unique object which corresponds to* x *under* f *is called the value of* f *at the element* x *of* A, *and is denoted by* $f(x)$ (or f_x, or $F(x)$, or F_x).

If f is a function, F its graph, and x an element of the domain of f, the relation $y = f(x)$ is then equivalent to $(x, y) \in F$ (Chapter I, § 5, no. 3, criterion C46).

Remark. The reader should beware of the confusion which may arise from the simultaneous use of the notations $f(x)$ and $f(X)$ (synonymous with $f\langle X\rangle$) introduced in Definition 3 (cf. Exercise 11).

Let A and B be two sets; a *mapping of* A *into* B is a function f whose source (which is equal to its domain) is equal to A and whose target is equal to B; such a function is also said to be *defined on* A and to *take its values in* B.

Instead of the phrase "let f be a mapping of A into B", the following phrases are often used : "let $f: A \to B$ be a mapping" or even "let $f: A \to B$". To simplify the presentation of an argument involving several mappings, we use *diagrams* such as

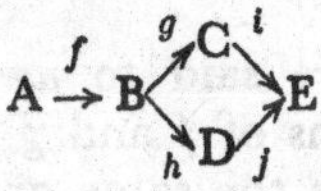

in which a group of signs such as $A \xrightarrow{f} B$ is to be interpreted as meaning that f is a mapping of A into B.

A function f defined on A is said to *transform* x *into* $f(x)$ (for all $x \in A$); $f(x)$ is called the *transform of* x *by* f or (by abuse of language) the *image* of x under f.

¶ Under certain circumstances, a functional graph is called a *family*; the domain is then called the *index set*, and the range is called (by abuse of language) the *set of elements* of the family. It is mainly in this connection that the indicial notation f_x is used to denote the value of f at the element x. When the index set is the product of two sets, we often speak of a *double family*.

¶ Likewise a function whose target is E is sometimes called a *family of elements of* E. If every element of E is a subset of a set F, we speak of a *family of subsets of* F.

Throughout this series we shall often use the word "function" in place of "functional graph".

Examples of functions

(1) The empty set is a functional graph. Every function whose graph is empty has domain and range equal to the empty set. Among such functions, the one whose target is empty (i.e., the function $(\emptyset, \emptyset, \emptyset)$) is called the *empty function*.

(2) Let A be a set. The identity correspondence of A (no. 3) is a function, called the *identity mapping of* A.

Thus with every set A there is associated a family, defined by the identity mapping of A, whose index set is A and whose set of elements is A. By abuse of language, a set is sometimes referred to as a "family", in which case it is the family thus associated with the set in question.

A function f is said to be *constant* if for all x and x' in the domain of f we have $f(x) = f(x')$.

¶ Let f be a mapping of a set E into E. An element x of E is said to be *fixed under* f if $f(x) = x$.

5. RESTRICTIONS AND EXTENSIONS OF FUNCTIONS

Two functions f and g are said to *agree* (or *coincide*) on a set E if E is contained in the domains of f and g and if $f(x) = g(x)$ for all $x \in E$. Two functions which have the same graph agree on their domain. To say that $f = g$ is to say that f and g have the same domain A and the same target B, and that they agree on A.

¶ Let $f = (F, A, B)$ and $g = (G, C, D)$ be two functions. To say that $F \subset G$ is to say that the domain A of f is contained in the domain C of g and that g agrees with f on A. If also $B \subset D$, then g is said to be an *extension* of f (more precisely, an extension of f to C), and g is said to extend f (to C). When g is called a family of elements of D, f is said to be a *subfamily* of g.

¶ Let f be a function and let A be a subset of the domain of f. It is immediate that the relation "$x \in A$ and $y = f(x)$" has a graph G with respect to x and y, that this graph is functional, and that A is its domain; the function whose graph is G, which has the same target as f, is called the *restriction of f to* A, and is sometimes denoted by $f|A$. A function is an extension of any of its restrictions. If two functions f, g have the same target and agree on a set E, then their restrictions to E are equal.

6. DEFINITION OF A FUNCTION BY MEANS OF A TERM

C54. *Let T, A be two terms and let x, y be distinct letters. Suppose that x does not appear in A and that y does not appear in either T or A. Let R be the relation "$x \in A$ and $y = T$". The relation R has a graph F with respect to the letters x and y. This graph is functional; its first projection is A, and its second projection is the set of objects of the form T for $x \in A$* (§ 1, no. 6). *For every $x \in A$ we have $F(x) = T$.*

Let B be the set of objects of the form T for $x \in A$. Then

$$R \Longrightarrow ((x, y) \in A \times B);$$

since the assembly denoted by $A \times B$ contains neither x nor y, R has a graph F with respect to the letters x and y (no. 1). It is clear that the relation

$$(x, y) \in F \text{ and } (x, y') \in F$$

implies $y = y'$, and hence F is a functional graph. The remaining statements are evident.

¶ If C is a set which contains the set B of objects of the form T for $x \in A$ (where y does not appear in C), then the function (F, A, C) is also denoted by the notation $x \to T$ $(x \in A,\ T \in C)$; the corresponding assembly in formal mathematics contains neither x nor y and does not depend on the choice of the letter y satisfying the above conditions. When the context is sufficiently explicit we shall permit ourselves the notations $x \to T(x \in A)$, $(T)_{x \in A}$, or $x \to T$, and sometimes simply T or (T). *Thus we may speak of "the function x^3", if the context indicates clearly that we mean the mapping $x \to x^3$ of the set of complex numbers into itself. *

Examples

(1) If f is a mapping of A into B, the function f is equal to the function $x \to f(x)$ $(x \in A,\ f(x) \in B)$, which is written simply as $x \to f(x)$ or also $(f_x)_{x \in A}$ (the latter notation is especially associated with the phrase "family of elements" instead of "function").

(2) Let G be a set of ordered pairs. The functions

$$z \to \mathrm{pr}_1 z \quad (z \in \mathrm{G},\ \mathrm{pr}_1 z \in \mathrm{pr}_1 \mathrm{G})$$

and

$$z \to \mathrm{pr}_2 z \quad (z \in \mathrm{G},\ \mathrm{pr}_2 z \in \mathrm{pr}_2 \mathrm{G})$$

are called respectively the *first and second coordinate functions on* G; they are denoted by pr_1 and pr_2 when there is no risk of confusion.

7. COMPOSITION OF TWO FUNCTIONS. INVERSE FUNCTION

PROPOSITION 6. *If f is a mapping of* A *into* B, *and g is a mapping of* B *into* C, *then $g \circ f$ is a mapping of* A *into* C.

Let F, G be the graphs of f, g, respectively, and let us show that $\mathrm{G} \circ \mathrm{F}$ is a functional graph. Let x, z, z' be objects such that $(x, z) \in \mathrm{G} \circ \mathrm{F}$, $(x, z') \in \mathrm{G} \circ \mathrm{F}$. There exist objects y, y' such that

$$(x, y) \in \mathrm{F}, \qquad (x, y') \in \mathrm{F}, \qquad (y, z) \in \mathrm{G}, \qquad (y', z') \in \mathrm{G}.$$

Since F is a functional graph, we have $y = y'$ and hence $(y, z') \in \mathrm{G}$. Since G is a functional graph, it follows that $z = z'$, which proves our assertion. Furthermore, the domain of $g \circ f$ is evidently A, and the proof is complete.

¶ The function $g \circ f$ may also be written as $x \to g(f(x))$, or as gf when there is no risk of confusion.

DEFINITION 10. *Let f be a mapping of* A *into* B. *The mapping f is said to be injective, or an injection, if any two distinct elements of* A *have distinct images under f. The mapping f is said to be surjective, or a surjection, if $f(\mathrm{A}) = \mathrm{B}$. If f is both injective and surjective, it is said to be bijective, or a bijection.*

Instead of saying that f is surjective, we sometimes say that f is a mapping of A *onto* B, or that it is a *parametric representation* of B by means of A (in which case A is called the *set of parameters* of the representation, and the elements of A are called *parameters*). If f is bijective, we sometimes say that f *puts* A *and* B *in one-to-one correspondence*. A bijection of A onto A is called a *permutation* of A.

Examples

(1) If $\mathrm{A} \subset \mathrm{B}$, the mapping of A into B whose graph is the diagonal of A is injective and is called the *canonical mapping* or the *canonical injection* (or simply the *injection*) of A into B.

(2) Let A be a set. The mapping $x \to (x, x)$ of A into the diagonal Δ_A of $\mathrm{A} \times \mathrm{A}$ is a bijective mapping, called the *diagonal mapping* of A.

(3) Let G be a set of ordered pairs. The mapping pr_1 (resp. pr_2) of G into $pr_1 G$ (resp. $pr_2 G$) is surjective; pr_1 is injective if and only if G is a functional graph.

(4) Let G be a set of ordered pairs. The mapping

$$z \to (pr_2 z, pr_1 z)$$

of G into $\overset{-1}{G}$ is a bijection (called the *canonical* bijection).

(5) Let A be a set and b an object. The mapping $x \to (x, b)$ of A into $A \times \{b\}$ is a bijection.

PROPOSITION 7. *Let f be a mapping of A into B. Then $\overset{-1}{f}$ is a function if and only if f is bijective.*

If $\overset{-1}{f}$ is a function, its source B is equal to its domain, i.e., to $f(A)$; hence f is surjective. To show that f is injective, let x and y be two elements of A such that $f(x) = f(y)$. If F denotes the graph of f, we have

$$(f(x), x) \in \overset{-1}{F} \quad \text{and} \quad (f(y), y) \in \overset{-1}{F},$$

hence

$$(f(x), y) \in \overset{-1}{F},$$

so that $x = y$, which proves the assertion.

¶ Conversely, if f is bijective, it is immediate that $\overset{-1}{F}$ is functional and that the domain of $\overset{-1}{f}$ is equal to B.

When f is bijective, $\overset{-1}{f}$ is called the *inverse mapping* of f; $\overset{-1}{f}$ is bijective, $\overset{-1}{f} \circ f$ is the identity mapping of A, and $f \circ \overset{-1}{f}$ is the identity mapping of B.

¶ If a permutation is the same as its inverse, it is said to be *involutory*.

Remark. Let f be a mapping of A into B. For each subset X of A we have (no. 3) $X \subset \overset{-1}{f}\langle f\langle X\rangle\rangle$. Furthermore, for each subset Y of B we have $f\langle \overset{-1}{f}\langle Y\rangle\rangle \subset Y$, for the relation $y \in f\langle \overset{-1}{f}\langle Y\rangle\rangle$ is equivalent to

$$(\exists x)((\exists z)(z \in Y \text{ and } z = f(x)) \text{ and } y = f(x))$$

and therefore implies the relation $(\exists z)(z \in Y \text{ and } y = z)$ and consequently also the relation $y \in Y$.

If f is *surjective*, we have $f\langle \overset{-1}{f}\langle Y\rangle\rangle = Y$ for every subset Y of B, for the relation $y \in Y \subset B$ implies by hypothesis the relation $(\exists x)(y = f(x))$ and therefore also $(\exists x)(y \in Y \text{ and } y = f(x))$; but "$y \in Y$ and $y = f(x)$" implies $(\exists z)(z \in Y \text{ and } z = f(x))$, and our assertion follows.

If f is *injective*, we have $\overset{-1}{f}\langle f\langle X\rangle\rangle = X$ for every subset X of A. For the relation $x \in \overset{-1}{f}\langle f\langle X\rangle\rangle$ is equivalent to $f(x) \in f\langle X\rangle$, hence to

$$(\exists z)(z \in X \text{ and } f(z) = f(x));$$

but the hypothesis means that $f(z) = f(x)$ implies $z = x$, hence $x \in \overset{-1}{f}\langle f\langle X\rangle\rangle$ implies $x \in X$.

8. RETRACTIONS AND SECTIONS

PROPOSITION 8. *Let f be a mapping of* A *into* B. *If there exists a mapping r* (resp. s) *of* B *into* A *such that $r \circ f$* (resp. $f \circ s$) *is the identity mapping of* A (resp. B), *then f is injective* (resp. *surjective*). *Conversely, if f is surjective, there exists a mapping s of* B *into* A *such that $f \circ s$ is the identity mapping of* B. *If f is injective and if* $A \neq \emptyset$, *there exists a mapping r of* B *into* A *such that $r \circ f$ is the identity mapping of* A.

If there exists a mapping r of B into A such that $r \circ f$ is the identity mapping of A, then the equality $f(x) = f(y)$, where $x \in A$ and $y \in A$, implies $x = r(f(x)) = r(f(y)) = y$, and so f is injective. If there exists a mapping s of B into A such that $f \circ s$ is the identity mapping of B, we have $B = f(s(B)) \subset f(A) \subset B$, so that f is surjective. If f is surjective, let T denote the term $\tau_y(y \in A \text{ and } f(y) = x)$. We have $f(T) = x$ for $x \in B$; if s denotes the mapping $x \to T$ $(x \in B,\ T \in A)$, then $f \circ s$ is the identity mapping of B. Finally, suppose that f is injective and that $A \neq \emptyset$, and let a be an element of A. The relation

$$\text{"}(y \in A \text{ and } x = f(y)) \text{ or } (y = a \text{ and } x \in B - f(A))\text{"}$$

implies $(x, y) \in B \times A$ and therefore has a graph R with respect to the letters x, y. This graph is functional by reason of the hypothesis on f, and has B as its domain; and we have $R(x) = a$ if $x \in B - f(A)$, and $f(R(x)) = x$ if $x \in f(A)$. Hence the function $r = (R, B, A)$ is such that $r \circ f$ is the identity mapping of A.

COROLLARY. *Let* A, B *be sets, let f be a mapping of* A *into* B, *and let g be a mapping of* B *into* A. *If $g \circ f$ is the identity mapping of* A *and if $f \circ g$ is the identity mapping of* B, *then f and g are bijective, and* $g = \overset{-1}{f}$.

DEFINITION 11. *Let f be an injective* (resp. *surjective*) *mapping of* A *into* B. *Any mapping r* (resp. s) *of* B *into* A *such that $r \circ f$* (resp. $f \circ s$) *is the identity mapping of* A (resp. B) *is called a retraction* (resp. *section*) *of f*.

Instead of retraction (resp. section) the phrase *left-inverse* (resp. *right-inverse*) is sometimes used.

¶ If f is injective (resp. surjective) and if r (resp. s) is a retraction (resp. section) of f, then f is a section (resp. retraction) of r (resp. s). Hence a retraction is surjective and a section is injective.

¶ If f is surjective and if s, s' are two sections of f such that $s(\mathrm{B}) = s'(\mathrm{B})$, then $s = s'$; for if $x \in \mathrm{B}$, there exists $y \in \mathrm{B}$ such that $s(x) = s'(y)$, and we have $x = f(s(x)) = f(s'(y)) = y$, so that $s(x) = s'(x)$ and consequently $s = s'$. Thus a section s is uniquely determined by the set $s(\mathrm{B})$. By abuse of language, the set $s(\mathrm{B})$ is sometimes called a *section* of f.

THEOREM 1. *Let f be a mapping of* A *into* B, *let f' be a mapping of* B *into* C, *and let $f'' = f' \circ f$. Then :*

(a) *If f and f' are injections, then f'' is an injection. If r, r' are retractions of f, f', respectively, then $r \circ r'$ is a retraction of f''.*

(b) *If f and f' are surjections, then f'' is a surjection. If s, s' are sections of f, f', respectively, then $s \circ s'$ is a section of f''.*

(c) *If f'' is an injection, then f is an injection. If r'' is a retraction of f'', then $r'' \circ f'$ is a retraction of f.*

(d) *If f'' is a surjection, then f' is a surjection. If s'' is a section of f'', then $f \circ s''$ is a section of f'.*

(e) *If f'' is a surjection and f' an injection, then f is a surjection. If s'' is a section of f'', then $s'' \circ f'$ is a section of f.*

(f) *If f'' is an injection and f a surjection, then f' is an injection. If r'' is a retraction of f'', then $f \circ r''$ is a retraction of f'.*

For every set E let $\mathrm{I_E}$ denote the identity mapping of E.

(a) We have $r \circ f = \mathrm{I_A}$ and $r' \circ f' = \mathrm{I_B}$, hence

$$(r \circ r') \circ (f' \circ f) = r \circ \mathrm{I_B} \circ f = r \circ f = \mathrm{I_A}.$$

If f and f' are injections, then f'' is an injection, by Proposition 8 if $\mathrm{A} \neq \emptyset$, and trivially if $\mathrm{A} = \emptyset$.

(b) We have $f \circ s = \mathrm{I_B}$ and $f' \circ s' = \mathrm{I_C}$, hence

$$(f' \circ f)(s \circ s') = f' \circ \mathrm{I_B} \circ s' = f' \circ s' = \mathrm{I_C}.$$

If f and f' are surjections, f'' is then a surjection by Proposition 8.

(c) We have $r'' \circ f'' = \mathrm{I_A}$, hence $(r'' \circ f') \circ f = r'' \circ f'' = \mathrm{I_A}$. If f'' is an injection, then f is an injection, by Proposition 8 if $\mathrm{A} \neq \emptyset$, and trivially if $\mathrm{A} = \emptyset$.

(d) We have $f'' \circ s'' = I_C$, hence $f' \circ (f \circ s'') = f'' \circ s'' = I_C$. If f'' is a surjection, then f' is a surjection by Proposition 8.

(e) We have $f'' \circ s'' = I_C$, and f' is a bijection by (d). Hence

$$f \circ (s'' \circ f') = (\overset{-1}{f'} \circ f') \circ f \circ (s'' \circ f') = \overset{-1}{f'} \circ (f'' \circ s'') \circ f'$$
$$= \overset{-1}{f'} \circ I_C \circ f' = \overset{-1}{f'} \circ f' = I_B.$$

If f'' is a surjection and f' an injection, f is then a surjection by Proposition 8.

(f) We have $r'' \circ f'' = I_A$, and f is a bijection by (c). Hence

$$(f \circ r'') \circ f' = (f \circ r'') \circ f' \circ (f \circ \overset{-1}{f}) = f \circ (r'' \circ f'') \circ \overset{-1}{f} = f \circ I_A \circ \overset{-1}{f}$$
$$= f \circ \overset{-1}{f} = I_B.$$

If f'' is an injection and f a surjection, then f' is an injection, by Proposition 8 if $A \neq \emptyset$, and trivially if $A = \emptyset$ (for then we have $B = f\langle A \rangle = \emptyset$).

PROPOSITION 9. (a) *Let* E, F, G *be sets, let* g *be a mapping of* E *onto* F *and* f *a mapping of* E *into* G. *Then there exists a mapping* h *of* F *into* G *such that* $f = h \circ g$ *if and only if the relation* $g(x) = g(y)$ (*where* $x \in E$, $y \in E$) *implies the relation* $f(x) = f(y)$. *The mapping* h *is then uniquely determined by* f; *if* s *is a section of* g, *we have* $h = f \circ s$.

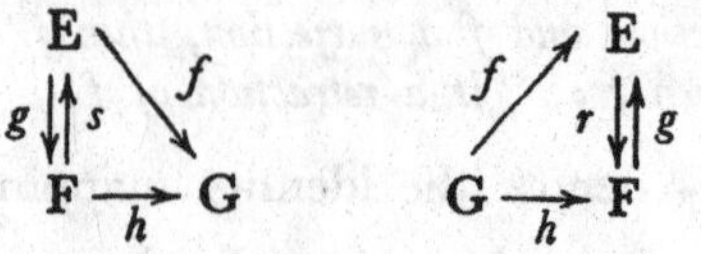

(a) (b)

(b) *Let* E, F, G *be sets, let* g *be an injection of* F *into* E, *and let* f *be a mapping of* G *into* E. *Then there exists a mapping* h *of* G *into* F *such that* $f = g \circ h$ *if and only if* $f(G) \subset f(F)$. *The mapping* h *is uniquely determined by* f; *if* r *is a retraction of* g, *we have* $h = r \circ f$.

(a) If $f = h \circ g$, the relation $g(x) = g(y)$ (where $x \in E$, $y \in E$) clearly implies $f(x) = f(y)$. And for every section s of g we have

$$h = h \circ (g \circ s) = f \circ s,$$

which shows that h is uniquely determined by f. Conversely, suppose that the relation $g(x) = g(y)$ implies $f(x) = f(y)$; let s be a section of g, and let $h = f \circ s$; then for every $x \in E$ we have $g(s(g(x))) = g(x)$, hence $f(s(g(x))) = f(x)$, that is, $h(g(x)) = f(x)$ and therefore $f = h \circ g$.

(b) If $f = g \circ h$, then clearly $f(G) \subset f(F)$, and for every retraction r of g we have $h = (r \circ g) \circ h = r \circ f$, which shows that h is uniquely determined by f. Conversely, suppose that $f(G) \subset f(F)$; let r be a retraction of g, and put $h = r \circ f$; for every $x \in G$, there exists $y \in F$ such that $f(x) = g(y)$, so that

$$g(h(x)) = g(r(f(x))) = g(r(g(y))) = g(y) = f(x)$$

and therefore $f = g \circ h$.

9. FUNCTIONS OF TWO ARGUMENTS

A *function of two arguments* is a function whose domain is a set of ordered pairs (or, equivalently, a subset of a product). Let f be such a function; if (x, y) is an element of the domain of f, the value $f((x, y))$ of f at the element (x, y) is generally denoted by $f(x, y)$.

¶ Let f be a function of two arguments, D its domain, and C its target. For each y let A_y be the set of all x such that $(x, y) \in D$ (that is, the section of $\overset{-1}{D}$ at y (no. 1)). The mapping

$$x \to f(x, y) \quad (x \in A_y, f(x, y) \in C)$$

is called *the partial mapping defined by f, with respect to the value y of the second argument*, and is denoted by $f(\bullet, y)$, or $f(\ , y)$ (or sometimes f_y); we have $f(\bullet, y)(x) = f(x, y)$ for all $(x, y) \in D$. Similarly, for each x let B_x be the set of all y such that $(x, y) \in D$. The mapping

$$y \to f(x, y) \quad (y \in B_x, f(x, y) \in C)$$

is called *the partial mapping defined by f, with respect to the value x of the first argument*, and is denoted by $f(x, \bullet)$, or $f(x, \ )$ (or sometimes f_x); we have $f(x, \bullet)(y) = f(x, y)$ for all $(x, y) \in D$.

¶ If for every y (resp. x) the partial mapping $f(\bullet, y)$ (resp. $f(x, \bullet)$) is a constant mapping, we say that f *does not depend on* its first (resp. second) argument; this means therefore that $f(x, y) = f(x', y)$ whenever (x, y) and (x', y) belong to D (resp. $f(x, y) = f(x, y')$ whenever (x, y) and (x, y') belong to D). For each y belonging to the second projection of D let $g(y)$ denote the common value of the $f(x, y)$ for $x \in A_y$; the mapping $y \to g(y)$ is then a mapping of $pr_2 D$ into C such that

$$g(y) = f(x, y) \quad \text{for all } (x, y) \in D.$$

¶ Conversely, let g be a mapping of a set B into a set C, and let A be any set. Then the mapping $(x, y) \to g(y)$ of $A \times B$ into C does not depend on its first argument.

¶ Let u be a mapping of A into C and v a mapping of B into D. The mapping $z \to (u(\mathrm{pr}_1 z), v(\mathrm{pr}_2 z))$ of $A \times B$ into $C \times D$ is called the *(canonical) extension of u and v to the product sets*, or simply *the product of u and v* (if there is no risk of confusion); its range is $u\langle A\rangle \times v\langle B\rangle$. It is denoted by $u \times v$. If u and v are injective (resp. surjective) then $u \times v$ is injective (resp. surjective). If u and v are bijective, then $u \times v$ is bijective, and its inverse mapping is $\overset{-1}{u} \times \overset{-1}{v}$. If u' is a mapping of C into a set E and if v' is a mapping of D into a set F, we have

$$(u' \times v') \circ (u \times v) = (u' \circ u) \times (v' \circ v).$$

If U, V are the graphs of u, v respectively, the graph W of $u \times v$ is the set of ordered pairs $((x, y), (z, t))$ of $(A \times B) \times (C \times D)$ such that $(x, z) \in U$ and $(y, t) \in V$; the mapping $((x, y), (z, t)) \to ((x, z), (y, t))$ puts W in one-to-one correspondence with the product $U \times V$ (a subset of $(A \times C) \times (B \times D)$) (cf. §5, no. 5).

4. UNION AND INTERSECTION OF A FAMILY OF SETS

1. DEFINITION OF THE UNION AND THE INTERSECTION OF A FAMILY OF SETS

Let X be a family (§3, no. 4) and I its index set. To help the intuitive interpretation of what follows, we shall refer to X as a *family of sets*. If $(X, I, \mathfrak{G})$ is a *family of subsets of a set* E (that is, a family whose target $\mathfrak{G}$ is such that the relation $Y \in \mathfrak{G}$ implies $Y \subset E$), we shall denote the family by $(X_\iota)_{\iota \in I}(X_\iota \in \mathfrak{G})$ or simply by $(X_\iota)_{\iota \in I}$ (§3, no. 6); by abuse of notation we shall denote any family of sets, which has I as its index set, by $(X_\iota)_{\iota \in I}$.

¶ Since the relation $(\forall x)((\iota \in I \text{ and } x \in X_\iota) \Longrightarrow (x \in X_\iota))$ is true, it follows from S5 (Chapter I, §4, no 2.) that the relation

$$(\forall \iota)(\exists Z)(\forall x)((\iota \in I \text{ and } x \in X_\iota) \Longrightarrow (x \in Z))$$

is true. By virtue of the scheme S8 (§1, no. 6) the relation $(\exists \iota)(\iota \in I$ and $x \in X_\iota)$ is *collectivizing in x*.

Definition 1. *Let* $(X_\iota)_{\iota \in I}$ *be a family of sets* (resp. *a family of subsets of a set* E). *The set* $\mathcal{E}_x((\exists \iota)(\iota \in I$ and $x \in X_\iota))$, *that is to say, the set of all x* *which belong to at least one set of the family* $(X_\iota)_{\iota\in I}$, *is called the union of the family, and is denoted by* $\bigcup_{\iota\in I} X_\iota$ (*).

If $(X_\iota)_{\iota\in I}$ is a family of subsets of a set E, then its union is a subset of E; notice that it does not depend on E, nor on the target $\mathfrak{G}$ of the mapping $\iota \to X_\iota$.

It is clear that if $I = \emptyset$, we have $\bigcup_{\iota\in I} X_\iota = \emptyset$, because the relation $(\exists\iota)(\iota \in I \text{ and } x \in X_\iota)$ is then false.

¶ Suppose now that $I \neq \emptyset$. If α is an element of I, the relation

$$(\forall\iota)((\iota \in I) \Longrightarrow (x \in X_\iota))$$

implies $x \in X_\alpha$ and therefore, by virtue of C52 (§1, no. 6), this relation is *collectivizing in* x.

DEFINITION 2. *Let* $(X_\iota)_{\iota\in I}$ *be a family of sets whose index set* I *is not empty. The set* $\mathcal{E}_x((\forall\iota)((\iota \in I) \Longrightarrow (x \in X_\iota))$, *that is to say, the set of all* x *which belong to every set of the family* $(X_\iota)_{\iota\in I}$, *is called the intersection of the family and is denoted by* $\bigcap_{\iota\in I} X_\iota$.

If $I = \emptyset$, the relation $(\forall\iota)((\iota \in I) \Longrightarrow (x \in X_\iota))$ is *not* collectivizing in x; for it is a true relation and there exists no set Y such that $x \in Y$ is a true relation, because Y would then be the set of all objects (cf. §1, no. 7, Remark).

If $(X_\iota)_{\iota\in I}$ is a family of subsets of a set E and if $I \neq \emptyset$, the relation "$x \in E$ and $(\forall\iota)((\iota \in I) \Longrightarrow (x \in X_\iota))$" is equivalent to

$$(\forall\iota)((\iota \in I) \Longrightarrow (x \in X_\iota));$$

consequently it is collectivizing in x, and the set of all x which satisfy this relation is equal to $\bigcap_{\iota\in I} X_\iota$. If $I = \emptyset$, the relation "$x \in E$ and $(\forall\iota)((\iota \in I) \Longrightarrow (x \in X_\iota))$" is equivalent to $x \in E$; it is therefore collectivizing in x, and the set of all x which satisfy this relation is E. Hence we may state the following definition:

DEFINITION 3. *Let* $(X_\iota)_{\iota\in I}$ *be a family of subsets of a set* E. *The set*

$$\mathcal{E}_x\ (x \in E \text{ and } (\forall\iota)((\iota \in I) \Longrightarrow (x \in X_\iota))),$$

(*) The scheme S8 therefore allows us to define the union of a family of sets without supposing *a priori* that these sets are subsets of the same set (which is the assumption made in the definition of union given in *Summary of Results*, §4, no. 2).

in other words, the set of all x which belong to E *and to each of the sets* X_ι, *is called the intersection of the family and is denoted by* $\bigcap_{\iota \in I} X_\iota$.

Hence for a family $(X_\iota)_{\iota \in \emptyset}$ of subsets of E we have

$$\bigcap_{\iota \in \emptyset} X_\iota = E.$$

But for a family $(X_\iota)_{\iota \in I}$ of subsets of E whose index set is not empty, the intersection $\bigcap_{\iota \in I} X_\iota$ depends neither on E nor on the target of $\iota \to X_\iota$; and this justifies the use of the same notation in Definitions 2 and 3.

PROPOSITION 1. *Let* $(X_\iota)_{\iota \in I}$ *be a family of sets, and let f be a mapping of a set* K *onto* I. *Then*

$$\bigcup_{\varkappa \in K} X_{f(\varkappa)} = \bigcup_{\iota \in I} X_\iota,$$

and, if $I \neq \emptyset$,

$$\bigcap_{\varkappa \in K} X_{f(\varkappa)} = \bigcap_{\iota \in I} X_\iota.$$

Let x be an element of $\bigcap_{\iota \in I} X_\iota$. There exists an index $\iota \in I$ such that $x \in X_\iota$. Since $f\langle K\rangle = I$, there exists an index $\varkappa \in K$ such that $\iota = f(\varkappa)$, whence $x \in X_{f(\varkappa)}$ and consequently

$$x \in \bigcup_{\varkappa \in K} X_{f(\varkappa)}.$$

Conversely, if $x \in \bigcup_{\varkappa \in K} X_{f(\varkappa)}$, there exists $\varkappa \in K$ such that $x \in X_{f(\varkappa)}$, and therefore, since $f(\varkappa) \in I$, $x \in \bigcup_{\iota \in I} X_\iota$. Hence

$$\bigcup_{\varkappa \in K} X_{f(\varkappa)} = \bigcup_{\iota \in I} X_\iota.$$

¶ Now suppose that $I \neq \emptyset$, and let x be an element of $\bigcap_{\iota \in I} X_\iota$. For each element $\varkappa$ of K we have $f(\varkappa) \in I$, hence $x \in X_{f(\varkappa)}$, and therefore

$$x \in \bigcap_{\varkappa \in K} X_{f(\varkappa)}.$$

Conversely, let x be an element of $\bigcap_{\varkappa \in K} X_{f(\varkappa)}$. If ι is any element of I, there exists an element $\varkappa$ of K such that $\iota = f(\varkappa)$, whence $x \in X_\iota$ and consequently $x \in \bigcap_{\iota \in I} X_\iota$. Hence

$$\bigcap_{\varkappa \in K} X_{f(\varkappa)} = \bigcap_{\iota \in I} X_\iota.$$

For families of subsets of a given set, it is clear that the second part of Proposition 1 remains valid without the restriction $I \neq \emptyset$.

COROLLARY. *Let $(X_\iota)_{\iota \in I}$ be a family of sets such that $X_\iota = X_\varkappa$ for each pair of indices $(\iota, \varkappa)$. Then for each $\alpha \in I$ we have*

$$\bigcup_{\iota \in I} X_\iota = X_\alpha, \quad \textit{and (if } I \neq \emptyset) \quad \bigcap_{\iota \in I} X_\iota = X_\alpha.$$

Apply Proposition 1 to the constant mapping $\iota \to \alpha$ of I onto $\{\alpha\}$.

DEFINITION 4. *Let $\mathfrak{F}$ be a set of sets and let Φ be the family of sets defined by the identity mapping of $\mathfrak{F}$. The union of the sets of Φ and (if $\mathfrak{F}$ is not empty) the intersection of the sets of Φ are called, respectively, the union and the intersection of the sets of $\mathfrak{F}$, and are denoted by $\bigcup_{X \in \mathfrak{F}} X$ and $\bigcap_{X \in \mathfrak{F}} X$.*

If follows immediately from Proposition 1 that if $(X_\iota)_{\iota \in I}$ is a family of sets, then the union and (if $I \neq \emptyset$) the intersection of the family are respectively equal to the union and the intersection of the sets of the set of elements of this family.

2. PROPERTIES OF UNION AND INTERSECTION

If $(X_\iota)_{\iota \in I}$ and $(Y_\iota)_{\iota \in I}$ are families of sets having the same index set I, and if $Y_\iota \subset X_\iota$ for each $\iota \in I$, then it is clear that

$$\bigcup_{\iota \in I} Y_\iota \subset \bigcup_{\iota \in I} X_\iota, \quad \text{and (if } I \neq \emptyset) \quad \bigcap_{\iota \in I} Y_\iota \subset \bigcap_{\iota \in I} X_\iota.$$

¶ Let $(X_\iota)_{\iota \in I}$ be a family of sets. If $J \subset I$, we have

$$\bigcup_{\iota \in J} X_\iota \subset \bigcup_{\iota \in I} X_\iota, \quad \text{and (if } J \neq \emptyset) \quad \bigcap_{\iota \in J} X_\iota \supset \bigcap_{\iota \in I} X_\iota.$$

PROPOSITION 2. *Let $(X_\iota)_{\iota \in I}$ be a family of sets whose index set I is the union of a family $(J_\lambda)_{\lambda \in L}$ of sets. Then*

$$\bigcup_{\iota \in I} X_\iota = \bigcup_{\lambda \in L} \Big(\bigcup_{\iota \in J_\lambda} X_\iota \Big),$$

and (*if* $L \neq \emptyset$ and $J_\lambda \neq \emptyset$ *for each* $\lambda \in L$)

$$\bigcap_{\iota \in I} X_\iota = \bigcap_{\lambda \in L} \Big(\bigcap_{\iota \in J_\lambda} X_\iota \Big)$$

("associativity" of union and intersection).

Let x be an element of $\bigcup_{\iota \in I} X_\iota$. There exists an index $\iota \in I$ such that $x \in X_\iota$. Since I is the union of the family $(J_\lambda)_{\lambda \in L}$, there exists an index $\lambda \in L$ such that $\iota \in J_\lambda$, whence $x \in \bigcup_{\iota \in J_\lambda} X_\iota$, and consequently

$$x \in \bigcup_{\lambda \in L} \Big(\bigcup_{\iota \in J_\lambda} X_\iota \Big).$$

Conversely, let x be an element of this set. There exists an index $\lambda \in L$ such that $x \in \bigcup_{\iota \in J_\lambda} X_\iota$, hence there exists an index $\iota \in J_\lambda$ (and therefore $\iota \in I$) such that $x \in X_\iota$; it follows that

$$x \in \bigcup_{\iota \in I} X_\iota.$$

Now suppose that $L \neq \emptyset$ and $J_\lambda \neq \emptyset$ for each $\lambda \in L$. Then $I \neq \emptyset$. Let x be an element of $\bigcap_{\iota \in I} X_\iota$. If $\lambda \in L$, we have $x \in X_\iota$ for each $\iota \in J_\lambda$ (since $J_\lambda \subset I$), whence $x \in \bigcap_{\iota \in J_\lambda} X_\iota$. Since the last inclusion is true for all $\lambda \in L$, x belongs to $\bigcap_{\lambda \in L} \Big(\bigcap_{\iota \in J_\lambda} X_\iota \Big)$. Conversely, let x be an element of this latter set, and let ι be any element of I. There exists $\lambda \in L$ such that $\iota \in J_\lambda$; since $x \in \bigcap_{\iota \in J_\lambda} X_\iota$, we have $x \in X_\iota$, which is true for all $\iota \in I$. Hence $x \in \bigcap_{\iota \in I} X_\iota$. This completes the proof.

For families of subsets of a given set the second part of Proposition 2 remains valid without restriction on L and J_λ.

3. IMAGES OF A UNION AND AN INTERSECTION

PROPOSITION 3. *Let* $(X_\iota)_{\iota \in I}$ *be a family of subsets of a set* A, *and let* Γ *be a correspondence between* A *and* B. *Then*

$$\Gamma\Big\langle \bigcup_{\iota \in I} X_\iota \Big\rangle = \bigcup_{\iota \in I} \Gamma\langle X_\iota \rangle, \qquad \Gamma\Big\langle \bigcap_{\iota \in I} X_\iota \Big\rangle \subset \bigcap_{\iota \in I} \Gamma\langle X_\iota \rangle.$$

The relation $(\exists x)\left(x \in \bigcup_{\iota \in I} X_\iota \text{ and } y \in \Gamma(x)\right)$ is equivalent to

$$(\exists x)(\exists \iota)(\iota \in I \text{ and } x \in X_\iota \text{ and } y \in \Gamma(x)),$$

hence to $(\exists \iota)(\iota \in I \text{ and } y \in \Gamma\langle X_\iota \rangle)$, hence is equivalent to $y \in \bigcup_{\iota \in I} \Gamma\langle X_\iota \rangle$; this proves the first formula. As to the second formula, we have $\bigcap_{\iota \in I} X_\iota \subset X_\iota$ for all $\iota \in I$, whence (§3, Proposition 2)

$$\Gamma\left\langle \bigcap_{\iota \in I} X_\iota \right\rangle \subset \Gamma\langle X_\iota \rangle,$$

and consequently

$$\Gamma\left\langle \bigcap_{\iota \in I} X_\iota \right\rangle \subset \bigcap_{\iota \in I} \Gamma\langle X_\iota \rangle.$$

¶ If Γ is an arbitrary correspondence (and in particular an arbitrary function), the formula

$$\Gamma\left\langle \bigcap_{\iota \in I} X_\iota \right\rangle = \bigcap_{\iota \in I} \Gamma\langle X_\iota \rangle$$

is usually *false*.

* For example, in the plane $\mathbf{R}^2$ the first projections of the lines $y = x$ and $y = x + 1$ are equal to $\mathbf{R}$, but the intersection of these lines is empty, and therefore so is the first projection of this intersection (*). *

However, we have the following important result :

Proposition 4. *Let f be a mapping of* A *into* B *and let* $(Y_\iota)_{\iota \in I}$ *be a family of subsets of* B. *Then* $\overset{-1}{f}\left\langle \bigcap_{\iota \in I} Y_\iota \right\rangle = \bigcap_{\iota \in I} \overset{-1}{f}\langle Y_\iota \rangle$.

To prove this, let x be an element of $\bigcap_{\iota \in I} \overset{-1}{f}\langle Y_\iota \rangle$. We have $f(x) \in Y_\iota$ for all $\iota \in I$, whence $f(x) \in \bigcap_{\iota \in I} Y_\iota$, and consequently

$$x \in \overset{-1}{f}\left\langle \bigcap_{\iota \in I} Y_\iota \right\rangle.$$

(*) A celebrated error arising from the application of this formula is that committed by H. Lebesgue in his attempt to prove that the projection on an axis of a Borel set in the plane is again a Borel set (this statement was subsequently shown to be incorrect, and the discussion it provoked was the origin of the theory of "Souslin" sets) : Lebesgue asserted that the projection of the intersection of a decreasing sequence of sets is equal to the intersection of their projections (*Journal de Mathématiques*, (6) **1** (1905), pp. 191-192).

Therefore $\bigcap_{\iota \in I} \overset{-1}{f}(Y_\iota) \subset \overset{-1}{f}\left\langle \bigcap_{\iota \in I} Y_\iota \right\rangle$; this relation, together with Proposition 3, gives the result.

COROLLARY. *If f is an injection of A into B and if $(X_\iota)_{\iota \in I}$ is a family of subsets of A whose index set I is not empty, then* $f\left\langle \bigcap_{\iota \in I} X_\iota \right\rangle = \bigcap_{\iota \in I} f\langle X_\iota \rangle$.

For we may write $f = i \circ g$, where i is the canonical injection of $f\langle A \rangle$ into B and g is a bijection of A onto $f\langle A \rangle$. If h denotes the inverse mapping of g, we have $f\langle X \rangle = \overset{-1}{h}\langle X \rangle$ for every subset X of A, and we are therefore brought back to Proposition 4.

4. COMPLEMENTS OF UNIONS AND INTERSECTIONS

PROPOSITION 5. *For every family $(X_\iota)_{\iota \in I}$ of subsets of a set* E, *we have*

$$\complement_E\left(\bigcup_{\iota \in I} X_\iota\right) = \bigcap_{\iota \in I} (\complement_E X_\iota), \qquad \complement_E\left(\bigcap_{\iota \in I} X_\iota\right) = \bigcup_{\iota \in I} (\complement_E X_\iota).$$

Let $x \in \complement_E\left(\bigcup_{\iota \in I} X_\iota\right)$. Then $x \in E$ and, for every $\iota \in I$, $x \notin X_\iota$, so that $x \in \complement_E X_\iota$; consequently

$$x \in \bigcap_{\iota \in I} (\complement_E X_\iota).$$

Conversely, let x be an element of $\bigcap_{\iota \in I} (\complement_E X_\iota)$; by the definition of intersection we have $x \in E$. Furthermore, if we had $x \in \bigcup_{\iota \in I} X_\iota$, there would exist an index $\varkappa \in I$ such that $x \in X_\varkappa$, which contradicts the hypothesi $x \in \bigcap_{\iota \in I} (\complement_E X_\iota)$; hence

$$x \in \complement_E\left(\bigcup_{\iota \in I} X_\iota\right).$$

This proves the first formula; the second one is an immediate consequence, in view of the relation $\complement_E(\complement_E X) = X$ for every subset X of E.

5. UNION AND INTERSECTION OF TWO SETS

If A, B are sets, we write

$$A \cup B = \bigcup_{X \in \{A, B\}} X, \qquad A \cap B = \bigcap_{X \in \{A, B\}} X.$$

It is clear that $A \cup B$ is the set of all objects which belong either to A or to B (or possibly to both), while $A \cap B$ is the set of all objects which belong to both A and B. In particular, $\{x, y\} = \{x\} \cup \{y\}$.

¶ Let $\{x, y, z\} = \{x, y\} \cup \{z\}$. The set $\{x, y, z\}$ is the set whose only elements are x, y, and z. Similarly we write

$$\{x, y, z, t\} = \{x, y, z\} \cup \{t\},$$

and so on.

¶ If now A, B, C, D are sets, we write

$$A \cup B \cup C = \bigcup_{X \in \{A, B, C\}} X, \qquad A \cap B \cap C = \bigcap_{X \in \{A, B, C\}} X;$$

$$A \cup B \cup C \cup D = \bigcup_{X \in \{A, B, C, D\}} X, \qquad A \cap B \cap C \cap D = \bigcap_{X \in \{A, B, C, D\}} X;$$

and so on.

¶ Let A, B, C be sets. From Propositions 1 and 2 we deduce the formulae

$$A \cup B = B \cup A, \qquad A \cap B = B \cap A,$$
$$A \cup (B \cup C) = (A \cup B) \cup C = A \cup B \cup C,$$
$$A \cap (B \cap C) = (A \cap B) \cap C = A \cap B \cap C.$$

These formulae are also immediate consequences of the theorems enunciated in the criterion C24 (Chapter I, §3, no. 5). Similarly one proves the formulae

$$A \cup (B \cap C) = (A \cup B) \cap (A \cup C), \qquad A \cap (B \cup C) = (A \cap B) \cup (A \cap C)$$

("distributivity" of union over intersection and of intersection over union; cf. §5, no. 6).

¶ The relation $A \subset B$ is equivalent to $A \cup B = B$ and to $A \cap B = A$. If A and B are subsets of a set E, we deduce from Proposition 5 (or from criterion C24) the formulae

$$\complement_E(A \cup B) = (\complement_E A) \cap (\complement_E B), \qquad \complement_E(A \cap B) = (\complement_E A) \cup (\complement_E B);$$

furthermore, we have

$$A \cup (\complement_E A) = E, \qquad A \cap (\complement_E A) = \emptyset.$$

¶ If Γ is a correspondence between E and F, and if A, B are subsets of E, it follows from Proposition 3 that

$$\Gamma\langle A \cup B\rangle = \Gamma\langle A\rangle \cup \Gamma\langle B\rangle, \qquad \Gamma\langle A \cap B\rangle \subset \Gamma\langle A\rangle \cap \Gamma\langle B\rangle,$$

and that, if f is a mapping of F into E,

$$\overset{-1}{f}\langle A \cap B\rangle = \overset{-1}{f}\langle A\rangle \cap \overset{-1}{f}\langle B\rangle$$

from Proposition 4.

¶ We record also the following Proposition on complements:

PROPOSITION 6. *Let f be a mapping of* A *into* B. *For every subset* Y *of* B, *we have*

$$\overset{-1}{f}\langle B - Y\rangle = \overset{-1}{f}\langle B\rangle - \overset{-1}{f}\langle Y\rangle.$$

For x belongs to $\overset{-1}{f}\langle B - Y\rangle$ if and only if $f(x)$ belongs to B but not to Y, i.e., if and only if x belongs to $\overset{-1}{f}\langle B\rangle$ but not to $\overset{-1}{f}\langle Y\rangle$.

COROLLARY. *Let f be an injection of* A *into* B. *For every subset* X *of* A, *we have* $f\langle A - X\rangle = f\langle A\rangle - f\langle X\rangle$.

Writing $f = i \circ g$, where i is the canonical injection of $f\langle A\rangle$ into B, we reduce the Corollary to Proposition 6 applied to $\overset{-1}{g}$.

¶ The intersection $X \cap A$ is sometimes called the *trace* of X on A. If $\mathfrak{F}$ is a family of sets, the set of traces on A of the sets belonging to $\mathfrak{F}$ is called the *trace* of $\mathfrak{F}$ on A.

6. COVERINGS

DEFINITION 5. *A family of sets* $(X_\iota)_{\iota \in I}$ *is said to be a covering of a set* E *(or to cover* E*) if* $E \subset \bigcup_{\iota \in I} X_\iota$. *If* $(X_\iota)_{\iota \in I}$ *and* $(Y_\varkappa)_{\varkappa \in K}$ *are coverings of* E, *the second of these coverings is said to be finer than the first (or to be a refinement of the first, or to refine the first) if, for each* $\varkappa \in K$, *there exists* $\iota \in I$ *such that*

$$Y_\varkappa \subset X_\iota.$$

A set of sets $\mathfrak{R}$ is a covering of E if the family of sets defined by the identity mapping of $\mathfrak{R}$ is a covering of E, in other words, if $E \subset \bigcup_{X \in \mathfrak{R}} X$.

If $\mathfrak{R}$, $\mathfrak{R}'$, $\mathfrak{R}''$ are three coverings of E such that $\mathfrak{R}'$ refines $\mathfrak{R}$ and $\mathfrak{R}''$ refines $\mathfrak{R}'$, it is clear that $\mathfrak{R}''$ refines $\mathfrak{R}$.

Let $(X_\iota)_{\iota \in I}$ be a covering of E. If J is a subset of I such that $(X_\iota)_{\iota \in J}$ is still a covering of E, then this covering clearly refines $(X_\iota)_{\iota \in I}$.

¶ Let $(X_\iota)_{\iota \in I}$ and $(Y_\varkappa)_{\varkappa \in K}$ be coverings of a set E. Then the family of sets $(X_\iota \cap Y_\varkappa)_{(\iota, \varkappa) \in I \times K}$ is a covering of E. For if $x \in E$, there exist indices $\iota \in I$ and $\varkappa \in K$ such that $x \in X_\iota$ and $x \in Y_\varkappa$, so that $x \in X_\iota \cap Y_\varkappa$ Moreover, it is clear that the covering $(X_\iota \cap Y_\varkappa)_{(\iota, \varkappa) \in I \times K}$ refines each o the coverings $(X_\iota)_{\iota\in I}$, $(Y_\varkappa)_{\varkappa\in K}$. Conversely, let $(Z_\lambda)_{\lambda\in L}$ be a covering of E which refines each of the coverings $(X_\iota)_{\iota\in I}$, $(Y_\varkappa)_{\varkappa\in K}$; if $\lambda\in L$, then there exist indices $\iota\in I$ and $\varkappa\in K$ such that $Z_\lambda\subset X_\iota$ and $Z_\lambda\subset Y_\varkappa$, so that $Z_\lambda\subset X_\iota\cap Y_\varkappa$; hence the covering $(Z_\lambda)_{\lambda\in L}$ is a refinement of

$$(X_\iota\cap Y_\varkappa)_{(\iota,\varkappa)\in I\times K}.$$

¶ Let $(X_\iota)_{\iota\in I}$ be a covering of a set A, and let f be a mapping of A *onto* a set B. The family $(f\langle X_\iota\rangle)_{\iota\in I}$ is then a covering of B (Proposition 3), called the *image under f of the covering* $(X_\iota)_{\iota\in I}$. If g is a mapping of a set C into the set A, the family $(\overset{-1}{g}\langle X_\iota\rangle)_{\iota\in I}$ is a covering of C, called the *inverse image under g of the covering* $(X_\iota)_{\iota\in I}$.

¶ Let E and F be sets, let $(X_\iota)_{\iota\in I}$ be a covering of E, and let $(Y_\varkappa)_{\varkappa\in K}$ be a covering of F. The family $(X_\iota\times Y_\varkappa)_{(\iota,\varkappa)\in I\times K}$ is then a covering of E × F, called the *product* of the coverings $(X_\iota)_{\iota\in I}$ of E and $(Y_\varkappa)_{\varkappa\in K}$ of F.

PROPOSITION 7. (1) *Let* E *be a set and* $(X_\iota)_{\iota\in I}$ *a covering of* E. *If f and g are two functions with domain* E *such that f and g agree on* $E\cap X_\iota$ *for each* $\iota\in I$, *then f and g agree on* E.

(2) *Let* $(X_\iota)_{\iota\in I}$ *be a family of sets and let* $(f_\iota)_{\iota\in I}$ *be a family of mappings with the same target* F *such that for each* $\iota\in I$ *the domain of* f_ι *is* X_ι, *and for each pair* $(\iota,\varkappa)\in I\times I$, f_ι *and* $f_\varkappa$ *agree on* $X_\iota\cap X_\varkappa$. *Then there is exactly one function f with domain* $A=\bigcup_{\iota\in I}X_\iota$ *and target* F *which extends each of the functions* f_ι $(i\in I)$.

(1) Let x be any element of E. Then there exists $\iota\in I$ such that $x\in X_\iota$, whence $f(x)=g(x)$ by hypothesis.

(2) Let G_ι be the graph of f_ι and let $G=\bigcup_{\iota\in I}G_\iota$; let us show that G is a functional graph. If $(x,y)\in G$ and $(x,y')\in G$, there exist two indices ι, $\varkappa$ in I such that $(x,y)\in G_\iota$ and $(x,y')\in G_\varkappa$. This implies $x\in X_\iota$, $x\in X_\varkappa$, $y=f_\iota(x)$, $y'=f_\varkappa(x)$; but since $x\in X_\iota\cap X_\varkappa$, we have

$$f_\iota(x)=f_\varkappa(x),$$

that is to say, $y=y'$. The graph G has domain $\mathrm{pr}_1G=\bigcup_{\iota\in I}\mathrm{pr}_1G_\iota=A$; the function $f=(G,A,F)$ therefore satisfies the required conditions. Its uniqueness follows from the first part of the Proposition.

7. PARTITIONS

DEFINITION 6. *Two sets* A *and* B *are said to be disjoint (or not to intersect) if* $A\cap B=\emptyset$. *If* $A\cap B\neq\emptyset$, *we say that* A *meets (or intersects)* B. *Let*

$(X_\iota)_{\iota\in I}$ *be a family of sets. The sets of this family are said to be mutually disjoint if the conditions* $\iota \in I$, $\varkappa \in I$, $\iota \neq \varkappa$ *imply* $X_\iota \cap X_\varkappa = \emptyset$.

Let f be a mapping of A into B, and let $(Y_\iota)_{\iota\in I}$ be a family of mutually disjoint subsets of B. Proposition 4 then shows that the sets of the family $(\overset{-1}{f}\langle Y_\iota\rangle)_{\iota\in I}$ of subsets of A are mutually disjoint. On the other hand, if $(X_\iota)_{\iota\in I}$ is a family of mutually disjoint subsets of A, the sets of the family $(f\langle X_\iota\rangle)_{\iota\in I}$ are not in general mutually disjoint.

PROPOSITION 8. *Let* $(X_\iota)_{\iota\in I}$ *be a family of mutually disjoint sets, and let* $(f_\iota)_{\iota\in I}$ *be a family of functions with the same target* F *such that the domain of* f_ι *is* X_ι *for each* $\iota \in I$. *Then there exists exactly one function f with domain* $\bigcup_{\iota\in I} X_\iota$ *and target* F *which extends each of the functions* f_ι $(\iota \in I)$.

This is a corollary of Proposition 7, since f_ι and $f_\varkappa$ clearly agree on $X_\iota \cap X_\varkappa = \emptyset$ whenever $\iota \neq \varkappa$.

DEFINITION 7. *A partition of a set* E *is a family of* non-empty *mutually disjoint subsets of* E *which covers* E.

Example. For every non-empty set A, the family $(\{x\})_{x\in A}$ is a partition of A.

If $(X_\iota)_{\iota\in I}$ is a partition of a set E, the mapping $\iota \to X_\iota$ of I onto the set $\mathfrak{F}$ of elements X_ι of the partition is bijective. Hence, if $\mathfrak{F}$ is given, the partition is determined up to a one-to-one correspondence between index sets. Usually when we speak of a partition, it is the set of elements of the partition with which we are concerned.

8. SUM OF A FAMILY OF SETS

PROPOSITION 9. *Let* $(X_\iota)_{\iota\in I}$ *be a family of sets. Then there exists a set* X *with the following property*: X *is the union of a family* $(X'_\iota)_{\iota\in I}$ *of mutually disjoint sets such that for each* $\iota \in I$ *there exists a one-to-one mapping of* X_ι *onto* X'_ι.

Let $A = \bigcup_{\iota\in I} X_\iota$. If $\iota \in I$, the mapping $x \to (x, \iota)$ $(x \in X_\iota)$ is a one-to-one mapping of X_ι onto a subset X'_ι of $A \times I$. Moreover, the image of X'_ι under the second coordinate function on $A \times I$ is contained in the set $\{\iota\}$; it follows that $X'_\iota \cap X'_\varkappa = \emptyset$ whenever $\iota \neq \varkappa$. We may then take $X = \bigcup_{\iota\in I} X'_\iota$.

DEFINITION 8. *Let* $(X_\iota)_{\iota\in I}$ *be a family of sets. The sum of this family is the union of the family of sets* $(X_\iota \times \{\iota\})_{\iota\in I}$.

Proposition 10. *Let* $(X_\iota)_{\iota \in I}$ *be a family of mutually disjoint sets. Let* A *be its union and* S *its sum. Then there exists a bijective mapping of* A *onto* S.

For each $\iota \in I$, let f_ι be a bijection of X_ι onto $X_\iota \times \{\iota\}$. By virtue of Proposition 8, there exists a mapping f of A into S which extends all the mappings f_ι. It is immediately verified that f is a bijective mapping of A onto S.

By abuse of language, a set E is said to be the *sum* of a family of sets $(X_\iota)_{\iota \in I}$ if there exists a bijection of E onto the sum of the family, as defined in Definition 8.

Note that if $(X_\iota)_{\iota \in I}$ is an arbitrary family of sets, the argument of Proposition 10 shows that there exists a mapping of the sum S *onto* the union A.

5. PRODUCT OF A FAMILY OF SETS

1. THE AXIOM OF THE SET OF SUBSETS

A4. $$(\forall X) \operatorname{Coll}_Y(Y \subset X).$$

This axiom means that for every set X there exists a set whose elements are all the subsets of X, namely the set $\mathcal{E}_Y(Y \subset X)$ (§1, no. 4); this set is denoted by $\mathfrak{P}(X)$ and is called the *set of subsets of* X. Clearly, if $X \subset X'$, we have $\mathfrak{P}(X) \subset \mathfrak{P}(X')$.

¶ Let A, B be two sets and let Γ be a correspondence between A and B. The *function* $X \to \Gamma\langle X\rangle$ $(X \subset \mathfrak{P}(A), \Gamma\langle X\rangle \in \mathfrak{P}(B))$ is called the *canonical extension* (or simply the *extension*) *of* Γ *to sets of subsets* and is denoted by $\hat{\Gamma}$; it is a mapping of $\mathfrak{P}(A)$ into $\mathfrak{P}(B)$. If Γ' is a correspondence between B and a set C, the formula $(\Gamma' \circ \Gamma)\langle X\rangle = \Gamma'\langle\Gamma\langle X\rangle\rangle$ shows that the extension of $\Gamma' \circ \Gamma$ to sets of subsets is the mapping $\hat{\Gamma}' \circ \hat{\Gamma}$.

Proposition 1. (1) *If f is a surjection of a set* E *onto a set* F, *the canonical extension $\hat{f}$ of f is a surjection of* $\mathfrak{P}(E)$ *onto* $\mathfrak{P}(F)$.

(2) *If f is an injection of* E *into* F, *the canonical extension $\hat{f}$ of f is an injection of* $\mathfrak{P}(E)$ *into* $\mathfrak{P}(F)$.

(1) If s is a section of f, then $f \circ s$ is the identity mapping of F, hence $\hat{f} \circ \hat{s}$ is the identity mapping of $\mathfrak{P}(F)$; therefore $\hat{f}$ is surjective and $\hat{s}$ is a section of $\hat{f}$ (§3, no. 8).

(2) The proposition is obvious if $E = \emptyset$, because then $\mathfrak{P}(E) = \{\emptyset\}$. If $E \neq \emptyset$ and if r is a retraction of f, then $r \circ f$ is the identity mapping of E, so that $\hat{r} \circ \hat{f}$ is the identity mapping of $\mathfrak{P}(E)$; therefore $\hat{f}$ is injective, and $\hat{r}$ is a retraction of $\hat{f}$ (§3, no. 8).

2. SET OF MAPPINGS OF ONE SET INTO ANOTHER

Let E, F be sets. The graph of a mapping of E into F is a subset of $E \times F$. The set of elements of $\mathfrak{P}(E \times F)$ which have the property of being graphs of mappings of E into F is therefore a subset of $\mathfrak{P}(E \times F)$, which is denoted by F^E. The set of triples $f = (G, E, F)$, where $G \in F^E$, is therefore the *set of mappings* of E into F; it is denoted by $\mathcal{F}(E, F)$. Clearly $G \to (G, E, F)$ is a bijection (called the *canonical bijection*) of F^E onto $\mathcal{F}(E, F)$. The existence of this bijection allows us to translate immediately every proposition relating to the set F^E into one relating to $\mathcal{F}(E, F)$, and *vice versa*.

¶ Let E, E′, F, F′ be sets. Let u be a mapping of E′ into E, and let v be a mapping of F into F′. Then the function $f \to v \circ f \circ u$ is a mapping of $\mathcal{F}(E, F)$ into $\mathcal{F}(E', F')$.

PROPOSITION 2. (1) *If u is a surjection of* E′ *onto* E *and v an injection of* F *into* F′, *then the mapping $f \to v \circ f \circ u$ is injective.*

(2) *If u is an injection of* E′ *into* E *and v a surjection of* F *onto* F′, *then the mapping $f \to v \circ f \circ u$ is surjective.*

We shall assume that the sets E, E′, F, F′ are all non-empty, since otherwise the proposition is trivially verified.

(1) Let s be a section of u and let r be a retraction of v (§3, Definition 11). Then $r \circ (v \circ f \circ u) \circ s = I_F \circ f \circ I_E = f$, so that

$$f \to v \circ f \circ u$$

is injective.

(2) Let r' be a retraction of u and let s' be a section of v. For every mapping $f' : E' \to F'$ we have $v \circ (s' \circ f' \circ r') \circ u = f'$, which shows that the mapping $f \to v \circ f \circ u$ is surjective.

COROLLARY. *If u is a bijection of* E′ *onto* E *and v is a bijection of* F *onto* F′, *then $f \to v \circ f \circ u$ is bijective.*

Let A, B, C be three sets and let f be a mapping of $B \times C$ into A. For every $y \in C$ let $f(\bullet, y)$ be the partial mapping $x \to f(x, y)$ of B into A (§ 3, no. 9); the function $y \to f(\bullet, y)$ is a mapping of C into $\mathcal{F}(B, A)$. Conversely, for every mapping g of C into $\mathcal{F}(B, A)$ there exists a unique mapping f of $B \times C$ into A such that $g(y) = f(\bullet, y)$ for each $y \in C$, namely the mapping $(x, y) \rightarrow (g(y))(x)$. Hence :

PROPOSITION 3. *If for every mapping f of $B \times C$ into A we denote by $\tilde{f}$ the mapping $y \rightarrow f(\bullet, y)$ of C into $\mathcal{F}(B, A)$, then the function $f \rightarrow \tilde{f}$ is a bijection* (called the *canonical bijection*) *of $\mathcal{F}(B \times C, A)$ onto $\mathcal{F}(C, \mathcal{F}(B, A))$.*

Similarly we define a *canonical bijection* of $\mathcal{F}(B \times C, A)$ onto $\mathcal{F}(B, \mathcal{F}(C, A))$. By reason of the one-to-one correspondence between mappings and functional graphs, these bijections give rise to *canonical bijections* of $A^{B \times C}$ onto $(A^B)^C$ (resp. $(A^C)^B$).

3. DEFINITIONS OF THE PRODUCT OF A FAMILY OF SETS

Let $(X_\iota)_{\iota \in I}$ be a family of sets and let F be a functional graph with domain I such that $F(\iota) \in X_\iota$ for each $\iota \in I$. Then for each $\iota \in I$ we have $F(\iota) \in A = \bigcup_{\iota \in I} X_\iota$, and therefore F is an element of $\mathfrak{P}(I \times A)$. The functional graphs with the above property therefore form a subset of $\mathfrak{P}(I \times A)$.

DEFINITION 1. *Let $(X_\iota)_{\iota \in I}$ be a family of sets. The set of functional graphs F with domain I such that $F(\iota) \in X_\iota$ for each $\iota \in I$ is called the product of the family of sets $(X_\iota)_{\iota \in I}$ and is denoted by $\prod_{\iota \in I} X_\iota$. For each $\iota \in I$, X_ι is called the factor of index ι in the product $\prod_{\iota \in I} X_\iota$. The mapping $F \rightarrow F(\iota)$ $\left(F \in \prod_{\iota \in I} X_\iota,\ F(\iota) \in X_\iota\right)$ is called the coordinate function* (or *projection*) *of index ι, and is denoted by* pr_ι.

$F(\iota)$ is called the *coordinate of index* ι (or *projection of index* ι) of F; the image $\mathrm{pr}_\iota\langle A\rangle$ of a subset A of $\prod_{\iota \in I} X_\iota$ under the coordinate function of index ι is called the *projection of index* ι of A. It is easily verified that $A \subset \prod_{\iota \in I} \mathrm{pr}_\iota\langle A\rangle$.

We shall often use the notation $(x_\iota)_{\iota \in I}$ to denote an element of $\prod_{\iota \in I} X_\iota$ (§3, no. 6).

If $I = \emptyset$, the set $\prod_{\iota \in I} X_\iota$ has only one element, namely the empty set (§3, no. 4, Example 1).

¶ If all the factors X_ι of the product $\prod_{\iota \in I} X_\iota$ are equal to the same set E, we have $\prod_{\iota \in I} X_\iota = E^I$; this follows immediately from the definitions.

¶ If $(X_\iota)_{\iota\in I}$ is an arbitrary family of sets and if E is a set such that

$$\bigcup_{\iota\in I} X_\iota \subset E,$$

then Definition 1 shows that $\prod_{\iota\in I} X_\iota \subset E^I$; there is therefore a one-to-one correspondence between $\prod_{\iota\in I} X_\iota$ and a set of mappings of I into E (i.e., a subset of $\mathcal{F}(I, E)$).

¶ If $I = \{\alpha\}$ is a set consisting of a single element, we have

$$\prod_{\iota\in I} X_\iota = X_\alpha^{\{\alpha\}};$$

the mapping $F \to F(\alpha)$ is then a bijection (called *canonical*) of $\prod_{\iota\in\{\alpha\}} X_\iota$ onto X_α.

¶ Let A, B be sets and let α, β be distinct objects (there exist two distinct objects, for example $\emptyset$ and $\{\emptyset\}$). Consider the graph $\{(\alpha, A), (\beta, B)\}$, which is clearly functional and is precisely the family $(X_\iota)_{\iota\in\{\alpha,\beta\}}$ such such that $X_\alpha = A$ and $X_\beta = B$. For each pair $(x, y) \in A \times B$, let $f_{x,y}$ be the functional graph $\{(\alpha, x), (\beta, y)\}$. It is evident that the function $(x, y) \to f_{x,y}$ is a bijective mapping of $A \times B$ onto $\prod_{\iota\in\{\alpha,\beta\}} X_\iota$; the inverse of this bijection is $g \to (g(\alpha), g(\beta))$. These two bijections are called *canonical*. In what follows we shall use this one-to-one correspondence to deduce properties of the product of two sets from properties of the product of a family of sets.

¶ Let $(X_\iota)_{\iota\in I}$ be a family of sets each of which consists of a single element, say $X_\iota = \{a_\iota\}$; then the product $\prod_{\iota\in I} X_\iota$ is a set consisting of the single element $(a_\iota)_{\iota\in I}$.

¶ Let E be a set. The graphs of the *constant* mappings $\iota \to x$ of I into E form a subset Δ of the product E^I, called the *diagonal*. If $\bar{x}$ denotes the graph of the mapping $\iota \to x$ (where $x \in E$), the mapping $x \to \bar{x}$ is a bijection of E onto Δ, called the *diagonal mapping*.

PROPOSITION 4. *Let $(X_\iota)_{\iota\in I}$ be a family of sets, and let u be a bijection of a set K onto the index set I. If U is the graph of u, the mapping $F \to F \circ U$ of $\prod_{\iota\in I} X_\iota$ into $\prod_{\varkappa\in K} X_{u(\varkappa)}$ is bijective.*

Let

$$A = \bigcup_{\iota\in I} X_\iota = \bigcup_{\varkappa\in K} X_{u(\varkappa)}$$

(§4, Proposition 1). The mapping $F \to F \circ U$ ($F \in A^I$) is a bijection of A^I onto A^K (Proposition 2). It is evident that the condition "for each $\iota \in I$, $F(\iota) \in X_\iota$" is equivalent to "for each $\varkappa \in K$, $(F \circ U)(\varkappa) \in X_{u(\varkappa)}$", and the result follows.

4. PARTIAL PRODUCTS

Let $(X_\iota)_{\iota \in I}$ be a family of sets, and let J be a subset of I. The product $\prod_{\iota \in J} X_\iota$ is called a *partial product* of $\prod_{\iota \in I} X_\iota$. If f is a function whose graph F is a member of $\prod_{\iota \in I} X_\iota$, then $F \circ \Delta_J$ (where Δ_J is the diagonal of $J \times J$) is the graph of the *restriction* of f to J. Clearly $F \circ \Delta_J \in \prod_{\iota \in J} X_\iota$; the mapping $F \to F \circ \Delta_J$ of $\prod_{\iota \in I} X_\iota$ into $\prod_{\iota \in I} X_\iota$ is called the *projection of index* J and is denoted by pr_J.

PROPOSITION 5. *Let $(X_\iota)_{\iota \in I}$ be a family of sets and let J be a subset of I. If for each $\iota \in I$ we have $X_\iota \neq \emptyset$, the projection pr_J is a mapping of $\prod_{\iota \in I} X_\iota$ onto $\prod_{\iota \in J} X_\iota$.*

In view of the remarks made above, it is enough to prove the following proposition :

PROPOSITION 6. *Let $(X_\iota)_{\iota \in I}$ be a family of sets such that $X_\iota \neq \emptyset$ for all $\iota \in I$. If g is a mapping of $J \subset I$ into $A = \bigcup_{\iota \in I} X_\iota$ such that $g(\iota) \in X_\iota$ for all $\iota \in J$, then there exists an extension f of g to I such that $f(\iota) \in X_\iota$ for all $\iota \in I$.*

For each $\iota \in I - J$ let T_ι denote the term $\tau_y(y \in X_\iota)$. Since $X_\iota \neq \emptyset$ by hypothesis, we have $T_\iota \in X_\iota$ for all $\iota \in I - J$ (Chapter I, §4, no. 1). If G is the graph of g, the graph $G \cup \left(\bigcup_{\iota \in I - J} \{(\iota, T_\iota)\}\right)$ is the graph of a function which has the required properties, as is immediately verified.

COROLLARY 1. *Let $(X_\iota)_{\iota \in I}$ be a family of sets such that for each $\iota \in I$ we have $X_\iota \neq \emptyset$. Then for each $\alpha \in I$ the projection pr_α is a mapping of $\prod_{\iota \in I} X_\iota$ onto X_α.*

Apply Proposition 5 to the subset $J = \{\alpha\}$ of I and note that pr_α is the composition of the canonical mapping of $X_\alpha^{\{\alpha\}}$ onto X_α and the mapping $pr_{\{\alpha\}}$.

COROLLARY 2. *Let* $(X_\iota)_{\iota\in I}$ *be a family of sets. Then* $\prod_{\iota\in I} X_\iota = \emptyset$ *if and only if there exists* $\iota \in I$ *such that* $X_\iota = \emptyset$.

If we have $X_\iota \neq \emptyset$ for each $\iota \in I$, then it follows from Corollary 1 that $\prod_{\iota\in I} X_\iota \neq \emptyset$; conversely, if $\prod_{\iota\in I} X_\iota \neq \emptyset$, the relation $\mathrm{pr}_\alpha\left(\prod_{\iota\in I} X_\iota\right) \subset X_\alpha$ shows that $X_\alpha \neq \emptyset$ for each $\alpha \in I$.

Hence, if we have a family $(X_\iota)_{\iota\in I}$ of non-empty sets, we may introduce (as an auxiliary constant) a function f with domain I such that $f(\iota) \in X_\iota$ for all $\iota \in I$. In practice one says: "take an element x_ι in each X_ι". Intuitively, we have thus "chosen" an element x_ι in each set X_ι; the introduction of the logical sign τ and the criteria governing its use absolve us from the necessity of formulating an "axiom of choice" to legalize this operation (cf. Summary of Results, §4, no. 10).

COROLLARY 3. *Let* $(X_\iota)_{\iota\in I}$ *and* $(Y_\iota)_{\iota\in I}$ *be two families of sets having the same index set* I. *If* $X_\iota \subset Y_\iota$ *for each* $\iota \in I$, *then*

$$\prod_{\iota\in I} X_\iota \subset \prod_{\iota\in I} Y_\iota.$$

Conversely, if $\prod_{\iota\in I} X_\iota \subset \prod_{\iota\in I} Y_\iota$, *and if* $X_\iota \neq \emptyset$ *for each* $\iota \in I$, *then* $X_\iota \subset Y_\iota$ *for each* $\iota \in I$.

The first assertion is obvious, and the second follows from Proposition 1, Corollary 5, because we have then, for each $\alpha \in I$,

$$X_\alpha = \mathrm{pr}_\alpha\left(\prod_{\iota\in I} X_\iota\right) \subset \mathrm{pr}_\alpha\left(\prod_{\iota\in I} Y_\iota\right) = Y_\alpha.$$

5. ASSOCIATIVITY OF PRODUCTS OF SETS

PROPOSITION 7. *Let* $(X_\iota)_{\iota\in I}$ *be a family of sets whose index set* I *is not the empty set. Let* $(J_\lambda)_{\lambda\in L}$ *be a partition of* I. *Then the mapping*

$$f \to (\mathrm{pr}_{J_\lambda} f)_{\lambda\in L}$$

of $\prod_{\iota\in I} X_\iota$ *into the product set* $\prod_{\lambda\in L}\left(\prod_{\iota\in J_\lambda} X_\iota\right)$ *is bijective* ("associativity" of products of sets).

From the interpretation of $\mathrm{pr}_{J_\lambda} f$ as the graph of the restriction of the function whose graph is f (no. 4), it follows that the statement that the mapping $f \to (\mathrm{pr}_{J_\lambda} f)_{\lambda\in L}$ is bijective means that, for each family $(v_\lambda)_{\lambda\in L}$, where v_λ is a mapping of J_λ into $\bigcup_{\iota\in I} X_\iota$, there exists a unique mapping u

of I into $\bigcup_{\iota \in I} X_\iota$ such that v_λ is the restriction of u to J_λ for each $\lambda \in L$. But this is a consequence of the hypothesis that $(J_\lambda)_{\lambda \in L}$ is a partition of I (§4, Proposition 8).

¶ The bijection defined in Proposition 7, and its inverse bijection are said to be *canonical*.

Remarks

(1) Let α, β be two distinct objects and let $(J_\lambda)_{\lambda \in \{\alpha, \beta\}}$ be a partition of I into two sets J_α, J_β. We thus obtain a one-to-one mapping (again called *canonical*) of the product $\prod_{\iota \in I} X_\iota$ onto $\left(\prod_{\iota \in J_\alpha} X_\iota\right) \times \left(\prod_{\iota \in J_\beta} X_\iota\right)$ by forming the composition of the canonical mapping of $\prod_{\lambda \in \{\alpha, \beta\}} \left(\prod_{\iota \in J_\lambda} X_\iota\right)$ onto $\left(\prod_{\iota \in J_\alpha} X_\iota\right) \times \left(\prod_{\iota \in J_\beta} X_\iota\right)$ and the canonical mapping of $\prod_{\iota \in I} X_\iota$ onto $\prod_{\lambda \in \{\alpha, \beta\}} \left(\prod_{\iota \in J_\lambda} X_\iota\right)$. If X_ι is a set consisting of a *single element* for each $\iota \in J_\beta$, then pr_{J_α} is a bijective mapping of $\prod_{\iota \in I} X_\iota$ onto $\prod_{\iota \in J_\alpha} X_\iota$.

(2) Let α, β, γ be three objects, no two of which are equal (three such objects exist; for example, $\emptyset$, $\{\emptyset\}$, $\{\{\emptyset\}\}$), and let A, B, C be sets. Consider the functional graph $\{(\alpha, A), (\beta, B), (\gamma, C)\}$, i.e., the family of sets $(X_\iota)_{\iota \in \{\alpha, \beta, \gamma\}}$ such that $X_\alpha = A$, $X_\beta = B$, $X_\gamma = C$. To the partition of $\{\alpha, \beta, \gamma\}$ formed by the two sets $\{\alpha, \beta\}$ and $\{\gamma\}$ there corresponds a canonical bijection of $\prod_{\iota \in \{\alpha, \beta, \gamma\}} X_\iota$ onto the product

$$\left(\prod_{\iota \in \{\alpha, \beta\}} X_\iota\right) \times X_\gamma^{\{\gamma\}},$$

and hence a bijection (again called *canonical*) of $\prod_{\iota \in \{\alpha, \beta, \gamma\}} X_\iota$ onto $A \times B \times C$ (§ 2, no. 2) which transforms each graph $f \in \prod_{\iota \in \{\alpha, \beta, \gamma\}} X_\iota$ into the element $(f(\alpha), f(\beta), f(\gamma))$ of $A \times B \times C$. By Proposition 4 we may therefore put any two of the six sets $A \times B \times C$, $B \times C \times A$, $C \times A \times B$, $B \times A \times C$, $A \times C \times B$, $C \times B \times A$ in one-to-one correspondence.

6. DISTRIBUTIVITY FORMULAE

PROPOSITION 8. *Let* $((X_{\lambda,\iota})_{\iota \in J_\lambda})_{\lambda \in L}$ *be a family (with index set* L*) of families of sets. Suppose that* $L \neq \emptyset$ *and that* $J_\lambda \neq \emptyset$ *for each* $\lambda \in L$. *Let*

$$I = \prod_{\lambda \in L} J_\lambda \neq \emptyset.$$

Then we have

$$\bigcup_{\lambda\in L}\Big(\bigcap_{\iota\in J_\lambda} X_{\lambda,\iota}\Big)=\bigcap_{f\in I}\Big(\bigcup_{\lambda\in L} X_{\lambda,f(\lambda)}\Big),$$
$$\bigcap_{\lambda\in L}\Big(\bigcup_{\iota\in J_\lambda} X_{\lambda,\iota}\Big)=\bigcup_{f\in I}\Big(\bigcap_{\lambda\in L} X_{\lambda,f(\lambda)}\Big)$$

("distributivity" of union over intersection, and of intersection over union).

Let x be an element of $\bigcup_{\lambda\in L}\Big(\bigcap_{\iota\in J_\lambda} X_{\lambda,\iota}\Big)$ and let f be any element of I. There exists an index λ such that $x\in\bigcap_{\iota\in J_\lambda} X_{\lambda,\iota}$; consequently $x\in X_{\lambda,f(\lambda)}$ and hence

$$x\in\bigcup_{\lambda\in L} X_{\lambda,f(\lambda)}.$$

Since this is true for each $f\in I$, we have

$$x\in\bigcap_{f\in I}\Big(\bigcup_{\lambda\in L} X_{\lambda,f(\lambda)}\Big).$$

Conversely, let x be an object which does not belong to the set

$$\bigcup_{\lambda\in L}\Big(\bigcap_{\iota\in J_\lambda} X_{\lambda,\iota}\Big).$$

Then for each $\lambda\in L$ we have $x\notin\bigcap_{\iota\in J_\lambda} X_{\lambda,\iota}$, which means that the set J'_λ of indices $\iota\in J_\lambda$ such that $x\notin X_{\lambda,\iota}$ is not empty for each $\lambda\in L$. By Proposition 5, Corollary 2, there exists a functional graph f with domain L such that for each $\lambda\in L$ we have $f(\lambda)\in J'_\lambda$. Consequently $f\in I$ and $x\notin X_{\lambda,f(\lambda)}$ for each $\lambda\in L$; hence

$$x\notin\bigcup_{\lambda\in L} X_{\lambda,f(\lambda)}$$

and thus

$$x\notin\bigcap_{f\in I}\Big(\bigcup_{\lambda\in L} X_{\lambda,f(\lambda)}\Big).$$

This completes the proof of the first formula. The second formula follows by applying the first to the family $((\complement_A X_{\lambda,\iota})_{\iota\in J_\lambda})_{\lambda\in L}$, where A denotes the union $\bigcap_{\lambda\in L}\Big(\bigcap_{\iota\in J_\lambda} X_\lambda\Big)$.

COROLLARY. *Let* $(X_\iota)_{\iota\in I}$ *and* $(Y_\varkappa)_{\varkappa\in K}$ *be two families of sets with non-empty index sets* I, K. *Then*

$$\Big(\bigcap_{\iota\in I} X_\iota\Big) \cup \Big(\bigcap_{\varkappa\in K} Y_\varkappa\Big) = \bigcap_{(\iota,\varkappa)\in I\times K} (X_\iota \cup Y_\varkappa),$$
$$\Big(\bigcup_{\iota\in I} X_\iota\Big) \cap \Big(\bigcup_{\varkappa\in K} Y_\varkappa\Big) = \bigcup_{(\iota,\varkappa)\in I\times K} (X_\iota \cap Y_\varkappa).$$

Let α, β be two distinct objects; apply the formulae of Proposition 8 (with $L = \{\alpha, \beta\}$, $J_\alpha = I$, $J_\beta = K$) to the family $((Z_{\lambda,\mu})_{\mu\in J_\lambda})_{\lambda\in L}$, where $Z_{\alpha,\iota} = X_\iota$ for all $\iota \in I$ and $Z_{\beta,\varkappa} = Y_\varkappa$ for each $\varkappa \in K$. By the existence of the canonical bijection of $\prod_{\lambda\in L} J_\lambda$ onto $I \times K$ (no. 3) and by Proposition 1 of § 4 we obtain the formulae stated.

PROPOSITION 9. *Let* $((X_{\lambda,\iota})_{\iota\in J_\lambda})_{\lambda\in L}$ *be a family (with index set* L) *of families of sets. Let* $I = \prod_{\lambda\in L} J_\lambda$. *Then*

$$\prod_{\lambda\in L}\Big(\bigcup_{\iota\in J_\lambda} X_{\lambda,\iota}\Big) = \bigcup_{f\in I}\Big(\prod_{\lambda\in L} X_{\lambda,f(\lambda)}\Big)$$

and (if $L \neq \emptyset$ *and* $J_\lambda \neq \emptyset$ *for each* $\lambda \in L$)

$$\prod_{\lambda\in L}\Big(\bigcap_{\iota\in J_\lambda} X_{\lambda,\iota}\Big) = \bigcap_{f\in I}\Big(\prod_{\lambda\in L} X_{\lambda,f(\lambda)}\Big)$$

("distributivity" of product over union and over intersection).

The first formula is trivially true if $L = \emptyset$ or if $J_\lambda = \emptyset$ for some $\lambda \in L$. If not, let g be an element of $\prod_{\lambda\in L}\Big(\bigcap_{\iota\in J_\lambda} X_{\lambda,\iota}\Big)$. For each $\lambda \in L$ there exists an index $\iota \in J_\lambda$ such that $g(\lambda) \in X_{\lambda,\iota}$; in other words, the set H_λ of indices $\iota \in J_\lambda$ such that $g(\lambda) \in X_{\lambda,\iota}$ is not empty. By Corollary 2 to Proposition 5 there is therefore a functional graph f with domain L such that

$$f(\lambda) \in H_\lambda$$

for each $\lambda \in L$, i.e., $g(\lambda) \in X_{\lambda,f(\lambda)}$. Hence we have $g \in \prod_{\lambda\in L} X_{\lambda,f(\lambda)}$ and consequently $g \in \bigcup_{f\in I}\Big(\prod_{\lambda\in L} X_{\lambda,f(\lambda)}\Big)$. Conversely, if

$$g \in \bigcup_{f\in I}\Big(\prod_{\lambda\in L} X_{\lambda,f(\lambda)}\Big),$$

there is a functional graph $f \in I$ such that for every $\lambda \in L$ we have

$$g(\lambda) \in X_{\lambda,f(\lambda)}$$

and, *a fortiori*, $g(\lambda) \in \bigcup_{\iota \in J_\lambda} X_{\lambda, \iota}$. This completes the proof of the first formula. The proof of the second formula is analogous but simpler, and we leave it to the reader.

COROLLARY 1. *Suppose that* $L \neq \emptyset$ *and that* $J_\lambda \neq \emptyset$ *for each* $\lambda \in L$. *If for e achindex* $\lambda \in L$ *the family* $(X_{\lambda, \iota})_{\iota \in J_\lambda}$ *is a partition of* $X_\lambda = \bigcup_{\iota \in J_\lambda} X_{\lambda, \iota}$, *then the family* $\left(\prod_{\lambda \in L} X_{\lambda, f(\lambda)}\right)_{f \in I}$ *is a partition of* $\prod_{\lambda \in L} X_\lambda$.

If we set

$$P_f = \prod_{\lambda \in L} X_{\lambda, f(\lambda)},$$

then, by virtue of the first formula of Proposition 9, it is sufficient to show that $P_f \neq \emptyset$ for all $f \in I$ and that $P_f \cap P_g = \emptyset$ whenever f and g are distinct elements of I. The first point follows from Proposition 5, Corollary 2. As to the second, if $f \neq g$, there exists $\lambda \in L$ such that

$$f(\lambda) \neq g(\lambda)$$

and therefore, by virtue of the hypothesis, $X_{\lambda, f(\lambda)} \cap X_{\lambda, g(\lambda)} = \emptyset$. It follows that there is no graph belonging to $P_f \cap P_g$; for if G were such a graph, we would have $G(\lambda) \in X_{\lambda, f(\lambda)} \cap X_{\lambda, g(\lambda)} = \emptyset$, which is absurd.

COROLLARY 2. *Let* $(X_\iota)_{\iota \in I}$ *and* $(Y_\varkappa)_{\varkappa \in K}$ *be two families of sets. Then*

$$\left(\bigcup_{\iota \in I} X_\iota\right) \times \left(\bigcup_{\varkappa \in K} Y_\varkappa\right) = \bigcup_{(\iota, \varkappa) \in I \times K} (X_\iota \times Y_\varkappa)$$

and, if I *and* K *are non-empty,*

$$\left(\bigcap_{\iota \in I} X_\iota\right) \times \left(\bigcap_{\varkappa \in K} Y_\varkappa\right) = \bigcap_{(\iota, \varkappa) \in I \times K} (X_\iota \times Y_\varkappa).$$

The proof follows the pattern of the proof of the Corollary to Proposition 8.

PROPOSITION 10. *Let* $(X_{\iota, \varkappa})_{(\iota, \varkappa) \in I \times K}$ *be a family of sets whose index set is the product of two sets* I *and* K. *If* $K \neq \emptyset$, *we have*

$$\bigcap_{\varkappa \in K}\left(\prod_{\iota \in I} X_{\iota, \varkappa}\right) = \prod_{\iota \in I}\left(\bigcap_{\varkappa \in K} X_{\iota, \varkappa}\right).$$

Both sides of the equality to be proved are functional graphs. A graph f belongs to the left-hand side if and only if, for each $\varkappa \in K$, $f \in \prod_{\iota \in I} X_{\iota, \varkappa}$; that is, if and only if $f(\iota) \in X_{\iota, \varkappa}$ for all $(\iota, \varkappa) \in I \times K$. For f to belong to the right-hand side it is necessary and sufficient that $f(\iota) \in \bigcap_{\varkappa \in K} X_{\iota, \varkappa}$ for each $\iota \in I$, i.e., that $f(\iota) \in X_{\iota, \varkappa}$ for each pair $(\iota, \varkappa) \in I \times K$. This completes the proof.

COROLLARY. *Let $(X_\iota)_{\iota \in I}$ and $(Y_\iota)_{\iota \in I}$ be two families of sets with the same index set $I \neq \emptyset$.* Then

$$\Big(\prod_{\iota \in I} X_\iota\Big) \cap \Big(\prod_{\iota \in I} Y_\iota\Big) = \prod_{\iota \in I} (X_\iota \cap Y_\iota),$$
$$\Big(\bigcap_{\iota \in I} X_\iota\Big) \times \Big(\bigcap_{\iota \in I} Y_\iota\Big) = \bigcap_{\iota \in I} (X_\iota \times Y_\iota).$$

Apply Proposition 10 to the case where K (resp. I) is a set consisting of two distinct elements.

7. EXTENSION OF MAPPINGS TO PRODUCTS

DEFINITION 2. *Let $(X_\iota)_{\iota \in I}$, $(Y_\iota)_{\iota \in I}$ be two families of sets, and let $(g_\iota)_{\iota \in I}$ be a family of functions with the same index set I such that g_ι is a mapping of X_ι into Y_ι for each $\iota \in I$. For each $f \in \prod_{\iota \in I} X_\iota$ let u_f be the graph of the function $\iota \to g_\iota(f(\iota))$ $(\iota \in I)$, which is an element of $\prod_{\iota \in I} Y_\iota$. The mapping $f \to u_f$ of $\prod_{\iota \in I} X_\iota$ into $\prod_{\iota \in I} Y_\iota$ is called the canonical extension* (or simply the *extension*) *to products of the family of mappings $(g_\iota)_{\iota \in I}$; it is also sometimes called the product of the family of mappings $(g_\iota)_{\iota \in I}$.*

When the index notation is used, the product of the family $(g_\iota)_{\iota \in I}$ is the function $(x_\iota)_{\iota \in I} \to (g_\iota(x_\iota))_{\iota \in I}$; it is sometimes denoted by $(g_\iota)_{\iota \in I}$.

If $I = \{\alpha, \beta\}$, where α and β are distinct, the extension to products of the family of mappings $(g_\iota)_{\iota \in I}$ is just $\psi \circ (g_\alpha \times g_\beta) \circ \varphi$, where φ denotes the canonical mapping of $\prod_{\iota \in I} X_\iota$ onto $X_\alpha \times X_\beta$ (no. 3) and ψ the canonical mapping of $Y_\alpha \times Y_\beta$ onto $\prod_{\iota \in I} Y_\iota$.

PROPOSITION 11. *Let $(X_\iota)_{\iota \in I}$, $(Y_\iota)_{\iota \in I}$, $(Z_\iota)_{\iota \in I}$ be three families of sets and let $(g_\iota)_{\iota \in I}$, $(g'_\iota)_{\iota \in I}$ be two families of functions, all having the same index set, such that g_ι is a mapping of X_ι into Y_ι and g'_ι a mapping of Y_ι into Z_ι, for each $\iota \in I$. Let g and g' be the extensions of the families $(g_\iota)_{\iota \in I}$ and $(g'_\iota)_{\iota \in I}$ to products. Then the extension of the family $(g'_\iota \circ g_\iota)_{\iota \in I}$ to products is equal to $g' \circ g$.*

This follows immediately from Definition 2.

COROLLARY. *Let* $(X_\iota)_{\iota\in I}$, $(Y_\iota)_{\iota\in I}$ *be two families of sets and let* $(g_\iota)_{\iota\in I}$ *be a family of functions. If* g_ι *is an injection* (resp. *surjection*) *of* X_ι *into* Y_ι *for each* $\iota \in I$, *then the extension* g *of* $(g_\iota)_{\iota\in I}$ *to products is an injection* (resp. *surjection*) *of* $\prod_{\iota\in I} X_\iota$ *into* $\prod_{\iota\in I} Y_\iota$.

(1) Let us assume that $X_\iota \neq \emptyset$ for each $\iota \in I$; otherwise the result is trivial. Suppose that g_ι is injective for each $\iota \in I$, and let r_ι be a retraction of g_ι (§ 3, no. 8, Definition 11), so that $r_\iota \circ g_\iota$ is the identity mapping of X_ι. Let r be the extension to products of the family $(r_\iota)_{\iota\in I}$; since $r \circ g$ is the extension to products of the family of identity mappings I_{X_ι}, $r \circ g$ is the identity mapping of $\prod_{\iota\in I} X_\iota$ and hence g is injective (§ 3, Proposition 8).

(2) Suppose that g_ι is a surjection of X_ι onto Y_ι for each $\iota \in I$, and let s_ι be a section of g_ι (§ 3, no. 8, definition 11), so that $g_\iota \circ s_\iota$ is the identity mapping of Y_ι. If s is the extension to products of the family $(s_\iota)_{\iota\in I}$, then $g \circ s$ is the extension to products of the family of identity mappings I_{Y_ι} and is therefore the identity mapping of $\prod_{\iota\in I} Y_\iota$; hence g is surjective (§ 3, Proposition 8).

Let $(X_\iota)_{\iota\in I}$ be a family of sets, and let E be a set. For every mapping f of E into $\prod_{\iota\in I} X_\iota$, $\text{pr}_\iota \circ f$ is a mapping of E into X_ι. If $\bar{f}$ is the extension to products of this family of mappings, and if d is the diagonal mapping of E into E^I, then it is immediate that $f = \bar{f} \circ d$. Conversely, let $(f_\iota)_{\iota\in I}$ be a family of functions such that f_ι is a mapping of E into X_ι for each $\iota \in I$, and let $\bar{f}$ be the extension to products of this family; then we have $\text{pr}_\iota \circ (\bar{f} \circ d) = f_\iota$ for each $\iota \in I$. By abuse of language, the mapping $\bar{f} \circ d$ is also written as $(f_\iota)_{\iota\in I}$. In this way we define a one-to-one mapping of the set $\prod_{\iota\in I} X_\iota^E$ onto the set $\left(\prod_{\iota\in I} X_\iota\right)^E$; this mapping and its inverse are said to be *canonical*.

6. EQUIVALENCE RELATIONS

In principle, from now on we shall stop using bold-face italic letters to denote undetermined assemblies; the reader will be able to discern easily from the context the assertions which apply to undetermined letters or relations.

1. DEFINITION OF AN EQUIVALENCE RELATION

Let $R\{x, y\}$ be a relation, x and y being distinct letters. The relation R is said to be *symmetric* (with respect to the letters x and y) if

$$R\{x, y\} \Longrightarrow R\{y, x\}.$$

If so, by substituting for x and y two letters x' and y', distinct from each other and from all the letters which appear in R, and then by substituting y and x for x' and y', respectively, we see that

$$R\{y, x\} \Longrightarrow R\{x, y\}.$$

Hence $R\{x, y\}$ and $R\{y, x\}$ are equivalent.

¶ Let z be a letter which does not appear in R. The relation R is said to be *transitive* (with respect to the letters x, y) if we have

$$(R\{x, y\} \text{ and } R\{y, z\}) \Longrightarrow R\{x, z\}.$$

Examples. The relation $x = y$ is symmetric and transitive. The relation $X \subset Y$ is transitive but not symmetric. The relation $X \cap Y = \emptyset$ is symmetric but not transitive.

If $R\{x, y\}$ is both symmetric and transitive it is said to be an *equivalence relation* (with respect to the letters x and y). In this case the notation $x \equiv y \pmod{R}$ is sometimes used as a synonym of $R\{x, y\}$; it is read "*x is equivalent to y modulo* R". If R is an equivalence relation, we have $R\{x, y\} \Longrightarrow (R\{x, x\} \text{ and } R\{y, y\})$, because $R\{x, y\}$ implies $R\{y, x\}$, and $(R\{x, y\}$ and $R\{y, x\})$ implies $(R\{x, x\}$ and $R\{y, y\})$ by virtue of the definitions.

¶ Let $R\{x, y\}$ be a relation; it is said to be *reflexive on* E (with respect to the letters x, y) if the relation $R\{x, x\}$ is equivalent to $x \in E$. If there is no possible ambiguity about E, one says simply, by abuse of language, that R is reflexive.

¶ An *equivalence relation on* E is defined to be an equivalence relation which is reflexive on E. If $R\{x, y\}$ is an equivalence relation on E, we have $R\{x, y\} \Longrightarrow ((x, y) \in E \times E)$, hence R has a graph (with respect to the letters x, y). Conversely, suppose that the equivalence relation $R\{x, y\}$ has a graph G. Observe that the relation $R\{x, x\}$ is equivalent to the relation $(\exists y)R\{x, y\}$; for the former implies the latter (Chapter I, §4, no. 2, scheme S5), and conversely, since $R\{x, y\}$ implies $R\{x, x\}$, $(\exists y)R\{x, y\}$ implies $(\exists y)R\{x, x\}$ and therefore also $R\{x, x\}$. Thus $R\{x, x\}$ is equivalent to $x \in \mathrm{pr}_1 G$, and hence R is an equivalence relation on $\mathrm{pr}_1 G$.

¶ An *equivalence* on a set E is a correspondence whose source and target are both equal to E, and whose graph F is such that the relation $(x, y) \in F$ is an equivalence relation on E.

Examples

(1) The relation $x = y$ is an equivalence relation which has no graph, for if it did, the first projection of this graph would be the set of all objects.
(2) The relation "$x = y$ and $x \in E$" is an equivalence relation on E whose graph is the diagonal of $E \times E$.
(3) The relation "there exists a bijection of X onto Y" is an equivalence relation which has no graph (cf. Chapter III, § 3).
(4) The relation "$x \in E$ and $y \in E$" is an equivalence relation on E, whose graph is $E \times E$.
(5) Suppose $A \subset E$; then the relation

$$(x \in E - A \text{ and } y = x) \text{ or } (x \in A \text{ and } y \in A)$$

is an equivalence relation on E.
(6) * The relation "$x \in \mathbf{Z}$ and $y \in \mathbf{Z}$ and $x - y$ is divisible by 4" is an equivalence relation on **Z**.*

PROPOSITION 1. *A correspondence* Γ *between* X *and* X *is an equivalence on* X *if and only if it satisfies the following conditions:* (a) X *is the domain of* Γ; (b) $\Gamma = \overset{-1}{\Gamma}$; (c) $\Gamma \circ \Gamma = \Gamma$.

Let Γ be a correspondence between X and X, and let G be its graph. If Γ is an equivalence on X, then $(x, x) \in G$ for all $x \in X$; hence X is the domain of Γ. The relation $(x, y) \in G$ is equivalent to $(y, x) \in G$, hence to $(x, y) \in \overset{-1}{G}$, so that $G = \overset{-1}{G}$ and therefore $\Gamma = \overset{-1}{\Gamma}$. The relations $(x, y) \in G$ and $(y, z) \in G$ imply $(x, z) \in G$, so that $G \circ G \subset G$; conversely, $(x, y) \in G$ implies $(x, x) \in G$ and therefore $(x, y) \in G \circ G$, so that $G \subset G \circ G$; hence $G = G \circ G$ and consequently $\Gamma = \Gamma \circ \Gamma$.

¶ Conversely, suppose that conditions (a), (b), and (c) are satisfied. The relation $(x, y) \in G$ is symmetric (by (b)) and transitive (by (c)); hence it is an equivalence relation, and by (a) it is an equivalence relation on X.

2. EQUIVALENCE CLASSES; QUOTIENT SET

Let f be a function, E its domain, F its graph. The relation "$x \in E$ and $y \in E$ and $f(x) = f(y)$" is an equivalence relation on E, called the equivalence relation *associated with* f. It is equivalent to the relation $(\exists z)((x, z) \in F \text{ and } (y, z) \in F)$, i.e., to $(\exists z)((x, z) \in F \text{ and } (z, y) \in \overset{-1}{F})$, and therefore its graph is $\overset{-1}{F} \circ F$.

¶ We shall now show that every equivalence relation R on a set E is of this type. Let G be the graph of R. For each $x \in E$ the (non-empty) set $G(x) \subset E$ is called the *equivalence class of x with respect to* R; it is thus the set of all $y \in E$ such that $R\{x, y\}$. Every set which can be written as $G(x)$ for some $x \in E$ is called an equivalence class (with respect to R). An element of an equivalence class is called a *representative* of this class. The set of equivalence classes with respect to R (that is, the set of all objects of the form $G(x)$ for some $x \in E$) is called the *quotient set* of E by R and is denoted by E/R. The mapping $x \to G(x)$ $(x \in E)$ whose domain is E and whose target is E/R is called the *canonical mapping* of E onto E/R.

C55. *Let* R *be an equivalence relation on a set* E, *and let* p *be the canonical mapping of* E *onto* E/R. *Then*

$$R\{x, y\} \Leftrightarrow (p(x) = p(y)).$$

With the notation above, let x and y be elements of E such that

$$(x, y) \in G.$$

Then $x \in E$ and $y \in E$; let us show that $G(x) = G(y)$. Since $y \in G(x)$, we have (Proposition 1) $G(y) \subset (G \circ G)(x) = G(x)$. On the other hand, we also have $(y, x) \in G$, whence $G(x) \subset G(y)$ and therefore

$$G(x) = G(y),$$

i.e., $p(x) = p(y)$. Conversely, if $G(x) = G(y)$, we have $y \in G(y) = G(x)$, so that $(x, y) \in G$. This completes the proof.

¶ A section of the canonical mapping p of E onto E/R (§ 3, no. 8, Definition 11) is called more briefly a *section* of E (with respect to the relation R).

Examples

(1) Let R be the equivalence relation "$x \in E$ and $y \in E$ and $x = y$" on a set E. The equivalence class of $x \in E$ is then the set $\{x\}$, and the canonical mapping $x \to \{x\}$ of E onto E/R is bijective.

(2) Let E, F be two sets and let R be the equivalence relation on $E \times F$ associated with the mapping pr_1 of $E \times F$ onto E. The equivalence classes with respect to R are the sets of the form $\{x\} \times F$, where $x \in E$; the mapping $x \to \{x\} \times F$ is a bijection of E onto $(E \times F)/R$.

Let R be an equivalence relation on a set E. The quotient set E/R is a subset of $\mathfrak{P}(E)$, and the identity mapping of E/R is a *partition* of E (§ 4, no. 7); for if G is the graph of R, we have $x \in G(x)$ for all $x \in E$, and if two equivalence classes $G(x)$ and $G(y)$ have a common element z, then $R\{x, z\}$ and $R\{y, z\}$, so that $G(x) = G(y)$. Furthermore, the relation

$$(\exists X)(X \in E/R \text{ and } x \in X \text{ and } y \in X)$$

is equivalent to $R\{x, y\}$.

¶ Conversely, let $(X_\iota)_{\iota \in I}$ be a partition of a set E. It is immediately verified that the relation $(\exists \iota)(\iota \in I \text{ and } x \in X_\iota \text{ and } y \in Y_\iota)$ is an equivalence relation R on E; the equivalence classes with respect to R are just the sets X_ι of the partition, and the mapping $\iota \to X_\iota$ is a bijection of I onto E/R. Every subset S of E such that, for each $\iota \in I$, the set $S \cap X_\iota$ consists of a single element is called a *system of representatives* (or a *transversal*) of the equivalence classes with respect to R. This name is also used to denote any injection of a set K into E such that the image of K under this injection is a system of representatives of the equivalence classes with respect to R; for example, any *section* of E with respect to R.

3. RELATIONS COMPATIBLE WITH AN EQUIVALENCE RELATION

Let $R\{x, x'\}$ be an equivalence relation and let $P\{x\}$ be a relation. The relation $P\{x\}$ is said to be *compatible with the equivalence relation* $R\{x, x'\}$ (with respect to x) if we have

$$(P\{x\} \text{ and } R\{x, y\}) \Longrightarrow P\{y\},$$

where y denotes a letter which appears neither in P nor in R.

For example, it follows from C43 (Chapter I, §5), no. 1) that every relation $P\{x\}$ is compatible with the equivalence relation $x = x'$.

C56. *Let $R\{x, x'\}$ be an equivalence relation on a set* E, *and let $P\{x\}$ be a relation which does not contain the letter x' and is compatible (with respect to x) with the equivalence relation $R\{x, x'\}$. Then, if t does not appear in $P\{x\}$, the relation* "$t \in E/R$ and $(\exists x)(x \in t$ and $P\{x\})$" *is equivalent to the relation* "$t \in E/R$ and $(\forall x)((x \in t) \Longrightarrow P\{x\})$".

Let $t \in E/R$. If there exists $a \in t$ such that $P\{a\}$, then for each $x \in t$ we have $R\{a, x\}$ and hence $P\{x\}$. Hence $(\exists x)(x \in t$ and $P\{x\})$ implies $(\forall x)((x \in t) \Longrightarrow P\{x\})$. The converse is obvious since $t \in E/R$ implies that $t \neq \varnothing$.

¶ The relation

$$t \in E/R \text{ and } (\exists x)(x \in t \text{ and } P\{x\})$$

is said to be *induced by* $P\{x\}$ *on passing to the quotient* (with respect to x) with respect to R. If this relation is denoted by $P'\{t\}$, and if f is the canonical mapping of E onto E/R, then the relation

$$y \in E \text{ and } P'\{f(y)\}$$

(where y does not appear in $P\{x\}$) is *equivalent* to ($y \in E$ and $P\{y\}$), as is immediately verified.

4. SATURATED SUBSETS

Let $R\{x, y\}$ be an equivalence relation on a set E, and let A be a subset of E. A is said to be *saturated with respect to* R if the relation $x \in A$ is compatible (with respect to x) with $R\{x, y\}$; or, equivalently, if *for each* $x \in A$ *the equivalence class of* x *is contained in* A. In other words, a set is saturated with respect to R if and only if it is the *union of a set of equivalence classes with respect to* R.

¶ Let f be the canonical mapping of E onto E/R. If A is saturated with respect to R, then the equivalence class of each $x \in A$, which is equal to $\overset{-1}{f}\langle\{f(x)\}\rangle$, is contained in A, and hence $\overset{-1}{f}\langle f\langle A\rangle\rangle \subset A$; since in any case we have $A \subset \overset{-1}{f}\langle f\langle A\rangle\rangle$, it follows that $\overset{-1}{f}\langle f\langle A\rangle\rangle = A$. Conversely, if $A = \overset{-1}{f}\langle f\langle A\rangle\rangle$, then for every $x \in A$ the equivalence class $K = f(x)$ of x with respect to R is an element of $f\langle A\rangle$; and since $K = \overset{-1}{f}\langle\{K\}\rangle$, we have $K \subset \overset{-1}{f}\langle f\langle A\rangle\rangle = A$. Thus the subsets of E which are saturated with respect to R are precisely the subsets A of E such that $A = \overset{-1}{f}\langle f\langle A\rangle\rangle$. Equivalently, they are the subsets of E of the form $\overset{-1}{f}\langle B\rangle$, where $B \subset E/R$; for the relation $A = \overset{-1}{f}\langle B\rangle$ implies $B = f\langle A\rangle$, hence $A = \overset{-1}{f}\langle f\langle A\rangle\rangle$.

¶ If $(X_\iota)_{\iota \in I}$ is a family of saturated subsets of E, then the sets $\bigcup_{\iota \in I} X_\iota$ and $\bigcap_{\iota \in I} X_\iota$ are saturated (§4, Propositions 3 and 4). If A is a saturated subset of E, then so is $\complement_E A$ (§4, Proposition 6).

¶ Let A now be an arbitrary subset of E. Then the set $\overset{-1}{f}\langle f\langle A\rangle\rangle$ contains A and is saturated. Conversely, if A′ is a saturated subset of E which contains A, we have $f\langle A'\rangle \supset f\langle A\rangle$, so that

$$A' = \overset{-1}{f}\langle f\langle A'\rangle\rangle \supset \overset{-1}{f}\langle f\langle A\rangle\rangle.$$

Hence we may say that $\overset{-1}{f}\langle f\langle A\rangle\rangle$ is the "smallest" saturated subset of E which contains A (cf. Chapter III); this set is called the *saturation* of A with respect to the relation R. It is immediately seen that the saturation of A is the union of the equivalence classes of the elements of A. If $(X_\iota)_{\iota\in I}$ is a family of subsets of E and if A_ι is the saturation of X_ι with respect to R, then the saturation of $\bigcup_{\iota\in I} X_\iota$ is $\bigcup_{\iota\in I} A_\iota$ (§4, Proposition 3).

5. MAPPINGS COMPATIBLE WITH EQUIVALENCE RELATIONS

Let R be an equivalence relation on a set E, and let f be a function with domain E. Then f is said to be *compatible with the relation* R if the relation $y = f(x)$ is compatible (with respect to x) with the relation $R\{x, x'\}$.

¶ It comes to the same thing to say that the restriction of f to each equivalence class is a constant mapping, in which case we say that *f is constant on each equivalence class with respect to* R. If g is the canonical mapping of E onto E/R, an equivalent statement is that the relation $g(x) = g(x')$ implies $f(x) = f(x')$; hence (§3, Proposition 9) we have the following criterion :

C57. *Let* R *be an equivalence relation on a set* E, *and let g be the canonical mapping of* E *onto* E/R. *Then a mapping f of* E *into* F *is compatible with* R *if and only if f can be put in the form $h \circ g$, where h is a mapping of* E/R *into* F. *The mapping h is uniquely determined by f; if s is any section of g, we have $h = f \circ s$.*

The mapping h is said to be *induced by f on passing to the quotient* with respect to R.

¶ Let f be a mapping of a set E into a set F, and let $A = f\langle E\rangle \subset F$. Let R be the equivalence relation associated with f (no. 2); it is clear that f is compatible with R. Moreover, the mapping h induced by f on passing to the quotient is an *injection* of E/R into F : for if t, t' are equivalence classes with respect to R such that $h(t) = h(t')$, we have $f(x) = f(x')$ for all $x \in t$ and $x' \in t'$, which implies $t = t'$ by the definition of R. Let k be the mapping of E/R onto A which has the same graph as h; then k is *bijective*. If j is the canonical injection of A into F and if g is the canonical mapping of E onto E/R, then we may write

$$f = j \circ k \circ g.$$

This relation is called the *canonical decomposition of f*.

¶ Let f be a mapping of a set E into a set F, let R be an equivalence relation on E, and let S be an equivalence relation on F. Let u be the canonical mapping of E onto E/R, and let v be the canonical mapping of F onto F/S. The mapping f is said to be ***compatible with the equivalence relations*** R ***and*** S if $v \circ f$ is compatible with R; this means that the relation $x \equiv x'$ (mod R) *implies* $f(x) \equiv f(x')$ (mod S). The mapping h of E/R into F/S induced by $v \circ f$ on passing to the quotient with respect to R is then called the ***mapping induced by f on passing to the quotients with respect to*** R ***and*** S; it is characterized by the relation $v \circ f = h \circ u$.

6. INVERSE IMAGE OF AN EQUIVALENCE RELATION; INDUCED EQUIVALENCE RELATION

Let φ be a mapping of a set E into a set F, and let S be an equivalence relation on F. If u is the canonical mapping of F onto F/S, the equivalence relation associated with the mapping $u \circ \varphi$ of E into F/S is called the *inverse image* of S under φ; if R is this relation, $R\{x, y\}$ is equivalent to $S\{\varphi(x), \varphi(y)\}$, and the equivalence classes with respect to R are the inverse images under φ of the equivalence classes with respect to S which meet $\varphi\langle E\rangle$.

¶ In particular, consider an equivalence relation R on a set E, and let A be a subset of E; then the inverse image of R under the injection j of A into E is called the equivalence relation *induced* by R on A, and is denoted by R_A.

¶ The equivalence classes with respect to R_A are the *traces* on A of the equivalence classes with respect to R which meet A. The injection j is obviously compatible with the relations R_A and R; the mapping h of A/R_A into E/R induced by j on passing to the quotient with respect to R_A and R is an *injective* mapping of A/R_A into E/R : for if f (resp. g) is the canonical mapping of E onto E/R (resp. A onto A/R_A), then the relation $h(g(x)) = h(g(x'))$, where $x \in A$ and $x' \in A$, is equivalent to $f(x) = f(x')$ and therefore to $g(x) = g(x')$. The image $h\langle A/R_A\rangle$ is equal to $f\langle A\rangle$. If k is the bijective mapping of A/R_A onto $f\langle A\rangle$ which has the same graph as h, then k and its inverse are said to be *canonical.*

7. QUOTIENTS OF EQUIVALENCE RELATIONS

Let R, S be two equivalence relations with respect to two letters x, y. We shall say that S is *finer* than R (or that R is *coarser* than S) if the relation $S \Longrightarrow R$ is true. If R and S are equivalence relations on the same set E, the statement that S is finer than R means that the graph of S is contained in that of R, or again that every equivalence class with respect to S is contained in an equivalence class with respect to R; or, equivalently, that every equivalence class with respect to R is saturated with respect to S.

Examples

(1) The relation "$x \in E$ and $y \in E$ and $x = y$" is finer than every equivalence relation on E. The relation "$x \in E$ and $y \in E$" is coarser than every equivalence relation on E.

* (2) The equivalence relation "$x \in \mathbf{Z}$ and $y \in \mathbf{Z}$ and $x - y$ is divisible by 4" is finer than the equivalence relation "$x \in \mathbf{Z}$ and $y \in \mathbf{Z}$ and $x - y$ is divisible by 2". *

Let R and S be two equivalence relations on the same set E, such that S is finer than R. Let f and g be the canonical mappings of E onto E/R and E onto E/S respectively; then the function f is compatible with S. Let h be the function induced by f on passing to the quotient with respect to S; then h is a mapping of E/S onto E/R. The equivalence relation associated with h on E/S is called the *quotient of* R *by* S and is denoted by R/S. The relation $x \equiv y \pmod{R}$ is equivalent to $g(x) \equiv g(y) \pmod{R/S}$, and the equivalence classes with respect to R/S are the images under g of the equivalence classes with respect to R. Let $h = j \circ h_2 \circ h_1$ be the canonical decomposition (no. 5) of the mapping h. Then h_1 is the canonical mapping of E/S onto (E/S)/(R/S), j is the identity mapping of E/R, and h_2 is a one-to-one mapping of (E/S)/(R/S) onto E/R. The mapping h_2 and its inverse are said to be *canonical*.

¶ Conversely, consider an arbitrary equivalence relation T on the set E/S, and let R be the equivalence relation on E which is the inverse image under g of the relation T (no. 6). Since the relation $x \equiv y \pmod{R}$ is equivalent to $g(x) \equiv g(y) \pmod{T}$, it follows that T is equivalent to R/S.

8. PRODUCT OF TWO EQUIVALENCE RELATIONS

Let $R\{x, y\}$ and $R'\{x', y'\}$ be two equivalence relations. Let $S\{u, v\}$ denote the relation

$$(\exists x)(\exists y)(\exists x')(\exists y')(u = (x, x') \text{ and } v = (y, y') \text{ and } R\{x, y\} \text{ and } R'\{x', y'\});$$

it is immediately verified that $S\{u, v\}$ is an equivalence relation, called the *product* of R and R′, and denoted by $R \times R'$. Suppose that R is an equivalence relation on a set E and that R′ is an equivalence relation on a set E′. Then the relation $S\{u, u\}$ is equivalent to

$$(\exists x)(\exists x')(u = (x, x') \text{ and } R\{x, x\} \text{ and } R'\{x', x'\})$$

i.e., to $(\exists x)(\exists x')(u = (x, x') \text{ and } x \in E \text{ and } x' \in E')$, hence to $u \in E \times E'$. It follows that $R \times R'$ is an equivalence relation on $E \times E'$. If

$$u = (x, x')$$

is an element of $E \times E'$, the relation $S\{u, v\}$ is equivalent to

$$(\exists y)(\exists y')(v = (y, y') \text{ and } R\{x, y\} \text{ and } R'\{x', y'\});$$

if G and G′ are the graphs of R and R′, respectively, this relation is in turn equivalent to $v \in G(x) \times G(x')$. Hence *every equivalence class with respect to* $R \times R'$ *is the product of an equivalence class with respect to* R *and an equivalence class with respect to* R′, *and conversely.*

¶ Let f and f' be the canonical mappings of E onto E/R and E′ onto E/′R′, respectively, and let $f \times f'$ be the canonical extension of f and f' to the product sets (§ 3, no. 9). Then $(f \times f')(x, x') = (f(x), f'(x'))$ for all $(x, x') \in E \times E'$. The inverse image under $f \times f'$ of an element (u, u') of $(E/R) \times (E'/R')$ is just the product $u \times u'$ of the equivalence class u with respect to R and the equivalence class u' with respect to R′. It follows that the equivalence relation associated with $f \times f'$ is equivalent to $R \times R'$. The mapping $f \times f'$ can therefore be written as $h \circ g$, where g is the canonical mapping of $E \times E'$ onto

$$(E \times E')/(R \times R')$$

and h is a one-to-one mapping of $(E \times E')/(R \times R')$ onto

$$(E/R) \times (E'/R');$$

this mapping h and its inverse are said to be *canonical*.

Remark. Let $P\{x, x'\}$ be a relation in which the letters y, y' do not occur. P is said to be *compatible* with the equivalence relations $R\{x, y\}$ and $R'\{x', y'\}$ (with respect to x and x') if the relation ($P\{x, x'\}$ and $R\{x, y\}$ and $R'\{x', y'\}$) implies $P\{y, y'\}$. Let $Q\{u\}$ be the relation $(\exists x)(\exists x')(u = (x, x') \text{ and } P\{x, x'\})$; then it comes to the same thing to say that $Q\{u\}$ is compatible (with respect to u) with the equivalence relation $S\{u, v\}$, the product of R and R′.

9. CLASSES OF EQUIVALENT OBJECTS

Let $R\{x, y\}$ be an equivalence relation, which need not have a graph. It is obvious that if x, x', y are three distinct letters, the relation $R\{x, x'\}$ implies "$R\{x, y\} \Longleftrightarrow R\{x', y\}$" and therefore also implies the relation $(\forall y)(R\{x, y\} \Longleftrightarrow R\{x', y\})$. By the scheme S7 (Chapter I, § 5, no. 1), if we put $\theta\{x\} = \tau_y(R\{x, y\})$, the relation $R\{x, x'\}$ implies

$$\theta\{x\} = \theta\{x'\}.$$

Note, on the other hand, that by definition $R\{x, \theta\{x\}\}$ is the relation $(\exists y)(R\{x,y\})$ and hence (no. 1) is equivalent to $R\{x, x\}$. It follows that the relation ($R\{x, x\}$ and $R\{x', x'\}$ and $\theta\{x\} = \theta\{x'\}$) is *equivalent* to $R\{x, x'\}$, for it implies, by S6 (Chapter I, §5, no. 1), the relation

$$(R\{x,x\} \text{ and } R\{x',x'\} \text{ and } (R\{x', \theta\{x\}\} \Longleftrightarrow R\{x', \theta\{x'\}\})),$$

hence also ($R\{x, \theta\{x\}\}$ and $R\{x', \theta\{x\}\}$), and finally $R\{x, x'\}$ by transitivity and symmetry. And since, conversely, $R\{x, x'\}$ implies $R\{x, x\}$ and $R\{x', x'\}$, the assertion is proved. The term $\theta\{x\}$ is called the *class of objects equivalent to* x (with respect to the relation R).

¶ Suppose now that T is a term such that the relation

$$(1) \qquad (\forall y)(R\{y, y\} \Longrightarrow (\exists x)(x \in T \text{ and } R\{x, y\}))$$

is true. Then the relation $(\exists x)(R\{x, x\}$ and $z = \theta\{x\})$ is *collectivizing in* z. For we may suppose that $x \in T$ implies $R\{x, x\}$; it is sufficient to replace T by the set of all $x \in T$ such that $R\{x, x\}$ (observing that $R\{x, y\}$ implies $R\{x, x\}$). Let Θ be the set of all objects of the form $\theta\{x\}$ for $x \in T$ (§1, no. 6). Suppose that $R\{y, y\}$ is true; then there exists $x \in T$ such that $R\{x, y\}$ and therefore $\theta\{y\} = \theta\{x\} \in \Theta$. The set Θ is called the *set of classes of equivalent objects* with respect to R; and for each x such that $R\{x, x\}$, $\theta\{x\}$ is the *unique* element $z \in \Theta$ such that $R\{x, z\}$.

Under the same hypothesis, let $A\{x\}$ be a term such that $R\{x, y\}$ implies $A\{x\} = A\{y\}$. Then the relation $(\exists x)(R\{x, x\}$ and $z = \{x\})$ is collectivizing in z, since $R\{x, x\}$, being equivalent to $R\{x, \theta\{x\}\}$, implies $A\{x\} = A\{\theta\{x\}\}$, and consequently if E is the set of objects of the form $A\{t\}$ for $t \in \Theta$, then $R\{x, x\}$ implies $A\{x\} \in E$. If f is the function $t \to A\{t\}$ ($t \in \Theta$, $A\{t\} \in E$), then the relation $R\{x, x\}$ implies $A\{x\} = f(\theta\{x\})$.

In particular, if R is an equivalence relation *on a set* F, we may take $A\{x\}$ to be the *equivalence class of* x *with respect to* R (no. 2), and the function f is then a *bijection* of Θ onto the quotient set F/R; this justifies the terminology introduced.

* *Example.* Let $R\{x, y\}$ be the equivalence relation "x and y are two vector spaces of the same finite dimension over **C**"; this relation has no graph. It satisfies condition (1) when T is the set of all vector subspaces of $\mathbf{C}^{(\mathbf{N})}$, or when the subset T' of T consists of the spaces $\mathbf{C}^n$ ($n \in \mathbf{N}$) with the conventions that $\mathbf{C}^0$ consists of the point 0 of $\mathbf{C}^{(\mathbf{N})}$ and that $\mathbf{C}^n$ ($n > 0$) is the sum of the first n components of the direct sum $\mathbf{C}^{(\mathbf{N})}$. With this second choice we have $\Theta = T'$.*

EXERCISES

§ 1

1. Show that the relation

$$(x = y) \Longleftrightarrow (\forall X)((x \in X) \Longrightarrow (y \in X))$$

is a theorem.

2. Show that $\emptyset \neq \{x\}$ is a theorem. Deduce that $(\exists x)(\exists y)(x \neq y)$ is a theorem.

3. Let A and B be two subsets of a set X. Show that the relation $B \subset \complement A$ is equivalent to $A \subset \complement B$ and that the relation $\complement B \subset A$ is equivalent to $\complement A \subset B$.

4. Prove that the relation $X \subset \{x\}$ is equivalent to

$$\text{"}X = \{x\} \text{ or } X = \emptyset\text{"}.$$

5. Prove that $\emptyset = \tau_X(\tau_x(x \in X) \notin X)$.

6. Let $\mathcal{T}$ be an equalitarian theory which contains the sign $\in$ and the following axiom:

A1′. $$(\forall y)(y = \tau_x((\forall z)(z \in x \Longleftrightarrow z \in y)))$$

(in other words, every term is equal to the set of its elements). Show that the axiom of extent A1 is a theorem in $\mathcal{T}$ (use the scheme S7).

§ 2

1. Let $\boldsymbol{R}\{x, y\}$ be a relation, the letters x and y being distinct; let z be a letter, distinct from x and y, which does not appear in

$R\{x, y\}$. Show that the relation $(\exists x)(\exists y)R\{x, y\}$ is equivalent to

$$(\exists z)(z \text{ is an ordered pair and } R\{\mathrm{pr}_1 z, \mathrm{pr}_2 z\})$$

and that the relation $(\forall x)(\forall y)R\{x, y\}$ is equivalent to

$$(\forall z)\ ((z \text{ is an ordered pair}) \Longrightarrow R\{\mathrm{pr}_1 z, \mathrm{pr}_2 z\}).$$

2. (a) Show that the relation $\{\{x\}, \{x, y\}\} = \{\{x'\}, \{x', y'\}\}$ is equivalent to "$x = x'$ and $y = y'$".

(b) Let $\mathcal{T}_0$ be the theory of sets and let $\mathcal{T}_1$ be the theory which has the same schemes and explicit axioms as $\mathcal{T}_0$, except for axiom A3. Show that if $\mathcal{T}_1$ is not contradictory, then $\mathcal{T}_0$ is not contradictory (use (a)).

§ 3

1. Show that the relations $x \in y$, $x \subset y$, $x = \{y\}$ have no graph with respect to x and y.

2. Let G be a graph. Show that the relation $X \subset \mathrm{pr}_1 G$ is equivalent to $X \subset \overset{-1}{G}\langle G\langle X\rangle\rangle$.

3. Let G, H be two graphs. Show that the the relation $\mathrm{pr}_1 H \subset \mathrm{pr}_1 G$ is equivalent to $H \subset H \circ \overset{-1}{G} \circ G$. Deduce that $G \subset G \circ \overset{-1}{G} \circ G$.

4. If G is a graph, show that $\varnothing \circ G = G \circ \varnothing = \varnothing$, and that $\overset{-1}{G} \circ G = \varnothing$ if and only if $G = \varnothing$.

5. Let A, B be two sets and let G be a graph. Show that

$$(A \times B) \circ G = \overset{-1}{G}\langle A\rangle \times B \qquad \text{and} \qquad G \circ (A \times B) = A \times G\langle B\rangle.$$

6. For each graph G, let G′ be the graph $((\mathrm{pr}_1 G) \times (\mathrm{pr}_2 G)) - G$. Show that

$$(\overset{-1}{G})' = \overset{-1}{\overline{G'}},$$

and that

$$G \circ (\overset{-1}{G})' \subset \Delta'_B, \qquad (\overset{-1}{G})' \circ G \subset \Delta'_A$$

if $A \supset \mathrm{pr}_1 G$ and $B \supset \mathrm{pr}_2 G$. Show that

$$G = (\mathrm{pr}_1 G) \times (\mathrm{pr}_2 G)$$

if and only if

$$G \circ \overset{-1}{G'} \circ G = \emptyset.$$

7. A graph G is functional if and only if for each set X we have

$$G\langle \overset{-1}{G}\langle X\rangle\rangle \subset X.$$

8. Let A, B be two sets, let Γ be a correspondence between A and B, and let Γ' be a correspondence between B and A. Show that if $\Gamma'(\Gamma(x)) = \{x\}$ for all $x \in A$ and if $\Gamma(\Gamma'(y)) = \{y\}$ for all $y \in B$, then Γ is a bijection of A onto B, and Γ' is the inverse mapping.

9. Let A, B, C, D be sets, f a mapping of A into B, g a mapping of B into C, h a mapping of C into D. If $g \circ f$ and $h \circ g$ are bijections, show that all of f, g, h are bijections.

10. Let A, B, C be sets, f a mapping of A into B, g a mapping of B into C, h a mapping of C into A. Show that if two of the three mappings $h \circ g \circ f$, $g \circ f \circ h$, $f \circ h \circ g$ are surjections and the third is an injection, then f, g, h are all bijections.

*11. Find the error in the following argument : Let N denote the set of natural numbers and let A denote the set of all integers $n > 2$ for which there exist three strictly positive integers x, y, z such that $x^n + y^n = z^n$. Then the set A is not empty (in other words, "Fermat's last theorem" is false). For let $B = \{A\}$ and $C = \{N\}$; B and C are sets consisting of a single element, hence there is a bijection f of B onto C. We have $f(A) = N$; if A were empty we should have $N = f(\emptyset) = \emptyset$, which is absurd. *

§ 4

1. Let G be a graph. Show that the following three propositions are equivalent :

(a) G is a functional graph.

(b) If X, Y are any two sets, then

$$\overset{-1}{G}(X \cap Y) = \overset{-1}{G}(X) \cap \overset{-1}{G}(Y).$$

(c) The relation $X \cap Y = \emptyset$ implies

$$\overset{-1}{G}(X) \cap \overset{-1}{G}(Y) = \emptyset.$$

2. Let G be a graph. Show that for each set X we have

$$G(X) = \mathrm{pr}_2(G \cap (X \times \mathrm{pr}_2 G)) \quad \text{and} \quad G(X) = G(X \cap \mathrm{pr}_1 G).$$

3. Let X, Y, Y′, Z be four sets. Show that

$$(Y' \times Z) \circ (X \times Y) = \varnothing \quad \text{if} \quad Y \cap Y' = \varnothing$$

and that

$$(Y' \times Z) \circ (X \times Y) = X \times Z \quad \text{if} \quad Y \cap Y' \neq \varnothing.$$

4. Let $(G_\iota)_{\iota \in I}$ be a family of graphs. Show that for every set X we have

$$\left(\bigcup_{\iota \in I} G_\iota\right)\langle X \rangle = \bigcup_{\iota \in I} G_\iota \langle X \rangle,$$

and that for every object x,

$$\left(\bigcap_{\iota \in I} G_\iota\right)\langle \{x\} \rangle = \bigcap_{\iota \in I} G_\iota \langle \{x\} \rangle.$$

Give an example of two graphs G, H and a set X such that

$$(G \cap H)\langle X \rangle \neq G\langle X \rangle \cap H\langle X \rangle.$$

5. Let $(G_\iota)_{\iota \in I}$ be a family of graphs and let H be a graph. Show that

$$\left(\bigcup_{\iota \in I} G_\iota\right) \circ H = \bigcup_{\iota \in I} (G_\iota \circ H) \quad \text{and} \quad H \circ \left(\bigcup_{\iota \in I} G_\iota\right) = \bigcup_{\iota \in I} (H \circ G_\iota).$$

6. A graph G is functional if and only if for each pair of graphs H, H′ we have $(H \cap H') \circ G = (H \circ G) \cap (H' \circ G)$.

7. Let G, H, K be three graphs. Prove the relation

$$(H \circ G) \cap K \subset (H \cap (K \circ \overset{-1}{G})) \circ (G \cap (\overset{-1}{H} \circ K)).$$

8. Let $\mathfrak{R} = (X_\iota)_{\iota \in I}$ and $\mathfrak{S} = (Y_\varkappa)_{\varkappa \in K}$ be two coverings of a set E.

(a) Show that if $\mathfrak{R}$ and $\mathfrak{S}$ are partitions of E and if $\mathfrak{R}$ is finer than $\mathfrak{S}$, then for every $\varkappa \in K$ there exists $\iota \in I$ such that $X_\iota \subset Y_\varkappa$.

(b) Give an example of two coverings $\mathfrak{R}$, $\mathfrak{S}$ such that $\mathfrak{R}$ is finer than $\mathfrak{S}$ but such that the property stated in (a) is not satisfied.

(c) Give an example of two partitions $\mathfrak{R}$, $\mathfrak{S}$ of E such that for every $\varkappa \in K$ there exists $\iota \in I$ such that $X_\iota \subset Y_\varkappa$ but such that $\mathfrak{R}$ is not a refinement of $\mathfrak{S}$.

§ 5

1. Let $(X_\iota)_{\iota \in I}$ be a family of sets. Show that if $(Y_\iota)_{\iota \in I}$ is a family of sets such that $Y_\iota \subset X_\iota$ for each $\iota \in I$, then

$$\prod_{\iota \in I} Y_\iota = \bigcap_{\iota \in I} \overset{-1}{\mathrm{pr}_\iota}(Y_\iota).$$

2. Let A, B be two sets. For each subset G of $A \times B$ let $\tilde{G}$ be the mapping $x \to G\langle\{x\}\rangle$ of A into $\mathfrak{P}(B)$. Show that the mapping $G \to \tilde{G}$ is a bijection of $\mathfrak{P}(A \times B)$ onto $(\mathfrak{P}(B))^A$.

*3. Let $(X_i)_{1 \leqslant i \leqslant n}$ be a finite family of sets. For each subset H of the index set $[1, n]$ let

$$P_H = \bigcup_{\iota \in H} X_i \quad \text{and} \quad Q_H = \bigcap_{\iota \in H} X_i.$$

Let $\mathfrak{F}_k$ be the set of subsets of $[1, n]$ which have k elements. Show that

$$\bigcup_{H \in \mathfrak{F}_k} Q_H \supset \bigcap_{H \in \mathfrak{F}_k} P_H \quad \text{if} \quad k \leqslant \frac{1}{2}(n+1),$$

and that

$$\bigcup_{H \in \mathfrak{F}_k} Q_H \subset \bigcap_{H \in \mathfrak{F}_k} P_H \quad \text{if} \quad k \geqslant \frac{1}{2}(n+1). \,*$$

§ 6

1. For a graph G to be the graph of an equivalence relation on a set E, it is necessary and sufficient that $\mathrm{pr}_1 G = E$, $G \circ \overset{-1}{G} \circ G = G$, and $\Delta_E \subset G$ (Δ_E being the diagonal of E).

2. If G is a graph such that

$$G \circ \overset{-1}{G} \circ G = G,$$

show that $\overset{-1}{G} \circ G$ and $G \circ \overset{-1}{G}$ are graphs of equivalences on $\mathrm{pr}_1 G$ and $\mathrm{pr}_2 G$ respectively.

3. Let E be a set, A a subset of E, and R the equivalence relation associated with the mapping $X \to X \cap A$ of $\mathfrak{P}(E)$ into $\mathfrak{P}(E)$. Show that there exists a bijection of $\mathfrak{P}(A)$ onto the quotient set $\mathfrak{P}(E)/R$.

4. Let G be the graph of an equivalence on a set E. Show that if A is a graph such that $A \subset G$ and $\mathrm{pr}_1 A = E$ (resp. $\mathrm{pr}_2 A = E$), then

$$G \circ A = G \quad (\text{resp. } A \circ G = G);$$

furthermore, if B is any graph, we have $(G \cap B) \circ A = G \cap (B \circ A)$ (resp. $A \circ (G \cap B) = G \cap (A \circ B)$).

5. Show that every intersection of graphs of equivalences on a set E is the graph of an equivalence on E. Give an example of two equivalences on a set E such that the union of their graphs is not the graph of an equivalence on E.

6. Let G, H be the graphs of two equivalences on E. Then $G \circ H$ is the graph of an equivalence on E if and only if $G \circ H = H \circ G$. The graph $G \circ H$ is then the intersection of all the graphs of equivalences on E which contain both G and H.

7. Let G_0, G_1, H_0, H_1 be the graphs of four equivalences on a set E such that $G_1 \cap H_0 = G_0 \cap H_1$ and $G_1 \circ H_0 = G_0 \circ H_1$. For each $x \in E$, let R_0 (resp. S_0) be the relation induced on $G_1(x)$ (resp. $H_1(x)$) by the equivalence relation $(x, y) \in G_0$ (resp. $(x, y) \in H_0$). Show that there exists a bijection of the quotient set $G_1(x)/R_0$ onto the quotient set

$$H_1(x)/S_0.$$

(If $A = G_1(x) \cap H_1(x)$, show that both quotient sets are in one-to-one correspondence with the quotient set of A by the equivalence relation induced by R_0 on A; this relation is equivalent to that induced by S_0 on A.)

8. Let E, F be two sets, let R be an equivalence relation on F, and let f be a mapping of E into F. If S is the equivalence relation which is the inverse image of R under f, and if $A = f\langle E\rangle$, define a canonical bijection of E/S onto A/R_A.

9. Let F, G be two sets, let R be an equivalence relation on F, let p be the canonical mapping of F onto F/R, and let f be a surjection of G onto F/R. Show that there exists a set E, a surjection g of E onto F, and a surjection h of E onto G, such that $p \circ g = f \circ h$.

10. (a) If $R\{x, y\}$ is any relation, then "$R\{x, y\}$ and $R\{y, x\}$" is a symmetric relation. Under what condition is it reflexive on a set E?

*(b) Let $R\{x, y\}$ be a reflexive and symmetric relation on a set E. Let $S\{x, y\}$ be the relation "there exists an integer $n > 0$ and a sequence $(x_i)_{0 \leqslant i \leqslant n}$ of elements of E such that $x_0 = x$, $x_n = y$, and for each index i such that $0 \leqslant i < n$, $R\{x_i, x_{i+1}\}$". Show that $S\{x, y\}$ is an equivalence relation on E and that its graph is the smallest of all graphs of equivalences on E which contain the graph of R. The equivalence classes with respect to S are called the *connected components* of E with respect to the relation R.

(c) Let $\mathfrak{F}$ be the set of subsets A of E such that, for each pair of elements (y, z) such that $y \in A$ and $z \in E - A$, we have "not $R\{y, z\}$".

For each $x \in E$, show that the intersection of the sets $A \in \mathfrak{F}$ such that $x \in A$ is the connected component of x with respect to the relation R.*

11. (a) Let $R\{x, y\}$ be a reflexive and symmetric relation on a set E. R is said to be *intransitive of order* 1 if for any four *distinct* elements x, y, z, t of E the relations $R\{x, y\}$, $R\{x, z\}$, $R\{x, t\}$, $R\{y, z\}$, and $R\{y, t\}$ imply $R\{z, t\}$. A subset A of E is said to be *stable* with respect to the relation R if $R\{x, y\}$ for all x and y in A. If a and b are two distinct elements of E such that $R\{a, b\}$, show that the set $C(a, b)$ of elements $x \in E$ such that $R\{a, x\}$ and $R\{b, x\}$ is stable and that

$$C(x, y) = C(a, b)$$

for each pair of distinct elements x, y of $C(a, b)$. The sets $C(a, b)$ (for each ordered pair (a, b) such that $R\{a, b\}$) and the connected components (Exercise 10) with respect to R which consist of a single element are called the *constituents* of E with respect to the relation R. Show that the intersection of two distinct constituents of E contains at most one element and that if A, B, C are three mutually distinct constituents, at least one of the sets $A \cap B$, $B \cap C$, $C \cap A$ is empty.

(b) Conversely, let $(X_\lambda)_{\lambda \in L}$ be a covering of a set E consisting of non-empty subsets of E and having the following properties: (1) if λ, μ are two distinct indices, $X_\lambda \cap X_\mu$ contains at most one element; (2) if λ, μ, ν are three distinct indices, then at least one of the three sets $X_\lambda \cap X_\mu$, $X_\mu \cap X_\nu$, $X_\nu \cap X_\lambda$ is empty. Let $R\{x, y\}$ be the relation "there exists $\lambda \in L$ such that $x \in X_\lambda$ and $y \in X_\lambda$"; show that R is reflexive on E, symmetric and intransitive of order 1, and that the X_λ are the constituents of E with respect to R.

(c) * Similarly, a relation $R\{x, y\}$ which is reflexive and symmetric on E is said to be *intransitive of order* $n - 3$ if, for every family $(x_i)_{1 \leqslant i \leqslant n}$ of distinct elements of E, the relations $R\{x_i, x_j\}$ for each pair $(i, j) \neq (n - 1, n)$ imply $R\{x_{n-1}, x_n\}$. Generalize the results of (a) and (b) to intransitive relations of arbitrary order. Show that a relation which is intransitive of order p is also intransitive of order q for all $q > p$. *

CHAPTER III

Ordered Sets, Cardinals, Integers

1. ORDER RELATIONS. ORDERED SETS

1. DEFINITION OF AN ORDER RELATION

Let $R\{x, y\}$ be a relation, x and y being distinct letters. R is said to be an *order relation with respect to the letters x and y* (or *between x and y*) if

$$(R\{x, y\} \text{ and } R\{y, z\}) \Longrightarrow R\{x, z\},$$
$$(R\{x, y\} \text{ and } R\{y, x\}) \Longrightarrow (x = y),$$
$$R\{x, y\} \Longrightarrow (R\{x, x\} \text{ and } R\{y, y\}).$$

The first of the above relations says that R is *transitive* with respect to the letters x and y (Chapter II, § 6, no. 1).

Examples

(1) The *relation of equality*, $x = y$, is an order relation.

(2) The relation $X \subset Y$ is an order relation between X and Y (Chapter II, § 1, no. 2, Propositions 1 and 2 and axiom A1) which is often called the *inclusion relation*.

(3) Let $R\{x, y\}$ be an order relation between x and y. The relation $R\{y, x\}$ is then an order relation *between x and y*, called the *opposite* of the order relation $R\{x, y\}$.

An *order relation on a set* E is an order relation $R\{x, y\}$ with respect to two distinct letters x, y such that the relation $R\{x, x\}$ is *equivalent to* $x \in E$ (in other words, is such that $R\{x, y\}$ is *reflexive* on E (Chapter II, § 6, no. 1)). Then the relation $R\{x, y\}$ implies "$x \in E$ and $y \in E$" and the relation ($R\{x, y\}$ and $R\{y, x\}$) is equivalent to "$x \in E$ and $y \in E$ and $x = y$".

Examples

(1) The relations of equality and inclusion are not order relations on a set, for the relations $x = x$ and $X \subset X$ are not collectivizing (Chapter II, § 1, no. 7).

(2) Let $R\{x, y\}$ be an order relation between x and y, and let E be a set such that $x \in E$ implies $R\{x, x\}$ (notice that the empty set satisfies this condition). The relation "$R\{x, y\}$ and $x \in E$ and $y \in E$" is then an order relation on E, as is immediately verified; it is called the order relation *induced* by $R\{x, y\}$ on E (cf. no. 4). By abuse of language, the phrase "the relation $S\{x, y\}$ is an order relation between elements of E" is often used in place of "the relation ($S\{x, y\}$ and $x \in E$ and $y \in E$) is an order relation on E". For example, if A is a set, the relation "$X \subset Y$ and $X \subset A$ and $Y \subset A$" is an order relation between subsets of A.

(3) Let E, F be sets. The relation "g extends f" is an order relation (between f and g) on the set of mappings of subsets of E into F.

(4) In the set $\mathfrak{P}(\mathfrak{P}(E))$ of sets of subsets of a set E, let $\mathfrak{L}$ be the set of *partitions* of E (Chapter II, § 4, no. 7). We recall that a partition ϖ is said to be *coarser* than a partition ϖ' if given any $Y \in \varpi'$ there exists $X \in \varpi$ such that $Y \subset X$ (Chapter II, § 4, no. 6). For each partition ϖ of E let $\tilde{\varpi}$ be the graph of the equivalence relation defined by ϖ on E (Chapter II, § 6, no. 2), that is to say, the union of the (mutually disjoint) sets $A \times A$, where A runs through ϖ. The relation "ϖ is coarser than ϖ'" is immediately seen to be equivalent to $\tilde{\varpi} \supset \tilde{\varpi}'$, and is therefore an order relation on the set $\mathfrak{L}$ between ϖ and ϖ'.

An *ordering* on a set E is a correspondence $\Gamma = (G, E, E)$ with E as source and as target, and such that the relation $(x, y) \in G$ is an order relation on E. By abuse of language we shall sometimes refer to the graph G of Γ as an ordering on E. If $R\{x, y\}$ is an order relation on E, it has a graph which is an ordering on E.

Proposition 1. *A correspondence* Γ *between* E *and* E *is an ordering on* E *if and only if its graph* G *satisfies the following conditions:*

(a) $G \circ G = G$.

(b) *The set* $G \cap \overset{-1}{G}$ *is the diagonal* Δ *of* $E \times E$.

For the relation $((x, y) \in G$ and $(y, z) \in G) \Longrightarrow ((x, z) \in G)$ can be written as $G \circ G \subset G$, and the relation

$$((x, y) \in G \text{ and } (y, x) \in G) \Longleftrightarrow (x = y \text{ and } x \in E \text{ and } y \in E)$$

can be written as $G \cap \overset{-1}{G} = \Delta$. From $G \cap \overset{-1}{G} = \Delta$ we then deduce $\Delta \subset G$, whence $G = \Delta \circ G \subset G \circ G$; since also $G \circ G \subset G$, we have

$$G \circ G = G.$$

2. PREORDER RELATIONS

Let R$\{x, y\}$ be a relation, x and y being distinct letters. If R is transitive and if we have R$\{x, y\} \Longrightarrow$ (R$\{x, x\}$ and R$\{y, y\}$), it does not necessarily follow that R is an order relation because the relation

$$(\mathrm{R}\{x, y\} \text{ and } \mathrm{R}\{y, x\})$$

does not necessarily imply $x = y$. R$\{x, y\}$ is said to be a *preorder relation* between x and y; R$\{y, x\}$ is then also a preorder relation between x and y, called the *opposite* of the relation R$\{x, y\}$.

For example, let $\mathfrak{K}$ be the set of subsets of $\mathfrak{P}$(E) which are coverings of E (Chapter II, § 4, no. 6). The relation "$\mathfrak{R}$ is coarser than $\mathfrak{R}'$" between elements $\mathfrak{R}$, $\mathfrak{R}'$ of $\mathfrak{K}$ (Chapter II, § 4, no. 6) is transitive and reflexive. But two distinct coverings can be such that each is coarser than the other; for example, this is the case when $\mathfrak{R}'$ is the union (in $\mathfrak{P}$(E)) of $\mathfrak{R}$ and a subset of E contained in a set of $\mathfrak{R}$ but not belonging to $\mathfrak{R}$.

But in any case the relation (R$\{x, y\}$ and R$\{y, x\}$) is an *equivalence relation* S$\{x, y\}$ with respect to x and y. Let x', y' be letters distinct from x, y which do not appear in R. Then R$\{x, y\}$ is *compatible* (with respect to x and y) with the equivalence relations S$\{x, x'\}$ and S$\{y, y'\}$; in other words (Chapter II, § 6, no. 8), the relation

$$(\mathrm{R}\{x, y\} \text{ and } \mathrm{S}\{x, x'\} \text{ and } \mathrm{S}\{y, y'\})$$

implies R$\{x', y'\}$.

¶ A *preorder relation on a set* E is a preorder relation R$\{x, y\}$ such that the relation R$\{x, x\}$ is equivalent to $x \in$ E. The relation R$\{x, y\}$ then implies "$x \in$ E and $y \in$ E".

¶ If R$\{x, y\}$ is a preorder relation on a set E, then the relation S$\{x, y\}$ defined above is an equivalence relation on E. Let R$'\{$X, Y$\}$ denote the relation

$$\mathrm{X} \in \mathrm{E/S} \text{ and } \mathrm{Y} \in \mathrm{E/S} \text{ and } (\exists x)(\exists y)(x \in \mathrm{X} \text{ and } y \in \mathrm{Y} \text{ and } \mathrm{R}\{x, y\}),$$

that is to say, the relation induced by R on passing to the quotient (with respect to x and y); we saw in Chapter II, § 6, no. 3, that it is equivalent to the relation

$$\mathrm{X} \in \mathrm{E/S} \text{ and } \mathrm{Y} \in \mathrm{E/S} \text{ and } (\forall x)(\forall y)((x \in \mathrm{X} \text{ and } y \in \mathrm{Y}) \Longrightarrow \mathrm{R}\{x, y\}).$$

¶ Let us show that $R'\{X, Y\}$ is an *order relation* between elements of E/S. The relation ($R'\{X, Y\}$ and $R'\{Y, Z\}$) is equivalent to

$$X \in E/S \text{ and } Y \in E/S \text{ and } Z \in E/S \text{ and}$$
$$(\forall x)(\forall y)(\forall z)((x \in X \text{ and } y \in Y \text{ and } z \in Z) \Rightarrow (R\{x, y\} \text{ and } R\{y, z\}))$$

(Chapter I, §4, criteria C40, C41). Since $R\{x, y\}$ is transitive and $Y \in E/S \Rightarrow Y \neq \emptyset$ (Chapter II, §6, no. 2), it follows immediately that $R'\{X, Y\}$ is transitive. Next, ($R'\{X, Y\}$ and $R'\{Y, X\}$) is equivalent to

$$X \in E/S \text{ and } Y \in E/S \text{ and}$$
$$(\forall x)(\forall y) \quad ((x \in X \text{ and } y \in Y) \Rightarrow (R\{x, y\} \text{ and } R\{y, x\})),$$

and hence equivalent to

$$X \in E/S \text{ and } Y \in E/S \text{ and } (\forall x)(\forall y)((x \in X \text{ and } y \in Y) \Rightarrow S\{x, y\}),$$

and therefore implies

$$X \in E/S \quad \text{and} \quad Y \in E/S \quad \text{and} \quad X = Y.$$

Moreover, $R\{x, y\}$ implies $R\{x, x\}$ and $R\{y, y\}$, and hence $R'\{X, Y\}$ implies each of the relations

$$X \in E/S \text{ and } (\forall x)((x \in X) \Rightarrow R\{x, x\}),$$
$$Y \in E/S \text{ and } (\forall y)((y \in Y) \Rightarrow R\{y, y\}),$$

whence $R'\{X, Y\}$ implies ($R'\{X, X\}$ and $R'\{Y, Y\}$). Finally, since $x \in E$ implies $R\{x, x\}$, $X \in E/S$ implies $R'\{X, X\}$, and the proof of our assertion is complete. $R'\{X, Y\}$ is said to be the order relation *associated* with $R\{x, y\}$.

¶ A *preordering* on a set E is a correspondence $\Gamma = (G, E, E)$ with E as source and as target, and such that $(x, y) \in G$ is a preorder relation on E. By abuse of language we refer sometimes to the graph G of Γ as a preordering on E. For this to be so it is necessary and sufficient that $\Delta \subset G$ and $G \circ G \subset G$ (which implies $G \circ G = G$). The equivalence relation S corresponding to the preorder relation $(x, y) \in G$ then has $G \cap \overset{-1}{G}$ as its graph; the order relation associated with $(x, y) \in G$ has as graph the subset G' of $(E/S) \times (E/S)$ which corresponds (Chapter II, §6, no. 8) to the image of G under the canonical mapping of $E \times E$ onto

$$(E \times E)/(S \times S).$$

Example. * Let A be a ring with an identity element. The relation $(\exists z)(z \in A \text{ and } y = zx)$ between two elements x, y of A is a preorder relation on A; it is read "x is a right divisor of y" or "y is a left multiple of x". *

3. NOTATION AND TERMINOLOGY

The definitions to be given in the remainder of this section apply to an arbitrary order (or preorder) relation $R\{x, y\}$ between x and y, but will be used mainly in the case where $R\{x, y\}$ is written $x \leqslant y$ *(by analogy with the usual order relation between integers or real numbers)* (or $x \subset y$, or by means of some analogous sign); we shall therefore state them only for the notation $x \leqslant y$, and leave to the reader the task of transcribing them into other notations. When $R\{x, y\}$ is written $x \leqslant y$, then $y \geqslant x$ is synonymous with $x \leqslant y$, and these relations are read "x is *smaller* than y", or "x is *less than* y", or "y is *larger* than x" or "y is *greater than* x". The relation $x \geqslant y$ is then the preorder relation *(between x and y) opposite to* $x \leqslant y$.

By abuse of language, we shall often speak of "the relation $\leqslant$" instead of "the relation $x \leqslant y$"; in this case "the relation $\geqslant$" is the opposite of "the relation $\leqslant$". We remark also that, in the same proof, we may often use the same sign $\leqslant$ to denote several different order relations when there is no risk of confusion.

The conditions for a relation written $x \leqslant y$ to be an order relation on a set E are as follows:

(RO_I) *The relation* "$x \leqslant y$ and $y \leqslant z$" *implies* $x \leqslant z$.
(RO_{II}) *The relation* "$x \leqslant y$ and $y \leqslant x$" *implies* $x = y$.
(RO_{III}) *The relation* $x \leqslant y$ *implies* "$x \leqslant x$ and $y \leqslant y$".
(RO_{IV}) *The relation* $x \leqslant x$ *is equivalent to* $x \in E$.

If we leave out condition (RO_{II}), we have the conditions for $x \leqslant y$ to be a preorder relation on E.

¶ When an order relation is written $x \leqslant y$ we shall write $x < y$ (or $y > x$) for the relation "$x \leqslant y$ and $x \neq y$"; these relations are read "x is *strictly smaller* than y", or "x is *strictly less* than y", or "y is *strictly larger* than x", or "y is *strictly greater* than x".

The example of the inclusion relation shows that the negation of $x \leqslant y$ (sometimes denoted by $x \nleqslant y$) is *not necessarily equivalent to* $y < x$ (cf. no. 12).

C58. *Let* $\leqslant$ *be an order relation, and let* x, y *be two distinct letters. The relation* $x \leqslant y$ *is equivalent to* "$x < y$ or $x = y$". *Each of the relations* "$x \leqslant y$ and $y < z$", "$x < y$ and $y \leqslant z$" *implies* $x < z$.

The first assertion follows from the criterion A $\Longrightarrow$ ((A and (not B)) or B) (Chapter I, § 3, criterion C24). To prove the second assertion, we remark that each of the hypotheses implies $x \leqslant z$, by transitivity; and the relation ($x = z$ and $x \leqslant y$ and $y \leqslant z$) would imply $x = y = z$, contrary to the hypothesis.

¶ In order to make matters easier and to replace metamathematical criteria by mathematical theorems we shall usually consider a theory $\mathcal{T}$ which contains the axioms and axiom schemes of the theory of sets, and in addition, two constants E and Γ satisfying the axiom

"Γ is an ordering on the set E" (no. 1).

We shall denote by $x \leqslant y$ the relation $y \in \Gamma\langle x \rangle$, and we shall say that the set E is *ordered by the ordering* Γ (or by the order relation $y \in \Gamma\langle x \rangle$) (cf. Chapter IV, § 1).

¶ Whenever, in $\mathcal{T}$, Γ is a preordering on E, we say likewise that E is *preordered by the preordering* Γ.

In some situations (for example in the following definition) the theories which we shall consider are a little more complicated. We shall leave it to the reader to make explicit the constants and axioms of such theories.

Let E, E′ be two sets ordered by orderings Γ, Γ'. An *isomorphism of* E *onto* E′ (for the orderings Γ and Γ') is a bijection f of E onto E′ such that the relations $x \leqslant y$ and $f(x) \leqslant f(y)$ are equivalent (cf. Chapter IV, § 1).

4. ORDERED SUBSETS. PRODUCT OF ORDERED SETS

Let E be a set ordered by an ordering Γ, with graph G. For each subset A of E, $G \cap (A \times A)$ is an ordering on A; the corresponding order relation is equivalent to "$x \leqslant y$ and $x \in A$ and $y \in A$", and we shall denote it simply by $x \leqslant y$ (by abuse of language). The ordering and the order relation thus defined on A are said to be *induced* by the ordering and order relation given on E; and the ordering and order relation on E are said to be *extensions* of the ordering and order relation which they induce on A. Whenever we consider A as an ordered set we have in mind the ordering induced on A by that on E, unless the contrary is expressly stated.

Examples. The relations induced by the inclusion relation $X \subset Y$ on various sets of subsets are of considerable importance. Here are some examples :

(1) Let E, F be two sets, and let $\Phi(E, F)$ be the set of all mappings of subsets of E into F. For each $f \in \Phi(E, F)$ let G_f be the graph of f, which is a subset of $E \times F$. If we endow $\Phi(E, F)$ with the order relation "g extends f" between f and g (no. 1, Example 3), then $f \to G_f$ is an isomorphism of the ordered set $\Phi(E, F)$ onto a subset of $\mathfrak{P}(E \times F)$, ordered by inclusion.

(2) For each partition ϖ of a set E, let $\tilde{\varpi}$ be the graph of the equivalence relation defined by ϖ on E. The mapping $\varpi \to \tilde{\varpi}$ is an isomorphism of the set $\mathfrak{L}$ of partitions of E, ordered by the relation "ϖ is finer than ϖ'" between ϖ and ϖ' (no. 1, Example 4) onto a subset of $\mathfrak{P}(E \times E)$, ordered by inclusion.

(3) Let E be a set and let $\Omega \subset \mathfrak{P}(E \times E)$ be the set of graphs of *preorderings* on E (no. 2) (or, by abuse of language, the set of all preorderings on E). The order relation $s \subset t$ between s and t, induced on Ω by the inclusion relation on $\mathfrak{P}(E \times E)$, is expressed by saying that "the preordering s is *finer* than t" (or that "t is *coarser* than s"). Let $x(s)y$ and $x(t)y$ respectively denote the preorder relations $(x, y) \in s$ and $(x, y) \in t$ on E; then to say that s is finer than t is equivalent to saying that the relation $x(s)y$ *implies* $x(t)y$.

Let $(E_\iota)_{\iota \in I}$ be a family of sets, and for each index $\iota \in I$ let Γ_ι be an ordering on E_ι; let $G_\iota \subset E_\iota \times E_\iota$ be its graph, and let $x_\iota \leqslant y_\iota$ denote the order relation $(x_\iota, y_\iota) \in G_\iota$ on E_ι. On the product set $F = \prod_{\iota \in I} E_\iota$, the relation

$$(\forall \iota)((\iota \in I) \Longrightarrow (x_\iota \leqslant y_\iota))$$

is an order relation between $x = (x_\iota)$ and $y = (y_\iota)$, as is immediately verified. The ordering and the order relation so defined on F are called the ***product of the orderings*** Γ_ι and the ***product of the order relations*** $x_\iota \leqslant y_\iota$; this relation is written $x \leqslant y$, and the set F, ordered by the product of the orderings Γ_ι, is called the ***product of the ordered sets*** E_ι.

It is immediately verified that the graph of the product ordering on F is the image of the product set $\prod_{\iota \in I} G_\iota$ under the canonical mapping of $\prod_{\iota \in I} (E_\iota \times E_\iota)$ onto $F \times F$ (Chapter II, § 5, no. 5).

An important example of a product of ordered sets is the set F^E of graphs of mappings of a set E into an ordered set F. There is a canonical bijection of F^E onto the set $\mathcal{F}(E, F)$ of mappings of E into F, and this mapping is an isomorphism of the ordered set F^E onto $\mathcal{F}(E, F)$ endowed with the ordering defined by the relation "for all $x \in E$, $f(x) \leqslant g(x)$" between two mappings f, g of E into F. This relation is written $f \leqslant g$.

It should be observed that in the ordered set $\mathcal{F}(E, F)$, the relation $f < g$ means

"for all $x \in E$, $f(x) \leqslant g(x)$, and there exists $y \in E$ such that $f(y) < g(y)$"

and not

"for all $x \in E$, $f(x) < g(x)$".

In order to avoid confusion it is better not to use the notation $f < g$ in this situation.

The definitions of this subsection apply without change to preordered sets when "ordering" is replaced by "preordering" throughout.

5. INCREASING MAPPINGS

DEFINITION 1. *Let* E *and* F *be preordered sets (the preorder relation in each being denoted by* $\leqslant$*). A mapping* f *of* E *into* F *is said to be increasing* (or *order-preserving*) *if the relation* $x \leqslant y$ *implies* $f(x) \leqslant f(y)$*; it is decreasing* (or *order-reversing*) *if the relation* $x \leqslant y$ *implies* $f(x) \geqslant f(y)$*. A mapping of* E *into* F *is said to be monotone if it is either increasing or decreasing.*

An increasing mapping of E into F becomes decreasing (and vice versa) when we replace *one* of the preorderings on E or on F by the opposite preordering. Every *constant* function is both increasing and decreasing; the converse of this statement is usually not true.

For example, if a set E is ordered by the equality relation, the identity mapping of E onto itself is both increasing and decreasing, but not constant if E has more than one element (cf. Exercise 7).

DEFINITION 2. *Let* E *and* F *be two ordered sets. A mapping* f *of* E *into* F *is said to be strictly increasing* (or *strictly order-preserving*) *if the relation* $x < y$ *implies* $f(x) < f(y)$*;* f *is said to be strictly decreasing* (or *strictly order-reversing*) *if the relation* $x < y$ *implies* $f(x) > f(y)$*. A mapping of* E *into* F *is said to be strictly monotone if it is either strictly increasing or strictly decreasing.*

Examples

(1) Let E be a set. The mapping $X \to E - X$ of $\mathfrak{P}(E)$ (ordered by inclusion) onto itself is strictly decreasing.

(2) Let E be an ordered set. For each $x \in E$ let U_x be the set of all $y \in E$ such that $y \geqslant x$. The mapping $x \rightarrow U_x$ is a strictly decreasing mapping of E into $\mathfrak{P}(E)$ (ordered by inclusion); indeed, the relation $x \leqslant y$ is equivalent to $U_x \supset U_y$.

An injective monotone mapping of an ordered set E into an ordered set F is *strictly monotone*; the converse is usually not true, because it may happen that $f(x) = f(y)$ when neither of the relations $x \leqslant y$, $x \geqslant y$ is true (cf. no. 12, Proposition 11).

¶ A bijective mapping f of an ordered set E onto an ordered set E′ is an isomorphism of E onto E′ (no. 3) if and only if both f and its inverse mapping are increasing.

¶ If I is an *ordered* index set, a *family of subsets* $(X_\iota)_{\iota \in I}$ of a set E is said to be *increasing* if $\iota \rightarrow X_\iota$ is an increasing mapping of I into $\mathfrak{P}(E)$, ordered by inclusion (in other words, if $\iota \leqslant \varkappa$ implies $X_\iota \subset X_\varkappa$). Similarly we define a *decreasing, strictly increasing,* or *strictly decreasing* family of subsets $(X_\iota)_{\iota \in I}$.

Proposition 2. *Let* E, E′ *be two ordered sets, and let* $u : E \rightarrow E'$ *and* $v : E' \rightarrow E$ *be two decreasing mappings such that for all* $x \in E$ *and all* $x' \in E'$ *we have* $v(u(x)) \geqslant x$ *and* $u(v(x')) \geqslant x'$. *Then*

$$u \circ v \circ u = u \quad \textit{and} \quad v \circ u \circ v = v.$$

For the relation $v(u(x)) \geqslant x$ implies $(u(v(u(x))) \leqslant u(x)$ because u is decreasing; on the other hand, we have $u(v(u(x))) \geqslant u(x)$ by replacing x' by $u(x)$ in the inequality $u(v(x')) \geqslant x'$. Hence the first equality; the proof of the second is similar.

6. MAXIMAL AND MINIMAL ELEMENTS

Definition 3. *Let* E *be an ordered set. An element* $a \in E$ *is said to be a minimal* (resp. *maximal*) *element of* E *if the relation* $x \leqslant a$ (resp. $x \geqslant a$) *implies* $x = a$.

Every minimal element of E is a maximal element with respect to the opposite ordering, and vice versa.

Examples

(1) Let A be a set. In the subset of $\mathfrak{P}(A)$ (ordered by inclusion) consisting of the non-empty subsets of A, the minimal elements are the subsets consisting of a single element.

(2) In the set $\Phi(E, F)$ of mappings of subsets of E into F (F being non-empty), ordered by the relation "v extends u" between u and v, the maximal elements are the mappings of the whole of E into F.

* (3) In the set of natural integers > 1, ordered by the relation "m divides n" between m and n, the minimal elements are the prime numbers. *

* (4) The set of real numbers has no maximal element and no minimal element. *

7. GREATEST ELEMENT AND LEAST ELEMENT

If there exists an element a in an ordered set E such that $a \leqslant x$ for all $x \in$ E, then a is the *only* element of E with this property; for if also $b \leqslant x$ for all $x \in$ E, then we should have $a \leqslant b$ and $b \leqslant a$, and consequently $a = b$.

DEFINITION 4. *Let* E *be an ordered set. An element* $a \in$ E *is said to be the least* (resp. *greatest*) *element of* E *if for all* $x \in$ E *we have* $a \leqslant x$ (resp. $x \leqslant a$).

An ordered set need not have a greatest element nor a least element. If E has a least element a, then a is the greatest element with respect to the opposite ordering.

¶ If E has a least element a, then a is the *unique minimal element* of E; for if $x \in$ E is distinct from a, we have $a < x$.

Examples

(1) Let $\mathfrak{S}$ be a non-empty subset of the set $\mathfrak{P}$(E) of subsets of a set E. If $\mathfrak{S}$ has a least (resp. greatest) element A with respect to the inclusion relation, then A is the intersection (resp. union) of the sets of $\mathfrak{S}$. Conversely, if the intersection (resp. union) of the sets of $\mathfrak{S}$ belongs to $\mathfrak{S}$, then it is the least (resp. greatest) element of $\mathfrak{S}$.

(2) In particular, $\varnothing$ is the least element and E the greatest element of $\mathfrak{P}$(E). In the set Φ(E, F) of mappings of subsets of E into F, ordered by extension of mappings (no. 1, Example 3), the empty mapping is the least element, and there is no greatest element unless F consists of a single element. The diagonal Δ of E $\times$ E is the least element of the set of graphs of equivalence relations on E (or of the set of preorderings on E).

PROPOSITION 3. *Let* E *be an ordered set and let* E′ *be the disjoint union of* E *and a set* $\{a\}$ *consisting of a single element. Then there exists a unique ordering on* E′ *which induces the given ordering on* E *and for which* a *is the greatest element of* E′.

For if G is the graph of the ordering on E, the graph of an ordering on E′ which satisfies the conditions of the Proposition must be the union G′ of G and the set of all pairs (x, a) where $x \in$ E′; conversely, it is clear that G′ is the graph of an ordering on E′ which satisfies the given conditions.

¶ The ordered set E′ is said to be obtained by *adjoining a greatest element* a *to* E (cf. Exercise 3).

¶ A subset A of a preordered set E is said to be *cofinal* (resp. *coinitial*) in E if for every $x \in E$ there exists $y \in A$ such that $x \leqslant y$ (resp. $y \leqslant x$). To say that an ordered set E has a greatest (resp. least) element therefore means that E has a cofinal (resp. coinitial) subset consisting of a single element.

8. UPPER AND LOWER BOUNDS

DEFINITION 5. *Let* E *be a preordered set and let* X *be a subset of* E. *Any element* $x \in E$ *such that* $x \leqslant y$ (resp. $x \geqslant y$) *for all* $y \in X$ *is called a lower* (resp. *upper*) *bound of* X *in* E.

Every upper bound of X is a lower bound of X with respect to the opposite ordering, and vice versa.
¶ If x is a lower bound of X, every element $z \leqslant x$ is also a lower bound of X. A lower bound of X is also a lower bound of every subset of X. An ordered set X has a least element if and only if there exists a lower bound of X which belongs to X.
¶ The set of lower bounds of a subset X of a preordered set E may be empty : this is the case when $X = E$ and E is an ordered set which has no least element.
¶ A subset X of E whose set of lower (resp. upper) bounds is not empty is said to be *bounded below* (resp. *bounded above*). A subset which is bounded both below and above is said to be *bounded*. If X is bounded below (resp. bounded above, bounded), the same is true of every subset of X.

Every subset consisting of a single element is bounded. But a subset consisting of two elements need not be bounded either above or below (no. 10).

Let E be a preordered set and let f be a mapping of an arbitrary set A into E. The mapping f is said (by abuse of language) to be *bounded below* (resp. *bounded above, bounded*) if the set $f(A)$ is bounded below (resp. bounded above, bounded) in E.

9. LEAST UPPER BOUND AND GREATEST LOWER BOUND

DEFINITION 6. *Let* E *be an ordered set and let* X *be a subset of* E. *An element of* E *is said to be the greatest lower bound or infimum* (resp. *least upper bound or supremum*) *of* X *in* E *if it is the greatest* (resp. *least*) *element of the set of lower* (resp. *upper*) *bounds of* X *in* E.

Given a subset X of an ordered set E, the least upper bound (resp. greatest lower bound) of X in E, when it exists, is denoted by

$$\sup_E X \text{ (resp. } \inf_E X)$$

or by sup X (resp. inf X) if there is no risk of ambiguity. The least upper bound (resp. greatest lower bound) of a set $\{x, y\}$ of two elements, when it exists, is denoted by $\sup(x, y)$ (resp. $\inf(x, y)$). Similarly for the least upper bound and greatest lower bound of a set of three elements, etc.
¶ If a subset X of E has a greatest element a, then a is the least upper bound of X in E.
¶ If X has a greatest lower bound a in E, then a is the least upper bound of X with respect to the opposite ordering on E. For this reason we may restrict ourselves for the most part, in what follows, to consideration of the properties of least upper bounds.

Examples

(1) The set of upper bounds of the empty set $\emptyset$ in an ordered set E is evidently E itself; hence $\emptyset$ has a supremum in E if and only if E has a *least* element, which is then the *least upper bound* of $\emptyset$.

(2) In the set $\mathfrak{P}(E)$ of subsets of a set E, ordered by inclusion, every subset $\mathfrak{S}$ of $\mathfrak{P}(E)$ has a least upper bound, namely the *union* of the sets of $\mathfrak{S}$, and a greatest lower bound, namely the *intersection* of the sets of $\mathfrak{S}$.

(3) Let E, F be two sets and let Θ be a subset of the $\Phi(E, F)$ of mappings of subsets of E into F, ordered by extension of mappings (no. 1, Example 3). For each $u \in \Phi(E, F)$ let $D(u)$ be the domain of u. The condition for the existence of a common extension of a family of mappings belonging to $\Phi(E, F)$ (Chapter II, § 4, no. 6, Proposition 7) shows that Θ has a least upper bound in $\Phi(E, F)$ if and only if for each pair (u, v) of elements of Θ we have $u(x) = v(x)$ whenever $x \in D(u) \cap D(v)$.

A mapping f of a set A into an ordered set E is said to have a least upper bound if the image $f(A)$ has a least upper bound in E; this bound is then called the *least upper bound of* f and is written $\sup_{x \in A} f(x)$. Similarly for the greatest lower bound.

In particular, if a subset A of E has a least upper bound in E, this bound is the least upper bound of the canonical injection of A into E, and may therefore be written as $\sup_{x \in A} x$.

PROPOSITION 4. *Let* E *be an ordered set and let* A *be a subset of* E *which has both a greatest lower bound and a least upper bound in* E. *Then* $\inf A \leqslant \sup A$ *if* $A \neq \emptyset$; *if* $A = \emptyset$, sup A *is the least and* inf A *the greatest element of* E.

This follows immediately from the definitions.

PROPOSITION 5. *Let* E *be an ordered set and let* A, B *be two subsets of* E, *each of which has a least upper bound* (resp. *greatest lower bound*) *in* E. *If* $A \subset B$, *then* $\sup A \leqslant \sup B$ (resp. $\inf A \geqslant \inf B$).

COROLLARY. *Let* $(x_\iota)_{\iota\in I}$ *be a family of elements of an ordered set* E *which has a least upper bound in* E. *If* J *is a subset of* I *such that the family* $(x_\iota)_{\iota\in J}$ *has a least upper bound in* E, *we have* $\sup_{\iota\in J} x_\iota \leqslant \sup_{\iota\in I} x_\iota$.

PROPOSITION 6. *Let* $(x_\iota)_{\iota\in I}$, $(y_\iota)_{\iota\in I}$ *be two families of elements of an ordered set* E, *indexed by the same set* I, *and such that* $x_\iota \leqslant y_\iota$ *for all* $\iota \in I$. *If both families have a least upper bound in* E, *then* $\sup_{\iota\in I} x_\iota \leqslant \sup_{\iota\in I} y_\iota$.

For $a = \sup_{\iota\in I} y_\iota$ is an upper bound of the set of the y_ι, so that $x_\iota \leqslant y_\iota \leqslant a$ for all $\iota \in I$, and hence $\sup_{\iota\in I} x_\iota \leqslant a$.

PROPOSITION 7. *Let* $(x_\iota)_{\iota\in I}$ *be a family of elements of an ordered set* E, *and let* $(J_\lambda)_{\lambda\in L}$ *be a covering of the index set* I. *Suppose that each of the subfamilies* $(x_\iota)_{\iota\in J_\lambda}$ *has a least upper bound in* E. *For the family* $(x_\iota)_{\iota\in I}$ *to have a least upper bound in* E, *it is necessary and sufficient that the family* $\left(\sup_{\iota\in J_\lambda} x_\iota\right)_{\lambda\in L}$ *should have a least upper bound in* E, *and then we have*

$$\sup_{\iota\in I} x_\iota = \sup_{\lambda\in L}\left(\sup_{\iota\in J_\lambda} x_\iota\right). \tag{1}$$

Let $b_\lambda = \sup_{\iota\in J_\lambda} x_\iota$. Suppose that $(x_\iota)_{\iota\in I}$ has a least upper bound a. Then $a \geqslant b_\lambda$ for each $\lambda \in L$ (Corollary to Proposition 5). On the other hand, if $c \geqslant b_\lambda$ for each $\lambda \in L$, then we have $c \geqslant x_\iota$ for each $\iota \in I$, because $(J_\lambda)_{\lambda\in L}$ is a covering of I; hence $c \geqslant a$, which proves that

$$a = \sup_{\lambda\in L} b_\lambda.$$

Conversely, suppose that the family $(b_\lambda)_{\lambda\in L}$ has a least upper bound a'. Then $a' \geqslant x_\iota$ for all $\iota \in I$. On the other hand, if $c' \geqslant x_\iota$ for all $\iota \in I$, then we have in particular

$$c' \geqslant \sup_{\iota\in J_\lambda} x_\iota = b_\lambda$$

for each $\lambda \in L$, so that $c' \geqslant a'$ and therefore

$$a' = \sup_{\iota\in I} x_\iota.$$

COROLLARY. *Let* $(x_{\lambda\mu})_{(\lambda,\mu)\in L\times M}$ *be a "double" family of elements of an ordered set* E *such that for each* $\mu \in M$ *the family* $(x_{\lambda\mu})_{\lambda\in L}$ *has a least upper bound in* E. *For the family* $(x_{\lambda\mu})_{(\lambda,\mu)\in L\times M}$ *to have a least upper bound in* E, *it is necessary and sufficient that the family* $\left(\sup_{\lambda\in L} x_{\lambda\mu}\right)_{\mu\in M}$ *should have a least upper bound in* E, *and then we have*

(2) $$\sup_{(\lambda,\mu)\in L\times M} x_{\lambda\mu} = \sup_{\mu\in M}\left(\sup_{\lambda\in L} x_{\lambda\mu}\right).$$

PROPOSITION 8. *Let* $(E_\iota)_{\iota\in I}$ *be a family of ordered sets. Let* A *be a subset of the product ordered set* $E = \prod_{\iota\in I} E_\iota$, *and let* $A_\iota = \mathrm{pr}_\iota A$ *for each* $\iota \in I$. *For* A *to have a least upper bound in* E *it is necessary and sufficient that, for each* $\iota \in I$, A_ι *should have a least upper bound in* E_ι, *and then we have*

$$\sup A = (\sup A_\iota)_{\iota\in I} = \left(\sup_{x\in A} \mathrm{pr}_\iota x\right)_{\iota\in I}.$$

Suppose that, for each $\iota \in I$, A_ι has a least upper bound b_ι in E_ι. To say that $c = (c_\iota)$ is an upper bound of A then means that $c_\iota \geqslant b_\iota$ for each $\iota \in I$, hence $(b_\iota)_{\iota\in I}$ is an upper bound of A. Conversely, suppose that A has a least upper bound $a = (a_\iota)_{\iota\in I}$; for each $\varkappa \in I$, $a_\varkappa$ is an upper bound of $A_\varkappa$, because if $x_\varkappa \in A_\varkappa$, there exists $x \in A$ such that $\mathrm{pr}_\varkappa x = x_\varkappa$, by the definition of $A_\varkappa$; on the other hand, if $a'_\varkappa$ is an upper bound of $A_\varkappa$ in $E_\varkappa$, the element $c' = (c'_\iota)_{\iota\in I}$ for which

$$c'_\iota = a_\iota \quad \text{for} \quad \iota \neq \varkappa \quad \text{and} \quad c'_\varkappa = a'_\varkappa$$

is an upper bound of A; consequently $c' \geqslant a$ and therefore $a'_\varkappa \geqslant a_\varkappa$; hence $a_\varkappa$ is the least upper bound of $A_\varkappa$ in $E_\varkappa$.

¶ Let F be a subset of an ordered set E, and let A be a subset of F. It can happen that one of the two elements $\sup_E A$, $\sup_F A$ exists but the other does not, or that both exist but are unequal.

Examples

* (1) In the ordered set $E = \mathbf{R}$ of real numbers, consider the subset $F = \mathbf{Q}$ of rational numbers and the set $A \subset F$ of rational numbers $< \sqrt{2}$; $\sup_E A$ exists but $\sup_F A$ does not.

(2) In the notation of Example 1, let G be the union of A and the set $\{2\}$; then $G \subset F$ and $\sup_G A$ exists, but $\sup_F A$ does not.

(3) With the same notation, $\sup_E A = \sqrt{2}$, $\sup_G A = 2$. *

However, we have the following result :

PROPOSITION 9. *Let* E *be an ordered set,* F *a subset of* E, A *a subset of* F. *If both* $\sup_E A$ *and* $\sup_F A$ *exist, we have* $\sup_E A \leqslant \sup_F A$. *If* $\sup_E A$ *exists and belongs to* F, *then* $\sup_F A$ *exists and is equal to* $\sup_E A$.

The first assertion follows from the fact that the set M of upper bounds of A in F is contained in the set N of upper bounds of A in E, and from Proposition 5. On the other hand, if the least element of N lies in F, then it belongs to M and is clearly the least element of M; this proves the second assertion.

10. DIRECTED SETS

Definition 7. *A preordered set* E *is said to be right directed* (resp. *left directed*) *if every subset of two elements of* E *is bounded above* (resp. *bounded below*).

In place of "right directed" we shall often use the expression "directed with respect to the relation $\leqslant$", and analogous expressions when the preorder relation is denoted by some other sign. For example, if $\mathfrak{S}$ is a set of subsets of a set A, we say that $\mathfrak{S}$ is *directed with respect to the relation* $\subset$ (resp. $\supset$) if, for each subset $\{X, Y\}$ consisting of two elements of $\mathfrak{S}$, there exists $Z \in \mathfrak{S}$ such that $X \subset Z$ and $Y \subset Z$ (resp. $X \supset Z$ and $Y \supset Z$).

Examples

(1) An ordered set which has a greatest element is right directed.

* (2) In a topological space, a fundamental system of neighbourhoods of a point is directed with respect to the relation $\supset$.

(3) The set of submodules of finite type of an arbitrary module is directed with respect to the relation $\subset$. *

Proposition 10. *In a right directed ordered set* E, *a maximal element* a *is the greatest element of* E.

For every $x \in E$ there exists by hypothesis $y \in E$ such that $x \leqslant y$ and $a \leqslant y$; since a is maximal, $y = a$.

¶ A right directed preordered set is left directed with respect to the opposite ordering. Any product of right directed sets is right directed. On the other hand, a subset of a right directed set is not necessarily right directed. However, a *cofinal* subset F of a right directed set E is always right directed; for given $x, y \in F$ there exists $z \in E$ such that $x \leqslant z$ and $y \leqslant z$, and then $t \in F$ such that $z \leqslant t$.

11. LATTICES

Definition 8. *An ordered set* E *is said to be a lattice if every subset consisting of two elements of* E *has a least upper bound and a greatest lower bound in* E.

Every product of lattices is a lattice; this follows from the condition for the existence of a least upper bound in a product of ordered sets (no. 9, Proposition 8). The set of subsets of a set A, ordered by inclusion, is a lattice because the union and intersection of two subsets of A are again subsets of A.

Examples

* (1) The set of integers $\geqslant 1$, ordered by the relation "m divides n" between m and n, is a lattice; the least upper bound of $\{m, n\}$ is the l.c.m. of m and n, and the greatest lower bound is their h.c.f.

(2) The set of subgroups of a group G, ordered by inclusion, is a lattice.

(3) The set of topologies on a set A, ordered by the relation "$\mathfrak{T}$ is coarser than $\mathfrak{T}'$" between $\mathfrak{T}$ and $\mathfrak{T}'$, is a lattice. (*General Topology*, Chapter I, § 2).

(4) The set $\mathcal{F}(\mathrm{I}, \mathbf{R})$ of all real-valued functions defined on an interval I of **R** is a lattice with respect to the order relation $f \leqslant g$ (no. 4), and as such is isomorphic to the product $\mathbf{R}^{\mathrm{I}}$. *

Remark. A lattice is obviously both left and right directed. But an ordered set which is both left and right directed is not necessarily a lattice. * An example of the latter is the set of mappings $x \to p(x)$ of **R** into itself, where p is a polynomial in $\mathbf{R}[\mathrm{X}]$, this set being ordered by the relation $p \leqslant q$ (no. 4). *

12. TOTALLY ORDERED SETS

DEFINITION 9. *Two elements x, y of a preordered set* E *are said to be comparable if the relation "$x \leqslant y$* or *$y \leqslant x$" is true. A set* E *is said to be totally ordered if it is ordered and if any two elements of* E *are comparable. The ordering on* E *is then said to be a total ordering, and the corresponding order relation a total order relation.*

If x and y are elements of a totally ordered set, then $x = y$ or $x < y$ or $x > y$; the negation of $x \leqslant y$ is thus $x > y$.

An ordering on E is a total ordering if and only if its graph G satisfies the relation $\mathrm{G} \cup \overset{-1}{\mathrm{G}} = \mathrm{E} \times \mathrm{E}$, as well as the relations $\mathrm{G} \circ \mathrm{G} = \mathrm{G}$ and $\mathrm{G} \cap \overset{-1}{\mathrm{G}} = \Delta$.

Examples

(1) Every subset of a totally ordered set is totally ordered by the induced ordering.

(2) Let E be an arbitrary ordered set. The empty subset of E is totally ordered, and so is every subset of E consisting of a single element.

* (3) The set **R** of real numbers is totally ordered. *

(4) If A is a set which has at least two distinct elements, the set $\mathfrak{P}(\mathrm{A})$ (ordered by inclusion) is not totally ordered, for if $x \neq y$, the subsets $\{x\}$ and $\{y\}$ are not comparable.

A totally ordered set is also totally ordered with respect to the opposite ordering; it is a lattice and *a fortiori* both left and right directed.

PROPOSITION 11. *Every strictly monotone mapping f of a totally ordered set* E *into an ordered set* F *is injective. If f is strictly increasing, f is an isomorphism of* E *onto* $f(\mathrm{E})$.

For $x \neq y$ implies that either $x < y$ or $x > y$; hence

$$f(x) < f(y) \quad \text{or} \quad f(x) > f(y),$$

so that $f(x) \neq f(y)$ in either case. It remains to be shown that if f is strictly increasing, $f(x) \leqslant f(y)$ implies $x \leqslant y$; if not, we should have $x > y$, and therefore $f(x) > f(y)$.

PROPOSITION 12. *Let* E *be a totally ordered set and let* X *be a subset of* E. *For an element b of* E *to be the least upper bound of* X *in* E, *it is necessary and sufficient that* (1) *b is an upper bound of* X, (2) *for every $c \in \mathrm{E}$ such that $c < b$, there exists $x \in \mathrm{X}$ such that $c < x \leqslant b$.*

The second condition says that no element $c < b$ is an upper bound of X, i.e., b is a minimal element of the set M of upper bounds of X; but this is the same as saying that b is the least element of M, since M is totally ordered (no. 10, Proposition 10).

13. INTERVALS

Let E be an ordered set and let a, b be two elements of E such that $a \leqslant b$. The subset of E consisting of elements x such that $a \leqslant x \leqslant b$ is called the *closed interval with left-hand endpoint a and right-hand endpoint b*, and is denoted by $[a, b]$. The set of all $x \in \mathrm{E}$ such that $a \leqslant x < b$ (resp. $a < x \leqslant b$) is called the *interval half-open on the right* (resp. *on the left*) with endpoints a and b, and is denoted by $[a, b[$ (resp. $]a, b]$). The set of all $x \in \mathrm{E}$ such that $a < x < b$ is called the *open interval* with endpoints a and b, and is denoted by $]a, b[$.

Note that a closed interval is never empty; the interval $[a, a]$ is the set $\{a\}$. On the other hand, the intervals $[a, a[$, $]a, a]$, $]a, a[$ are all empty; and an open interval $]a, b[$ may be empty even when $a < b$.

Let a be an element of E. The set of all $x \in \mathrm{E}$ such that $x \leqslant a$ (resp. $x < a$) is called the *closed* (resp. *open*) *interval unbounded on the left, with right-hand endpoint a*, and is denoted by $]\leftarrow, a]$ (resp. $]\leftarrow, a[$); likewise, the set of all $x \in \mathrm{E}$ such that $x \geqslant a$ (resp. $x > a$) is called the *closed* (resp. *open*) *interval with left-hand endpoint a, unbounded on the right*, and is denoted by $[a, \leftarrow[$ (resp. $]a, \leftarrow[$). Finally, E itself may be regarded as the *open interval unbounded on the left and on the right*, denoted by]←, →[.

PROPOSITION 13. *In a lattice, the intersection of two intervals is an interval.* Consider for example the intersection of two closed intervals $[a, b]$ and $[c, d]$, and let $\alpha = \sup(a, c)$, $\beta = \inf(b, d)$. If both $a \leqslant x \leqslant b$ and $c \leqslant x \leqslant d$, then we have $\alpha \leqslant x \leqslant \beta$, and conversely; if $\alpha \not\leqslant \beta$, the intersection of $[a, b]$ and $[c, d]$ is empty; if $\alpha \leqslant \beta$, this intersection is $[\alpha, \beta]$. We leave it to the reader to carry through the proof for the other cases.

2. WELL-ORDERED SETS

1. SEGMENTS OF A WELL-ORDERED SET

A relation $R\{x, y\}$ is said to be a *well-ordering relation between x and y* if R is an order relation between x and y and if, for each non-empty set E on which $R\{x, y\}$ induces an order relation (i.e., such that $x \in E$ implies $R\{x, x\}$; cf. § 1, no. 1), E, ordered by this relation, has a *least element.*

¶ A set E ordered by an ordering Γ is said to be *well-ordered* if the relation $y \in \Gamma\langle x\rangle$ is a well-ordering relation between x and y; Γ is then said to be a *well-ordering* on E. The following definition is equivalent to this:

DEFINITION 1. *A set* E *is said to be well-ordered if it is ordered and if each non-empty subset of* E *has a least element.*

A well-ordered set E is totally ordered because every subset $\{x, y\}$ of E has a least element. Every subset A of E which is bounded above in E has a least upper bound in E.

Examples

(1) Let $E = \{\alpha, \beta\}$ be a set whose elements are distinct. It is easily verified that the subset $\{(\alpha, \alpha), (\beta, \beta), (\alpha, \beta)\}$ of $E \times E$ is the graph of a well-ordering on E.

(2) Every subset (in particular, the empty subset) of a well-ordered set is well-ordered by the induced ordering.

* (3) The existence of totally ordered sets which are not well-ordered is equivalent to the axiom of infinity (§ 4, no. 4, Corollary 1 to Proposition 3, and Exercise 3).

(4) If Γ is a well-ordering on E, the ordering opposite to Γ is a well-ordering on E only if E is finite (§ 4, Exercise 3). *

(5) Let E be a well-ordered set. The set E_1 obtained by adjoining to E a greatest element b (§ 1, no. 7) is well-ordered, for if H is any

non-empty subset of E_1 other than $\{b\}$, the least element of $H \cap E$ is also the least element of H.

Remark. * As a consequence of the axiom of infinity (§ 6, no. 1), there exist well-ordered sets which have no greatest element, for example the set **N** of natural integers. *

DEFINITION 2. *In an ordered set* E, *a subset of* E *such that the relations* $x \in S$, $y \in E$, *and* $y \leqslant x$ *imply* $y \in S$ *is called a segment of* E.

Clearly, every intersection and union of segments of E is a segment of E. If S is a segment of E, every segment of S is also a segment of E. The set E itself and the empty set are segments of E.

PROPOSITION 1. *In a well-ordered set* E, *every segment of* E *other than* E *itself is an interval* $]\leftarrow, a[$, *where* $a \in E$.

Let S be a segment of E such that $S \neq E$. Since $E - S$ is not empty, it has a least element a. By virtue of Definition 2, the relation $x \geqslant a$ implies $x \notin S$; otherwise we would have $a \in S$, which is absurd. Hence $E - S$ is the interval $[a, \rightarrow[$, and S is the interval $]\leftarrow, a[$.

¶ For every element x in a totally ordered set E the segment $]\leftarrow, x[$ is called the *segment with endpoint* x, and is denoted by S_x.

Note that if E is well-ordered and not empty, S_x has a least element α and consequently is also the interval $[\alpha, x[$.

Let E be a totally ordered set. The union A of the S_x, as x runs through E, is E if E has no greatest element; and if E has a greatest element b, we have $A = E - \{b\}$.

PROPOSITION 2. *The set* E* *of segments of a well-ordered set* E *is well-ordered by inclusion. The mapping* $x \rightarrow S_x$ *is an isomorphism of the well-ordered set* E *onto the set of segments of* E *other than* E *itself.*

It is clear that if $x \in E$ and $y \in E$, the relation $x \leqslant y$ implies $S_x \subset S_y$, and that $x < y$ implies $S_x \neq S_y$; the mapping $x \rightarrow S_x$ is therefore an isomorphism of E onto the set S(E) of segments of E distinct from E itself (§ 1, no. 12, Proposition 11), and consequently S(E) is well-ordered. Moreover, E* is isomorphic to the well-ordered set obtained from S(E) by adjoining a greatest element.

PROPOSITION 3. *Let* $(X_\iota)_{\iota \in I}$ *be a family of well-ordered sets such that for each pair of indices* $(\iota, \varkappa)$ *one of the sets* X_ι, $X_\varkappa$ *is a segment of the other. Then there exists a unique ordering on the set* $E = \bigcup_{\iota \in I} X_\iota$ *which induces the given ordering on each of the* X_ι. *Endowed with this ordering,* E *is a well-ordered set. Every segment of* X_ι *is a segment of* E; *for each* $x \in X_\iota$, *the segment with endpoint* x *in* X_ι *is equal to the segment with endpoint* x *in* E; *and each segment of* E *is either* E *itself or a segment of one of the* X_ι.

The first assertion is a consequence of the following general lemma :

LEMMA 1. *Let* $(X_\alpha)_{\alpha \in A}$ *be a family of ordered sets, directed with respect to the relation* $\subset$ (in other words, such that for each pair of indices α, β there exists an index γ such that $X_\alpha \subset X_\gamma$ and $X_\beta \subset X_\gamma$). *Suppose that, for each pair of indices* (α, β) *such that* $X_\alpha \subset X_\beta$, *the ordering induced on* X_α *by that of* X_β *is identical with the given ordering on* X_α. *Under these conditions there exists a unique ordering on the set* $E = \bigcup_{\alpha \in A} X_\alpha$ *which induces the given ordering on each* X_α.

Let G_α be the graph of the given ordering on X_α. If G is the graph of an ordering on E which induces on each X_α the ordering whose graph is G_α, then we must have $G_\alpha \subset G$ for each $\alpha \in A$; hence G contains $\bigcup_{\alpha \in A} G_\alpha$. On the other hand, for each pair (x, y) of elements of E there exists by hypothesis an index $\alpha \in A$ such that $x \in X_\alpha$ and $y \in X_\alpha$; if $(x, y) \in G$, we have $(x, y) \in G_\alpha$, so that $G \subset \bigcup_{\alpha \in A} G_\alpha$. Hence if the required ordering on E exists, its graph is necessarily $G = \bigcup_{\alpha \in A} G_\alpha$. It remains to be shown that this set satisfies the conditions of the lemma. Since $G_\beta \cap (X_\alpha \times X_\alpha) = G_\alpha$ if $X_\alpha \subset X_\beta$, we have $G \cap (X_\alpha \times X_\alpha) = G_\alpha$ for all $\alpha \in A$; on the other hand, it follows from the hypothesis that any three elements x, y, z of E belong to the same X_α. Hence $(x, y) \in G$ is an order relation on E, and the lemma is proved.

¶ We now take up the proof of Proposition 3. Let us begin by showing that each X_ι is a segment of E. Indeed, if $x \in X_\iota$, $y \in E$, and $y \leqslant x$, there exists an index $\varkappa$ such that $X_\iota \subset X_\varkappa$ and $y \in X_\varkappa$; since by hypothesis X_ι is a segment of $X_\varkappa$, we have $y \in X_\iota$, which proves the assertion. The same reasoning proves that for each $x \in X_\iota$ the segment with endpoint x in X_ι is identical with the interval $]\leftarrow, x[$ in E. Next let us show that E is well-ordered. If H is a non-empty subset of E, there is an index $\iota \in I$ such that $H \cap X_\iota \neq \emptyset$; if a is the least element of $H \cap X_\iota$ in X_ι, then a is also the least element of H in E. That is, if $x \in H$, there exists an index $\varkappa \in I$ such that $X_\iota \subset X_\varkappa$ and $x \in X_\varkappa$; we cannot have $x < a$, because the interval $]\leftarrow, a[$ is contained in X_ι, and therefore we have $x \geqslant a$ since $X_\varkappa$ is totally ordered.

¶ Finally, we must show that a segment of E, other than E itself, is a segment of one of the X_ι; this is an immediate consequence of the preceding arguments, since such a segment is of the form $]\leftarrow, x[$ (Proposition 1) and since x belongs to some X_ι.

2. THE PRINCIPLE OF TRANSFINITE INDUCTION

LEMMA 2. *Let* E *be a well-ordered set and let* $\mathfrak{S}$ *be a set of segments of* E *with the following properties :* (1) *every union of segments belonging to* $\mathfrak{S}$ *belongs to* $\mathfrak{S}$; (2) *if* $S_x \in \mathfrak{S}$, *then* $S_x \cup \{x\} \in \mathfrak{S}$. *Then every segment of* E *belongs to* $\mathfrak{S}$.

Suppose that there are segments of E which do not belong to $\mathfrak{S}$, and let S be the smallest of them (no. 1, Proposition 2). If S has no greatest element, then S is the union of the segments of S distinct from S itself, and these segments belong to $\mathfrak{S}$ by virtue of the definition of S; hence $S \in \mathfrak{S}$, which is absurd. If, on the other hand, S has a greatest element a, then $S = S_a \cup \{a\}$, and since S_a is a segment of S distinct from S, we have $S_a \in \mathfrak{S}$; but then also $S \in \mathfrak{S}$, which again is absurd.

¶ For greater convenience we shall place ourselves in a theory $\mathcal{T}$ in which E is a set *well-ordered* by a relation written $x \leqslant y$. We have then the following criteria :

C59. (Principle of transfinite induction). *Let* $R\{x\}$ *be a relation in* $\mathcal{T}$ *(x not being a constant of* $\mathcal{T}$*) such that the relation*

$$(x \in E \text{ and } (\forall y)((y \in E \text{ and } y < x) \Longrightarrow R\{y\})) \Longrightarrow R\{x\}$$

is a theorem in $\mathcal{T}$. *Under these conditions the relation* $(x \in E) \Longrightarrow R\{x\}$ *is a theorem in* $\mathcal{T}$.

Let $\mathfrak{S}$ be the set of segments S of E such that $(y \in S) \Longrightarrow R\{y\}$. It is clear that every union of segments belonging to $\mathfrak{S}$ also belongs to $\mathfrak{S}$. On the other hand, if $S_x \in \mathfrak{S}$, we have $R\{x\}$ by hypothesis; hence $(y \in S_x \cup \{y\}) \Longrightarrow R\{y\}$ by the method of disjunction of cases. Hence (Lemma 2) $E \in \mathfrak{S}$, which proves the criterion.

¶ In the applications of C59, the relation

$$x \in E \text{ and } (\forall y)((y \in E \text{ and } y < x) \Longrightarrow R\{y\})$$

is usually called the "inductive hypothesis".

¶ In what follows, for every mapping g of a segment S of E into a set F, and for each $x \in S$ we shall denote by $g^{(x)}$ the mapping of the segment $S_x =]\leftarrow, x[$ of E *onto* $g(S_x)$ which coincides with g on S_x. With this notation we have

C60. (Definition of a mapping by transfinite induction.) *Let u be a letter,* $T\{u\}$ *a term in the theory* $\mathcal{T}$. *There exists a set* U *and a mapping* f *of* E *onto* U *such that for all* $x \in E$ *we have* $f(x) = T\{f^{(x)}\}$. *Furthermore, the set* U *and the mapping* f *are uniquely determined by these conditions.*

Let us first prove the uniqueness. Suppose that f' and U′ also satisfy the conditions of the criterion. Let $\mathfrak{S}$ be the set of segments S of E such that f and f' coincide on S. It is clear that every union of segments belonging to $\mathfrak{S}$ also belongs to $\mathfrak{S}$. On the other hand, if $S_x \in \mathfrak{S}$, then f and f' agree on S_x and therefore $f^{(x)} = f'^{(x)}$; consequently

$$f(x) = T\{f^{(x)}\} = T\{f'^{(x)}\} = f'(x),$$

which shows that $S_x \cup \{x\} \in \mathfrak{S}$. It follows that $E \in \mathfrak{S}$ (Lemma 2), hence $f' = f$ and $U' = f'(E) = f(E) = U$.

¶ Now let $\mathfrak{S}_1$ denote the set of segments S of E for which there exists a set U_S and a mapping f_S of S *onto* U_S such that for all $x \in S$ we have $f_S(x) = T\{f^{(x)}\}$. For each $S \in S_1$, f_S and U_S are uniquely determined, by the first part of the proof; in particular, if S′ and S″ are two segments belonging to $\mathfrak{S}_1$ such that $S' \subset S''$, then $f_{S'}$ is the mapping of S′ onto $f_{S''}(S')$ which agrees with $f_{S''}$ on S′. From this remark it follows that every union of segments belonging to $\mathfrak{S}_1$ also belongs to $\mathfrak{S}_1$ (Chapter II, § 4, no. 6, Proposition 7). On the other hand, if $S_x \in \mathfrak{S}_1$, we define on $S = S_x \cup \{x\}$ a function f_S extending f_{S_x} by putting

$$f_S(x) = T\{f_{S_x}\}$$

(Chapter II, § 4, no. 7, Proposition 8); since $f_S^{(x)} = f_{S_x}$, it is obvious that $S_x \cup \{x\} \in \mathfrak{S}_1$. Hence (Lemma 2) $E \in \mathfrak{S}_1$, and the proof is complete.

¶ Usually this criterion is applied in situations where there exists a set F such that *for every mapping h of a segment of* E *onto a subset of* F *we have* $T\{h\} \in F$. Then the set U obtained by applying C60 is a *subset of* F. For, with the notation used above, let $\mathfrak{S}_2$ be the subset of $\mathfrak{S}_1$ consisting of segments S of E such that $U_S \subset F$. It is evident that every union of segments belonging to $\mathfrak{S}_2$ also belongs to $\mathfrak{S}_2$; on the other hand, the hypothesis on F implies that if $S_x \in \mathfrak{S}_2$, we have $S_x \cup \{x\} \in \mathfrak{S}_2$. The assertion now follows from Lemma 2.

3. ZERMELO'S THEOREM

LEMMA 3. *Let* E *be a set, let* $\mathfrak{S}$ *be a subset of* $\mathfrak{P}(E)$, *and let* p *be a mapping of* $\mathfrak{S}$ *into* E *such that* $p(X) \notin X$ *for all* $X \in \mathfrak{S}$. *Then there exists a subset* M *of* E *and a well-ordering* Γ *on* M *such that, if* $x \leqslant y$ *denotes the relation* $y \in \Gamma\langle x\rangle$ *and* S_x *denotes the segment* $]\leftarrow, x]$,

(1) *for all* $x \in M$ *we have* $S_x \in \mathfrak{S}$ *and* $p(S_x) = x$;

(2) $M \notin \mathfrak{S}$.

Let $\mathfrak{M}$ be the set of subsets G of $E \times E$ satisfying the following conditions :

(a) G is the graph of a well-ordering on $\mathrm{pr}_1 G = U$;

(b) if $x \leqslant y$ denotes the relation $(x, y) \in G$ on U, then for each $x \in U$ the segment S_x is such that $S_x \in \mathfrak{S}$ and $p(S_x) = x$.

¶ We shall show that if G and G′ are two elements of M and if U, U′ denote their first projections, then one of the two sets U, U′ is contained in the other and that if, for example, $U \subset U'$, then $G = G' \cap (U \times U)$ (in other words, that the order relation on U is induced by the order relation on U′) and U is a *segment* of U′.

¶ Consider the set V of elements $x \in U \cap U'$ such that the segments with endpoint x are the same in U and U′, and such that the orderings induced on this segment by the orderings on U and U′ are identical. It is clear that V is a *segment* in both U and U′ and that the orderings induced on V are the same; our assertion will therefore be proved if we show that either $V = U$ or $V = U'$. Let us argue by contradiction and suppose that $V \neq U$ and $V \neq U'$. Let x be the least element of $U - V$ in U and let x' be the least element of $U' - V$ in U′; we have $V = S_x$ in U, and $V = S_{x'}$ in U′. But by hypothesis, $V \in \mathfrak{S}$ and

$$x = p(S_x), \quad x' = p(S_{x'}),$$

so that $x = x'$. Hence by definition $x \in V$, which is absurd.

¶ We may therefore apply Proposition 3 of no. 1 to the set of first projections $U = \mathrm{pr}_1 G$ (where $G \in \mathfrak{M}$) and thus obtain a well-ordered set

$$M = \bigcup_{G \in \mathfrak{M}} \mathrm{pr}_1 G.$$

It is easily seen that the graph of the ordering on M belongs to $\mathfrak{M}$. If we had $M \in S$, then, putting $a = p(M)$, we should have $a \notin M$. We could therefore adjoin to M the element a as greatest element, and the set $M' = M \cup \{a\}$ would be well-ordered. Since $M = S_a$ in M′, we should have $S_a \in \mathfrak{S}$ and $p(S_a) = a$; the graph of the ordering on M′ would therefore belong to $\mathfrak{M}$, which is absurd.

Note that if $\varnothing \notin \mathfrak{S}$ (and in particular if $\mathfrak{S}$ is empty), the set M whose existence is asserted by Lemma 3 is the empty set; this follows from condition 1 of Lemma 3.

Theorem 1 (Zermelo). *Every set* E *can be well-ordered.*

Let $\mathfrak{S} = \mathfrak{P}(E) - \{E\}$ be the set of all subsets of E other than E itself. For each $X \in \mathfrak{S}$ let $p(X) = \tau_x (x \in E - X)$; since the relation $X \in \mathfrak{S}$

implies $(\exists x)(x \in \mathrm{E} - \mathrm{X})$, we have $p(\mathrm{X}) \in \mathrm{E} - \mathrm{X}$ (Chapter I, § 4, no. 1) and therefore $p(\mathrm{X}) \notin \mathrm{X}$. We may therefore apply Lemma 3, and consequently there exists a well-ordering on a subset M of E such that $\mathrm{M} \notin \mathfrak{S}$; but the only subset of E which does not belong to $\mathfrak{S}$ is E itself, and the theorem is proved.

4. INDUCTIVE SETS

Definition 3. *An ordered set* E *is said to be inductive if every totally ordered subset of* E *has an upper bound in* E.

Examples

(1) Let $\mathfrak{F}$ be a set of subsets of a set A, ordered by inclusion, and such that for every totally ordered subset $\mathfrak{G}$ of $\mathfrak{F}$ the union of the sets of $\mathfrak{G}$ belongs to $\mathfrak{F}$. Then $\mathfrak{F}$ is inductive with respect to the relation $\subset$ because the union of the sets of $\mathfrak{G}$ is the least upper bound of $\mathfrak{G}$ in $\mathfrak{P}(\mathrm{A})$.

(2) An important example of a set of subsets which is inductive with respect to the relation $\subset$ is the set $\mathfrak{F}$ of graphs of mappings of subsets of a set A into a set B. For $\mathfrak{F}$ is a subset of $\mathfrak{P}(\mathrm{A} \times \mathrm{B})$, and to say that a subset $\mathfrak{G}$ of $\mathfrak{F}$ is totally ordered by inclusion means that the elements of $\mathfrak{G}$ are graphs of mappings such that, given any two of these mappings, one is an extension of the other. It follows immediately that the union of the sets of $\mathfrak{G}$ is an element of $\mathfrak{F}$ (Chapter II, § 4, no. 6, Proposition 7). Hence the set $\Phi(\mathrm{A}, \mathrm{B})$ of mappings of subsets of A into B is inductive with respect to the order relation "v extends u" between u and v.

* (3) It follows from the axiom of infinity (§ 6, no. 1) that the well-ordered set of natural integers is not inductive with respect to the relation $\leqslant$. *

Theorem 2 ("Zorn's lemma"). *Every inductive ordered set has a maximal element.*

This theorem is a particular case of the following result :

Proposition 4. *Let* E *be an ordered set in which every well-ordered subset is bounded above; then* E *has a maximal element.*

We shall say that an element $v \in \mathrm{E}$ is a *strict upper bound* of a subset X of E if v is an upper bound of X and $v \notin \mathrm{X}$. Let $\mathfrak{S}$ be the set of subsets of E which have a strict upper bound in E, and for each $\mathrm{S} \in \mathfrak{S}$ put $p(\mathrm{S}) = \tau_v(v$ is a strict upper bound of S); then $p(\mathrm{S})$ is a strict upper bound of S. Applying Lemma 3 of no. 3 to $\mathfrak{S}$ and p, we see that there exists a subset M of E and a well-ordering Γ on M which satisfies the conditions of Lemma 3; in particular, M has no strict upper bound in E. Furthermore, the ordering Γ is identical with that induced on M by the ordering on E. For in M the relation "$y \in \Gamma\langle x \rangle$ and $x \neq y$" is equivalent to $x \in S_y$; and since $p(S_y) = y$ is an upper bound of S_y (with respect to the ordering on E), it implies $x < y$ in E. But this means that the injection of M into E is a strictly increasing mapping (M being endowed with the ordering Γ); and since M is totally ordered, it follows that the relations $y \in \Gamma\langle x \rangle$ and $x \leqslant y$ are equivalent in M (§ 1, no. 12, Proposition 11). This being so, there exists by hypothesis an upper bound m of M in E; but since M has no strict upper bound, it follows that m is a maximal element of E.

COROLLARY 1. *Let* E *be an inductive ordered set and let* a *be an element of* E. *Then there exists a maximal element* m *of* E *such that* $m \geqslant a$.

For it follows from Definition 3 that the set F of elements $x \geqslant a$ in E is inductive, and a maximal element of F is also maximal in E.

COROLLARY 2. *Let* $\mathfrak{F}$ *be a set of subsets of a set* E *such that, for every subset* $\mathfrak{G}$ *of* $\mathfrak{F}$ *which is totally ordered by inclusion, the union* (resp. *intersection*) *of the sets of* $\mathfrak{G}$ *belongs to* $\mathfrak{F}$; *then* $\mathfrak{F}$ *has a maximal* (resp. *minimal*) *element.*

5. ISOMORPHISMS OF WELL-ORDERED SETS

THEOREM 3. *Let* E *and* F *be two well-ordered sets. Then at least one of the following two statements is true :*

(1) *there exists a unique isomorphism of* E *onto a segment of* F;

(2) *there exists a unique isomorphism of* F *onto a segment of* E.

Let $\mathfrak{F}$ be the set of mappings of subsets of E into F such that each mapping is defined on a segment of E and is an isomorphism of this segment onto a segment of F. Then the set $\mathfrak{F}$, ordered by the relation "v extends u" between u and v, is *inductive*. For if $\mathfrak{G}$ is a totally ordered subset of $\mathfrak{F}$, the union S of the domains of the mappings $u \in \mathfrak{G}$ is a union of segments of E and is therefore itself a segment of E. If v is the least upper bound of $\mathfrak{G}$ in $\Phi(E, F)$ (no. 4, Example 2), then $v(S)$ is the union of the ranges of the mappings $u \in \mathfrak{G}$ and is therefore a segment of F. Finally, for each pair of elements x, y of S such that $x < y$ there exists $u \in \mathfrak{G}$ whose domain contains both x and y (because $\mathfrak{G}$ is totally ordered); and since $v(x) = u(x) < u(y) = v(y)$, v is an isomorphism of S onto $v(S)$, and our assertion is proved.

¶ Now let u_0 be a maximal element of $\mathfrak{F}$ (no. 4, Theorem 2) and let S_0 be the segment of E which is the domain of u_0. If we show that either $S_0 = E$ or $u_0(S_0) = F$, the theorem will be proved. Let us argue by contradiction and suppose that $S_0 \neq E$ and $u_0(S_0) \neq F$. There will then be an element $a \in E$ and an element $b \in F$ such that $S_0 =]\leftarrow, a[$ and $u_0(S_0) =]\leftarrow, b[$ (no. 1, Proposition 1). Extend u_0 to a mapping u_1 of the segment $]\leftarrow, a]$ into F by putting $u_1(a) = b$; since u_1 is an isomorphism of $]\leftarrow, a]$ onto the segment $]\leftarrow, b]$, this contradicts the maximality of u_0 in $\mathfrak{F}$.

¶ The uniqueness asserted in Theorem 3 is a consequence of the following Lemma :

Lemma 4. *Let* E, F *be two well-ordered sets and let* f, g *be two increasing mappings of* E *into* F *such that* $f(E)$ *is a segment of* F *and* g *is strictly increasing; then* $f(x) \leqslant g(x)$ *for all* $x \in E$.

Suppose, on the contrary, that the set of elements $y \in E$ such that $f(y) > g(y)$ is not empty; then this set will have a least element a. If $x < a$, we have then $f(x) \leqslant g(x) < g(a) < f(a)$ since g is strictly increasing. Since $f(E)$ is a segment of F, there exists $z \in E$ such that $g(a) = f(z)$; f is increasing, so that $f(z) < f(a)$ implies $z < a$. Hence

$$f(z) \leqslant g(z) < g(a) = f(z),$$

which is absurd.

Corollary 1. *The only isomorphism of a well-ordered set* E *onto a segment of* E *is the identity mapping of* E *onto itself.*

Put F = E in Theorem 3.

Corollary 2. *Let* E, F *be two well-ordered sets. If there exists an isomorphism* f *of* E *onto a segment* T *of* F *and an isomorphism* g *of* F *onto a segment* S *of* E, *then we must have* S = E, T = F, *and* g, f *are inverses of each other.*

For $g \circ f$ is an isomorphism of E onto the segment $g(T) \subset S$ of E; by Corollary 1 we have $g(T) = S = E$, and $g \circ f$ is the identity mapping of E. Similarly, $f \circ g$ is the identity mapping of F, whence the result.

Corollary 3. *Every subset* A *of a well-ordered set* E *is isomorphic to a segment of* E.

By virtue of Theorem 3 it is enough to prove that there exists no isomorphism g of E onto a segment of A of the form S_a. If there were, g would then be a strictly increasing mapping of E into E such that $g(a) \in S_a$, in other words such that $g(a) < a$; but this inequality contradicts Lemma 4 (with f as the identity mapping).

6. LEXICOGRAPHIC PRODUCTS

Let $(E_\iota)_{\iota \in I}$ be a family of ordered sets, indexed by a *well-ordered* set I. Consider the product set $E = \prod_{\iota \in I} E_\iota$ and the relation

"$x \in E$ and $y \in E$, and for the least index $\iota \in I$ such that $pr_\iota x \neq pr_\iota y$, we have $pr_\iota x < pr_\iota y$",

which we shall denote by $R\{x, y\}$. It is evident that $R\{x, x\}$ is equivalent to $x \in E$, that $R\{x, y\}$ implies $R\{x, x\}$ and $R\{y, y\}$, and that ($R\{x, y\}$ and $R\{y, x\}$) implies $x = y$. Also it is easily verified that ($R\{x, y\}$ and $R\{y, z\}$) implies $R\{x, z\}$ (consider the least index $\iota \in I$ for which at least two of the three elements $pr_\iota x$, $pr_\iota y$, $pr_\iota z$ are unequal); hence $R\{x, y\}$ is an *order relation on the product set* E. This relation and the ordering it defines are called the *lexicographic order relation* and the *lexicographic ordering* on E (induced by the given orderings on I and on the E_ι); the set E with this ordering is called the *lexicographic product* of the family of ordered sets $(E_\iota)_{\iota \in I}$. If each E_ι is *totally ordered* the lexicographic product is also *totally ordered.*

3. EQUIPOTENT SETS. CARDINALS

1. THE CARDINAL OF A SET

DEFINITION 1. *A set* X *is said to be equipotent to a set* Y *if there exists a bijection of* X *onto* Y. *The relation "*X *is equipotent to* Y*" is denoted by* Eq(X, Y).

The relations Eq(X, Y) and Eq(Y, X) are clearly equivalent, so that the relation Eq(X, Y) is *symmetric*; when it is true, we say that X *and* Y *are equipotent.* Next, Eq(X, X) is true. Finally, the relation Eq(X, Y) is *transitive* since the composition of two bijections is a bijection (Chapter II, § 3, no. 8, Theorem 1); it is therefore an *equivalence relation*, reflexive on every set.

¶ From what has been said it follows that if X and Y are equipotent, the relation

$$(\forall Z)(\mathrm{Eq}(X, Z) \Longleftrightarrow \mathrm{Eq}(Y, Z))$$

is true. Now the axiom scheme S7 (Chapter I, § 5, no. 1) gives us the following axiom :

$$((\forall Z)(\mathrm{Eq}(X, Z) \Longleftrightarrow \mathrm{Eq}(Y, Z)) \Longrightarrow (\tau_Z(\mathrm{Eq}(X, Z)) = \tau_Z(\mathrm{Eq}(Y, Z))).$$

Hence, if X and Y are equipotent, we have

$$\tau_Z(\mathrm{Eq}(X, Z)) = \tau_Z(\mathrm{Eq}(Y, Z)),$$

which justifies the following definition :

DEFINITION 2. *The set* $\tau_Z(\mathrm{Eq}(X, Z))$ *is called the cardinal of* X (or the *power* of X) *and is written* Card(X).

Since Eq(X, X) is true, Card(X) is *equipotent* to X (Chapter I, § 4, Scheme S5). We have therefore proved the following result :

PROPOSITION 1. *Two sets* X *and* Y *are equipotent if and only if their cardinals are equal.*

Examples

(1) Card(∅) is denoted by 0. Since the only set equipotent to ∅ is ∅ (Chapter II, § 3, nos. 1 and 4), we have $0 = \mathrm{Card}(\emptyset) = \emptyset$.

(2) All sets consisting of a single element are equipotent, since $\{(a, b)\}$ is the graph of a bijection of $\{a\}$ onto $\{b\}$; in particular, they are all equipotent to $\{\emptyset\}$. The cardinal $\mathrm{Card}(\{\emptyset\}) = \tau_Z(\mathrm{Eq}(\{\emptyset\}, Z))$ is denoted by 1. (*)

(3) $\mathrm{Card}(\{\emptyset, \{\emptyset\}\})$ is denoted by 2; this is the cardinal of every set consisting of two distinct elements.

*(4) A Hilbert space of countable type is equipotent to the set of real numbers. *

2. ORDER RELATION BETWEEN CARDINALS

The relation "X is equipotent to a subset of Y" is equivalent to "there exists an injection of X into Y"; it is also equivalent to the relation "Card (X) is equipotent to a subset of Card (Y)" (Chapter II, § 3, no. 8, Theorem 1).

(*) The mathematical *term denoted* (Chapter I, § 1, no. 1) by the symbol "1" is of course not to be confused with the *word* "one" in ordinary language. The term denoted by "1" is equal, by virtue of the definition above, to the term denoted by the symbol

$$\begin{aligned}\tau_Z((\exists u)(\exists U)(u = (U, \{\emptyset\}, Z) \text{ and } U \subset \{\emptyset\} \times Z \\ \text{and } (\forall x)((x \in \{\emptyset\}) \Rightarrow (\exists y)((x, y) \in U)) \\ \text{and } (\forall x)(\forall y)(\forall y')(((x, y) \in U \text{ and } (x, y') \in U) \Rightarrow (y = y')) \\ \text{and } (\forall y)((y \in Z) \Rightarrow (\exists x)((x, y) \in U)))).\end{aligned}$$

As a rough estimate, the term so *denoted* is an assembly of several tens of thousands of signs (each of which is one of τ, $\square$, $\vee$, $\neg$, $=$, $\in$, $\supset$).

THEOREM 1. *The relation* $R\{\mathfrak{x}, \mathfrak{y}\}$: "*$\mathfrak{x}$ and $\mathfrak{y}$ are cardinals and $\mathfrak{x}$ is equipotent to a subset of $\mathfrak{y}$" is a well-ordering relation* (§ 2, no. 1).

Since $R\{\mathfrak{x}, \mathfrak{x}\}$ is true for every cardinal $\mathfrak{x}$, what must be proved is that, for every set E of cardinals the relation "$\mathfrak{x} \in E$ and $\mathfrak{y} \in E$ and $R\{\mathfrak{x}, \mathfrak{y}\}$" is a well-ordering relation on E. Consider the set $A = \bigcup_{\mathfrak{x} \in E} \mathfrak{x}$. Every cardinal $\mathfrak{x} \in E$ is then a subset of A. By § 2, Theorem 1 there exists a well-ordering relation on A, which we shall denote by $\mathfrak{x} \leqslant y$, and every subset of A is equipotent to a segment of A (§ 2, no. 5, Theorem 3, Corollary x). For every cardinal $\mathfrak{x} \in E$ consider the set of segments of A which are equipotent to $\mathfrak{x}$; this set of segments is not empty and therefore has a least element (§ 2, no. 1, Proposition 2); let $\varphi(\mathfrak{x})$ denote this least element. The relation

"$\mathfrak{x} \in E$ and $\mathfrak{y} \in E$ and $\mathfrak{x}$ is equipotent to a subset of $\mathfrak{y}$"

is then *equivalent* to

"$\mathfrak{x} \in E$ and $\mathfrak{y} \in E$ and $\varphi(\mathfrak{x}) \subset \varphi(\mathfrak{y})$".

For clearly the second relation implies the first. Conversely, if $\mathfrak{x}$ is equipotent to a subset of $\varphi(\mathfrak{y})$, we cannot have $\varphi(\mathfrak{y}) \subset \varphi(\mathfrak{x})$ and $\varphi(\mathfrak{y}) \neq \varphi(\mathfrak{x})$; otherwise there would exist a segment of $\varphi(\mathfrak{y})$ equipotent to $\mathfrak{X}$ (§ 2, no. 5, Theorem 3, Corollary 3), contrary to the definition of $\varphi(\mathfrak{x})$. Since the set of segments of A is well-ordered by inclusion (§ 2, no. 1, Proposition 2), the theorem follows.

¶ We shall denote the relation $R\{\mathfrak{x}, \mathfrak{y}\}$ by $\mathfrak{x} \leqslant \mathfrak{y}$. A set X is equipotent to a subset of a set Y if and only if Card (X) $\leqslant$ Card (Y).

Clearly we have $0 \leqslant \mathfrak{x}$ for every cardinal $\mathfrak{x}$, and $1 \leqslant \mathfrak{x}$ for every cardinal $\mathfrak{x} \neq 0$.

COROLLARY 1. *Given any two sets, one of them is equipotent to a subset of the other.*

COROLLARY 2. *Two sets each of which is equipotent to a subset of the other are equipotent.*

Remark. Given any set A, there exists a set whose elements are the cardinals Card(X) for all the subsets X of A, namely, the set of objects of the form Card (X) for $X \in \mathfrak{P}(A)$ (Chapter II, § 1, no. 6). For every cardinal $\mathfrak{a}$ the relation "$\mathfrak{x}$ is a cardinal and $\mathfrak{x} \leqslant \mathfrak{a}$" is therefore collectivizing in $\mathfrak{x}$ (Chapter II, § 1, no. 4), because it is equivalent to the relation "$\mathfrak{x}$ is of the form Card (X) for $X \subset \mathfrak{a}$"; the set of all $\mathfrak{x}$ satisfying this relation is called the *set of cardinals* $\leqslant \mathfrak{a}$.

PROPOSITION 2. *For every family* $(\mathfrak{a}_\iota)_{\iota \in I}$ *of cardinals, there exists a unique cardinal* $\mathfrak{b}$ *such that* $\mathfrak{a}_\iota \leqslant \mathfrak{b}$ *for all* $\iota \in I$ *and such that every cardinal* $\mathfrak{c}$ *for which* $\mathfrak{a}_\iota \leqslant \mathfrak{c}$ *for all* $\iota \in I$ *is* $\geqslant \mathfrak{b}$.

There exists a set E containing all the sets $\mathfrak{a}_\iota$ (e.g., the sum of these sets (Chapter II, § 4, no. 8)), whence $\mathfrak{a}_\iota \leqslant \mathfrak{a} = \text{Card (E)}$ for all $\iota \in I$. The set F of cardinals $\leqslant \mathfrak{a}$ is well-ordered and contains all the $\mathfrak{a}_\iota$, and therefore the family $(\mathfrak{a}_\iota)_{\iota \in I}$ has a least upper bound $\mathfrak{b}$ in F. Let $\mathfrak{c}$ be a cardinal $\geqslant \mathfrak{a}_\iota$ for all $\iota \in I$; if $\mathfrak{c} < \mathfrak{b} \leqslant \mathfrak{a}$, then $\mathfrak{c} \in F$, and the inequality $\mathfrak{a}_\iota \leqslant \mathfrak{c}$ contradicts the definition of the least upper bound of the family $(\mathfrak{a}_\iota)$ in the ordered set F; hence the result.

¶ By abuse of language, the cardinal $\mathfrak{b}$ is called the *least upper bound* of the family $(\mathfrak{a}_\iota)_{\iota \in I}$ and is denoted by $\sup_{\iota \in I} \mathfrak{a}_\iota$.

PROPOSITION 3. *Let* X *and* Y *be sets. If there exists a surjection f of* X *onto* Y, *then* Card (Y) $\leqslant$ Card (X).

For there exists a section s associated with f (Chapter II, § 3, no. 8, Proposition 8) and s is an injection of Y into X.

3. OPERATIONS ON CARDINALS

DEFINITION 3. *Let* $(\mathfrak{a}_\iota)_{\iota \in I}$ *be a family of cardinals. The cardinal of the* product (resp. sum) *of the sets* $\mathfrak{a}_\iota$ *is called the* cardinal product (resp. cardinal sum) *of the cardinals* $\mathfrak{a}_\iota$ *and is denoted by* $\mathop{\mathsf{P}}_{\iota \in I} \mathfrak{a}_\iota$ $\left(\text{resp. } \sum_{\iota \in I} \mathfrak{a}_\iota\right)$.

Whenever there is no risk of confusion we shall say simply "product" and "sum" in place of "cardinal product" and "cardinal sum", and we shall write $\prod_{\iota \in I} \mathfrak{a}_\iota$ in place of $\mathop{\mathsf{P}}_{\iota \in I} \mathfrak{a}_\iota$ (cf. Exercise 2).

PROPOSITION 4. *Let* $(E_\iota)_{\iota \in I}$ *be a family of sets,* P *their product, and* S *their sum, and let* $\mathfrak{a}_\iota$ *be the cardinal of* E_ι. *Then the cardinal of* P (resp. S) *is the cardinal product* (resp. *cardinal sum*) *of the family* $(\mathfrak{a}_\iota)_{\iota \in I}$.

For there exists a bijection of P (resp. S) onto the product (resp. sum) of the sets $(\mathfrak{a}_\iota)$ (Chapter II, § 4, no. 8, Proposition 10, and § 5, no. 7, Corollary to Proposition 11).

COROLLARY. *If* $(E_\iota)_{\iota \in I}$ *is any family of sets, the cardinal of the union* $\bigcup_{\iota \in I} E_\iota$ *is at most equal to the sum* $\sum_{\iota \in I} \text{Card}(E_\iota)$.

For there exists a mapping of the sum S of the E_ι onto the union of the E_ι (Chapter II, § 4, no. 8); the Corollary therefore follows from Propositions 3 and 4.

PROPOSITION 5. (a) *Let* $(\mathfrak{a}_\iota)_{\iota \in I}$ *be a family of cardinals, and let* f *be a bijection of a set* K *onto the index set* I. *Then*

$$\sum_{\varkappa \in K} \mathfrak{a}_{f(\varkappa)} = \sum_{\iota \in I} \mathfrak{a}_\iota, \qquad \mathop{\mathsf{P}}_{\varkappa \in K} \mathfrak{a}_{f(\varkappa)} = \mathop{\mathsf{P}}_{\iota \in I} \mathfrak{a}_\iota.$$

(b) *Let* $(\mathfrak{a}_\iota)_{\iota \in I}$ *be a family of cardinals and let* $(J_\lambda)_{\lambda \in L}$ *be a partition of* I. *Then*

$$\sum_{\iota \in I} \mathfrak{a}_\iota = \sum_{\lambda \in L} \Big(\sum_{\iota \in J_\lambda} \mathfrak{a}_\iota \Big), \qquad \mathop{\mathsf{P}}_{\iota \in I} \mathfrak{a}_\iota = \mathop{\mathsf{P}}_{\lambda \in L} \Big(\mathop{\mathsf{P}}_{\iota \in J_\lambda} \mathfrak{a}_\iota \Big)$$

("associativity of the sum and product").

(c) *Let* $((\mathfrak{a}_{\lambda\iota})_{\iota \in J_\lambda})_{\lambda \in L}$ *be a family (indexed by* L*) of families of cardinals, and let* $I = \prod_{\lambda \in L} J_\lambda$. *Then*

$$\mathop{\mathsf{P}}_{\lambda \in L} \Big(\sum_{\iota \in J_\lambda} \mathfrak{a}_{\lambda\iota} \Big) \sum_{f \in I} = \Big(\mathop{\mathsf{P}}_{\lambda \in L} \mathfrak{a}_{\lambda, f(\lambda)} \Big)$$

("distributivity of product over sum").

The relations in (a) follow from the analogous formulae for the union and product of sets, for the fact that f is a bijection implies that if $(X_\iota)_{\iota \in I}$ is a family of mutually disjoint sets, the elements of the family $(X_{f(\varkappa)})_{\varkappa \in K}$ are also mutually disjoint (cf. Chapter II, § 4, no. 1, Proposition 1 and § 5, no. 3, Proposition 4).

¶ The relations in (b) are immediate consequences of the associativity formulae for unions and products (Chapter II, § 4, no. 2, Proposition 2 and § 5, no. 5, Proposition 7) and the distributivity of intersection over union (Chapter II, § 5, no. 6, Proposition 8), which shows that if $(X_\iota)_{\iota \in I}$ is a family of mutually disjoint sets, then the elements of the family

$$\Big(\bigcup_{\iota \in J_\lambda} X_\iota \Big)_{\lambda \in L}$$

are also mutually disjoint.

¶ Finally, (c) follows from the distributivity of the product over union and intersection (Chapter II, § 5, no. 6, Proposition 9 and Corollary 1).

¶ Let $\mathfrak{a}$ and $\mathfrak{b}$ be two cardinals. If I is a set consisting of two distinct elements (e.g., the cardinal 2), there exists a mapping f of I onto $\{\mathfrak{a}, \mathfrak{b}\}$ which defines a family of cardinals. The sum and product of this family depend only on $\mathfrak{a}$ and $\mathfrak{b}$ (by reason of Proposition 5(a)); these cardinals are called respectively the *sum* and the *product* of $\mathfrak{a}$ and $\mathfrak{b}$, and are denoted by $\mathfrak{a}+\mathfrak{b}$ and $\mathfrak{a}\mathfrak{b}$. Similarly for the sum and product of three or more cardinals. Proposition 5 then implies the following corollary:

COROLLARY. *Let $\mathfrak{a}$, $\mathfrak{b}$, $\mathfrak{c}$ be cardinals. Then*

$$(1) \qquad \mathfrak{a}+\mathfrak{b}=\mathfrak{b}+\mathfrak{a}, \qquad \mathfrak{a}\mathfrak{b}=\mathfrak{b}\mathfrak{a};$$

$$(2) \qquad \mathfrak{a}+(\mathfrak{b}+\mathfrak{c})=(\mathfrak{a}+\mathfrak{b})+\mathfrak{c}, \qquad \mathfrak{a}(\mathfrak{b}\mathfrak{c})=(\mathfrak{a}\mathfrak{b})\mathfrak{c};$$

$$(3) \qquad \mathfrak{a}(\mathfrak{b}+\mathfrak{c})=\mathfrak{a}\mathfrak{b}+\mathfrak{a}\mathfrak{c}.$$

4. PROPERTIES OF THE CARDINALS 0 AND 1

PROPOSITION 6. *Let $(\mathfrak{a}_\iota)_{\iota\in I}$ be a family of cardinals, and let* J (resp. K) *be a subset of* I *such that $\mathfrak{a}_\iota=0$ for all $\iota \notin J$* (resp. *$\mathfrak{a}_\iota=1$ for all $\iota\notin K$*). *Then*

$$\sum_{\iota\in I}\mathfrak{a}_\iota=\sum_{\iota\in J}\mathfrak{a}_\iota \qquad \left(\text{resp. } \mathbf{P}_{\iota\in I}\mathfrak{a}_\iota=\mathbf{P}_{\iota\in K}\mathfrak{a}_\iota\right).$$

The proposition is obvious as regards the sum, for the sum S_I of the family of sets $(\mathfrak{a}_\iota)_{\iota\in I}$ is equipotent to the union of the sum S_J of the family $(\mathfrak{a}_\iota)_{\iota\in I}$ and the empty set, and hence equipotent to S_J. The assertion concerning products follows from the fact that the projection pr_K of the product set $\prod_{\iota\in I}\mathfrak{a}_\iota$ onto the partial product $\prod_{\iota\in K}\mathfrak{a}_\iota$ is a bijection (Chapter II, § 5, no. 5, Remark 1).

COROLLARY 1. *For every cardinal $\mathfrak{a}$ we have $\mathfrak{a}+0=\mathfrak{a}.1=\mathfrak{a}$.*

COROLLARY 2. *Let $\mathfrak{a}$ and $\mathfrak{b}$ be cardinals and let* I *be a set equipotent to $\mathfrak{b}$. For each $\iota\in I$ let $\mathfrak{a}_\iota=\mathfrak{a}$, $\mathfrak{c}_\iota=1$. Then*

$$\mathfrak{a}\mathfrak{b}=\sum_{\iota\in I}\mathfrak{a}_\iota, \qquad \mathfrak{b}=\sum_{\iota\in I}\mathfrak{c}_\iota.$$

The second formula is a consequence of the fact that any set is the union of its one-element subsets. The first formula follows from the second by multiplying by $\mathfrak{a}$ and using Corollary 1.

PROPOSITION 7. *Let $(\mathfrak{a}_\iota)_{\iota\in I}$ be a family of cardinals. Then $\mathbf{P}_{\iota\in I}\mathfrak{a}_\iota\neq 0$ if and only if $\mathfrak{a}_\iota\neq 0$ for all $\iota\in I$.*

This is merely a translation of the condition that a product set should be non-empty (Chapter II, § 5, no. 4, Proposition 5, Corollary 2).

PROPOSITION 8. *If $\mathfrak{a}$ and $\mathfrak{b}$ are cardinals such that $\mathfrak{a}+1=\mathfrak{b}+1$, then $\mathfrak{a}=\mathfrak{b}$.*

Let $X = \mathfrak{a} + 1 = \mathfrak{b} + 1$. Then there exist subsets A, B of X with cardinals $\mathfrak{a}$, $\mathfrak{b}$, respectively, such that the complements $X - A$, $X - B$ each consist of a single element. Let u, v be these elements. The intersection $C = A \cap B$ has as complement in X the set $\{u, v\}$. If $u = v$, then $A = B = C$, so that $\mathfrak{a} = \mathfrak{b}$. If $u \neq v$, then $A = C \cup \{v\}$, $B = C \cup \{u\}$, and therefore $\mathfrak{a} = 1 + \text{Card}(C) = \mathfrak{b}$.

The reader should beware of assuming that $\mathfrak{a} + \mathfrak{m} = \mathfrak{b} + \mathfrak{m}$ implies $\mathfrak{a} = \mathfrak{b}$ for all cardinals $\mathfrak{m}$ (cf. § 6); * we shall see later, however, that this implication is true if $\mathfrak{m}$ is *finite* (§ 5, no. 2, Proposition 3, Corollary 4 and § 6, no. 3, Theorem 3, Corollary 4). *

5. EXPONENTIATION OF CARDINALS

Definition 4. *Let $\mathfrak{a}$ and $\mathfrak{b}$ be cardinals. The cardinal of the set of mappings of $\mathfrak{b}$ into $\mathfrak{a}$ is denoted by $\mathfrak{a}^{\mathfrak{b}}$, by abuse of notation.*

The abuse of notation here lies in the the fact that $\mathfrak{a}^{\mathfrak{b}}$ already denotes the set of graphs of mappings of $\mathfrak{b}$ into $\mathfrak{a}$ (Chapter II, § 5, no. 3), and this set is not necessarily a cardinal (Exercise 2). It will always be clear from the context which meaning is to be attached to the symbol $\mathfrak{a}^{\mathfrak{b}}$.

Proposition 9. *Let X and Y be two sets, $\mathfrak{a}$ and $\mathfrak{b}$ their respective cardinals. Then the set X^Y has cardinal $\mathfrak{a}^{\mathfrak{b}}$.*

For there exists a bijection of X^Y onto the set of mappings of $\mathfrak{b}$ into $\mathfrak{a}$ (Chapter II, § 5, no. 2, Proposition 2, Corollary).

Proposition 10. *Let $\mathfrak{a}$ and $\mathfrak{b}$ be cardinals and let I be a set such that* $\text{Card}(I) = \mathfrak{b}$. *If $\mathfrak{a}_\iota = \mathfrak{a}$ for all $\iota \in I$, we have* $\mathfrak{a}^{\mathfrak{b}} = \prod_{\iota \in I} \mathfrak{a}_\iota$.

This follows from the definition of the product of a family of sets as a set of functional graphs (Chapter II, § 5, no. 3).

Corollary 1. *Let $\mathfrak{a}$ be a cardinal and let $(\mathfrak{b}_\iota)_{\iota \in I}$ be a family of cardinals. Then*

$$\mathfrak{a}^{\sum_{\iota \in I} \mathfrak{b}_\iota} = \prod_{\iota \in I} \mathfrak{a}^{\mathfrak{b}_\iota}.$$

Let S be the sum of the sets $\mathfrak{b}_\iota$, and put $\mathfrak{a}_s = \mathfrak{a}$ for all $s \in S$. Both sides of the equality to be proved are then equal to $\prod_{s \in S} \mathfrak{a}_s$, by virtue of Proposition 10 and the associativity formula for products (no. 3, Proposition 5(b)).

COROLLARY 2. *Let* $(\mathfrak{a}_\iota)_{\iota\in I}$ *be a family of cardinals and let* $\mathfrak{b}$ *be a cardinal. Then*

$$\left(\prod_{\iota\in I}\mathfrak{a}_\iota\right)^{\mathfrak{b}} = \prod_{\iota\in I}\mathfrak{a}_\iota^{\mathfrak{b}}.$$

Let $\mathfrak{a}_{\iota\beta} = \mathfrak{a}_\iota$ for each pair $(\iota, \beta) \in I \times \mathfrak{b}$. Then, by associativity of the product, we have

$$\left(\prod_{\iota\in I}\mathfrak{a}_\iota\right)^{\mathfrak{b}} = \prod_{\beta\in\mathfrak{b}}\left(\prod_{\iota\in I}\mathfrak{a}_{\iota\beta}\right) = \prod_{\iota\in I}\left(\prod_{\beta\in\mathfrak{b}}\mathfrak{a}_{\iota\beta}\right) = \prod_{\iota\in I}\mathfrak{a}_\iota^{\mathfrak{b}}.$$

COROLLARY 3. *Let* $\mathfrak{a}, \mathfrak{b}, \mathfrak{c}$ *be cardinals. Then* $\mathfrak{a}^{\mathfrak{b}\mathfrak{c}} = (\mathfrak{a}^{\mathfrak{b}})^{\mathfrak{c}}$.

Let $\mathfrak{b}_\gamma = \mathfrak{b}$ for all $\gamma \in \mathfrak{c}$. Then

$$\mathfrak{a}^{\mathfrak{b}\mathfrak{c}} = \mathfrak{a}^{\sum_{\gamma\in\mathfrak{c}} \mathfrak{b}_\gamma} = \prod_{\gamma\in\mathfrak{c}}\mathfrak{a}^{\mathfrak{b}_\gamma} = (\mathfrak{a}^{\mathfrak{b}})^{\mathfrak{c}}$$

by virtue of Corollary 1.

PROPOSITION 11. *Let* $\mathfrak{a}$ *be a cardinal. Then* $\mathfrak{a}^0 = 1$, $\mathfrak{a}^1 = \mathfrak{a}$, $1^{\mathfrak{a}} = 1$; *and* $0^{\mathfrak{a}} = 0$ *if* $\mathfrak{a} \neq 0$.

For there exists a unique mapping of $\varnothing$ into any given set (namely, the mapping whose graph is the empty set); the set of mappings of a set consisting of a single element into an arbitrary set X is equipotent to X (Chapter II, § 5, no. 3); there exists a unique mapping of an arbitrary set into a set consisting of a single element; and, finally, there is no mapping of a non-empty set into $\varnothing$.

¶ Note in particular that $0^0 = 1$.

PROPOSITION 12. *Let* X *be a set and let* $\mathfrak{a}$ *be its cardinal. Then the cardinal of the set* $\mathfrak{P}(X)$ *of all subsets of* X *is* $2^{\mathfrak{a}}$.

Let α and β be the elements of the cardinal 2. For each subset Y of X let f_Y be the mapping of X into 2 defined by $f_Y(x) = \alpha$ if $x \in Y$ and $f_Y(x) = \beta$ if $x \in X - Y$. Let u be the mapping $Y \to f_Y$ of $\mathfrak{P}(X)$ into 2^X. Conversely, with each mapping g of X into 2 let us associate the subset $\overset{-1}{g}(\alpha)$ of X, and let v be the mapping $g \to \overset{-1}{g}(\alpha)$ of 2^X into $\mathfrak{P}(X)$. It is clear that $u \circ v$ and $v \circ u$ are the identity mappings of 2^X and $\mathfrak{P}(X)$; hence u and v are bijections (Chapter II, § 3, no. 8, Proposition 8, Corollary) and therefore Card $(\mathfrak{P}(X)) = 2^{\mathfrak{a}}$.

6. ORDER RELATION AND OPERATIONS ON CARDINALS

PROPOSITION 13. *Let* $\mathfrak{a}$ *and* $\mathfrak{b}$ *be cardinals. Then* $\mathfrak{a} \geqslant \mathfrak{b}$ *if and only if there exists a cardinal* $\mathfrak{c}$ *such that* $\mathfrak{a} = \mathfrak{b} + \mathfrak{c}$.

For the relation $\mathfrak{a} \geqslant \mathfrak{b}$ means that there exists a subset B of $\mathfrak{a}$ which is equipotent to $\mathfrak{b}$ (no. 2), i.e., $\mathfrak{a}$ is equipotent to the set which is the sum of $\mathfrak{b}$ and a set c.

If $\mathfrak{a} \geqslant \mathfrak{b}$, there usually exist many cardinals $\mathfrak{c}$ such that $\mathfrak{a} = \mathfrak{b} + \mathfrak{c}$ (cf. § 6); in general, therefore, it is not possible to define the "difference" $\mathfrak{a} - \mathfrak{b}$ of two such cardinals (cf. § 5, no. 2).

Proposition 14. *Let $(\mathfrak{a}_\iota)_{\iota \in I}$ and $(\mathfrak{b}_\iota)_{\iota \in I}$ be two families of cardinals, both indexed by the same set* I, *and such that $\mathfrak{a}_\iota \geqslant \mathfrak{b}_\iota$ for all $\iota \in I$. Then*

$$\sum_{\iota \in I} \mathfrak{a}_\iota \geqslant \sum_{\iota \in I} \mathfrak{b}_\iota, \qquad \prod_{\iota \in I} \mathfrak{a}_\iota \geqslant \prod_{\iota \in I} \mathfrak{b}_\iota.$$

The second inequality follows from the inclusion relations between products of sets (Chapter II, § 5, no. 4, Proposition 6, Corollary 3). As to the first inequality, if a set E is the union of a family $(A_\iota)_{\iota \in I}$ of mutually disjoint subsets and if $B_\iota \subset A_\iota$ for all $\iota \in I$, then the B_ι are also mutually disjoint and $\bigcup_\iota B_\iota \subset \bigcup_\iota A_\iota$ (Chapter II, § 4, no. 2).

Corollary 1. *Let $(\mathfrak{a}_\iota)_{\iota \in I}$ be a family of cardinals. For each subset* J *of* I *we have $\sum_{\iota \in J} \mathfrak{a}_\iota \leqslant \sum_{\iota \in I} \mathfrak{a}_\iota$. If also $\mathfrak{a}_\iota \neq 0$ for all $\iota \in I - J$, then $\prod_{\iota \in J} \mathfrak{a}_\iota \leqslant \prod_{\iota \in I} \mathfrak{a}_\iota$.*

Put $\mathfrak{b}_\iota = \mathfrak{a}_\iota$ if $\iota \in J$, and $\mathfrak{b}_\iota = 0$ (resp. $\mathfrak{b}_\iota = 1$) if $\iota \in I - J$. Then apply Proposition 14, observing that the relation $\mathfrak{a} \neq 0$ implies $\mathfrak{a} \geqslant 1$.

Corollary 2. *If $\mathfrak{a}$, $\mathfrak{a}'$, $\mathfrak{b}$, $\mathfrak{b}'$ are cardinals such that $\mathfrak{a} \leqslant \mathfrak{a}'$, $\mathfrak{b} \leqslant \mathfrak{b}'$, and $\mathfrak{a}' > 0$, then $\mathfrak{a}^{\mathfrak{b}} \leqslant \mathfrak{a}'^{\mathfrak{b}'}$.*

For $\mathfrak{a}^{\mathfrak{b}} \leqslant \mathfrak{a}'^{\mathfrak{b}}$ by Propositions 10 and 14, and $\mathfrak{a}'^{\mathfrak{b}} \leqslant \mathfrak{a}'^{\mathfrak{b}'}$ by Proposition 10 and Corollary 1 to Proposition 14.

Theorem 2 (Cantor). *For each cardinal $\mathfrak{a}$, we have $2^{\mathfrak{a}} > \mathfrak{a}$.*

We have $\text{Card}(\mathfrak{P}(\mathfrak{a})) = 2^{\mathfrak{a}}$ (no. 5, Proposition 12). The mapping $x \to \{x\}$ $(x \in \mathfrak{a})$ is an injection of $\mathfrak{a}$ into $\mathfrak{P}(\mathfrak{a})$, whence $\mathfrak{a} \leqslant 2^{\mathfrak{a}}$. Hence it is enough to show that $\mathfrak{a} \neq 2^{\mathfrak{a}}$, i.e., that for every mapping f of $\mathfrak{a}$ into $\mathfrak{P}(\mathfrak{a})$, the image $f(\mathfrak{a})$ is distinct from $\mathfrak{P}(\mathfrak{a})$. Let X be the set of all $x \in \mathfrak{a}$ such that $x \notin f(x)$. If $x \in X$, we have $x \notin f(x)$, whence $f(x) \neq X$; if $x \in \mathfrak{a} - X$, we have $x \in f(x)$ and $x \notin X$, whence $f(x) = X$. This shows that $X \notin f(\mathfrak{a})$ and proves the theorem.

Corollary. *There does not exists a set that has every cardinal as an element.*

If U were such a set, there would exist a set S, the sum of the family of sets $(X)_{X \in U}$, so that every cardinal is equipotent to a subset of S. In particular, let $\mathfrak{S} = \text{Card}(S)$; since $2^{\mathfrak{S}}$ is a cardinal, we would have $2^{\mathfrak{S}} \leqslant \mathfrak{S}$, in contradiction to Theorem 2.

4. NATURAL INTEGERS. FINITE SETS

1. DEFINITION OF INTEGERS

DEFINITION 1. *A cardinal* $\mathfrak{a}$ *is said to be finite if* $\mathfrak{a} \neq \mathfrak{a} + 1$. *A finite cardinal is also called a natural integer* (or simply an *integer* if there is no risk of confusion (*)). *A set* E *is said to be finite if* Card (E) *is a finite cardinal; and* Card (E) *is then called the number of elements of* E.

A family (Chapter II, § 3, no. 4) is said to be *finite* if its index set is finite.

When we say that the number of objects of a certain type is an integer *m*, we mean that these objects are elements of a finite set whose number of elements is *m*. A set whose number of elements is *m* is also called a *set of m elements*.

PROPOSITION 1. *A cardinal* $\mathfrak{a}$ *is finite if and only if* $\mathfrak{a} + 1$ *is finite.*

For the relations $\mathfrak{a} = \mathfrak{b}$ and $\mathfrak{a} + 1 = \mathfrak{b} + 1$ between cardinals $\mathfrak{a}$ and $\mathfrak{b}$ are equivalent (§ 3, no. 4, Proposition 8); the relations $\mathfrak{a} \neq \mathfrak{a} + 1$ and $\mathfrak{a} + 1 \neq (\mathfrak{a} + 1) + 1$ are therefore equivalent.

¶ It is clear that $0 \neq 1$; hence 0 is an integer. It follows that 1 and 2 are integers. The cardinals $2 + 1$ and $(2 + 1) + 1$ are integers, denoted by 3 and 4, respectively.

2. INEQUALITIES BETWEEN INTEGERS

PROPOSITION 2. *Let n be an integer. Then every cardinal* $\mathfrak{a}$ *such that* $\mathfrak{a} \leqslant n$ *is an integer. If* $n \neq 0$, *there exists a unique integer m such that* $n = m + 1$, *and the relation* $\mathfrak{a} < n$ *is equivalent to* $\mathfrak{a} \leqslant m$.

If $\mathfrak{a} \leqslant n$, there exists a cardinal $\mathfrak{b}$ such that $n = \mathfrak{a} + \mathfrak{b}$ (§ 3, no. 6, Proposition 13). Then $(\mathfrak{a} + 1) + \mathfrak{b} = (\mathfrak{a} + \mathfrak{b}) + 1 = n + 1$ (§ 3, no. 3, Proposition 5, Corollary); and since $n \neq n + 1$, we have

$$(\mathfrak{a} + 1) + \mathfrak{b} \neq \mathfrak{a} + \mathfrak{b}.$$

Hence $\mathfrak{a} + 1 \neq \mathfrak{a}$, which means that $\mathfrak{a}$ is an integer. If $n \neq 0$, we have $n \geqslant 1$ (§ 3, no. 2), and therefore there exists a unique cardinal *m* such

(*) The notion of "integer" will be generalized later, in Algebra, where we shall define the *rational integers* and the *algebraic integers.*

that $n = m + 1$ (§ 3, no. 6, Proposition 13 and no. 4, Proposition 8). Since $m \leqslant n$, m is an integer, from what has already been proved. Finally, if an integer a is such that $a < n$, we have $n = a + b$, with $b \neq 0$ (§ 3, no. 6, Proposition 13); since b is an integer, we have $b = c + 1$ and $n = m + 1 = (a + c) + 1$. It follows that $m = a + c$ (§ 3, no. 4, Proposition 8), hence $a \leqslant m$. Conversely, if $a \leqslant m$, we have

$$a \leqslant m + 1 = n;$$

and if $a = n = m + 1$, we would have $a > m$, contrary to hypothesis.

COROLLARY 1. *Every subset of a finite set is finite.*

COROLLARY 2. *If* X *is a subset of a finite set* E *and* $X \neq E$, *then*

$$\text{Card}(X) < \text{Card}(E).$$

For X is contained in the complement X′ of a subset of E consisting of a single element; we have $\text{Card}(X) \leqslant \text{Card}(X')$ and $\text{Card}(E) = \text{Card}(X') + 1$, hence (Proposition 2) $\text{Card}(X') < \text{Card}(E)$ and *a fortiori* $\text{Card}(X) < \text{Card}(E)$.

Definition 1 shows that, conversely, if E is a set such that

$$\text{Card}(X) < \text{Card}(E)$$

for every subset X of E such that $X \neq E$, then E is finite.

COROLLARY 3. *If f is a mapping of a finite set* E *into a set* F, *then* $f(E)$ *is a finite subset of* F.

For $\text{Card}(f(E)) \leqslant \text{Card}(E)$ (§ 3, no. 2, Proposition 3).

COROLLARY 4. *Let* E *and* F *be two finite sets with the same number of elements, and let f be a mapping of* E *into* F. *Then the following statements are equivalent:*

(a) *f is an injection;*
(b) *f is a surjection;*
(c) *f is a bijection.*

It is enough to prove that (a) and (b) are equivalent. If f is injective, then $\text{Card}(f(E)) = \text{Card}(E) = \text{Card}(F)$, whence $f(E) = F$ (Corollary 2). If f is not injective, let x and x' be two elements of E such that $x \neq x'$ and $f(x) = f(x')$. Then, putting $E' = E - \{x\}$, we have $f(E') = f(E)$, whence $\text{Card}(f(E)) \leqslant \text{Card}(E') < \text{Card}(E)$ by virtue of Corollary 2; but since $\text{Card}(F) = \text{Card}(E)$, it follows that $f(E) \neq F$.

3. THE PRINCIPLE OF INDUCTION

C61 (Principle of Induction). *Let* $R\{n\}$ *be a relation in a theory* $\mathcal{T}$ *(where n is not a constant of* $\mathcal{T}$*). Suppose that the relation*

$$R\{0\} \text{ and } (\forall n)((n \text{ is an integer and } R\{n\}) \Longrightarrow R\{n+1\})$$

is a theorem in $\mathcal{T}$*. Under these conditions the relation*

$$(\forall n)((n \text{ is an integer}) \Longrightarrow R\{n\})$$

is a theorem in $\mathcal{T}$.

We shall argue by contradiction. Suppose that the relation

$$(\exists n)(n \text{ is an integer and } (\text{not } R\{n\}))$$

is true. Let q be an integer such that "not $R\{q\}$" (method of the auxiliary constant; cf. Chapter I, § 3, no. 3 and § 4, no. 1). The integers n for which "$n \leqslant q$ and (not $R\{n\}$)" form a well-ordered non-empty set (§ 3, no. 2, Remark), which therefore has a least element s. If $s = 0$, then "not $R\{0\}$", contrary to hypothesis. If $s > 0$, then $s = s' + 1$, where s' is an integer such that $s' < s$ (no. 2, Proposition 2). By definition of s, we have $R\{s'\}$, but then the hypothesis implies that $R\{s\}$ is true, contrary to the definition of s.

In order to apply the principle of induction it is necessary in particular to prove the relation

$$(n \text{ is an integer and } R\{n\}) \Longrightarrow R\{n+1\}.$$

For this purpose the method of the auxiliary hypothesis (Chapter I, § 3, no. 3) is commonly used, and it is for this reason that the relation "n is an integer and $R\{n\}$" (or even $R\{n\}$) is called the *inductive hypothesis*.

Remark. There are various criteria which are frequently used under the name of "principle of induction". They can all be easily deduced from C61, and we indicate here the most important of them :

(1) Let $S\{n\}$ be the relation

$$(\forall p)((n \text{ is an integer and } p \text{ is an integer and } p < n) \Longrightarrow R\{p\})$$

and suppose that $S\{n\}$ *implies* $R\{n\}$. Then the relation

$$(\forall n)((n \text{ is an integer}) \Longrightarrow R\{n\})$$

is true. For the relation S$\{0\}$ is true, and by hypothesis S$\{n\}$ implies R$\{n\}$; since the relation $m < n + 1$ is equivalent to $m \leqslant n$ (no. 2, Proposition 2), the relation S$\{n+1\}$ is equivalent to "S$\{n\}$ and R$\{n\}$", consequently, S$\{n\}$ implies S$\{n+1\}$. The criterion C61 therefore proves that the relation

$$(\forall n)\,((n \text{ is an integer}) \Longrightarrow \mathrm{S}\{n\})$$

is true, and since S$\{n\}$ implies R$\{n\}$, the relation

$$(\forall n)\,((n \text{ is an integer}) \Longrightarrow \mathrm{R}\{n\})$$

is true.

(2) Let k be an integer and let R$\{n\}$ be a relation such that the relation

$$\mathrm{R}\{k\} \text{ and } (\forall n)\,((n \text{ is an integer} \geqslant k \text{ and } \mathrm{R}\{n\}) \Longrightarrow \mathrm{R}\{n+1\})$$

is true. Then the relation

$$(\forall n)\,((n \text{ is an integer} \geqslant k) \Longrightarrow \mathrm{R}\{n\})$$

is true (*"induction starting at k"*). For let S$\{n\}$ be the relation

$$(n \geqslant k) \Longrightarrow \mathrm{R}\{n\}.$$

Then by the method of disjunction of cases we see that S$\{0\}$ is true. On the other hand, it is easily verified that the relation

$$(n \text{ is an integer and } \mathrm{S}\{n\}) \Longrightarrow \mathrm{S}\{n+1\}$$

is true. It follows from C61 that the relation

$$(n \text{ is an integer}) \Longrightarrow \mathrm{S}\{n\}$$

is true, which proves our assertion.

(3) Let a and b be two integers such that $a \leqslant b$, and let R$\{n\}$ be a relation such that

$$\mathrm{R}\{a\} \text{ and } (\forall n)((n \text{ is an integer and } a \leqslant n < b \text{ and } \mathrm{R}\{n\}) \Longrightarrow \mathrm{R}\{n+1\}).$$

Then the relation

$$(\forall n)((n \text{ is an integer and } a \leqslant n \leqslant b) \Longrightarrow \mathrm{R}\{n\})$$

is true. The proof is similar to that of the preceding case; we take S$\{n\}$ to be the relation "$(a \leqslant n < b) \Longrightarrow \mathrm{R}\{n\}$" (*"induction restricted to an interval"*).

(4) Let a, b be two integers such that $a \leqslant b$, and let $\mathrm{R}\{n\}$ be a relation such that

$$\mathrm{R}\{b\} \text{ and } (\forall n)((n \text{ is an integer and } a \leqslant n < b \text{ and } \mathrm{R}\{n+1\}) \Longrightarrow \mathrm{R}\{n\}).$$

Then the relation

$$(\forall n)((n \text{ is an integer and } a \leqslant n \leqslant b) \Longrightarrow \mathrm{R}\{n\})$$

is true. For we have the relation

$$(n \text{ is an integer and } a \leqslant n < b \text{ and } (\text{not } \mathrm{R}\{n\})) \Longrightarrow \text{not } \mathrm{R}\{n+1\}.$$

If for some n such that $a \leqslant n \leqslant b$ we had (not $\mathrm{R}\{n\}$), it would follow from (3) that (not $\mathrm{R}\{b\}$), contrary to the hypothesis, whence the result ("*descending induction*").

4. FINITE SUBSETS OF ORDERED SETS

PROPOSITION 3. *Let* E *be a right directed preordered set* (resp. *a lattice*, resp. *a totally ordered set*). *Then every non-empty finite subset of* E *is bounded above* (resp. *has a least upper bound and a greatest lower bound*, resp. *has a greatest element and a least element*).

The proof is by induction on the number n of elements of the subset under consideration. The result is trivial for $n = 1$. Let X be a subset of $n + 1$ elements of E (with $n \geqslant 1$), and put $\mathrm{X} = \mathrm{Y} \cup \{x\}$, where Y has n elements and is therefore not empty. The inductive hypothesis implies the existence of an upper bound (resp. least upper bound, resp. greatest element) y of Y. Since E is right directed (resp. a lattice, resp. totally ordered), $\{x, y\}$ has an upper bound (resp. least upper bound, resp. greatest element) which is evidently an upper bound (resp. least upper bound, resp. greatest element) of X.

COROLLARY 1. *Every totally ordered finite set is well-ordered and has a greatest element.*

COROLLARY 2. *Every finite ordered set has a maximal element.*

For such a set is inductive by Corollary 1 (cf. § 2, no. 4, Theorem 2).

5. PROPERTIES OF FINITE CHARACTER

DEFINITION 2. *Let* E *be a set. A set* $\mathfrak{S}$ *of subsets of* E *is said to be of finite character if the relation* $\mathrm{X} \in \mathfrak{S}$ *is equivalent to the relation "every finite subset of* X *belongs to* $\mathfrak{S}$".

A property $P\{X\}$ of a subset X of a set E is said to be *of finite character* if the set of subsets X of E for which $P\{X\}$ is true is of finite character.

Examples

(1) The set of totally ordered subsets of an ordered set E is of finite character. Indeed, a subset X of E is totally ordered if and only if every subset of X consisting of two elements is totally ordered.

* (2) The set of all free subsets of a module is of finite character. The same is true of the set of all algebraically free subsets of an extension of a field.

(3) The set of submodules of a module E is not of finite character, because a finite subset of a submodule of E is not necessarily a submodule of E. *

THEOREM 1. *Every set $\mathfrak{S}$ of subsets of a set* E *which is of finite character has a maximal element (when ordered by inclusion).*

By Theorem 2 of § 2, no. 4 it is enough to show that $\mathfrak{S}$ is inductive. To do this we shall show that if $\mathfrak{G}$ is any subset of $\mathfrak{S}$ which is totally ordered by inclusion, the union X of the sets of $\mathfrak{G}$ belongs to $\mathfrak{S}$ (§ 2, no. 4, Theorem 2, Corollary 2). Since $\mathfrak{S}$ is of finite character, it is enough to show that every finite subset Y of X belongs to $\mathfrak{S}$. Now, for each $y \in Y$, there exists a set $Z_y \in \mathfrak{G}$ such that $y \in Z_y$. Since the set of sets Z_y $(y \in Y)$ is finite and totally ordered by inclusion, it has a greatest element S (no. 4, Corollary 1 to Proposition 3); in other words, there exists a set $S \in \mathfrak{G}$ such that $Y \subset S$. But since $S \in \mathfrak{S}$ and since Y is a finite subset of S, we have $Y \in \mathfrak{S}$ because $\mathfrak{S}$ is of finite character; this completes the proof.

5. PROPERTIES OF INTEGERS

1. OPERATIONS ON INTEGERS AND FINITE SETS

PROPOSITION 1. *Let $(a_i)_{i \in I}$ be a finite family of integers. Then the cardinals $\sum_{i \in I} a_i$ and $\prod_{i \in I} a_i$ are integers.*

Let us begin by showing that if a and b are integers, then so is $a + b$. We proceed by induction on b. The assertion is true for $b = 0$ since $a + 0 = a$. If $a + b$ is an integer, then so is $(a + b) + 1$ (§ 4, no. 1, Proposition 1). But $(a + b) + 1 = a + (b + 1)$ (§ 3, no. 3, Corollary to Proposition 5); hence $a + (b + 1)$ is an integer, and consequently $a + b$ is an integer for all integers b.

Next we show by induction on $n = \text{Card}(I)$ that $\sum_{i \in I} a_i$ is an integer. This is clear if $n = 0$, for then $I = \emptyset$ and $\sum_{i \in I} a_i = 0$. If $\text{Card}(I) = n + 1$, we may write $I = J \cup \{k\}$, where $\text{Card}(J) = n$ and $k \notin J$. Then

$$\sum_{i \in I} a_i = a_k + \sum_{i \in J} a_i$$

(§ 3, no. 3, Proposition 5). The inductive hypothesis is that $\sum_{i \in J} a_i$ is an integer; hence so is $a_k + \sum_{i \in J} a_i$, from the first paragraph of the proof. This shows that $\sum_{i \in I} a_i$ is an integer for all n.

Since the product ab of two integers a and b is the sum of a finite family of integers all equal to a (§ 3, no. 4, Proposition 6, Corollary 2), ab is an integer. We shall prove by induction on $n = \text{Card}(I)$ that $\prod_{i \in I} a_i$ is an integer. This is true for $n = 0$, because then

$$\prod_{i \in I} a_i = 1.$$

If $\text{Card}(I) = n + 1$, we have (with the same notation as above)

$$\prod_{i \in I} a_i = a_k \cdot \prod_{i \in J} a_i$$

(§ 3, no. 3, Proposition 5), and the inductive hypothesis therefore implies that $\prod_{i \in I} a_i$ is an integer. Consequently $\prod_{i \in I} a_i$ is an integer for all n.

Corollary 1. *The union* E *of a finite family* $(X_i)_{i \in I}$ *of finite sets is a finite set.*

For the sum S of the family (X_i) is finite; and since there exists a mapping of S onto E (Chapter II, § 4, no. 8), the set E is finite (§ 4, no. 2, Proposition 2, Corollary 3).

Corollary 2. *The product of a finite family of finite sets is a finite set.*

Corollary 3. *If a and b are integers, a^b is an integer.*

For a^b is the product of a finite family of integers all equal to a (§ 3, no. 5, Proposition 10).

Corollary 4. *The set of subsets of a finite set* E *is finite.*

For its cardinal is $2^{\text{Card}(E)}$ (§ 3, no. 5, Proposition 12).

2. STRICT INEQUALITIES BETWEEN INTEGERS

PROPOSITION 2. *Let a and b be two integers. Then $a < b$ if and only if there exists an integer $c > 0$ such that $b = a + c$.*

If $a < b$, there exists a cardinal $c \leqslant b$ (so that c is an integer (§ 4, no. 2, Proposition 2)) such that $b = a + c$ (§ 3, no. 6, Proposition 13); if $a \neq b$, we must have $c \neq 0$. Conversely, if $b = a + c$ and $c \neq 0$, then $c \geqslant 1$ and hence $a < a + 1 \leqslant a + c = b$.

PROPOSITION 3. *Let $(a_i)_{i \in I}$ and $(b)_{i \in I}$ be two finite families of integers such that $a_i \leqslant b_i$ for each $i \in I$ and $a_i < b_i$ for at least one index i. Then*

$$\sum_{i \in I} a_i < \sum_{i \in I} b_i.$$

If also $b_i > 0$ for each $i \in I$, then

$$\prod_{i \in I} a_i < \prod_{i \in I} b_i.$$

Let j be an index such that $a_j < b_j$, and let $J = I - \{j\}$. Then

$$b_j = a_j + c_j$$

with $c_j > 0$ (Proposition 2), and therefore (§ 3, no. 6, Proposition 14)

$$\sum_{i \in I} b_i = a_j + c_j + \sum_{i \in J} b_i \geqslant c_j + a_j + \sum_{i \in J} a_i = c_j + \sum_{i \in I} a_i.$$

Since $c_j > 0$, the first part of the Proposition follows from Proposition 2. Likewise

$$\prod_{i \in I} b_i = (a_j + c_j) \prod_{i \in J} b_i = a_j . \prod_{i \in J} b_i + c_j . \prod_{i \in J} b_i \geqslant \prod_{i \in I} a_i + c_j . \prod_{i \in J} b_i.$$

Since c_j and all the b_i are $\neq 0$, the product $c_j . \prod_{i \in J} b_i$ is $\neq 0$ (§ 3, no. 4, Proposition 7); hence the second part of the Proposition.

COROLLARY 1. *Let a, a', and b be integers such that $a < a'$ and $b > 0$. Then $a^b < a'^b$.*

We need only express a^b and a'^b as products of finite families of integers (§ 3, no. 5, Proposition 10) and apply Proposition 3, observing that the relation $a < a'$ implies $a' > 0$.

COROLLARY 2. *Let* a, b, *and* b' *be integers such that* $a > 1$ *and* $b < b'$; *then* $a^b < a^{b'}$.

For there exists an integer $c > 0$ such that $b' = b + c$ (Proposition 2); since $c \geqslant 1$, we have $a^c \geqslant a > 1$, whence $a^{b'} = a^b a^c > a^b$.

COROLLARY 3. *Let* a, b, b' *be integers* (resp. *integers such that* $a > 0$). *Then* $a + b = a + b'$ (resp. $ab = ab'$) *if and only if* $b = b'$.

COROLLARY 4. *If* a *and* b *are integers such that* $a \leqslant b$, *then there exists a unique integer* c *such that* $b = a + c$.

The existence of c follows from Proposition 13 of § 3, no. 6, and its uniqueness from Corollary 3 above.

¶ The integer c such that $b = a + c$ (where $a \leqslant b$) is called the *difference* of the integers b and a, and is written $b - a$. It is easily verified that if a, b, a', b' are integers such that $a \leqslant b$ and $a' \leqslant b'$, then

$$(b - a) + (b' - a') = (b + b') - (a + a').$$

3. INTERVALS IN SETS OF INTEGERS

Every set of integers, being a set of cardinals, is well-ordered (§ 3, no. 2, Theorem 1). Furthermore, for each integer a the relation "x is a cardinal and $x \leqslant a$" is collectivizing in x (§ 3, no. 2, Remark after Theorem 1), and the set of x which satisfy this relation is a set of integers (§ 4, no. 2, Proposition 2), which may therefore be denoted by $[0, a]$.

PROPOSITION 4. *Let* a *and* b *be integers. Then the mapping* $x \rightarrow a + x$ *is a strictly inrceasing isomorphism of the interval* $[0, b]$ *onto the interval* $[a, a + b]$, *and* $y \rightarrow y - a$ *is the inverse isomorphism.*

Clearly the relations $0 \leqslant x \leqslant b$ imply $a \leqslant a + x \leqslant a + b$. The mapping $x \rightarrow a + x$ is strictly increasing (and therefore injective) by Proposition 3 of no. 2. Finally, the relations $a \leqslant y \leqslant a + b$ imply $y = a + x$ with $x \geqslant 0$ and $a + x \leqslant a + b$, whence $x \leqslant b$ (no. 2, Proposition 3). This completes the proof.

PROPOSITION 5. *If* a *and* b *are integers such that* $a \leqslant b$, *then the interval* $[a, b]$ *is a finite set whose number of elements is* $b - a + 1$.

By Proposition 4 we may restrict ouselves to the case where $a = 0$. The proof is by induction on b. If $b = 0$, the result is clear. The relation $0 \leqslant x \leqslant b + 1$ is equivalent to "$0 \leqslant x < b + 1$ or $x = b + 1$", and the relation $0 \leqslant x < b + 1$ is equivalent to $0 \leqslant x \leqslant b$ (§ 4, no. 2, Proposition 2); in other words, the interval $[0, b + 1]$ is the union of $[0, b]$

and $\{b+1\}$, and these two sets are disjoint. By the inductive hypothesis the number of elements of $[0, b+1]$ is equal to $(b+1)+1$, and the Proposition is proved.

PROPOSITION 6. *For every finite set* E *which is totally ordered and has* n *elements* ($n \geqslant 1$), *there exists a unique isomorphism of* E *onto the interval* $[1, n]$.

Since E and $[1, n]$ are well-ordered (§ 4, no. 4, Proposition 3, Corollary 1) and have the same number of elements (Proposition 5), the result follows from Theorem 3 of § 2, no. 5 and Corollary 2 to Proposition 2 of § 4, no. 2.

4. FINITE SEQUENCES

A *finite sequence* (resp. *finite sequence of elements of a set* E) is a family (resp. a family of elements of E) whose index set I is a finite set of integers. The number of elements of I is called the *length* of the sequence.

Let $(t_i)_{i \in I}$ be a finite sequence of length n. By Proposition 6 of no. 3 there exists a unique isomorphism f of the interval $[1, n]$ onto the set of integers I. For each $k \in [1, n]$, $t_{f(k)}$ is said to be the *kth term of the sequence*; $t_{f(1)}$ (resp. $t_{f(n)}$) is the *first* (resp. *last*) *term* of the sequence.

Let $P\{i\}$ be a relation such that the elements i for which $P\{i\}$ is true form a finite set of integers. A finite sequence $(t_i)_{i \in I}$ is then often written $(t_i)_{P\{i\}}$. For example, when $I = [a, b]$, the notation $(t_i)_{a \leqslant i \leqslant b}$ is often used. Under the same conditions, to denote the product of a family of sets $(X_i)_{i \in I}$ the notations

$$\prod_{P\{i\}} X_i \quad \text{and} \quad \prod_{i=a}^{b} X_i$$

are used; and analogous notations for union, intersection, cardinal product, cardinal sum, * composition laws in Algebra *, and so on.

5. CHARACTERISTIC FUNCTIONS OF SETS

Let E be a non-empty set and A a subset of E. The *characteristic function* of the subset A of E is the mapping φ_A of E into the set $\{0, 1\}$ defined by

$$\varphi_A(x) = 1 \quad \text{if} \quad x \in A; \qquad \varphi_A(x) = 0 \quad \text{if} \quad x \in E - A.$$

¶ Clearly the relation $\varphi_A = \varphi_B$ is equivalent to $A = B$. We have $\varphi_E(x) = 1$ for all $x \in E$ and $\varphi_\varnothing(x) = 0$ for all $x \in E$; these are the only constant characteristic functions on E. The following Proposition is an immediate consequence of the definitions:

PROPOSITION 7. *For each pair of subsets* A, B *of a non-empty set* E, *we have*

$$\varphi_{E-A}(x) = 1 - \varphi_A(x), \tag{1}$$

$$\varphi_{A \cap B}(x) = \varphi_A(x)\varphi_B(x), \tag{2}$$

$$\varphi_{A \cup B}(x) + \varphi_{A \cap B}(x) = \varphi_A(x) + \varphi_B(x) \tag{3}$$

for all $x \in E$.

6. EUCLIDEAN DIVISION

THEOREM 1. *Let a and b be integers such that $b > 0$. Then there exist integers q and r such that $a = bq + r$ and $r < b$, and the integers q and r are uniquely determined by these conditions.*

The conditions on q and r are equivalent to $bq \leqslant a < b(q+1)$ and $r = a - bq$ (no. 2, Proposition 2). Hence we have to find q such that $bq \leqslant a < b(q+1)$; in other words, q must be the smallest integer such that $a < b(q+1)$, which shows that q and $r = a - bq$ are uniquely determined. To prove their existence, we note that there exist integers p such that $a < bp$, for example $a + 1$ (since $b > 0$). Let m be the least of these integers. We have $m \neq 0$, and we may therefore write $m = q + 1$ with $q \leqslant m$ (§ 4, no. 2, Proposition 2); it follows that $bq \leqslant a < b(q+1)$.

DEFINITION 1. *With the notation of Theorem 1, r is called the remainder of the division of a by b. If $r = 0$, we say that a is a multiple of b, or that a is divisible by b, or that b is a divisor of a, or that b divides a, or that b is a factor of a. The number q is then called the quotient of a by b and is denoted by $\frac{a}{b}$ or a/b.*

If a is not a multiple of b, the number q is called the *integral part of the quotient of a by b* (cf. *General Topology*, Chapter IV, § 8, no. 2).

In this chapter, writing a/b or $\frac{a}{b}$ will imply that b divides a.

The relations $a = bq$ and $q = a/b$ are equivalent (if $b > 0$). Every multiple a' of a multiple a of b is a multiple of b, and

$$a'/b = (a'/a)(a/b) \quad \text{if} \quad a \neq 0.$$

Also, if c and d are multiples of b, then $c + d$ and $c - d$ (if $d \leqslant c$) are multiples of b, and we have

$$\frac{c+d}{b} = \frac{c}{b} + \frac{d}{b}, \quad \frac{c-d}{b} = \frac{c}{b} - \frac{d}{b}.$$

The integers which are multiples of 2 are said to be *even*, and the others *odd*. By Theorem 1 the odd integers are of the form $2n + 1$.

7. EXPANSION TO BASE b

PROPOSITION 8. *Let b be an integer > 1. For each integer $k > 0$ let E_k be the lexicographic product* (§ 2, no. 6) *of the family $(J_h)_{0 \leqslant h \leqslant k-1}$ of intervals all identical with $[0, b-1]$. For each $r = (r_0, r_1, \ldots, r_{k-1}) \in E_k$, let $f_k(r) = \sum_{h=0}^{k-1} r_h b^{k-h-1}$; then the mapping f_k is an isomorphism of the ordered set E_k onto the interval $[0, b^k - 1]$.*

The proof is by induction on k. For $k = 1$ it is an immediate consequence of the definitions. For each $r = (r_0, \ldots, r_{k-1}, r_k) \in E_{k+1}$, put

$$\varphi(r) = (r_0, \ldots, r_{k-1}) \in E_k.$$

Then the mapping $r \to (\varphi(r), r_k)$ is an isomorphism of E_{k+1} onto the lexicographical product of E_k and $J = [0, b-1]$; this is immediate from the definitions. We may write

$$f_{k+1}(r) = b \cdot f_k(\varphi(r)) + r_k;$$

let us show that the relation $r < r'$ in E_{k+1} implies $f_{k+1}(r) < f_{k+1}(r')$. Indeed, we have either $\varphi(r) < \varphi(r')$, or else $\varphi(r) = \varphi(r')$ and $r_k < r'_k$. In the first case, the inductive hypothesis implies that $f_k(\varphi(r)) < f_k(\varphi(r'))$, and therefore (§ 4, no. 2, Proposition 2) $f_k(q(r')) \geqslant f_k(\varphi(r)) + 1$; consequently

$$f_{k-1}(r') \geqslant b \cdot f_k(\varphi(r)) + b > f_{k+1}(r),$$

since $r_k \leqslant b - 1$ (no. 2, Proposition 3). If, on the other hand, $\varphi(r) = \varphi(r')$ and $r_k < r'_k$, it is clear that $f_{k+1}(r) < f_{k+1}(r')$. Now the inductive hypothesis shows that $f_k(\varphi(r)) \leqslant b^k - 1$, whence

$$f_{k+1}(r) \leqslant b(b^k - 1) + b - 1 = b^{k+1} - 1.$$

It follows that f_{k+1} is an isomorphism of E_{k+1} onto a subset of the interval $[0, b^{k+1} - 1]$; but this interval and E_{k+1} have the same number of elements, namely b^{k+1} (no. 3, Proposition 5); therefore f_{k+1} is a bijection (§ 4, no. 2, Proposition 2, Corollary 4), and the proof is complete.

¶ We note now that for every integer a we have $a < b^a$. This is proved by induction on a; the result is evident for $a = 0$, and the hypothesis $a < b^a$ implies $a + 1 \leqslant b^a < b.b^a = b^{a+1}$ (no. 2, Proposition 3 and § 4, no. 2, Proposition 2). There is therefore a least integer k such that $a < b^k$, and Proposition 8 then shows that there exists a unique finite sequence $(r_h)_{0 \leqslant h \leqslant k-1}$ such that $0 \leqslant r_h \leqslant b - 1$ for $0 \leqslant h \leqslant k - 1$ and

$$a = \sum_{h=0}^{k-1} r_h b^{k-h-1}.$$

Furthermore, we must have $r_0 > 0$, for otherwise $a < b^{k-1}$ by virtue of Proposition 8. The expression

$$\sum_{h=0}^{k-1} r_h b^{k-h-1}$$

is called the *expansion to base* b of the integer a.

* In all parts of mathematics which do not involve numerical computations, Proposition 8 is useful mainly when applied to a *prime* number b. *

When the integer b is small enough for this to be practicable, we may represent each integer $< b$ by a distinctive symbol called a *digit*. The digits which represent 0 and 1 are usually 0 and 1. Let a be an integer and let $\sum_{h=0}^{k-1} r_h b^{k-h-1}$ be its expansion to base b. If the integer k which appears in this expansion is sufficiently small for this to be practicable, it is usual to associate with the integer a the succession of symbols obtained by writing $r_0 r_1 \ldots r_{k-2} r_{k-1}$ from left to right and then replacing each integer r_i by the digit which represents it; the symbol so obtained is called the *numerical symbol* associated with a. One then often replaces a by its numerical symbol in the terms and relations in which it appears.

For example, if C, Q, F, D are digits, the numerical symbols CQ, CQF, CQFD are respectively associated with $\mathrm{C}b + \mathrm{Q}$, $\mathrm{C}b^2 + \mathrm{Q}b + \mathrm{F}$, $\mathrm{C}b^3 + \mathrm{Q}b^2 + \mathrm{F}b + \mathrm{D}$.

It follows from Proposition 8 that the numerical symbol associated with an integer a is unique, and that if $a < b^k$ it contains at most k digits. Notice that the numerical symbol associated with the integer b^k consists of the digit 1 followed by k digits 0.

This system of representation of integers by numerical symbols is called the *system of numeration to base* b. In practical numerical computations, the following systems are used: (*a*) the system to base 2, or the *dyadic system*, in which the digits are 0 and 1; (*b*) the *decimal system*, in which the digits are $0, 1, 2, 3, 4, 5 = 4 + 1, 6 = 5 + 1, 7 = 6 + 1, 8 = 7 + 1, 9 = 8 + 1$, and in which b is the integer $9 + 1$ (whose numerical symbol in this system is therefore 10).

Since the Middle Ages the decimal system has been traditionally used in numerical calculations, and we shall use it in this series whenever we have occasion to write down an integer explicitly. We refer the reader to the part of this series devoted to numerical calculation for an account of methods for obtaining the numerical symbols associated with the sum, difference, product, and integral part of the quotient of two integers given by their numerical symbols.

8. COMBINATORIAL ANALYSIS

PROPOSITION 9. *Let* E *and* F *be two sets,* $\mathfrak{a}$ *and* $\mathfrak{b}$ *their cardinals,* f *a surjection of* E *onto* F *such that the sets* $\overset{-1}{f}(y)$, *where* $y \in \mathrm{F}$, *all have the same cardinal* $\mathfrak{c}$. *Then* $\mathfrak{a} = \mathfrak{bc}$.

For the family $(\overset{-1}{f}(y))_{y \in \mathrm{F}}$ is a partition of E, each element of which partition is a set of cardinal $\mathfrak{c}$; hence the result (§ 3, no. 4, Proposition 6, Corollary 2).

DEFINITION 2. *Let* n *be an integer. The product* $\prod_{i<n} (i + 1)$ *is denoted by* $n!$ (read "factorial n").

We have $0! = 1$ (Chapter II, § 5, no. 3) and $1! = 1$. For each integer n we have $(n + 1)! = n!\,(n+1)$. This relation, together with the relation $0! = 1$, characterizes the term $n!$, as is easily seen by induction on n.

PROPOSITION 10. *Let* m *and* n *be integers such that* $m \leqslant n$. *Then* $n!/(n - m)!$ *is the number of injective mappings of a set* A *with* m *elements into a set* B *with* n *elements.*

The proof is by induction on the number $m \leqslant n$ of elements of A. If $m = 0$, the result is evident. Suppose that $m + 1 \leqslant n$. Let A be a set with $m + 1$ elements, let A′ be a subset of A with m elements, and let $\{a\} = \mathrm{A} - \mathrm{A}'$. Let F, F′ be the sets of injective mappings of A, A′ respectively into B, and let φ be the mapping $f \to f|\mathrm{A}'$ which maps each function $f \in \mathrm{F}$ to its restriction to A′. For each $f' \in \mathrm{F}'$ an element f of $\overset{-1}{\varphi}(f')$ is uniquely determined by its value $f(a)$; since f is injective, we must have $f(a) \in \mathrm{B} - f'(\mathrm{A}')$. It follows that $\overset{-1}{\varphi}(f')$ has the same number $n - m$ of elements as $\mathrm{B} - f'(\mathrm{A}')$. Hence, by Proposition 9, F has

$$(n - m)\frac{n!}{(n - m)!} = \frac{n!}{(n - m - 1)!}$$

elements by virtue of the inductive hypothesis. This completes the proof.

COROLLARY. *The number of permutations of a finite set with n elements is equal to* $n!$.

For this number is equal to the number of injections of the set into itself (§ 4, no. 2, Proposition 2, Corollary 4).

PROPOSITION 11. *Let* E *be a finite set with n elements, and let* $(p_i)_{1 \leqslant i \leqslant h}$ *be a finite sequence of integers such that* $\sum_{i=1}^{h} p_i = n$. *Then the number of coverings* $(X_i)_{1 \leqslant i \leqslant h}$ *of* E *by mutually disjoint sets* X_i *such that* Card $(X_i) = p_i$ *for* $1 \leqslant i \leqslant h$ *is equal to*

$$n! \Big/ \prod_{i=1}^{h} p_i!.$$

Let G be the set of permutations of E and let P be the set of coverings $(X_i)_{1 \leqslant i \leqslant h}$ which satisfy the conditions of the Proposition. Since $\sum_{i=1}^{h} p_i = n$, P is not empty. Let $(A_i)_{1 \leqslant i \leqslant h}$ be an element of P. For each permutation $f \in G$ the family $(f(A_i))_{1 \leqslant i \leqslant h}$ again belongs to P; let us denote it by $\varphi(f)$. For each element $(X_i)_{1 \leqslant i \leqslant h}$ let us calculate the number of permutations $f \in G$ such that $\varphi(f) = (X_i)$. We shall have $\varphi(f) = (X_i)$ if and only if for each index i we have $f(A_i) = X_i$. Hence the set of permutations f under consideration is equipotent to the product of the sets of bijections of A_i onto X_i (Chapter II, § 4, no. 7, Proposition 8); consequently the set $\overset{-1}{\varphi}((X_i)_{1 \leqslant i \leqslant h})$ has $\prod_{i=1}^{h} p_i!$ elements (Corollary to Proposition 10). Since G has $n!$ elements, the result now follows from Proposition 9.

COROLLARY 1. *Let* A *be a set with n elements and let p be an integer* $\leqslant n$. *Then the number of subsets of* A *which have p elements is* $\dfrac{n!}{p!(n-p)!}$.

Put $h = 2$, $p_1 = p$, $p_2 = n - p$ in Proposition 11.

¶ The number of subsets containing p elements in a set of n elements (where $p \leqslant n$) is denoted by $\binom{n}{p}$ and is called the *binomial coefficient with indices n and p*. From the relation $\binom{n}{p} = \dfrac{n!}{p!(n-p)!}$ it follows immediately that $\binom{n}{p} = \binom{n}{n-p}$.

This is also a consequence of the fact that, if E is a set with n elements, $X \to E - X$ is a bijection of the set of subsets of E consisting of p elements onto the set of subsets of $n - p$ elements.

We put $\binom{n}{p} = 0$ for each pair of natural integers such that $p > n$. With this convention the number of subsets of p elements in a set of n elements is $\binom{n}{p}$ for *every* natural integer p.

Corollary 2. *Let* E *and* F *be totally ordered finite sets with p and n elements, respectively. Then the number of strictly increasing mappings of* E *into* F *is* $\binom{n}{p}$.

For such a mapping is an injection of E into F (§ 1, no. 12, Proposition 11), and since E and F are well-ordered (§ 4, no. 4, Corollary 1 to Proposition 3), for each subset X of p elements of F there is exactly one strictly increasing mapping of E onto X (§ 2, no. 5, Theorem 3).

Proposition 12. *For each integer n, we have*

$$\sum_{p} \binom{n}{p} = 2^n.$$

For if E is a set of n elements, the left-hand side of the equality is the number of subsets of E. Now apply Proposition 12 of § 3, no. 5.

Proposition 13. *If n and p are integers, then*

$$\binom{n+1}{p+1} = \binom{n}{p+1} + \binom{n}{p}.$$

Let E be a set with $n + 1$ elements, let P be the set of all subsets of E containing $p + 1$ elements, let a be an element of E, and put

$$E' = E - \{a\}.$$

Let P′ (resp. P″) denote the set of subsets of $p + 1$ elements of E which contain a (resp. do not contain a). The set P″ is the set of subsets of $p + 1$ elements of E′ and therefore has $\binom{n}{p+1}$ elements. The mapping $X \to X \cap E'$ is a bijection of P′ onto the set of subsets of p elements of E′, and P′ therefore has $\binom{n}{p}$ elements. The result now follows from the fact that P is the union of the disjoint sets P′ and P″.

Proposition 13 can also be proved by means of a simple calculation from the formula $\binom{n}{p} = \dfrac{n!}{p!(n-p)!}$ for $p \leqslant n$.

Proposition 14. *Let n be an integer* > 0. *Then the number a_n* (resp. b_n) *of ordered pairs (i, j) of integers such that* $1 \leqslant i \leqslant j \leqslant n$ (resp. $1 \leqslant i < j \leqslant n$) *is* $\frac{1}{2}n(n+1)$ ((resp. $\frac{1}{2}n(n-1)$)).

For b_n is the number of subsets of 2 elements in $[1, n]$; hence

$$b_n = \frac{n!}{2!(n-2)!} = \frac{1}{2}n(n-1).$$

The value of a_n is deduced from this by noting that the set of ordered pairs (i, j) such that $1 \leqslant i \leqslant j \leqslant n$ is the union of the set of ordered pairs (i, j) such that $1 \leqslant i \leqslant j < n$ and the set of pairs (i, i) where $1 \leqslant i \leqslant n$. Thus $a_n = n + b_n = \frac{1}{2}n(n+1)$.

COROLLARY. *For each integer $n > 0$, we have*

$$\sum_{i=1}^{n} i = \frac{1}{2}n(n+1).$$

In the set A of ordered pairs of integers (i, j) such that $1 \leqslant i \leqslant j \leqslant n$, let A_k denote the subset of pairs (i, k), where $1 \leqslant i \leqslant k$ (for an arbitrary integer $k \leqslant n$). Then A_k has k elements. But $(A_k)_{1 \leqslant k \leqslant n}$ is a partition of A; hence the result.

PROPOSITION 15. *Let n and h be integers and let E be a set with h elements. Then the number of mappings u of E into $[0, n]$ such that $\sum_{x \in E} u(x) \leqslant n$ (resp. $\sum_{x \in E} u(x) = n$, for $h > 0$) is*

$$\binom{n+h}{h} \left(\text{resp. } \binom{n+h-1}{h-1}\right).$$

Let $A(h, n)$ (resp. $B(h, n)$) denote the number of mappings u of E into $[0, n]$ such that $\sum_{x \in E} u(x) \leqslant n$ (resp. $\sum_{x \in E} u(x) = n$ for $h > 0$). We show first that $A(h-1, n) = B(h, n)$. For this, let E′ be a subset of E with $h-1$ elements, and let $\{a\} = E - E'$. If u is a mapping of E into $[0, n]$ such that $\sum_{x \in E} u(x) = n$, then its restriction u' to E′ is such that $\sum_{x \in E'} u'(x) \leqslant n$, and moreover $u(a) = n - \sum_{x \in E'} u'(x)$. Conversely, every mapping u' of E′ into $[0, n]$ which satisfies $\sum_{x \in E'} u'(x) \leqslant n$ defines a unique mapping u of E into $[0, n]$, of which u' is the restriction, and which is such that $\sum_{x \ni E} u(x) = n$.

We note next that if $\sum_{x \in E} u(x) \leqslant n$, then either $\sum_{x \in E} u(x) = n$, or else $\sum_{x \in E} u(x) \leqslant n-1$, and these two possibilities are mutually exclusive. Consequently

$$A(h, n) = A(h, n-1) + B(h, n) = A(h, n-1) + A(h-1, n).$$

Since $A(0, 0) = 1 = \binom{0}{0}$, the formula $A(h, n) = \binom{n+h}{h}$ follows from above and from Proposition 13, by induction on $n + h$.

* The number of monomials $X_1^{\alpha_1} X_2^{\alpha_2} \ldots X_h^{\alpha_n}$ in h indeterminates with total degree $\leqslant n$ is evidently equal to the number of mappings $i \rightarrow \alpha_i$ of $[1, h]$ into $[0, n]$ such that $\sum_{i=0}^{h} \alpha_i \leqslant n$, and is therefore equal to $\binom{n+h}{h}$ by Proposition 15; this number is also the number of monomials in $h + 1$ indeterminates with total degree n. *

6. INFINITE SETS

1. THE SET OF NATURAL INTEGERS

DEFINITION 1. *A set is said to be infinite if it is not finite.*

In particular, a cardinal is infinite if it is not an integer.

The relation "there exists an infinite set" implies that the relation "x is an integer" is *collectivizing* (Chapter II, § 1, no. 4); for if $\mathfrak{a}$ is an infinite cardinal and n an arbitrary integer, we cannot have $\mathfrak{a} \leqslant n$ (§ 4, no. 2, Proposition 2). We have therefore $n < \mathfrak{a}$ for all integers n, which shows that the set of integers $< \mathfrak{a}$ (§ 3, no. 2, Remark following Theorem 1) contains all integers. Conversely, if the relation "x is an integer" is collectivizing, then the set E of integers is an *infinite* set. For each integer n, the interval $[0, n]$ is a subset of $n + 1$ elements of E (§ 5, no. 3, Proposition 5). Therefore $\mathrm{Card}(E) \geqslant n + 1 > n$. But to say that $\mathrm{Card}(E) \neq n$ for every integer n means that E is infinite.

¶ We now introduce the following axiom :

A5 ("Axiom of infinity".) *There exists an infinite set.*

It is not known whether or not this axiom can be deduced from the axioms and axiom schemes introduced previously; although the question has not been definitively settled, it is to be presumed that this axiom is independent of the others.

The preceding remarks then prove the following theorem :

THEOREM 1. *The relation "x is an integer" is collectivizing.*

We shall denote by **N** the set of integers (also called "the set of natural integers" when necessary to avoid ambiguity). The cardinal of **N** is denoted by $\aleph_0$. Whenever **N** is considered as an ordered set, it is always the ordering (called the *usual* ordering) defined in § 3, no. 2 that is under consideration, unless the contrary is expressly stated.

DEFINITION 2. *A sequence* (resp. *a sequence of elements of a set* E) *is a family* (resp. *a family of elements of* E) *whose index set is a subset of* **N**. *The sequence is said to be infinite if its index set is an infinite subset of* **N**.

Let $P\{n\}$ be a relation and let I denote the set of integers n such that $P\{n\}$ is true. I is then a subset of **N**. A sequence $(x_n)_{n \in I}$ is then sometimes written $(x_n)_{P\{n\}}$, and x_n is called the *nth term* in the sequence. A sequence whose index set is the set of integers $n \geqslant k$ is often written $(x_n)_{k \leqslant n}$ or $(x_n)_{n \geqslant k}$, or even just (x_n) if $k = 0$ or $k = 1$. Under the same conditions, for example, the notations $\prod_{P\{n\}} X_n$ and $\prod_{n=k}^{\infty} X_n$ are used to denote the product of a sequence of sets $(X_n)_{n \in I}$, and there are analogous notations for unions, intersections, cardinal products, and cardinal sums.

Every subfamily of a sequence is a sequence, called a *subsequence* of the given sequence.

Two sequences $(x_n)_{n \in I}$, $(y_n)_{n \in I}$ with the same index set are said to *differ only in the order of their terms* if there exists a permutation f of the index set I such that $x_{f(n)} = y_n$ for all $n \in I$.

A *multiple sequence* is a family whose index set is a subset of a product $\mathbf{N}^p$ (p an integer) ("double sequence" for $p = 2$, "triple sequence" for $p = 3$, and so on).

Let I be a set equipotent to **N** and let f be a bijection of **N** onto I. For each family $(x_\iota)_{\iota \in I}$ indexed by the set I, the sequence $n \to x_{f(n)}$ is said to be obtained by *arranging the family* $(x_\iota)_{\iota \in I}$ *in the order defined by* f. The sequences which correspond in this way to two distinct bijections of **N** onto I differ only in the order of their terms. For a finite family indexed by a set I of n elements we may similarly define a finite sequence with $[1, n]$ or $[0, n-1]$ as index set, by arranging the family in the order defined by a bijection of one or the other of these intervals onto I.

2. DEFINITION OF MAPPINGS BY INDUCTION

Since the set **N** is well-ordered, we may apply criterion C60 (§ 2, no. 2), which now takes the following form (with the same notation) :

C62. *Let u be a letter and let* $T\{u\}$ *be a term. Then there exists a set* U *and a mapping* f *of* **N** *onto* U *such that for each integer n we have* $f(n) = T\{f^{(n)}\}$, *where* $f^{(n)}$ *denotes the mapping of* $[0, n[$ *onto* $f([0, n[)$ *which agrees with* f *on* $[0, n[$. *Moreover, the set* U *and the mapping* f *are uniquely determined by this condition.*

We shall deduce from this the following criterion :

C63. *Let* $S\{v\}$ *and* a *be two terms. Then there exists a set* V *and a mapping* f *of* **N** *onto* V *such that* $f(0) = a$ *and* $f(n) = S\{f(n-1)\}$ *for each integer* $n \geqslant 1$. *Moreover, the set* V *and the mapping* f *are uniquely determined by these conditions.*

To deduce C63 from C62 (*), let

$$D(u) = \mathcal{E}_x(x \in \mathbf{N} \text{ and } (\exists y)((x, y) \in \mathrm{pr}_1(\mathrm{pr}_1(u))))$$

for each letter u. If u is a mapping of a subset of **N** into a set, then $D(u)$ is just the domain of u (Chapter II, § 3, no. 1). Let $M(u)$ be the least upper bound of $D(u)$ in **N** (†). Let φ be the empty mapping, with $\emptyset$ as source and target, i.e. (Chapter II, § 3, nos. 1 and 4), the triple $(\emptyset, \emptyset, \emptyset))$ and consider the relation

$$(u = \varphi \text{ and } y = a) \text{ or } (u \neq \varphi \text{ and } y = S\{u(M(u))\})$$

which we denote by $R\{y, u\}$; finally, let $T\{u\}$ be the term $\tau_y(R\{y, u\})$. Apply C62 to the term $T\{u\}$. Since $f^{(0)}$ is equal to φ, we have $T\{f^{(0)}\} = a$; hence $f(0) = a$. If on the other hand $n > 0$, we have $D(f^{(n)}) = [0, n-1]$ and $M(f^{(n)}) = n - 1$, whence

$$T\{f^{(n)}\} = S\{f^{(n)}(n-1)\} = S\{f(n-1)\}.$$

Examples

(1) Suppose that a is an element of a set E and that $S\{u\}$ is the term $g(u)$, where g is a mapping of E into itself (**). Then it is immediately seen by induction on n that for all $n \in \mathbf{N}$ we have $f(n) \in E$; consequently f is a mapping of **N** into E such that $f(0) = a$ and $f(n+1) = g(f(n))$ for all integers n.

Likewise, let h be a mapping of $\mathbf{N} \times E$ into E, and let ψ be the mapping of $\mathbf{N} \times E$ into itself defined by $\psi(n, x) = (n + 1, h(n, x))$. By the preceding discussion there exists a unique mapping $g = (\theta, f)$ of **N** into $\mathbf{N} \times E$ such that $g(0) = (0, a)$ and $g(n+1) = \psi(g(n))$ for all n, from which follows the existence and uniqueness of a mapping f

---

(*) **It is also possible to give a direct proof of C63, analogous to the proof of C60 (§ 2, no. 2).**

(†) **The definition of the least upper bound (§ 1, nos. 7, 8, and 9) can be formulated in such a way that it has a meaning even for a set which is not bounded above (it denotes a term, in the formalized language, of the form $\tau_x(R\{x\})$, which the reader will have no difficulty in writing down).**

(**) **If $g = (G, E, E)$, the term $g(u)$ is the term denoted by $\tau_y((u, y) \in G)$.**

of **N** into E such that $f(0) = a$ and $f(n+1) = h(n, f(n))$ for each integer n.

(2) Let X be a set and let E be the set of mappings of X into itself. Let e denote the identity mapping of X into itself, and let f be any element of E. Take $S\{u\}$ to be the term $f \circ u$ (*). By applying C63 we see that there exists a unique mapping of **N** into E, denoted by $n \to f^n$, such that $f^0 = e$ and $f^{n+1} = f \circ f^n$. The mapping f^n is called the *nth iterate* of the mapping f.

(3) If we take $S\{u\}$ to be the term $\mathfrak{P}(u)$, and a to be a set E, it follows likewise that there exists a mapping, denoted by $n \to \mathfrak{P}^n(E)$, of **N** into a set V(E) such that $\mathfrak{P}^0(E) = E$, $\mathfrak{P}^1(E) = \mathfrak{P}(E)$, and $\mathfrak{P}^{n+1}(E) = \mathfrak{P}(\mathfrak{P}^n(E))$ for every integer n.

Remark. Let E be a set, let A be a subset of E, let g be a mapping of A into E, and let a be an element of A. Take $S\{u\}$ to be the term $g(u)$. Criterion C63 is applicable and proves the existence of a mapping f of **N** onto a set V such that $f(0) = a$ and $f(n+1) = g(f(n))$ for every integer n. It may happen that $V \subset A$; if not, let p be the largest integer such that $f([0, p]) \subset A$. Then $f(p+1) = g(p) \notin A$, and $g(g(p))$ is a term about which nothing can be said. Hence in this case f is considered to be defined only on the interval $[0, p+1]$ ("restricted induction").

3. PROPERTIES OF INFINITE CARDINALS

THEOREM 2. *For every infinite cardinal* $\mathfrak{a}$ *we have* $\mathfrak{a}^2 = \mathfrak{a}$.

We shall use the following two lemmas:

LEMMA 1. *Every infinite set* E *contains a set equipotent to* **N**.

There exists a well-ordering relation on E (§ 2, no. 3, Theorem 1), which we shall denote by $x \leqslant y$. The hypothesis implies that the well-ordered set E cannot be isomorphic to a segment of **N** distinct from **N**, for such a segment is of the form $[0, n]$ (§ 2, no. 1, Proposition 1) and is therefore finite. It follows that **N** is isomorphic to a segment of E (§ 2, no. 3, Theorem 3), whence the result.

LEMMA 2. *The set* $\mathbf{N} \times \mathbf{N}$ *is equipotent to* **N**.

Since $\mathbf{N} \times \mathbf{N}$ contains the set $\{0\} \times \mathbf{N}$, which is equipotent to **N**, we have $\mathrm{Card}(\mathbf{N}) \leqslant \mathrm{Card}(\mathbf{N} \times \mathbf{N})$. To complete the proof it is enough to define an injection f of $\mathbf{N} \times \mathbf{N}$ into **N**. For this purpose we note that

(*) Here we mean the term denoted by (T, X, X), where T is the term denoted by $\mathcal{E}_z(z$ is an ordered pair and $(\exists y)((\mathrm{pr}_1 z, y) \in \mathrm{pr}_1(\mathrm{pr}_1(u))$ and $(y, \mathrm{pr}_2 z) \in \mathrm{pr}_1(\mathrm{pr}_1(f)))$.

there exists an injection φ of $\mathbf{N}$ into the set of mappings of $\mathbf{N}$ into $\mathrm{I} = \{0, 1\}$, obtained as follows : if r is the least integer such that $n > 2^r$, and if $\sum_{k=0}^{r-1} \varepsilon_k 2^{r-k-1}$ is the dyadic expansion of n (§ 5, no. 7), $\varphi(n)$ is defined to be the sequence $(u_m)_{m \in \mathbf{N}}$ such that $u_m = \varepsilon_{r-m-1}$ for $m < r$ and $u_m = 0$ for $m \geqslant r$. Proposition 8 of § 5, no. 7 shows that φ is injective. For each pair $(n, n') \in \mathbf{N} \times \mathbf{N}$ we define $f(n, n')$ as follows : if $\varphi(n) = (u_m)$ and $\varphi(n') = (v_m)$, let $f(n, n')$ be the integer s such that $\varphi(s) = w_m$, where $w_{2m} = u_m$ and $w_{2m+1} = v_m$ for all $m \in \mathbf{N}$. It is clear that the relation $f(n, n') = f(n_1, n_1')$ implies $\varphi(n) = \varphi(n_1)$ and $\varphi(n') = \varphi(n_1')$; hence $(n, n') = (n_1, n_1')$, and therefore f is injective.

¶ We come now to the proof of Theorem 2. Let E be a set such that $\mathrm{Card}(\mathrm{E}) = \mathfrak{a}$. Let D be a subset of E equipotent to $\mathbf{N}$ (Lemma 1). Then there exists a bijection ψ_0 of D onto $\mathrm{D} \times \mathrm{D}$ (Lemma 2). Let $\mathfrak{M}$ be the set of pairs (X, ψ), where X is a subset of E containing D and ψ is a bijection of X onto $\mathrm{X} \times \mathrm{X}$ which extends ψ_0. Order the set $\mathfrak{M}$ by means of the relation

"$\mathrm{X} \subset \mathrm{X}'$ and ψ' is an extension of ψ"

between (X, ψ) and (X', ψ'). Then it is immediately seen that $\mathfrak{M}$ is *inductive* (cf. § 2, no. 4, Example 2). Hence, by Zorn's Lemma (§ 2, no. 4, Theorem 2), $\mathfrak{M}$ has a maximal element (F, f). We shall show that $\mathrm{Card}(\mathrm{F}) = \mathfrak{a}$, which will suffice to prove the theorem. If $\mathrm{Card}(\mathrm{F})$ is not equal to $\mathfrak{a}$, let $\mathrm{Card}(\mathrm{F}) = \mathfrak{b} < \mathfrak{a}$. Then $\mathfrak{b} = \mathfrak{b}^2$ and $\mathfrak{b}$ is infinite, and $\mathfrak{b} \leqslant 2\mathfrak{b} \leqslant 3\mathfrak{b} \leqslant \mathfrak{b}^2 = \mathfrak{b}$ (§ 3, no. 6, Proposition 14); hence $2\mathfrak{b} = \mathfrak{b}$ and $3\mathfrak{b} = \mathfrak{b}$. From the hypothesis $\mathfrak{b} < \mathfrak{a}$ it follows that $\mathrm{Card}(\mathrm{E} - \mathrm{F}) > \mathfrak{b}$, for otherwise we should have $\mathrm{Card}(\mathrm{E}) \leqslant 2\mathfrak{b} = \mathfrak{b}$, and we have assumed that $\mathfrak{b} < \mathrm{Card}(\mathrm{E})$. Hence there is a subset $\mathrm{Y} \subset \mathrm{E} - \mathrm{F}$ equipotent to F. Let $\mathrm{Z} = \mathrm{F} \cup \mathrm{Y}$; we shall show that there exists a bijection g of Z onto $\mathrm{Z} \times \mathrm{Z}$ which extends f. We have

$$\mathrm{Z} \times \mathrm{Z} = (\mathrm{F} \times \mathrm{F}) \cup (\mathrm{F} \times \mathrm{Y}) \cup (\mathrm{Y} \times \mathrm{F}) \cup (\mathrm{Y} \times \mathrm{Y}),$$

and the four products on the right-hand side are mutually disjoint. Since F and Y are equipotent, we have

$$\mathrm{Card}(\mathrm{F} \times \mathrm{Y}) = \mathrm{Card}(\mathrm{Y} \times \mathrm{F}) = \mathrm{Card}(\mathrm{Y} \times \mathrm{Y}) = \mathfrak{b}^2 = \mathfrak{b},$$

whence

$$\mathrm{Card}((\mathrm{F} \times \mathrm{Y}) \cup (\mathrm{Y} \times \mathrm{F}) \cup (\mathrm{Y} \times \mathrm{Y})) = 3\mathfrak{b} = \mathfrak{b}.$$

There is therefore a bijection f_1 of Y onto the set

$$(\mathrm{F} \times \mathrm{Y}) \cup (\mathrm{Y} \times \mathrm{F}) \cup (\mathrm{Y} \times \mathrm{Y});$$

the mapping g of Z into Z $\times$ Z, which is equal to f on F and to f_1 on Y, is therefore a bijection which extends f, contrary to the definition of f. Hence Card(F) $= \mathfrak{a}$, and the proof is complete.

COROLLARY 1. *If $\mathfrak{a}$ is an infinite cardinal, then $\mathfrak{a}^n = \mathfrak{a}$ for every integer $n \geqslant 1$.*

By induction on n.

COROLLARY 2. *The product of a finite family $(\mathfrak{a}_i)_{i \in I}$ of non-zero cardinals, of which the greatest is an infinite cardinal $\mathfrak{a}$, is equal to $\mathfrak{a}$.*

Let $\mathfrak{b}$ denote the product and let n be the number of elements in I. Then $\mathfrak{b} \leqslant \mathfrak{a}^n = \mathfrak{a}$ (§ 3, no. 6, Proposition 14). On the other hand, since $\mathfrak{a}_i \geqslant 1$ for all $i \in I$, we have $\mathfrak{b} \geqslant \mathfrak{a}$ (§ 3, no. 6, Proposition 14).

COROLLARY 3. *Let $\mathfrak{a}$ be an infinite cardinal and let $(\mathfrak{a}_\iota)_{\iota \in I}$ be a family of cardinals $\leqslant \mathfrak{a}$ whose index set I has a cardinal $\leqslant \mathfrak{a}$. Then $\sum_{\iota \in I} \mathfrak{a}_\iota \leqslant \mathfrak{a}$; and if $\mathfrak{a}_\iota = \mathfrak{a}$ for at least one index $\iota \in I$,* then $\sum_{\iota \in I} \mathfrak{a}_\iota = \mathfrak{a}$.

Let $\mathfrak{b}$ be the cardinal of I; then we have $\sum_{\iota \in I} \mathfrak{a}_\iota \leqslant \mathfrak{a}\mathfrak{b} \leqslant \mathfrak{a}^2 = \mathfrak{a}$ (§ 3, no. 6, Proposition 14), and $\sum_{\iota \in I} \mathfrak{a}_\iota \geqslant \mathfrak{a}_\varkappa$ for all $\varkappa \in I$.

COROLLARY 4. *If $\mathfrak{a}$ and $\mathfrak{b}$ are two non-zero cardinals, one of which is infinite, we have $\mathfrak{a}\mathfrak{b} = \mathfrak{a} + \mathfrak{b} = \sup(\mathfrak{a}, \mathfrak{b})$.*

This follows directly from Corollaries 2 and 3.

4. COUNTABLE SETS

DEFINITION 3. *A set is said to be countable if it is equipotent to a subset of the set* N *of integers.*

PROPOSITION 1. *Every subset of a countable set is countable. The product of a finite family of countable sets is countable. The union of a sequence of countable sets is countable.*

The first assertion is obvious. The others follow from the Corollaries to Theorem 2, no. 3.

¶ We have proved (no. 3, Lemma 1) that if $\mathfrak{a}$ is any infinite cardinal, then Card(N) $\leqslant \mathfrak{a}$. This has the following consequences :

PROPOSITION 2. *Every countable infinite set* E *is equipotent to* N.

For Card(E) $\leqslant$ Card(N) by definition; and Card(N) $\leqslant$ Card(E) since E is infinite.

PROPOSITION 3. *Every infinite set has a partition* $(X_\iota)_{\iota \in I}$ *formed of countable infinite sets* X_ι, *the index set* I *being equipotent to* E.

For $\text{Card}(E) = \text{Card}(E)\,\text{Card}(N)$ (no. 3, Theorem 2, Corollary 4).

PROPOSITION 4. *Let* f *be a mapping of a set* E *onto an infinite set* F *such that, for each* $y \in F$, $\overset{-1}{f}(y)$ *is countable. Then* F *is equipotent to* E.

For the sets $\overset{-1}{f}(y)$ $(y \in F)$ form a partition of E; hence

$$\text{Card}(E) \leqslant \text{Card}(F)\text{Card}(N) = \text{Card}(F),$$

and $\text{Card}(F) \leqslant \text{Card}(E)$ by Proposition 3 of § 3, no. 2.

PROPOSITION 5. *The set* $\mathfrak{F}(E)$ *of finite subsets of an infinite set* E *is equipotent to* E.

For each integer n, let $\mathfrak{F}_n$ denote the set of subsets of E which have n elements. For each $X \in \mathfrak{F}_n$ there exists a bijection of $[1, n]$ onto X. Hence the cardinal of $\mathfrak{F}_n$ is at most equal to that of the set of mappings of $[1, n]$ into E, i.e., to $\text{Card}(E^n) = \text{Card}(E)$ (no. 3, Theorem 2, Corollary 1). Therefore

$$\text{Card}(\mathfrak{F}(E)) = \sum_{n \in N} \text{Card}(\mathfrak{F}_n) \leqslant \text{Card}(E)\text{Card}(N) = \text{Card}(E).$$

On the other hand, since $x \to \{x\}$ is an injective mapping of E into $\mathfrak{F}(E)$, we have $\text{Card}(E) \leqslant \text{Card}(\mathfrak{F}(E))$.

DEFINITION 4. *A set is said to have the power of the continuum if it is equipotent to the set of all subsets of* N.

A set which has the power of the continuum is not countable (§ 3, no. 6, Theorem 2).

* The name "power of the continuum" arises from the fact that the set of real numbers is equipotent to $\mathfrak{P}(N)$ (*General Topology*, Chapter IV, § 8). * The *continuum hypothesis* is the assertion that every uncountable set contains a subset which has the power of the continuum; the *generalized continuum hypothesis* is the assertion that, for every infinite cardinal $\mathfrak{a}$, every cardinal $> \mathfrak{a}$ is $\geqslant 2^{\mathfrak{a}}$.

5. STATIONARY SEQUENCES

DEFINITION 5. *A sequence* $(x_n)_{n \in N}$ *of elements of a set* E *is said to be stationary if there exists an integer* m *such that* $x_n = x_m$ *for all integers* $n \geqslant m$.

Proposition 6. *Let E be an ordered set. Then the following statements are equivalent :*

(a) *Every non-empty subset of* E *has a maximal element.*
(b) *Every increasing sequence* (x_n) *of elements of* E *is stationary.*

We show first that (a) implies (b). Let X be the set of elements of the sequence (x_n), and let x_m be a maximal element of X. If $n \geqslant m$, we have by hypothesis $x_n \geqslant x_m$, and therefore $x_n = x_m$ by the maximality of x_m. Conversely, suppose that there exists a non-empty subset A of E which has no maximal element. For each $x \in A$, let T_x be the set of all $y \in A$ such that $y > x$. By assumption, $T_x \neq \emptyset$ for all $x \in A$; hence there exists a mapping f of A into A such that $f(x) > x$ for all $x \in A$ (Chapter II, § 5, no. 4, Proposition 6). If $a \in A$, then the sequence $(x_n)_{n \in N}$ defined inductively by the conditions $x_0 = a$, $x_{n+1} = f(x_n)$ is evidently increasing and not stationary.

Corollary 1. *A totally ordered set* E *is well-ordered if and only if every decreasing sequence of elements of* E *is stationary.*

For to say that E is well-ordered is equivalent to saying that every non-empty subset of E has a minimal element (§ 1, no. 10, Proposition 10), and the assertion therefore follows from Proposition 6.

Corollary 2. *Every increasing sequence of elements of a finite ordered set is stationary.*

For every finite ordered set has a maximal element (§ 4, no. 4, Proposition 3, Corollary 2).

¶ An ordered set E which satisfies the equivalent conditions of Proposition 6 is sometimes said to be *Noetherian.*

Proposition 7 ("Principle of Noetherian induction"). *Let* E *be a Noetherian set, and let* F *be a subset of* E *with the following property : if* $a \in E$ *is such that the relation* $x > a$ *implies* $x \in F$, *then* $a \in F$. *Under these conditions,* $F = E$.

Indeed, suppose $E \neq F$; then $E - F$ has a maximal element b. By definition we have $x \in F$ for all $x > b$; but this implies $b \in F$, which is absurd.

7. INVERSE LIMITS AND DIRECT LIMITS

1. INVERSE LIMITS

Let I be a preordered set and let $(E_\alpha)_{\alpha\in I}$ be a family of sets indexed by I. For each pair (α, β) of elements of I such that $\alpha \leqslant \beta$, let $f_{\alpha\beta}$ be a mapping of E_β into E_α. Suppose that the $f_{\alpha\beta}$ satisfy the following conditions :

(LP_I) *The relations* $\alpha \leqslant \beta \leqslant \gamma$ *imply* $f_{\alpha\gamma} = f_{\alpha\beta} \circ f_{\beta\gamma}$.
(LP_{II}) *For each* $\alpha \in I$, $f_{\alpha\alpha}$ *is the identity mapping of* E_α.

Let $G = \prod_{\alpha\in I} E_\alpha$ be the *product* of the family of sets $(E_\alpha)_{\alpha\in I}$, and let E denote the subset of G consisting of all x which satisfy *each* of the relations

$$(1) \qquad \mathrm{pr}_\alpha x = f_{\alpha\beta}(\mathrm{pr}_\beta x)$$

for each pair of indices (α, β) such that $\alpha \leqslant \beta$. E is said to be the *inverse limit of the family* $(E_\alpha)_{\alpha\in I}$ *with respect to the family of mappings* $(f_{\alpha\beta})$, and we write $E = \varprojlim (E_\alpha, f_{\alpha\beta})$ or simply $E = \varprojlim E_\alpha$ when there is no risk of ambiguity. By abuse of language, the pair $((E_\alpha), (f_{\alpha\beta}))$ (usually denoted by $(E_\alpha, f_{\alpha\beta})$) is called an *inverse system of sets*, relative to the index set I. The *restriction* f_α of the projection pr_α to E is called the *canonical mapping* of E into E_α, and we have the relation

$$(2) \qquad f_\alpha = f_{\alpha\beta} \circ f_\beta$$

whenever $\alpha \leqslant \beta$; this is merely a transcription of the relations (1) which define E.

Examples

(1) Suppose that the order relation on I is the relation of *equality*. Then the only pairs (α, β) such that $\alpha \leqslant \beta$ are the pairs (α, α) where $\alpha \in I$; and since $f_{\alpha\alpha}$ is the identity mapping, the relation (1) is satisfied for *all* $x \in G$; in other words, $\varprojlim E_\alpha$ is then the *product* $\prod_{\alpha\in I} E_\alpha$.

(2) Suppose that I is *right directed*, that E_α is the same set F for all $\alpha \in I$, and that $f_{\alpha\beta}$ is the identity mapping of F onto itself whenever $\alpha \leqslant \beta$. Then $E = \varprojlim E_\alpha$ is the *diagonal* Δ of the product $\prod_{\alpha\in I} E_\alpha = F^I$. Indeed, it is clear that each $x \in \Delta$ satisfies the relations (1). Conversely, let x be an element of E, and let us show that for each pair of indices

(α, β) we have $\mathrm{pr}_\alpha x = \mathrm{pr}_\beta x$. By hypothesis, there exists an index $\gamma \in \mathrm{I}$ such that $\alpha \leqslant \gamma$ and $\beta \leqslant \gamma$; hence by (1) we have $\mathrm{pr}_\alpha x = f_{\alpha\gamma}(\mathrm{pr}_\gamma x) = \mathrm{pr}_\gamma x$, and similarly $\mathrm{pr}_\beta x = \mathrm{pr}_\gamma x$, which proves our assertion.

It should be noted that $\mathrm{E} = \varprojlim \mathrm{E}_\alpha$ can be *empty* even when all the E_α are non-empty and all the mappings $f_{\alpha\beta}$ are *surjective* (Exercise 4; see no. 4).

¶ It is clear that for each subset J of I the pair consisting of the subfamily $(\mathrm{E}_\alpha)_{\alpha \in \mathrm{J}}$ and the family $(f_{\alpha\beta})$, where $\alpha \in \mathrm{J}$, $\beta \in \mathrm{J}$, and $\alpha \leqslant \beta$, is again an inverse system of sets relative to J; it is said to be obtained by *restricting* the index set to J. Let E and E′ respectively denote the inverse limits of the families $(\mathrm{E}_\alpha)_{\alpha \in \mathrm{I}}$ and $(\mathrm{E}_\alpha)_{\alpha \in \mathrm{J}}$. For each $x \in \mathrm{E}$ the element

$$g(x) = (f_\alpha(x))_{\alpha \in \mathrm{J}} \tag{3}$$

belongs to E′ by virtue of (2); the mapping $g : \mathrm{E} \to \mathrm{E}'$ so defined is called *canonical*. If J′ is a subset of J, and E″ the inverse limit of the family $(\mathrm{E}_\alpha)_{\alpha \in \mathrm{J}'}$, and if $g' : \mathrm{E}' \to \mathrm{E}''$ and $g'' : \mathrm{E} \to \mathrm{E}''$ are the canonical mappings, then by definition we have

$$g'' = g' \circ g. \tag{4}$$

2. INVERSE SYSTEMS OF MAPPINGS

PROPOSITION 1. *Let* I *be an ordered set, let* $(\mathrm{E}_\alpha, f_{\alpha\beta})$ *be an inverse system of sets relative to* I, *let* $\mathrm{E} = \varprojlim \mathrm{E}_\alpha$ *be its inverse limit, and for each* $\alpha \in \mathrm{I}$ *let*

$$f_\alpha : \mathrm{E} \to \mathrm{E}_\alpha$$

be the canonical mapping. For each $\alpha \in \mathrm{I}$, *let* u_α *be a mapping of a set* F *into* E_α *such that*

$$f_{\alpha\beta} \circ u_\beta = u_\alpha \qquad \textit{whenever } \alpha \leqslant \beta. \tag{5}$$

Then:

(a) *there exists a unique mapping* u *of* F *into* E *such that*

$$u_\alpha = f_\alpha \circ u \qquad \textit{for all } \alpha \in \mathrm{I}; \tag{6}$$

(b) *the mapping* u *is injective if and only if, for each pair of distinct elements* y, z *of* F, *there exists* $\alpha \in \mathrm{I}$ *such that* $u_\alpha(y) \neq u_\alpha(z)$.

For the relation $u_\alpha = f_\alpha \circ u$ means that for each $y \in F$ we have

$$\mathrm{pr}_\alpha(u(y)) = u_\alpha(y);$$

the element $u(y) \in \prod_{\alpha \in I} E_\alpha$ is uniquely determined by $u(y) = (u_\alpha(y))_{\alpha \in I}$. It remains to be shown that $u(y) \in E$ for all $y \in F$, in other words, that

$$\mathrm{pr}_\alpha(u(y)) = f_{\alpha\beta}(\mathrm{pr}_\beta(u(y)))$$

whenever $\alpha \leqslant \beta$. But this can be written in the form

$$u_\alpha(y) = f_{\alpha\beta}(u_\beta(y))$$

and therefore follows from (5). The second part of the Proposition follows immediately from the definitions.

COROLLARY 1. *Let* $(E_\alpha, f_{\alpha\beta})$ *and* $(F_\alpha, g_{\alpha\beta})$ *be two inverse systems of sets relative to the same index set* I; *let* $E = \varprojlim E_\alpha$, $F = \varprojlim F_\alpha$, *and let* f_α (resp. g_α) *be the canonical mapping of* E *into* E_α (resp. *of* F *into* F_α) *for each* $\alpha \in I$, *For each* $\alpha \in I$, *let* u_α *be a mapping of* E *into* F_α *such that the diagram*

$$\begin{array}{ccc} E_\beta & \xrightarrow{u_\beta} & F_\beta \\ {\scriptstyle f_{\alpha\beta}}\downarrow & & \downarrow{\scriptstyle g_{\alpha\beta}} \\ E_\alpha & \xrightarrow[u_\alpha]{} & F_\alpha \end{array}$$

is commutative (*) *whenever* $\alpha \leqslant \beta$. *Then there exists a unique mapping* $u : E \to F$ *such that for each* $\alpha \in I$ *the diagram*

$$\begin{array}{ccc} E & \xrightarrow{u} & F \\ {\scriptstyle f_\alpha}\downarrow & & \downarrow{\scriptstyle g_\alpha} \\ E_\alpha & \xrightarrow[u_\alpha]{} & F_\alpha \end{array}$$

is commutative.

Put $v_\alpha = u_\alpha \circ f_\alpha$. If $\alpha \leqslant \beta$, we have, by (2),

$$g_{\alpha\beta} \circ v_\beta = g_{\alpha\beta} \circ u_\beta \circ f_\beta = u_\alpha \circ f_{\alpha\beta} \circ f_\beta = u_\alpha \circ f_\alpha = v_\alpha,$$

and we may therefore apply Proposition 1 to the mappings v_α; hence the

(*) This means that $u_\alpha \circ f_{\alpha\beta} = g_{\alpha\beta} \circ u_\beta$.

existence and uniqueness of a mapping $u : E \to F$ such that

$$g_\alpha \circ u = v_\alpha = u_\alpha \circ f_\alpha$$

for each $\alpha \in I$.

¶ A family of mappings $u_\alpha : E_\alpha \to F_\alpha$ which satisfies the conditions of Corollary 1 is called an *inverse system of mappings* of $(E_\alpha, f_{\alpha\beta})$ into $(F_\alpha, g_{\alpha\beta})$. The mapping u defined in Corollary 1 is called the *inverse limit* of the family (u_α) and is written $u = \varprojlim u_\alpha$ when there is no risk of confusion.

COROLLARY 2. *Let* $(E_\alpha, f_{\alpha\beta})$, $(F_\alpha, g_{\alpha\beta})$, $(G_\alpha, h_{\alpha\beta})$ *be three inverse systems of sets relative to the same index set* I; *let* $E = \varprojlim E_\alpha$, $F = \varprojlim F_\alpha$, $G = \varprojlim G_\alpha$, *and let* f_α (resp. g_α, h_α) *be the canonical mapping of* E (resp. F, G) *into* E_α (resp. F_α, G_α). *If* (u_α) *and* (v_α) *are two inverse systems of mappings,* $u_\alpha : E_\alpha \to F_\alpha$, $v_\alpha : F_\alpha \to G_\alpha$, *then the composite mappings* $v_\alpha \circ u_\alpha : E_\alpha \to G_\alpha$ *form an inverse system of mappings, and we have*

$$\varprojlim (v_\alpha \circ u_\alpha) = (\varprojlim v_\alpha) \circ (\varprojlim u_\alpha). \tag{7}$$

For if we put $w_\alpha = v_\alpha \circ u_\alpha$, then if $\alpha \leqslant \beta$ we have

$$w_\alpha \circ f_{\alpha\beta} = v_\alpha \circ (u_\alpha \circ f_{\alpha\beta}) = v_\alpha \circ (g_{\alpha\beta} \circ u_\beta) = (h_{\alpha\beta} \circ v_\beta) \circ u_\beta = h_{\alpha\beta} \circ w_\beta,$$

which shows that (w_α) is an inverse system of mappings. Furthermore, if $u = \varprojlim u_\alpha$ and $v = \varprojlim v_\alpha$, then $h_\alpha \circ (v \circ u) = (v_\alpha \circ g_\alpha) \circ u = (v_\alpha \circ u_\alpha) \circ f_\alpha$ for each $\alpha \in I$, and therefore, by the uniqueness of the inverse limit, we have $v \circ u = \varprojlim w_\alpha$.

¶ Let $(E_\alpha, f_{\alpha\beta})$ be an inverse system of sets, and for each $\alpha \in I$ let M_α be a subset of E_α. If $f_{\alpha\beta}(M_\beta) \subset M_\alpha$ whenever $\alpha \leqslant \beta$, the M_α are said to form an *inverse system of subsets* of the E_α. Let $g_{\alpha\beta}$ be the mapping of M_β into M_α (where $\alpha \leqslant \beta$) whose graph is the same as that of the restriction of $f_{\alpha\beta}$ to M_β. Then it is clear that $(M_\alpha, g_{\alpha\beta})$ is an inverse system of sets and that

$$\varprojlim M_\alpha = (\varprojlim E_\alpha) \cap \prod_{\alpha \in I} M_\alpha. \tag{8}$$

PROPOSITION 2. *Let* $(E_\alpha, f_{\alpha\beta})$ *and* $(E'_\alpha, f'_{\alpha\beta})$ *be two inverse systems of sets relative to* I, *and let* u_α *be a mapping of* E_α *into* E'_α *for each* $\alpha \in I$, *such that the* u_α *form an inverse system of mappings. Let* $u = \varprojlim u_\alpha$. *Then for each* $x' = (x'_\alpha) \in E' = \varprojlim E'_\alpha$, *the* $\overset{-1}{u}_\alpha(x'_\alpha)$ *form an inverse system of subsets of the* E_α, *and* $\overset{-1}{u}(x') = \varprojlim \overset{-1}{u}_\alpha(x'_\alpha)$.

For if $\alpha \leqslant \beta$ and $x_\beta \in \overset{-1}{u}_\beta(x'_\beta)$, we have

$$u_\alpha(f_{\alpha\beta}(x_\beta)) = f'_{\alpha\beta}(u_\beta(x_\beta)) = f'_{\alpha\beta}(x'_\beta) = x'_\alpha,$$

from which the first assertion follows; and to say that $x = (x_\alpha) \in \mathrm{E} = \varprojlim \mathrm{E}_\alpha$ is such that $u(x) = x'$ means, by definition, that $u_\alpha(x_\alpha) = x'_\alpha$ for each $\alpha \in \mathrm{I}$.

COROLLARY. *If u_α is injective* (resp. *bijective*) *for each $\alpha \in \mathrm{I}$, then u is injective* (resp. *bijective*).

With the notation of Proposition 2, the images $u_\alpha(\mathrm{E}_\alpha)$ also form an inverse system of subsets of the E'_α, and we have

$$u(\mathrm{E}) \subset \varprojlim u_\alpha(\mathrm{E}_\alpha); \tag{9}$$

but the two sides of this relation are *not necessarily equal* (Exercise 4).

PROPOSITION 3. *Let* I *be a preordered set, let* $(\mathrm{E}_\alpha, f_{\alpha\beta})$ *be an inverse system of sets relative to* I, *and let* $\mathrm{E} = \varprojlim \mathrm{E}_\alpha$. *Let* J *be a cofinal subset of* I *such that* J *is right directed, and let* E′ *be the inverse limit of the inverse system of sets obtained from* $(\mathrm{E}_\alpha, f_{\alpha\beta})$ *by restricting the index set to* J. *Then the canonical mapping g of* E *into* E′ (no. 1, *formula* (3)) *is bijective.*

For each $\alpha \in \mathrm{J}$, let f'_α be the canonical mapping $\mathrm{E}' \to \mathrm{E}_\alpha$. Then, by (2) and (5), g is the unique mapping of E into E′ such that $f_\alpha = f'_\alpha \circ g$ for all $\alpha \in \mathrm{J}$ (Proposition 1). We shall show that g is injective by using the criterion of Proposition 1. If x, y are distinct elements of E, then by definition there exists $\alpha \in \mathrm{I}$ such that $f_\alpha(x) \neq f_\alpha(y)$; since J is cofinal in I, there exists $\lambda \in \mathrm{J}$ such that $\alpha \leqslant \lambda$; since $f_{\alpha\lambda}(f_\lambda(x)) \neq f_{\alpha\lambda}(f_\lambda(y))$, we have $f_\lambda(x) \neq f_\lambda(y)$. It remains to be shown that g is surjective. Let $x' = (x'_\lambda)_{\lambda \in \mathrm{J}}$ be an element of E′. For each $\alpha \in \mathrm{I}$ there exists $\lambda \in \mathrm{J}$ such that $\alpha \leqslant \lambda$, and the element $f_{\alpha\lambda}(x'_\lambda)$ does not depend on the index $\lambda \in \mathrm{J}$ such that $\alpha \leqslant \lambda$; for if $\mu \in \mathrm{J}$ is such that $\alpha \leqslant \mu$, there exists $\nu \in \mathrm{J}$ such that $\lambda \leqslant \nu$ and $\mu \leqslant \nu$, hence $f_{\alpha\lambda}(x'_\lambda) = f_{\alpha\lambda}(f_{\lambda\nu}(x'_\nu)) = f_{\alpha\nu}(x'_\nu)$, and similarly $f_{\alpha\mu}(x'_\mu) = f_{\alpha\nu}(x'_\nu)$. Let x_α be the common value of the $f_{\alpha\lambda}(x'_\lambda)$ for the $\lambda \in \mathrm{J}$ such that $\alpha \leqslant \lambda$, and let $x = (x_\alpha)_{\alpha \in \mathrm{I}}$. Then $x \in \mathrm{E}$, because if $\alpha \leqslant \beta$ and if $\lambda \in \mathrm{J}$ is such that $\beta \leqslant \lambda$, we have

$$f_{\alpha\beta}(x_\beta) = f_{\alpha\beta}(f_{\beta\lambda}(x'_\lambda)) = f_{\alpha\lambda}(x'_\lambda) = x_\alpha.$$

Finally, we have $x'_\lambda = f_{\lambda\lambda}(x'_\lambda)$ for all $\lambda \in \mathrm{J}$, hence $x_\lambda = x'_\lambda$ for all $\lambda \in \mathrm{J}$; in other words, $f(x_\lambda) = x'_\lambda$, so that $g(x) = x'$. Therefore g is surjective and the proof is complete.

In particular, if I has a *greatest element* ω, we may take $\mathrm{J} = \{\omega\}$, so that $\varprojlim \mathrm{E}_\alpha$ is canonically identified with E_ω.

Remarks

(1) For each $\alpha \in I$ put $E'_\alpha = f_\alpha(E)$. Then the sets E'_α form an *inverse system of subsets* of the E_α by reason of (2), and it is immediately clear that $\varprojlim E'_\alpha = E = \varprojlim E_\alpha$. The mapping $f'_{\alpha\beta} : E'_\beta \to E'_\alpha$ (where $\alpha \leqslant \beta$), whose graph is the same as that of the restriction of $f_{\alpha\beta}$ to E_β, is *surjective*, and we have

$$E'_\alpha = f_\alpha(E) \subset \bigcap_{\beta \geqslant \alpha} f_{\alpha\beta}(E_\beta) \tag{10}$$

for all $\alpha \in I$.

(2) Let I be a (*right*) *directed* ordered set, let $(E_\alpha, f_{\alpha\beta})$ be an inverse system of sets relative to I, and for each $\alpha \in I$ let $u_\alpha : F \to E_\alpha$ be a mapping such that the family (u_α) satisfies formula (5). Consider the inverse system $(F_\alpha, i_{\alpha\beta})$ indexed by I, where $F_\alpha = F$ for all $\alpha \in I$ and $i_{\alpha\beta}$ is the identity mapping of F. Then (no. 1, Example 2) F is canonically identified with $\varprojlim F_\alpha$. If we consider u_α as a mapping of F_α into E_α, then (u_α) is an inverse system of mappings, and the mapping $u : F \to E$ defined by (6) is identified with the inverse limit of this system of mappings. Hence, by abuse of language, we write $u = \varprojlim u_\alpha$.

(3) Let I be an ordered set and let $(E_\alpha, f_{\alpha\beta})$ be an inverse system of sets relative to I. For each finite subset J of I, let F_J be the inverse limit of the (finite) inverse system obtained from $(E_\alpha, f_{\alpha\beta})$ by restricting the index set to J. If J and K are any two finite subsets of I such that $J \subset K$, let g_{JK} denote the canonical mapping (3) of F_K into F_J. Then the relation (4) shows that (F_J, g_{JK}) is an *inverse system* of sets relative to the *directed* set (with respect to the relation $\subset$) $\mathfrak{F}(I)$ of finite subsets of I. Next, for each $J \in \mathfrak{F}(I)$ let $h_J : E \to F_J$ be the canonical mapping (3). By virtue of (4) and with the abuse of language mentioned in Remark (2), (h_J) is an *inverse system* of mappings. Put $h = \varprojlim h_J : E \to F = \varprojlim F_J$, and let us show that h is a *bijection* (called *canonical*). Indeed, let $y = (y_J) \in F$. By definition we have $y_J = (x_{\alpha, J})_{\alpha \in J}$, where $x_{\alpha, J} \in E_\alpha$ for all $\alpha \in J$. If $J \subset K$, then by definition of the mapping g_{JK} and because $y_J = g_{JK}(y_K)$, we have $x_{\alpha, J} = x_{\alpha, K}$ for all $\alpha \in J$. Hence, given $\alpha \in I$, there is a unique element $x_\alpha \in E_\alpha$ such that $x_\alpha = x_{\alpha, J}$ for all finite subsets J of I which contain α. If $\alpha \leqslant \beta$, there is a finite subset J of I which contains both α and β; hence $x_\alpha = f_{\alpha\beta}(x_\beta)$ by definition. Consequently $x = (x_\alpha)$ is the unique element of E such that $h(x) = y$.

3. DOUBLE INVERSE LIMIT

Let I, L be two preordered sets, and $I \times L$ their product (§ 1, no. 4). Consider an inverse system of sets $(E^\lambda_\alpha, f^{\lambda\mu}_{\alpha\beta})$ relative to the index set $I \times L$.

We have

(11) $$f_{\alpha\gamma}^{\lambda\nu} = f_{\alpha\beta}^{\lambda\mu} \circ f_{\beta\gamma}^{\mu\nu} \quad \text{whenever } \alpha \leqslant \beta \leqslant \gamma \text{ and } \lambda \leqslant \mu \leqslant \nu.$$

Let E or $\varprojlim_{\alpha,\lambda} E_\alpha^\lambda$ denote the inverse limit of this inverse system.

For each $\lambda \in L$ put $g_{\alpha\beta}^\lambda = f_{\alpha\beta}^{\lambda\lambda} : E_\beta^\lambda \to E_\alpha^\lambda$. It follows from (11) that

(12) $$g_{\alpha\gamma}^\lambda = g_{\alpha\beta}^\lambda \circ g_{\beta\gamma}^\lambda \quad \text{whenever } \alpha \leqslant \beta \leqslant \gamma,$$

so that $(E_\alpha^\lambda, g_{\alpha\beta}^\lambda)$ is an inverse system of sets relative to I. Let $F^\lambda = \varprojlim_\alpha E_\alpha^\lambda$ denote its inverse limit. For fixed $\lambda \leqslant \mu$ in L it follows from (11) that the $h_\alpha^{\lambda\mu} = f_{\alpha\alpha}^{\lambda\mu} : E_\alpha^\mu \to E_\alpha^\lambda$ form an inverse system of mappings, whose inverse limit we denote by $h^{\lambda\mu} = \varprojlim_\alpha h_\alpha^{\lambda\mu} : F^\mu \to F^\lambda$. If $\lambda \leqslant \mu \leqslant \nu$ in L, we have

(13) $$h^{\lambda\nu} = h^{\lambda\mu} \circ h^{\mu\nu}$$

(no. 2, Proposition 2, Corollary 2); hence $(F^\lambda, h^{\lambda\mu})$ is an inverse system of sets relative to L. Let $F = \varprojlim F^\lambda$ be its inverse limit. We shall define a *canonical bijection* $F \to E$. To do this we note that F is by definition a subset of $\prod_{\lambda \in L} F^\lambda$, and F^λ is a subset of $\prod_{\alpha \in I} E_\alpha^\lambda$; hence F may be canonically identified with a subset of $\prod_{(\alpha,\lambda) \in I \times L} E_\alpha^\lambda = G$ (Chapter II, § 5, no. 5, Proposition 7). For each $z \in G$, let $\mathrm{pr}^\lambda(z)$ denote the element $(\mathrm{pr}_\alpha^\lambda(z))_{\alpha \in I}$ of $\prod_{\alpha \in I} E_\alpha^\lambda$. Then $z \in F$ if and only if

(14) $$\mathrm{pr}^\lambda(z) = h^{\lambda\mu}(\mathrm{pr}^\mu(z)) \qquad \text{whenever } \lambda \leqslant \mu \text{ in L}$$

and $\mathrm{pr}^\lambda(z) \in F^\lambda$ for all $\lambda \in L$; that is to say, whenever $\alpha \leqslant \beta$ in I we have

(15) $$\mathrm{pr}_\alpha^\lambda(z) = f_{\alpha\beta}^{\lambda\lambda}(\mathrm{pr}_\beta^\lambda(z)).$$

But $h^{\lambda\mu}(\mathrm{pr}^\mu(z)) = (f_{\alpha\alpha}^{\mu\lambda}(\mathrm{pr}_\alpha^\mu(z)))_{\alpha \in I}$; it therefore follows from (14) and (15) that if $\alpha \leqslant \beta$ and $\lambda \leqslant \mu$, we have

$$\mathrm{pr}_\alpha^\lambda(z) = f_{\alpha\alpha}^{\lambda\mu}(f_{\alpha\beta}^{\mu\mu}(\mathrm{pr}_\beta^\mu(z))) = f_{\alpha\beta}^{\lambda\mu}(\mathrm{pr}_\beta^\mu(z)),$$

which implies that $z \in E$. The converse is obvious, and we have therefore proved

PROPOSITION 4. *If* $(E_\alpha^\lambda, f_{\alpha\beta}^{\lambda\mu})$ *is an inverse system of sets relative to a product* $I \times L$ *of preordered sets, then (up to a canonical bijection) we have*

$$\varprojlim_{\alpha,\lambda} E_\alpha^\lambda = \varprojlim_{\lambda} (\varprojlim_{\alpha} E_\alpha^\lambda). \tag{16}$$

COROLLARY 1. *Let* $(E'^\lambda_\alpha, f'^{\mu\lambda}_{\alpha\beta})$ *be another inverse system of sets relative to* $I \times L$, *and for each* $(\alpha, \lambda) \in I \times L$ *let* u_α^λ *be a mapping of* E_α^λ *into* E'^λ_α *such that the* u_α^λ *form an inverse system of mappings. Then*

$$\varprojlim_{\alpha,\lambda} u_\alpha^\lambda = \varprojlim_{\lambda} (\varprojlim_{\alpha} u_\alpha^\lambda). \tag{17}$$

The verification is similar to that of Proposition 4.

COROLLARY 2. *Let* $(E_\alpha^\lambda, f_{\alpha\beta}^\lambda)_{\lambda \in L}$ *be a family of inverse systems of sets relative to* I. *If* $\prod_{\lambda \in L} f_{\alpha\beta}^\lambda$ *denotes the extension to products* (Chapter II, § 5, no. 7, Definition 2) *of the family of mappings* $(f_{\alpha\beta}^\lambda)_{\lambda \in L}$, *then* $\left(\prod_{\lambda \in L} E_\alpha^\lambda, \prod_{\lambda \in L} f_{\alpha\beta}^\lambda\right)$ *is an inverse system of sets relative to* I, *and (up to a canonical bijection) we have*

$$\varprojlim_{\alpha} \prod_{\lambda \in L} E_\alpha^\lambda = \prod_{\lambda \in L} (\varprojlim_{\alpha} E_\alpha^\lambda). \tag{18}$$

Consider the double inverse system $(E_\alpha^\lambda, g_{\alpha\beta}^{\lambda\mu})$ relative to $I \times L$, where the order relation on L is equality (no. 1, Example 1), and apply Proposition 4.

4. CONDITIONS FOR AN INVERSE LIMIT TO BE NON-EMPTY

In this subsection we shall give the two most frequently used sufficient conditions for an inverse limit to be non-empty (see also Exercise 5).

PROPOSITION 5. *Let* $(E_\alpha, f_{\alpha\beta})$ *be an inverse system of sets relative to a* directed *set* I *which has a* countable *cofinal subset, and suppose furthermore that the* $f_{\alpha\beta}$ *are* surjective. *Then, if* $E = \varprojlim E_\alpha$, *the canonical mapping* $f_\alpha : E \to E_\alpha$ *is surjective for each* $\alpha \in I$ *(and, a fortiori,* E *is not empty provided that none of the* E_α *is empty).*

Let (α_n) be a sequence of elements of I which form a cofinal subset of I. Since I is directed, we can define a sequence (β_n) of elements of I inductively by the conditions $\beta_0 = \alpha_0$, $\beta_n \geqslant \beta_i$ for $i < n$ and $\beta_n \geqslant \alpha_n$. Clearly the sequence (β_n) is increasing and forms a cofinal subset of I. In view of Proposition 1 of no. 1 and the relations $f_\alpha = f_{\alpha\beta_n} \circ f_{\beta_n}$ for

$\alpha \leqslant \beta_n$, we need only prove the Proposition for the case $I = \mathbf{N}$. Moreover, it is clear that it suffices to prove that f_0 is surjective. Let $x_0 \in E_0$. Define $x_n \in E_n$ (for $n \geqslant 1$) inductively to be an element of the set $\overset{-1}{f}_{n-1,n}(x_{n-1})$, which is possible because the latter set is non-empty by hypothesis. We then show by induction on $n - m$ that $x_m = f_{mn}(x_n)$ for $m \leqslant n$, and it follows that $x = (x_n)$ belongs to E.

The second criterion concerns inverse systems $(E_\alpha, f_{\alpha\beta})$ relative to an index set I such that for each $\alpha \in I$ we are given a set $\mathfrak{S}_\alpha$ of subsets of E_α which satisfy the following conditions :

(i) *Every intersection of sets belonging to* $\mathfrak{S}_\alpha$ *also belongs to* $\mathfrak{S}_\alpha$.

It follows in particular (by considering the intersection of the empty family) that $E_\alpha \in \mathfrak{S}_\alpha$.

(ii) *If a set of subsets* $\mathfrak{F} \subset \mathfrak{S}_\alpha$ *is such that every finite intersection of sets belonging to* $\mathfrak{F}$ *is non-empty, then* $\bigcap_{M \in \mathfrak{F}} M$ *is non-empty.*

In view of (i), it is clear that (ii) is equivalent to the following condition :

(ii)′ *If* $\mathfrak{G} \subset \mathfrak{S}_\alpha$ *is left directed (with respect to inclusion) and does not contain the empty set, then* $\bigcap_{M \in \mathfrak{G}} M$ *is non-empty.*

THEOREM 1. *Suppose that* I *is directed, that the sets* $\mathfrak{S}_\alpha$ *satisfy conditions* (i) *and* (ii), *and that the inverse system* $(E_\alpha, f_{\alpha\beta})$ *has the following properties:*

(iii) *For each pair of indices* α, β *such that* $\alpha \leqslant \beta$, *and each* $x_\alpha \in E_\alpha$, *we have* $\overset{-1}{f}_{\alpha\beta}(x_\alpha) \in \mathfrak{S}_\beta$.

(iv) *For each pair of indices* α, β *such that* $\alpha \leqslant \beta$, *and each* $M_\beta \in \mathfrak{S}_\beta$, *we have* $f_{\alpha\beta}(M_\beta) \in \mathfrak{S}_\alpha$.

Let $E = \varprojlim E_\alpha$ *and let* $f_\alpha : E \to E_\alpha$ *be the canonical mapping for each* $\alpha \in I$. *Then:*

(a) *For each* $\alpha \in I$ *we have*

$$f_\alpha(E) = \bigcap_{\beta \geqslant \alpha} f_{\alpha\beta}(E_\beta). \tag{19}$$

(b) *If* E_α *is non-empty for each* $\alpha \in I$, *then* E *is non-empty.*

Let Σ be the set of families $\mathfrak{A} = (A_\alpha)_{\alpha \in I}$ which satisfy the following conditions :

$$A_\alpha \neq \emptyset \quad \textit{and} \quad A_\alpha \in \mathfrak{S}_\alpha \quad \textit{for all} \quad \alpha \in I; \tag{20}$$

$$f_{\alpha\beta}(A_\beta) \subset A_\alpha \quad \textit{whenever} \quad \alpha \leqslant \beta. \tag{21}$$

If $\mathfrak{A} = (A_\alpha)$ and $\mathfrak{A}' = (A'_\alpha)$ are any two elements of Σ, let the relation $\mathfrak{A} \leqslant \mathfrak{A}'$ mean that $A_\alpha \supset A'_\alpha$ for all α. Clearly Σ is *ordered* by this relation.

(1) Let us first show that the ordered set Σ is *inductive*. Let L be a totally ordered set and let $\lambda \to \mathfrak{A}^\lambda = (A^\lambda_\alpha)_{\alpha \in I}$ be a strictly increasing mapping of L into Σ. For each $\alpha \in I$, put $B_\alpha = \bigcap_{\lambda \in L} A^\lambda_\alpha$. Then it is immediately seen that the family $\mathfrak{B} = (B_\alpha)_{\alpha \in I}$ satisfies (21); by reason of (i) and (ii), it also satisfies (20), hence belongs to Σ; and it is clear that $\mathfrak{B}$ is an upper bound of the set of the $\mathfrak{A}^\lambda$.

(2) Let $\mathfrak{A} = (A_\alpha)$ be a *maximal* element of Σ. We shall show that $A_\alpha = f_{\alpha\beta}(A_\beta)$ whenever $\alpha \leqslant \beta$. Let $A'_\alpha = \bigcap_{\beta \geqslant \alpha} f_{\alpha\beta}(A_\beta)$ for all $\alpha \in I$, and let us show that $\mathfrak{A}' = (A'_\alpha)$ belongs to Σ. Note first that if $\alpha \leqslant \beta \leqslant \gamma$, we have $f_{\alpha\gamma}(A_\gamma) = f_{\alpha\beta}(f_{\beta\gamma}(A_\gamma)) \subset f_{\alpha\beta}(A_\beta)$ by (21); moreover, $f_{\alpha\beta}(A_\beta) \in \mathfrak{S}_\alpha$ by (iv), and $f_{\alpha\beta}(A_\beta) \neq \emptyset$ by (20). Hence conditions (i) and (ii) show that $\mathfrak{A}'$ satisfies (20). Also, if $\alpha \leqslant \beta$, we have

$$f_{\alpha\beta}(A'_\beta) \subset \bigcap_{\gamma \geqslant \beta} f_{\alpha\beta}(f_{\beta\gamma}(A_\gamma)) = \bigcap_{\gamma \geqslant \beta} f_{\alpha\gamma}(A_\gamma);$$

and for each $\delta \geqslant \alpha$ there exists $\gamma \in I$ such that $\gamma \geqslant \delta$ and $\gamma \geqslant \beta$, so that $f_{\alpha\gamma}(A_\gamma) \subset f_{\alpha\delta}(A_\delta)$ and consequently

$$\bigcap_{\gamma \geqslant \beta} f_{\alpha\gamma}(A_\gamma) = \bigcap_{\delta \geqslant \alpha} f_{\alpha\delta}(A_\delta) = A'_\alpha.$$

Hence $\mathfrak{A}'$ satisfies (21) and therefore belongs to Σ. Since $A'_\alpha \subset A_\alpha$ for all α, the maximality of $\mathfrak{A}$ in Σ implies $\mathfrak{A}' = \mathfrak{A}$, and our assertion is proved.

(3) We shall establish next that if $\mathfrak{A} = (A_\alpha)$ is a *maximal* element of Σ, then each of the A_α *consists of a single element*. Let $x_\alpha \in A_\alpha$. For each $\beta \geqslant \alpha$, put $B_\beta = A_\beta \cap \overset{-1}{f}_{\alpha\beta}(x_\alpha)$; if β is not $\geqslant \alpha$, put $B_\beta = A_\beta$; then we shall see that $\mathfrak{B} = (B_\beta)$ belongs to Σ. If β is not $\geqslant \alpha$, the relation $\beta \leqslant \gamma$ implies $f_{\beta\gamma}(B_\gamma) \subset f_{\beta\gamma}(A_\gamma) \subset A_\beta = B_\beta$. If, on the other hand, $\alpha \leqslant \beta \leqslant \gamma$, then since

$$\overset{-1}{f}_{\alpha\gamma}(x_\alpha) = \overset{-1}{f}_{\beta\gamma}(\overset{-1}{f}_{\alpha\beta}(x_\alpha)),$$

we have

$$f_{\beta\gamma}(\overset{-1}{f}_{\alpha\gamma}(x_\alpha)) \subset \overset{-1}{f}_{\alpha\beta}(x_\alpha),$$

and since $f_{\beta\gamma}(A_\gamma) \subset A_\beta$, we again have $f_{\beta\gamma}(B_\gamma) \subset B_\beta$, so that the family $\mathfrak{B}$ satisfies (21). Since $A_\alpha = f_{\alpha\beta}(A_\beta)$ whenever $\alpha \leqslant \beta$ by part (2) of the proof, it is clear that $B_\beta \neq \emptyset$ for all $\beta \in I$. Finally, by virtue of (i) and (iii), we have $B_\beta \in \mathfrak{S}_\beta$ for all $\beta \in I$, and hence $\mathfrak{B} \in \Sigma$. Since $B_\beta \subset A_\beta$ for all $\beta \in I$, the maximality of $\mathfrak{A}$ implies that $B_\beta = A_\beta$ for all β, and in particular $A_\alpha = \{x_\alpha\}$.

(4) We are now in a position to prove Theorem 1. Let us start with (a). We have

$$f_\alpha(E) \subset \bigcap_{\beta \geqslant \alpha} f_{\alpha\beta}(E_\beta).$$

Conversely, let

$$x_\alpha \in \bigcap_{\beta \geqslant \alpha} f_{\alpha\beta}(E_\beta),$$

and put

$$B_\beta = \overset{-1}{f}_{\alpha\beta}(x_\alpha)$$

if $\beta \geqslant \alpha$, and $B_\beta = E_\beta$ otherwise. By the definition of x_α, the B_β are non-empty, and we have $B_\beta \in \mathfrak{S}_\beta$ for all $\beta \in I$ by virtue of (iii) and (i); moreover, it is evident that $f_{\beta\gamma}(B_\gamma) \subset B_\beta$ whenever $\beta \leqslant \gamma$. Hence $\mathfrak{B} = (B_\beta) \in \Sigma$. Let $\mathfrak{A} = (A_\alpha)$ be a maximal element of Σ such that $\mathfrak{A} \geqslant \mathfrak{B}$ (the existence of $\mathfrak{A}$ follows from (1) and § 2, no. 4, Theorem 2, Corollary 1). Since, by (3), A_β is of the form $\{y_\beta\}$ for all $\beta \in I$, it follows that $y = (y_\beta)$ belongs to E, and $f_\alpha(y) = y_\alpha = x_\alpha$ by definition.

Finally, (a) implies (b). We may as well assume that I is not empty (otherwise there is nothing to prove). The hypothesis that the E_α are non-empty implies that $f_{\alpha\beta}(E_\beta) \neq \emptyset$ for all $\beta \geqslant \alpha$. Since the $f_{\alpha\beta}(E_\beta)$, for fixed α and variable $\beta \geqslant \alpha$, form a left directed set of subsets of E_α belonging to $\mathfrak{S}_\alpha$, condition (ii)′ proves that

$$\bigcap_{\beta \geqslant \alpha} f_{\alpha\beta}(E_\beta) \neq \emptyset.$$

Hence $f_\alpha(E) \neq \emptyset$ by (a), and *a fortiori* $E \neq \emptyset$. Q.E.D.

Remark. Suppose that condition (iii) in Theorem 1 is replaced by the following weaker condition :

(iii)′ *For each* $\alpha \in I$ *and each non-empty set* $M_\alpha \in \mathfrak{S}_\alpha$ *there exists* $x_\alpha \in M_\alpha$ *such that* $\overset{-1}{f}_{\alpha\beta}(x_\alpha) \in \mathfrak{S}_\beta$ *for each* $\beta \geqslant \alpha$.

Then the conclusion (b) of Theorem 1 remains valid; the proofs of parts (1) and (2) of Theorem 1 remain unchanged and the proof of part (3) remains valid provided that we are careful to take $x_\alpha \in A_\alpha$ such that $\overset{-1}{f}_{\alpha\beta}(x_\alpha)$ belongs to $\mathfrak{S}_\beta$ for all $\beta \geqslant \alpha$. Finally, the proof of (4) shows that if

$$\bigcap_{\beta \geqslant \alpha} f_{\alpha\beta}(E_\beta) \neq \emptyset$$

and if we choose an x_α in this set such that $\overset{-1}{f_{\alpha\beta}}(x_\alpha) \in \mathfrak{S}_\beta$ whenever $\beta \geqslant \alpha$, there exists $y \in E$ such that $f_\alpha(y) = x_\alpha$, which proves our assertion.

Examples

(1) If the E_α are *finite* sets, Theorem 1 can be applied by taking $\mathfrak{S}_\alpha$ to be the set of *all* subsets of E_α. * This example is generalized in *General Topology* to the situation in which the E_α are *compact* topological spaces, the $f_{\alpha\beta}$ *continuous* maps, and $\mathfrak{S}_\alpha$ the set of *closed* subsets of E_α (*General Topology*, Chapter I, § 9, no. 6). *

* (2) Let A be a ring with an identity element, and for each $\alpha \in I$ let T_α be an *Artinian* left A-module. Let E_α be a *homogeneous space* for T_α on which T_α operates faithfully (so that E_α is an *affine space* attached to T_α). For $\beta \geqslant \alpha$, suppose that $f_{\alpha\beta} : E_\beta \to E_\alpha$ is an *affine mapping*. Take $\mathfrak{S}_\alpha$ to be the set consisting of the empty set and the *affine linear varieties* in E_α. Then condition (i) is trivially satisfied, and (ii) follows from the fact that T_α is Artinian; for this implies that there exists a minimal element in the set of finite intersections of sets $M \in \mathfrak{F}$, and this minimal element must be equal to $\bigcap_{M \in \mathfrak{F}} M$. Finally, since $f_{\alpha\beta}$ is affine, conditions (iii) and (iv) are trivially satisfied. *

5. DIRECT LIMITS

Let I be a (*right*) *directed* preordered set and let $(E_\alpha)_{\alpha \in I}$ be a family of sets indexed by I. For each pair (α, β) of elements of I such that $\alpha \leqslant \beta$, let $f_{\beta\alpha}$ be a *mapping of* E_α *into* E_β. Suppose that the $f_{\beta\alpha}$ satisfy the following conditions:

(LI_I) *The relations* $\alpha \leqslant \beta \leqslant \gamma$ *imply* $f_{\gamma\alpha} = f_{\gamma\beta} \circ f_{\beta\alpha}$.
(LI_{II}) *For each* $\alpha \in I$, $f_{\alpha\alpha}$ *is the identity mapping of* E_α.

Let G be the set which is the *sum* of the family of sets $(E_\alpha)_{\alpha \in I}$ (Chapter II, § 4, no. 8); by abuse of language, we shall identify the E_α with their canonical images in G, and for each $x \in G$ we shall denote by $\lambda(x)$ the unique index $\alpha \in I$ such that $x \in E_\alpha$. Let $R\{x, y\}$ denote the following relation between two elements x, y of G: "there exists an element $\gamma \in I$ such that $\gamma \geqslant \alpha = \lambda(x)$ and $\gamma \geqslant \beta = \lambda(y)$ for which $f_{\gamma\alpha}(x) = f_{\gamma\beta}(y)$". Then R is an *equivalence relation on* G. It is clear that R is reflexive and symmetric on G. To show that R is transitive, let $x \in E_\alpha$, $y \in E_\beta$, $z \in E_\gamma$, and suppose that there exist $\lambda \in I$ such that $\lambda \geqslant \alpha$, $\lambda \geqslant \beta$ and $f_{\lambda\alpha}(x) = f_{\lambda\beta}(y)$, and $\mu \in I$ such that $\mu \geqslant \beta$, $\mu \geqslant \gamma$, and $f_{\mu\beta}(y) = f_{\mu\gamma}(z)$. Since I is a directed set, there exists $\nu \in I$ such that $\nu \geqslant \lambda$ and $\nu \geqslant \mu$;

by (LI_I) we then have

$$f_{\nu\alpha}(x) = f_{\nu\lambda}(f_{\lambda\alpha}(x)) = f_{\nu\lambda}(f_{\lambda\beta}(y)) = f_{\nu\beta}(y)$$
$$= f_{\nu\mu}(f_{\mu\beta}(y)) = f_{\nu\mu}(f_{\mu\gamma}(z)) = f_{\nu\gamma}(z),$$

which establishes our assertion. The quotient set $E = G/R$ is called the *direct limit of the family* $(E_\alpha)_{\alpha\in I}$ *with respect to the family of mappings* $(f_{\beta\alpha})$, and is written $E = \varinjlim (E_\alpha, f_{\beta\alpha})$, or simply $E = \varinjlim E_\alpha$ when there is no risk of ambiguity. By abuse of language, the pair $((E_\alpha), (f_{\beta\alpha}))$ (which is usually written $(E_\alpha, f_{\beta\alpha})$) is called a *direct system of sets*, relative to the directed set I.

¶ Clearly E is not empty provided at least one of the E_α is not empty. We denote by f_α the restriction to E_α of the canonical mapping f of G onto $E = G/R$; f_α is called the *canonical mapping* of E_α into E. If $\alpha \leqslant \beta$, we have the relation

(22) $$f_\beta \circ f_{\beta\alpha} = f_\alpha$$

since for each $x \in E_\alpha$ we have $f_{\beta\beta}(f_{\beta\alpha}(x)) = f_{\beta\alpha}(x)$ by (LI_I); and therefore the elements $x \in E_\alpha$ and $f_{\beta\alpha}(x) \in E_\beta$ are congruent mod R.

Examples

(1) Let A, B be two sets, and let $(V_\alpha)_{\alpha\in I}$ be a family of subsets of A whose index set I is directed, and such that the relation $\alpha \leqslant \beta$ implies $V_\beta \subset V_\alpha$. Let E_α denote the set of all mappings of V_α into B, and for each pair of indices (α, β) such that $\alpha \leqslant \beta$ let $f_{\beta\alpha}$ be the mapping of E_α into E_β which sends each function $u \in E_\alpha$ to its *restriction* to V_β. It is obvious that the conditions (LI_I) and (LI_{II}) are satisfied, and the set $E = \varinjlim E_\alpha$ is called the set of *germs of mappings* of the V_α into B. * The most frequent case is that in which (V_α) is the family of *neighbourhoods* of a subset of a topological space A (*General Topology*, Chapter I, §6, no. 10). *

(2) Suppose that, for each $\alpha \in I$, E_α is the same set F and that whenever $\alpha \leqslant \beta$, $f_{\beta\alpha}$ is the identity mapping of F onto itself. Then there exists a *canonical bijection* of $\varinjlim E_\alpha$ onto F. In order to define $\varinjlim E_\alpha$, we have to form the set G which is the sum of the family (E_α); G is therefore the union of a family (G_α) of mutually disjoint sets, and for each $\alpha \in I$ there is a canonical bijection $h_\alpha : F \to G_\alpha$. We have next to consider the equivalence relation R on G corresponding to the partition $(P_y)_{y\in F}$, where P_y is the set of all $h_\alpha(y)$ as α runs through I. Clearly $y \to P_y$ is a bijection whose inverse is the bijection required. We shall identify F with $\varinjlim E_\alpha$ by means of this canonical bijection.

Lemma 1. *Let* $(E_\alpha, f_{\beta\alpha})$ *be a direct system of sets,* $E = \varinjlim E_\alpha$ *its direct limit, and for each* $\alpha \in I$ *let* $f_\alpha \colon E_\alpha \to E$ *the canonical mapping.*

(i) *Let* $(x^{(i)})_{1 \leqslant i \leqslant n}$ *be a finite system of elements of* E. *Then there exists* $\alpha \in I$ *and a finite system* $(x_\alpha^{(i)})_{1 \leqslant i \leqslant n}$ *of elements of* E_α *such that* $x^{(i)} = f_\alpha(x_\alpha^{(i)})$ *for* $1 \leqslant i \leqslant n$.

(ii) *Let* $(y_\alpha^{(i)})_{1 \leqslant i \leqslant n}$ *be a finite system of elements of some* E_α. *If* $f_\alpha(y_\alpha^{(i)}) = f_\alpha(y_\alpha^{(j)})$ *for each pair of indices* (i, j), *then there exists* $\beta \geqslant \alpha$ *such that* $f_{\beta\alpha}(y_\alpha^{(i)}) = f_{\beta\alpha}(y_\alpha^{(j)})$ *for each pair* (i, j).

(i) By the definition of E there exists for each i an index $\beta_i \in I$ and an element $z_{\beta_i} \in E_{\beta_i}$ such that $x^{(i)} = f_{\beta_i}(z_{\beta_i})$. Take α such that $\alpha \geqslant \beta_i$ for $1 \leqslant i \leqslant n$, and $x_\alpha^{(i)} = f_{\alpha\beta_i}(z_{\beta_i})$.

(ii) By the definition of E, for each pair (i, j) there exists $\gamma_{ij} \in I$ such that $\gamma_{ij} \geqslant \alpha$ and $f_{\gamma_{ij}\alpha}(y_\alpha^{(i)}) = f_{\gamma_{ij}\alpha}(y_\alpha^{(j)})$. Take β such that $\beta \geqslant \gamma_{ij}$ for all pairs (i, j), and use the relations $f_{\beta\alpha} = f_{\beta\gamma_{ij}} \circ f_{\gamma_{ij}\alpha}$.

6. DIRECT SYSTEMS OF MAPPINGS

PROPOSITION 6. *Let* I *be a directed set, let* $(E_\alpha, f_{\beta\alpha})$ *be a direct system of sets relative to* I, *let* $E = \varinjlim E_\alpha$ *be the direct limit, and for each* $\alpha \in I$ *let*

$$f_\alpha : E_\alpha \to E$$

be the canonical mapping. For each $\alpha \in I$, *let* u_α *be a mapping of* E_α *into a set* F *such that*

$$u_\beta \circ f_{\beta\alpha} = u_\alpha \quad \textit{whenever } \alpha \leqslant \beta. \tag{23}$$

Then:

(a) *There exists a unique mapping u of* E *into* F *such that*

$$u_\alpha = u \circ f_\alpha \quad \textit{for all } \alpha \in I. \tag{24}$$

(b) *u is surjective if and only if* F *is the union of the sets* $u_\alpha(E_\alpha)$.

(c) *u is injective if and only if for each* $\alpha \in I$ *the relations* $x \in E_\alpha$, $y \in E_\alpha$, $u_\alpha(x) = u_\alpha(y)$ *imply that there exists* $\beta \geqslant \alpha$ *such that* $f_{\beta\alpha}(x) = f_{\beta\alpha}(y)$.

(a) With the notation of no. 5, let v be the mapping of G into F which agrees with u_α on E_α for each $\alpha \in I$ (Chapter II, §4, no. 7, Proposition 8). The hypothesis implies that v is compatible with the equivalence relation R (Chapter II, §6, no. 5); hence there exists a unique mapping u of $E = G/R$ into F such that $v = u \circ f$ (*loc. cit.*).

(b) Since E is the union of the $f_\alpha(E_\alpha)$, the relation $F = \bigcup_{\alpha \in I} u_\alpha(E_\alpha)$ is clearly necessary and sufficient for u to be surjective.

(c) By Lemma 1 of no. 5 any two elements of E can always be written in the form $f_\alpha(x)$ and $f_\alpha(y)$, where $x \in E_\alpha$ and $y \in E_\alpha$, for a suitable choice of $\alpha \in I$. It follows also from the lemma that the relation $f_\alpha(x) = f_\alpha(y)$ is equivalent to the existence of $\beta \geqslant \alpha$ such that $f_{\beta\alpha}(x) = f_{\beta\alpha}(y)$. Since $u_\alpha(x) = u(f_\alpha(x))$ and $u_\alpha(y) = u(f_\alpha(y))$, this completes the proof.

¶ If the mapping u is *bijective*, it is sometimes said, by abuse of language, that F is the direct limit of the family (E_α).

Remark. Suppose that each of the mappings $f_{\beta\alpha}$ is *injective*. Then each of the f_α is *injective*, by the definition of the relation R. In this case we generally identify E_α and $f_\alpha(E_\alpha)$ and consider E therefore as the *union* of the E_α. Conversely, let $(F_\alpha)_{\alpha \in I}$ be an increasing family of subsets of a set F and suppose that F is the *union* of this family. If $j_{\beta\alpha}$ denotes the canonical injection of F_α into F_β for $\alpha \leqslant \beta$, then it follows from Proposition 6 that we may identify F with the direct limit of the family F_α with respect to the family of mappings $(j_{\beta\alpha})$, and the canonical mapping of F_α into $\varinjlim F_\alpha$ with the canonical injection of F_α into F, for each $\alpha \in I$.

COROLLARY 1. *Let* $(E_\alpha, f_{\beta\alpha})$ *and* $(F_\alpha, g_{\beta\alpha})$ *be two direct systems of sets relative to the same index set* I; *let* $E = \varinjlim E_\alpha$, $F = \varinjlim F_\alpha$, *and for each* $\alpha \in I$ *let* f_α (resp. g_α) *be the canonical mapping of* E_α (resp. F_α) *into* E (resp. F). *For each* $\alpha \in I$ *let* u_α *be a mapping of* E_α *into* F_α *such that, whenever* $\alpha \leqslant \beta$, *the diagram*

$$\begin{array}{ccc} E_\alpha & \xrightarrow{u_\alpha} & F_\alpha \\ {\scriptstyle f_{\beta\alpha}}\downarrow & & \downarrow{\scriptstyle g_{\beta\alpha}} \\ E_\beta & \xrightarrow[u_\beta]{} & F_\beta \end{array}$$

is commutative. Then there exists a unique mapping $u : E \to F$ *such that, for each* $\alpha \in I$, *the diagram*

$$\begin{array}{ccc} E_\alpha & \xrightarrow{u_\alpha} & F_\alpha \\ {\scriptstyle f_\alpha}\downarrow & & \downarrow{\scriptstyle g_\alpha} \\ E & \xrightarrow[u]{} & F \end{array}$$

is commutative.

Put $v_\alpha = g_\alpha \circ u_\alpha$. If $\alpha \leqslant \beta$, then by (22) we have

$$v_\beta \circ f_{\beta\alpha} = g_\beta \circ u_\beta \circ f_{\beta\alpha} = g_\beta \circ g_{\beta\alpha} \circ u_\alpha = g_\alpha \circ u_\alpha = v_\alpha.$$

We may therefore apply Proposition 6 to the mappings v_α, whence the existence and uniqueness of a mapping $u : E \to F$ such that

$$u \circ f_\alpha = v_\alpha = g_\alpha \circ u_\alpha$$

for all $\alpha \in I$.

¶ A family of mappings $u_\alpha : E_\alpha \to F_\alpha$ which satisfies the conditions of Corollary 1 is called a *direct system of mappings* of $(E_\alpha, f_{\beta\alpha})$ into $(F_\alpha, g_{\beta\alpha})$. The mapping defined in Corollary 1 is called the *direct limit* of the family (u_α) and is written $u = \varinjlim u_\alpha$ when there is no risk of ambiguity.

Corollary 2. *Let* $(E_\alpha, f_{\beta\alpha})$, $(F_\alpha, g_{\beta\alpha})$, $(G_\alpha, h_{\beta\alpha})$ *be three direct systems of sets relative to* I. *Let* $E = \varinjlim E_\alpha$, $F = \varinjlim F_\alpha$, $G = \varinjlim G_\alpha$, *and let* f_α (resp. g_α, h_α) *be the canonical mapping of* E_α (resp. F_α, G_α) *into* E (resp. F, G). *If* (u_α) *and* (v_α) *are two direct systems of mappings* $u_\alpha : E_\alpha \to F_\alpha$, $v_\alpha : F_\alpha \to G_\alpha$, *then the mappings* $v_\alpha \circ u_\alpha : E_\alpha \to G_\alpha$ *form a direct system of mappings, and we have*

$$\varinjlim (v_\alpha \circ u_\alpha) = (\varinjlim v_\alpha) \circ (\varinjlim u_\alpha). \tag{25}$$

For if we put $w_\alpha = v_\alpha \circ u_\alpha$, then for $\alpha \leqslant \beta$ we have

$$h_{\beta\alpha} \circ w_\alpha = (h_{\beta\alpha} \circ v_\alpha) \circ u_\alpha = (v_\beta \circ g_{\beta\alpha}) \circ u_\alpha = v_\beta \circ (u_\beta \circ f_{\beta\alpha}) = w_\beta \circ f_{\beta\alpha},$$

which shows that (w_α) is a direct system of mappings. Furthermore, if $u = \varinjlim u_\alpha$ and $v = \varinjlim v_\alpha$, then for all $\alpha \in I$ we have

$$(v \circ u) \circ f_\alpha = v \circ (g_\alpha \circ u_\alpha) = h_\alpha \circ (v_\alpha \circ u_\alpha),$$

and by virtue of the uniqueness of the direct limit, we have $v \circ u = \varinjlim w_\alpha$.

Proposition 7. *Let* $(E_\alpha, f_{\beta\alpha})$ *and* $(E'_\alpha, f'_{\beta\alpha})$ *be two direct systems of sets relative to* I, *and for each* $\alpha \in I$ *let* u_α *be a mapping of* E_α *into* E'_α *such that the* u_α *form a direct system of mappings. Let* $u = \varinjlim u_\alpha$. *If each* u_α *is injective* (resp. *surjective*) *then* u *is injective* (resp. *surjective*).

Let $E = \varinjlim E_\alpha$, $E' = \varinjlim E'_\alpha$, and let $f_\alpha : E_\alpha \to E$, $f'_\alpha : E'_\alpha \to E'$ be the canonical mappings. Suppose that each u_α is injective. To show that u is injective it is enough, by Proposition 6, to verify that if $x \in E_\alpha$ and $y \in E_\alpha$ are such that $f'_\alpha(u_\alpha(x)) = f'_\alpha(u_\alpha(y))$, then there exists $\beta \geqslant \alpha$ such that $f_{\beta\alpha}(x) = f_{\beta\alpha}(y)$. Now the hypothesis implies (no. 6, Lemma 1) that there exists $\beta \geqslant \alpha$ such that

$$f'_{\beta\alpha}(u_\alpha(x)) = f'_{\beta\alpha}(u_\alpha(y)), \quad \text{i.e.,} \quad u_\beta(f_{\beta\alpha}(x)) = u_\beta(f_{\beta\alpha}(y)),$$

and hence $f_{\beta\alpha}(x) = f_{\beta\alpha}(y)$ since u_β is injective.

Now suppose that the u_α are surjective. Then we have

$$E' = \bigcup_\alpha f'_\alpha(E'_\alpha) = \bigcup_\alpha f'_\alpha(u_\alpha(E_\alpha)) = \bigcup_\alpha u(f_\alpha(E_\alpha)) = u\Big(\bigcup_\alpha f_\alpha(E_\alpha)\Big) = u(E).$$

With the notation of Proposition 7, let M_α be a subset of E_α for each $\alpha \in I$; if we have $f_{\beta\alpha}(M_\alpha) \subset M_\beta$ whenever $\alpha \leqslant \beta$, the family $(M_\alpha)_{\alpha \in I}$ is said to be a *direct system of subsets of the* E_α. Let $g_{\beta\alpha}$ (where $\alpha \leqslant \beta$) be the mapping of M_α into M_β whose graph is the same as that of the restriction of $f_{\beta\alpha}$ to M_α. Then it is clear that $(M_\alpha, g_{\beta\alpha})$ is a direct system of sets; and Proposition 7, applied to the canonical injections $j_\alpha : M_\alpha \to E_\alpha$, allows us to *identify* $M = \varinjlim M_\alpha$ with a subset of E by means of the injection $j = \varinjlim j_\alpha$.

Corollary. *Let* $(E_\alpha, f_{\beta\alpha})$ *and* $(E'_\alpha, f'_{\beta\alpha})$ *be two direct systems of sets, let* (u_α) *be a direct system of mappings* $u_\alpha : E_\alpha \to E'_\alpha$, *and let* $u = \varinjlim u_\alpha$.

(i) *Let* (M_α) *be a direct system of subsets of the* E_α. *Then* $(u_\alpha(M_\alpha))$ *is a direct system of subsets of the* E'_α, *and we have*

$$\varinjlim u_\alpha(M_\alpha) = u(\varinjlim M_\alpha). \tag{26}$$

(ii) *Let* $(a'_\alpha)_{\alpha \in I}$ *be a family such that* $a'_\alpha \in E'_\alpha$ *for each* $\alpha \in I$ *and* $f'_{\beta\alpha}(a'_\alpha) = a'_\beta$ *whenever* $\alpha \leqslant \beta$. *Then the sets* $\overset{-1}{u}_\alpha(a'_\alpha)$ *form a direct system of subsets of the* E_α, *and we have*

$$\varinjlim \overset{-1}{u}_\alpha(a'_\alpha) = \overset{-1}{u}(a'), \tag{27}$$

where a' *is the unique element of* $\varinjlim E'_\alpha$ *which is the canonical image of* a'_α *for each* $\alpha \in I$.

(i) It is evident that the $u_\alpha(M_\alpha)$ form a direct system of subsets of the E'_α, and we may write $u_\alpha(M_\alpha) = v_\alpha(M_\alpha)$, where v_α is the mapping of M_α onto $u_\alpha(M_\alpha)$ whose graph is the same as that of the restriction of u_α to M_α. The formula (26) then follows from Proposition 7 because the v_α are surjective.

(ii) Let $N_\alpha = \overset{-1}{u}_\alpha(a'_\alpha)$. If $\alpha \leqslant \beta$ and $x_\alpha \in N_\alpha$, then

$$u_\beta(f_{\beta\alpha}(x_\alpha)) = f'_{\beta\alpha}(u_\alpha(x_\alpha)) = f'_{\beta\alpha}(a'_\alpha) = a'_\beta;$$

hence $f_{\beta\alpha}(x_\alpha) \in N_\beta$, and the N_α therefore form a direct system of subsets of the E_α. With the notation of the proof of Proposition 7, consider an element $x \in \varinjlim N_\alpha$. There exists $\alpha \in I$ and $x_\alpha \in N_\alpha$ such that $x = f_\alpha(x_\alpha)$, so that $u(x) = u(f_\alpha(x_\alpha)) = f'_\alpha(u_\alpha(x_\alpha)) = f'_\alpha(a'_\alpha) = a'$. Conversely, if

$x \in \overset{-1}{u}(a')$ and if $x = f_\alpha(x_\alpha)$ for some $\alpha \in \mathrm{I}$ and some $x_\alpha \in \mathrm{E}_\alpha$, then we have $a' = u(f_\alpha(x_\alpha)) = f'_\alpha(u_\alpha(x_\alpha)) = f'_\alpha(a'_\alpha)$. Hence (no. 5, Lemma 1) there exists $\beta \geqslant \alpha$ such that $f'_{\beta\alpha}(u'_\alpha(x_\alpha)) = f'_{\beta\alpha}(a'_\alpha) = a'_\beta$; i.e., $u_\alpha(f_{\beta\alpha}(x_\alpha)) = a'_\beta$, and therefore $f_{\beta\alpha}(x_\alpha) \in \mathrm{N}_\beta$. Since $x = f_\beta(f_{\beta\alpha}(x_\alpha))$, it follows that $x \in \varinjlim \mathrm{N}_\alpha$.

Remark. Suppose that, for each $\alpha \in \mathrm{I}$, $u_\alpha : \mathrm{E}_\alpha \to \mathrm{E}'$ is a mapping such that the family (u_α) satisfies (23). Consider the direct system $(\mathrm{E}_\alpha, i_{\beta\alpha})$ relative to I, where $\mathrm{E}'_\alpha = \mathrm{E}'$ for all $\alpha \in \mathrm{I}$, and $i_{\beta\alpha}$ is the identity mapping of E'. Then (no. 5, Example 2) E' can be identified canonically with $\varinjlim \mathrm{E}'_\alpha$. If u_α is considered as a mapping of E_α into E'_α, then (u_α) is a direct system of mappings, and the mapping $u : \mathrm{E} \to \mathrm{E}'$ defined by (24) is identified with the direct limit of this system of mappings. Hence, by abuse of language, we write $u = \varinjlim u_\alpha$.

If J is a subset of I which is directed (with respect to the induced preordering), it is clear that the pair consisting of the subfamily $(\mathrm{E}_\alpha)_{\alpha \in \mathrm{J}}$ and the family $(f_{\beta\alpha})$, where $\alpha \leqslant \beta$ and $\alpha \in \mathrm{J}$ and $\beta \in \mathrm{J}$, is a direct system of sets relative to J; it is said to be obtained by *restricting* the index set to J. Let E, E′ respectively denote the direct limits of the families $(\mathrm{E}_\alpha)_{\alpha \in \mathrm{I}}$ and $(\mathrm{E}_\alpha)_{\alpha \in \mathrm{J}}$, and for each $\alpha \in \mathrm{I}$ let $f_\alpha : \mathrm{E}_\alpha \to \mathrm{E}$ denote the canonical mapping. Then $(f_\alpha)_{\alpha \in \mathrm{J}}$ is a direct system of mappings, and consequently $g = \varinjlim f_\alpha$ is a mapping of E′ into E, called *canonical*. Moreover, if J′ is a directed subset of J, if E″ is the direct limit of the family $(\mathrm{E}_\alpha)_{\alpha \in \mathrm{J}'}$, and if

$$g' : \mathrm{E}'' \to \mathrm{E}' \qquad \text{and} \qquad g'' : \mathrm{E}'' \to \mathrm{E}$$

are the canonical mappings, then it follows immediately from Proposition 6 that

$$g'' = g \circ g'. \tag{28}$$

Proposition 8. *Let* I *be a directed set, let* $(\mathrm{E}_\alpha, f_{\beta\alpha})$ *be a direct system of sets relative to* I, *and let* $\mathrm{E} = \varinjlim \mathrm{E}_\alpha$ *be its direct limit. Let* J *be a cofinal subset of* I *and let* E′ *be the direct limit of the direct system of sets obtained from* $(\mathrm{E}_\alpha, f_{\beta\alpha})$ *by restricting the index set to* J. *Then the canonical mapping* g *of* E′ *into* E *is bijective.*

J is necessarily a directed set (§ 1, no. 10). We shall use the criteria of Proposition 6 to show that g is bijective. The condition for injectivity follows immediately from the definitions and from Lemma 1 of no. 5. To show that g is surjective, we note that for each $\alpha \in \mathrm{J}$ we have

$$g(\mathrm{E}_\alpha) = f_\alpha(\mathrm{E}_\alpha).$$

Now, for each $\beta \in I$, there exists $\gamma \in J$ such that $\beta \leqslant \gamma$, from which it follows that $g(E_\gamma) \supset g(f_{\gamma\beta}(E_\beta)) = f_\beta(E_\beta)$. E is therefore the union of the sets $g(E_\alpha)$ as α runs through J.

7. DOUBLE DIRECT LIMIT. PRODUCT OF DIRECT LIMITS

Let I, L be two directed sets, and let $I \times L$ be their product (§ 1, no. 4) which, with the product preordering, is again a directed set. Consider a direct system of sets $(E_\alpha^\lambda, f_{\beta\alpha}^{\mu\lambda})$ relative to $I \times L$. We have then

$$f_{\gamma\alpha}^{\nu\lambda} = f_{\gamma\beta}^{\mu\nu} \circ f_{\beta\alpha}^{\mu\lambda} \tag{29}$$

whenever $\alpha \leqslant \beta \leqslant \gamma$ and $\lambda \leqslant \mu \leqslant \nu$.

Let E or $\varinjlim_{\alpha,\lambda} E_\alpha^\lambda$ denote the direct limit of this direct system. For each $\lambda \in L$ put $g_{\beta\alpha}^\lambda = f_{\beta\alpha}^{\lambda\lambda} : E_\alpha^\lambda \to E_\beta^\lambda$. Then from (29) we have

$$g_{\gamma\alpha}^\lambda = g_{\gamma\beta}^\lambda \circ g_{\beta\alpha}^\lambda \quad \text{whenever } \alpha \leqslant \beta \leqslant \gamma; \tag{30}$$

in other words, $(E_\alpha^\lambda, g_{\beta\alpha}^\lambda)$ is a direct system of sets relative to I. Let $F^\lambda = \varinjlim_\alpha E_\alpha^\lambda$ denote its direct limit. If $\lambda \leqslant \mu$ are fixed elements of L, it follows from (29) that the mappings $h_\alpha^{\mu\lambda} = f_{\alpha\alpha}^{\lambda\mu} : E_\alpha^\lambda \to E_\alpha^\mu$ form a direct system of mappings. Let $h^{\mu\lambda} = \varinjlim_\alpha h_\alpha^{\mu\lambda} : F^\lambda \to F^\mu$ denote the direct limit of this system of mappings. If $\lambda \leqslant \mu \leqslant \nu$ in L, then

$$h^{\nu\lambda} = h^{\nu\mu} \circ h^{\mu\lambda} \tag{31}$$

(no. 6, Proposition 6, Corollary 2), and therefore $(F^\lambda, h^{\mu\lambda})$ is a direct system of sets relative to L. Let $F = \varinjlim F^\lambda$ be its direct limit. We shall define a *canonical bijection* $E \to F$. For this purpose, let g_α^λ denote the canonical mapping $E_\alpha^\lambda \to F^\lambda$, and h^λ the canonical mapping $F^\lambda \to F$, and put $u_\alpha^\lambda = h^\lambda \circ g_\alpha^\lambda$. If $\alpha \leqslant \beta$ and $\lambda \leqslant \mu$, then we have

$$\begin{aligned} u_\beta^\mu \circ f_{\beta\alpha}^{\mu\lambda} &= h^\mu \circ g_\beta^\mu \circ f_{\beta\alpha}^{\mu\lambda} = h^\mu \circ g_\beta^\mu \circ f_{\beta\alpha}^{\mu\mu} \circ f_{\alpha\alpha}^{\mu\lambda} = h^\mu \circ g_\alpha^\mu \circ f_{\alpha\alpha}^{\mu\lambda} \\ &= h^\mu \circ h^{\mu\lambda} \circ g_\alpha^\lambda = h^\lambda \circ g_\alpha^\lambda = u_\alpha^\lambda \end{aligned}$$

from (29) and the definition of the $h^{\mu\lambda}$. Hence the u_α^λ form a direct system of mappings relative to $I \times L$. Put $u = \varinjlim_{\alpha,\lambda} u_\alpha^\lambda : E \to F$. We shall show that u is bijective by applying the criteria of no. 6, Proposition 6. In the first place, F is the union of the sets $h^\lambda(F^\lambda)$, and each F^λ is the union of the sets $g_\alpha^\lambda(E_\alpha^\lambda)$; hence F is the union of the sets

$$h^\lambda(g_\alpha^\lambda(E_\alpha^\lambda)) = u_\alpha^\lambda(E_\alpha^\lambda).$$

Next, let x, y be two elements of E_α^λ such that $u_\alpha^\lambda(x) = u_\alpha^\lambda(y)$, i.e., $h^\lambda(g_\alpha^\lambda(x)) = h^\lambda(g_\alpha^\lambda(y))$. Then (no. 5, Lemma 1) there exists $\mu \geqslant \lambda$ such that $h^{\mu\lambda}(g_\alpha^\lambda(x)) = h^{\mu\lambda}(g_\alpha^\lambda(y))$, i.e., $g_\alpha^\mu(f_{\alpha\alpha}^{\mu\lambda}(x)) = g_\alpha^\mu(f_{\alpha\alpha}^{\mu\lambda}(y))$; likewise there exists $\beta \geqslant \alpha$ such that $g_{\beta\alpha}^\mu(f_{\alpha\alpha}^{\mu\lambda}(x)) = g_{\beta\alpha}^\mu(f_{\alpha\alpha}^{\mu\lambda}(y))$ (no. 5, Lemma 1), i.e., $f_{\beta\alpha}^{\mu\lambda}(x) = f_{\beta\alpha}^{\mu\lambda}(y)$; and this shows (no. 6, Proposition 6) that u is injective. We have therefore proved:

Proposition 9. *If $(E_\alpha^\lambda, f_{\beta\alpha}^{\mu\lambda})$ is a direct system of sets relative to a product $I \times L$ of two directed sets, then (up to a canonical bijection) we have*

$$(32)\qquad \varinjlim_{\alpha,\lambda} E_\alpha^\lambda = \varinjlim_{\lambda} (\varinjlim_{\alpha} E_\alpha^\lambda).$$

Corollary. *Let $(E'^\lambda_\alpha, f'^{\mu\lambda}_{\beta\alpha})$ be another direct system of sets relative to $I \times L$, and for each $(\alpha, \lambda) \in I \times L$ let u_α^λ be a mapping of E_α^λ into E'^λ_α, such that the u_α^λ form a direct system of mappings. Then we have*

$$(33)\qquad \varinjlim_{\alpha,\lambda} u_\alpha^\lambda = \varinjlim_{\lambda} (\varinjlim_{\alpha} u_\alpha^\lambda).$$

We leave the verification to the reader.

Proposition 10. *Let $(E_\alpha, f_{\beta\alpha})$ and $(E'_\alpha, f'_{\beta\alpha})$ be two direct systems of sets, both relative to the same directed set I. Let $E = \varinjlim E_\alpha$, $E' = \varinjlim E'_\alpha$, and let $f_\alpha : E_\alpha \to E$, $f'_\alpha : E'_\alpha \to E'$ denote the canonical mappings, for each $\alpha \in I$. Then $(E_\alpha \times E'_\alpha, f_{\beta\alpha} \times f'_{\beta\alpha})$ is a direct systems of sets, $(f_\alpha \times f'_\alpha)$ is a direct system of mappings, and $\varinjlim (f_\alpha \times f'_\alpha)$ is a bijection*

$$(34)\qquad \varinjlim (E_\alpha \times E'_\alpha) \to (\varinjlim E_\alpha) \times (\varinjlim E'_\alpha).$$

The first two assertions of the Proposition are immediately verified. To show that $g = \varinjlim (f_\alpha \times f'_\alpha)$ is bijective, we apply Proposition 6 of no. 6. Clearly $E \times E'$ is the union of the sets $f_\alpha(E_\alpha) \times f'_\alpha(E'_\alpha)$; hence g is surjective. If (x, x') and (y, y') are two elements of $E_\alpha \times E'_\alpha$ such that $f_\alpha(x) = f_\alpha(y)$ and $f'_\alpha(x') = f'_\alpha(y')$, then (no. 5, Lemma 1) there exist two elements β, γ of I such that $\beta \geqslant \alpha$, $\gamma \geqslant \alpha$, and $f_{\beta\alpha}(x) = f_{\beta\alpha}(y)$, $f'_{\gamma\alpha}(x') = f'_{\gamma\alpha}(y')$; since I is directed, there exists $\delta \in I$ such that $\delta \geqslant \beta$ and $\delta \geqslant \gamma$; hence $f_{\delta\alpha}(x) = f_{\delta\alpha}(y)$ and $f'_{\delta\alpha}(x') = f'_{\delta\alpha}(y')$. This completes the proof.

The bijection g is called *canonical*.

COROLLARY. *Let* $(F_\alpha, g_{\beta\alpha})$ *and* $(F'_\alpha, g'_{\beta\alpha})$ *be two direct systems of sets relative to* I, *and for each* $\alpha \in I$ *let* $u_\alpha : E_\alpha \to F_\alpha$, $u'_\alpha : E'_\alpha \to F'_\alpha$ *be mappings such that* (u_α) *and* (u'_α) *are two direct systems of mappings. Then* $(u_\alpha \times u'_\alpha)$ *is a direct system of mappings, and (up to canonical bijections) we have*

$$\varinjlim (u_\alpha \times u'_\alpha) = (\varinjlim u_\alpha) \times (\varinjlim u'_\alpha). \tag{35}$$

We leave the verification to the reader.

EXERCISES

§ 1

1. Let E be an ordered set in which there exists at least one pair of distinct comparable elements. Show that, if $R\{x, y\}$ denotes the relation "$x \in E$ and $y \in E$ and $x < y$", then R satisfies the first two conditions of no. 1 but not the third.

2. (a) Let E be a preordered set and let $S\{x, y\}$ be an equivalence relation on E. Let $R\{X, Y\}$ denote the relation "$X \in E/S$ and $Y \in E/S$ and for each $x \in X$ there exists $y \in Y$ such that $x \leqslant y$". Show that R is a preorder relation on E/S, called the *quotient* by S of the relation $x \leqslant y$. The quotient set E/S, endowed with this preorder relation, is called (by abuse of language; cf. Chapter IV, § 2, no. 6) the *quotient* by S of the preordered set E.

(b) Let φ be the canonical mapping of E onto E/S. Show that if g is a mapping of the preordered quotient set E/S into a preordered set F such that $g \circ \varphi$ is an increasing mapping, then g is an increasing mapping. The mapping φ is increasing if and only if S satisfies the following condition :

(C) The relations $x \leqslant y$ and $x \equiv x' \pmod{S}$ in E imply that there exists $y' \in E$ such that $y \equiv y' \pmod{S}$ and $x' \leqslant y'$.

If this condition is satisfied, the equivalence relation S is said to be *weakly compatible* (in x and y) with the preorder relation $x \leqslant y$. Every equivalence relation S which is *compatible* (in x) with the preorder relation $x \leqslant y$ (Chapter II, § 6, no. 3) is *a fortiori* weakly compatible (in x and y) with this relation.

(c) Let E_1 and E_2 be two preordered sets. Show that if S_1 is the equivalence relation $\mathrm{pr}_1 z = \mathrm{pr}_1 z'$ on $E_1 \times E_2$, then S_1 is weakly compatible in z and t with the product preorder relation $z \leqslant t$ on $E_1 \times E_2$ (but is not usually compatible with this relation in z or t separately); moreover, if φ_1 is the canonical mapping of $E_1 \times E_2$ onto $(E_1 \times E_2)/S_1$, and if $pr_1 = f_1 \circ \varphi_1$ is the canonical decomposition of pr_1 with respect to the equivalence relation S_1, then f_1 is an isomorphism of $(E_1 \times E_2)/S_1$ onto E_1.

(d) With the hypothesis of (a), suppose that E is an *ordered* set and that the following condition is satisfied :

(C') The relations $x \leqslant y \leqslant z$ and $x \equiv z \pmod{S}$ in E imply $x \equiv y \pmod{S}$.

Show that $R\{X, Y\}$ is then an *order* relation between X and Y on E/S.

(e) Give an example of a totally ordered set E with four elements and an equivalence relation S on E such that neither of the conditions (C) and (C') is satisfied, but such that E/S is an ordered set.

(f) Let E be an ordered set, let f be an increasing mapping of E into an ordered set F, and let $S\{x, y\}$ be the equivalence relation $f(x) = f(y)$ on E. Then the condition (C') is satisfied. Moreover, the condition (C) is satisfied if and only if the relations $x \leqslant y$ and $f(x) = f(x')$ imply that there exists $y' \in E$ such that $x' \leqslant y'$ and $f(y) = f(y')$. Let $f = g \circ \varphi$ be the canonical decomposition of f. Then g is an isomorphism of E/S onto $f(E)$ if and only if this condition is satisfied and, in addition, the relation $f(x) \leqslant f(y)$ implies that there exist x', y' such that $f(x) = f(x')$, $f(y) = f(y')$, and $x' \leqslant y'$.

3. Let I be an ordered set and let $(E_\iota)_{\iota \in I}$ be a family of non-empty ordered sets indexed by I.

(a) Let F be the *sum* (Chapter II, § 4, no. 8) of the family $(E_\iota)_{\iota \in I}$; for each $x \in F$, let $\lambda(x)$ be the index ι such that $x \in E_\iota$; and let G be the graph consisting of all pairs $(x, y) \in F \times F$ such that either $\lambda(x) < \lambda(y)$ or else $\lambda(x) = \lambda(y)$ and $x \leqslant y$ in $E_{\lambda(x)}$. Show that G is the graph of an ordering on F. The set F endowed with this ordering is called the *ordinal sum* of the family $(E_\iota)_{\iota \in I}$ (relative to the ordering on I) and is denoted by $\sum_{\iota \in I} E_\iota$. Show that the equivalence relation corresponding to the partition $(E_\iota)_{\iota \in I}$ of F satisfies conditions (C) and (C') of Exercise 2, and that the quotient ordered set (Exercise 2) is canonically isomorphic to I.

(b) If the set I is the ordinal sum of a family $(J_\lambda)_{\lambda \in L}$ of ordered sets, where L is an ordered set, show that the ordered set $\sum_{\iota \in I} E_\iota$ is canonically isomorphic to the ordinal sum $\sum_{\lambda \in L} F_\lambda$, where $F_\lambda = \sum_{\iota \in I_\lambda} E_\iota$ ("associativity" of the ordinal sum). If I is the linearly ordered set $\{1, 2\}$, we write $E_1 + E_2$ for the ordinal sum of E_1 and E_2. Show that $E_2 + E_1$ and $E_1 + E_2$ are not necessarily isomorphic.

(c) An ordinal sum $\sum_{\iota \in I} E_\iota$ is right directed if and only if I is right directed and E_ω is right directed for each maximal element ω of I.

(d) An ordinal sum $\sum_{\iota \in I} E_\iota$ is totally ordered if and only if I and each E_ι is totally ordered.

(e) An ordinal sum $\sum_{\iota \in I} E_\iota$ is a lattice if and only if the following conditions are satisfied :

(I) The set I is a lattice, and for each pair (λ, μ) of non-comparable indices in I, $E_{\sup(\lambda, \mu)}$ (resp. $E_{\inf(\lambda, \mu)}$) has a least (resp. greatest) element.

(II) For each $\alpha \in I$ and each pair (x, y) of elements of E_α such that the set $\{x, y\}$ is bounded above (resp. bounded below) in E_α, the set $\{x, y\}$ has a least upper bound (resp. greatest lower bound) in E_α.

(III) For each $\alpha \in I$ such that E_α contains a set of two elements which has no upper bound (resp. lower bound) in E_α, the set of indices $\lambda \in I$ such that $\lambda > \alpha$ (resp. $\lambda < \alpha$) has a least element (resp. greatest element) β, and E_β has a least element (resp. greatest element).

* 4. Let E be an ordered set, and let $(E_\iota)_{\iota \in I}$ be the partition of E formed by the connected components of E (Chapter II, § 6, Exercise 10) with respect to the reflexive and symmetric relation "either $x = y$ or x and y are not comparable".

(a) Show that if $\iota \neq \varkappa$ and $x \in E_\iota$ and $y \in E_\varkappa$, then x, y are comparable; and that if, for example, $x \leqslant y$, $y' \in E_\varkappa$, and $y' \neq y$, then also $x \leqslant y'$ (use the fact that there exists no partition of $E_\varkappa$ into two sets A and B such that every element of A is comparable with every element of B).

(b) Deduce from (a) that the equivalence relation S corresponding to the partition (E_ι) of E is compatible (in x and in y) with the order relation $x \leqslant y$ on E, and that the quotient ordered set E/S (Exercise 2) is totally ordered.

(c) What are the connected components of an ordered set $E = F \times G$ which is the product of two totally ordered sets? *

5. Let E be an ordered set. A subset X of E is said to be *free* if no two distinct elements of X are comparable. Let $\mathfrak{I}$ be the set of free subsets of E. Show that, on $\mathfrak{I}$, the relation "given any $x \in X$, there exists $y \in Y$ such that $x \leqslant y$" is an order relation between X and Y, written $X \leqslant Y$. The mapping $x \to \{x\}$ is an isomorphism of E onto a

subset of the ordered set $\mathfrak{J}$. If $X \subset Y$, where $X \in \mathfrak{J}$ and $Y \in \mathfrak{J}$, show that $X \leqslant Y$. The ordered set $\mathfrak{J}$ is totally ordered if and only if E is totally ordered, and then $\mathfrak{J}$ is canonically isomorphic to E.

6. Let E and F be two ordered sets, and let $\mathcal{A}(E, F)$ be the subset of the product ordered set F^E consisting of the *increasing* mappings of E into F.

(a) Show that if E, F, G are there ordered sets, then the ordered set $\mathcal{A}(E, F \times G)$ is isomorphic to the product ordered set $\mathcal{A}(E, F) \times \mathcal{A}(E, G)$.

(b) Show that if E, F, G are three ordered sets, then the ordered set $\mathcal{A}(E \times F, G)$ is isomorphic to the ordered set $\mathcal{A}(E, \mathcal{A}(F, G))$.

(c) If $E \neq \emptyset$, then $\mathcal{A}(E, F)$ is a lattice if and only if F is a lattice.

(d) Suppose that E and F are both non-empty. Then $\mathcal{A}(E, F)$ is totally ordered if and only if one of the following conditions is satisfied :

(α) F consists of a single element;

(β) E consists of a single element and F is totally ordered;

(γ) E and F are both totally ordered and F has two elements.

7. In order that every mapping of an ordered set E into an ordered set F with at least two elements, which is both an increasing and a decreasing mapping, should be constant on E, it is necessary and sufficient that E should be connected with respect to the reflexive and symmetric relation "x and y are comparable" (Chapter II, § 6, Exercise 10). This condition is satisfied in particular if E is either left or right directed.

8. Let E and F be two ordered sets, let f be an increasing mapping of E into F, and g an increasing mapping of F into E. Let A (resp. B) be the set of all $x \in E$ (resp. $y \in F$) such that $g(f(x)) = x$ (resp. $f(g(y)) = y$). Show that the two ordered sets A and B are canonically isomorphic.

* 9. If E is a lattice, prove that

$$\sup_j (\inf_i x_{ij}) \leqslant \inf_i (\sup_j x_{ij})$$

for every finite "double" family (x_{ij}). *

10. Let E and F be two lattices. Then a mapping f of E into F is increasing if and only if

$$f(\inf(x, y)) \leqslant \inf(f(x), f(y))$$

for all $x \in E$ and all $y \in E$.

* Give an example of an increasing mapping f of the product ordered set $\mathbf{N} \times \mathbf{N}$ into the ordered set $\mathbf{N}$ such that the relation

$$f(\inf(x, y)) = \inf(f(x, y))$$

is false for at least one pair $(x, y) \in \mathbf{N} \times \mathbf{N}$.*

11. A lattice E is said to be *complete* if every subset of E has a least upper bound and a greatest lower bound in E; this means, in particular, that E has a greatest and a least element.

(a) Show that if an ordered set E is such that every subset of E has a least upper bound in E, then E is a complete lattice.

(b) A product of ordered sets is a complete lattice if and only if each of the factors is a complete lattice.

(c) An ordinal sum (Exercise 3) $\sum_{\iota \in I} E_\iota$ is a complete lattice if and only if the following conditions are satisfied:

(I) I is a complete lattice.

(II) If J is a subset of I which has no greatest element, and if $\sigma = \sup J$, then E_σ has a least element.

(III) For each $\iota \in I$ every subset of E_ι which has an upper bound in E_ι has a least upper bound in E_ι.

(IV) For each $\iota \in I$ such that E_ι has no greatest element, the set of all $\varkappa > \iota$ has a least element α, and E_α has a least element.

(d) The ordered set $\mathcal{A}(E, F)$ of increasing mappings of an ordered set E into an ordered set F (Exercise 6) is a complete lattice if and only if F is a complete lattice.

12. Let Φ be a set of mappings of a set A into itself. Let $\mathfrak{F}$ be the subset of $\mathfrak{P}(A)$ consisting of all $X \subset A$ such that $f(X) \subset X$ for each $f \in \Phi$. Show that $\mathfrak{F}$ is a complete lattice with respect to the relation of inclusion.

13. Let E be an ordered set. A mapping f of E into itself is said to be a *closure* if it satisfies the following conditions: (1) f is increasing; (2) for each $x \in E$, $f(x) \geqslant x$; (3) for each $x \in E$, $f(f(x)) = f(x)$. Let F be the set of elements of E which are invariant under f.

(a) Show that for each $x \in E$ the set F_x of elements $y \in F$ such that $x \leqslant y$ is not empty and has a least element, namely $f(x)$. Conversely, if G is a subset of E such that, for each $x \in E$, the set of all $y \in G$ such that $x \leqslant y$ has a least element $g(x)$, then g is a closure and G is the set of elements of E which are invariant under g.

(b) Suppose that E is a complete lattice. Show that the greatest lower bound in E of any non-empty subset of F belongs to F.

(c) Show that if E is a lattice, then $f(\sup(x, y)) = f(\sup(f(x), y))$ for each pair of elements x, y of E.

14. Let A and B be two sets, and let R be any subset of A × B. For each subset X of A (resp. each subset Y of B) let $\rho(X)$ (resp. $\sigma(Y)$) denote the set of all $y \in B$ (resp. $x \in A$) such that $(x, y) \in R$ for all $x \in A$ (resp. $(x, y) \in R$ for all $y \in B$). Show that ρ and σ are decreasing mappings, and that the mappings $X \to \sigma(\rho(X))$ and $Y \to \rho(\sigma(Y))$ are closures (Exercise 13) in $\mathfrak{P}(A)$ and $\mathfrak{P}(B)$ respectively (ordered by inclusion).

15. (a) Let E be an ordered set, and for each subset X of E let $\rho(X)$ (resp. $\sigma(X)$) denote the set of upper (resp. lower) bounds of X in E. Show that, in $\mathfrak{P}(E)$, the set $\tilde{E}$ of subsets X such that $X = \sigma(\rho(X))$ is a complete lattice, and that the mapping $i : x \to \sigma(\{x\})$ is an isomorphism (called *canonical*) of E onto an ordered subset E′ of $\tilde{E}$ such that, if a family (x_ι) of elements of E has a least upper bound (resp. greatest lower bound) in E, the image of this least upper bound (resp. greatest lower bound) is the least upper bound (resp. greatest lower bound) in $\tilde{E}$ of the family of images of the x_ι. $\tilde{E}$ is called the *completion* of the ordered set E.

(b) Show that, for every subset X of E, $\sigma(\rho(X))$ is the least upper bound in $\tilde{E}$ of the subset $i(X)$ of $\tilde{E}$. If f is any increasing mapping of E into a complete lattice F, there exists a unique increasing mapping $\bar{f}$ of $\tilde{E}$ into F such that $f = \bar{f} \circ i$ and $\bar{f}(\sup Z) = \sup(\bar{f}(Z))$ for every subset Z of $\tilde{E}$.

(c) If E is totally ordered, show that $\tilde{E}$ is totally ordered.

¶ 16. A lattice E is said to be *distributive* if it satisfies the following two conditions:

(D′) $\sup(x, \inf(y, z)) = \inf(\sup(x, y), \sup(x, z)),$

(D″) $\inf(x, \sup(y, z)) = \sup(\inf(x, y) \inf(x, z))$

for all x, y, z in E. A totally ordered set is a distributive lattice.

(a) Show that each of the conditions (D), (D′) separately implies the condition

(D) $\sup(\inf(x, y), \inf(y, z), \inf(z, x)) = \inf(\sup(x, y), \sup(y, z), \sup(z, x))$

for all x, y, z in E.

(b) Show that the condition (D) implies the condition

(M) If $x \geqslant z$, then $\sup(z, \inf(x, y)) = \inf(x, \sup(y, z))$.

Deduce that (D) implies each of (D′) and (D″), and hence that the three axioms (D), (D′), and (D″) are equivalent (to show, for example, that (D) implies (D′), take the least upper bound of x and each side of (D), and use (M)).

(c) Show that each of the two conditions

$$(\mathrm{T}') \qquad \inf(z, \sup(x,y)) \leqslant \sup(x, \inf(y,z)),$$
$$(\mathrm{T}'') \quad \inf(\sup(x,y), \sup(z, \inf(x,y))) = \sup(\inf(x,y), \inf(y,z), \inf(z,x))$$

for all x, y, z in E is necessary and sufficient for E to be distributive. (To show that (T′) implies (D″), consider the element

$$\inf(z, \sup(x, \inf(y,z))).)$$

¶ 17. A lattice E which has a least element α is said to be ***relatively complemented*** if, for each pair of elements x, y of E such that $x \leqslant y$, there exists an element x' such that $\sup(x, x') = y$ and $\inf(x, x') = \alpha$. Such an element x' is called a ***relative complement*** of x with respect to y.

* (a) Show that the set E of vector subspaces of a vector space of dimension $\geqslant 2$, ordered by inclusion, is a relatively complemented lattice, but that if x, y are two elements of E such that $x \leqslant y$, there exist in general several distinct relative complements of x with respect to y. *

(b) If E is distributive and relatively complemented, show that if $x \leqslant y$ in E, there exists a unique relative complement of x with respect to y. E is said to be a *Boolean lattice* if it is distributive and relatively complemented and if, moreover, it has a greatest element ω. For each $x \in \mathrm{E}$, let x^* be the complement of x with respect to ω. Then the mapping $x \to x^*$ is an isomorphism of E onto the ordered set obtained by endowing E with the opposite ordering, and we have $(x^*)^* = x$. If A is any set, then the set $\mathfrak{P}(\mathrm{A})$ of all subsets of A, ordered by inclusion, is a Boolean lattice.

(c) If E is a *complete* Boolean lattice (Exercise 11), show that for each family (x_λ) of elements of E and each $y \in \mathrm{E}$ we have

$$\inf(y, \sup_\lambda(x_\lambda)) = \sup_\lambda(\inf(y, x_\lambda)).$$

(Reduce to the case $y = \alpha$, and use the fact that if $\inf(z, x_\lambda) = \alpha$ for every index λ, then $z^* \geqslant x_\lambda$ for every λ).

¶ * 18. Let A be a set with at least three elements, let $\mathfrak{L}$ be the set of all partitions of A, ordered by the relation "ϖ is finer than ϖ'" between ϖ and ϖ' (no. 1, Example 4). Show that $\mathfrak{L}$ is a complete lattice (Exercise 11), is not distributive (Exercise 17), but is relatively complemented. (To prove the last assertion, well-order the sets belonging to a partition.) *

19. An ordered set E is said to be *without gaps* if it contains two distinct comparable elements and if, for each pair of elements x, y such that $x < y$, the open interval $]x, y[$ is not empty. Show that an ordinal sum $\sum_{\iota \in I} E_\iota$ (Exercise 3) is without gaps if and only if the following conditions are satisfied:

(I) Either I contains two distinct comparable elements, or else there exists $\iota \in I$ such that E_ι contains two distinct comparable elements.

(II) Each E_ι which contains at least two distinct comparable elements is without gaps.

(III) If α, β are two elements of I such that $\alpha < \beta$, and if the interval $]\alpha, \beta[$ in I is empty, then either E_α has no maximal element or else E_β has no minimal element.

In particular, every ordinal sum $\sum_{\iota \in I} E_\iota$ of sets without gaps is itself without gaps provided that no E_ι has a maximal element (or provided that no E_ι has a minimal element). If I is without gaps, and if each E_ι is either without gaps or contains no two distinct comparable elements, then $\sum_{\iota \in I} E_\iota$ is without gaps.

¶ 20. An ordered set E is said to be *scattered* if no ordered subset of E is without gaps (Exercise 19). Every subset of a scattered set is scattered. * Every well-ordered set of more than one element is scattered. *

(a) Suppose that E is scattered. Then, if x, y are any two elements of E such that $x < y$, there exist two elements x', y' of E such that $x \leqslant x' < y' \leqslant y$ and such that the interval $]x', y'[$ is empty. * Give an example of a totally ordered set which satisfies this condition and is not scattered (consider Cantor's triadic set). *

(b) An ordinal sum $E = \sum_{\iota \in I} E_\iota$ (where neither I nor any E_ι is empty) is scattered if and only if I and each E_ι is scattered. (Note that E contains a subset isomorphic to I and that every subset F of E is the ordinal sum of those sets $F \cap E_\iota$ which are non-empty; finally use Exercise 19.)

21. Let E be a non-empty totally ordered set, and let $S\{x, y\}$ be the relation "the closed interval with endpoints x, y is scattered" (Exercise 20). Show that S is an equivalence relation which is weakly compatible (Exercise 2) in x and y with the order relation on E, that the equivalence classes with respect to S are scattered sets, and that the quotient ordered set E/S (Exercise 2) is either without gaps or else consists of a single element. Deduce that E is isomorphic to an ordinal sum of scattered sets whose index set is either without gaps or else consists of a single element.

¶ 22. (a) Let E be an ordered set. A subset U of E is said to be *open* if, for each $x \in U$, U contains the interval $[x, \rightarrow[$. An open set U is said to be *regular* if there exists no open set $V \supset U$, distinct from U, such that U is cofinal in V. Show that every open set U is cofinal in exactly one regular open set $\tilde{U}$ (*). The mapping $U \rightarrow \tilde{U}$ is increasing. If U, V are two open sets such that $U \cap V = \emptyset$, then also $\tilde{U} \cap \tilde{V} = \emptyset$.

(b) Show that the set R(E) of regular open subsets of E, ordered by inclusion, is a complete Boolean lattice (Exercise 17). For R(E) to consist of two elements it is necessary and sufficient that E should be non-empty and *right directed.*

(c) If F is a cofinal subset of E, show that the mapping $U \rightarrow U \cap F$ is an isomorphism of R(E) onto R(F).

(d) If E_1, E_2 are two ordered sets, then every open set in $E_1 \times E_2$ is of the form $U_1 \times U_2$, where U_i is open in E_i $(i=1, 2)$. The set $R(E_1 \times E_2)$ is isomorphic to $R(E_1) \times R(E_2)$.

¶ 23. Let E be an ordered set and let $R_0(E) = R(E) - \{\emptyset\}$ (Exercise 22). For each $x \in E$, let $r(x)$ denote the unique regular open set in which the interval $[x, \rightarrow[$ (which is an open set) is cofinal. The mapping r so defined is called the canonical mapping of E into $R_0(E)$. Endow $R_0(E)$ with the order relation *opposite* to the relation of inclusion.

(a) Show that the mapping r is increasing and that $r(E)$ is cofinal in $R_0(E)$.

(b) An ordered set E is said to be *antidirected* if the canonical mapping $r: E \rightarrow R_0(E)$ is injective. For this to be so it is necessary and sufficient that the following two conditions should be satisfied:

(I) If x and y are two elements of E such that $x < y$, there exists $z \in E$ such that $x < z$ and such that the intervals $[y, \rightarrow[$ and $[z, \rightarrow[$ do not intersect.

(II) If x and y are two non-comparable elements of E, then either there exists $x' \geqslant x$ such that the intervals $[x', \rightarrow[$ and $[y, \rightarrow[$ do not intersect, or else there exists $y' \geqslant y$ such that the intervals $[x, \rightarrow[$ and $[y', \rightarrow[$ do not intersect.

(c) Show that, for every ordered set E, $R_0(E)$ is antidirected and that the canonical mapping of $R_0(E)$ into $R_0(R_0(E))$ is bijective (use Exercise 22 (a)).

* 24. (a) An ordered set E is said to be *branched* (on the right) if for each $x \in E$ there exist y, z in E such that $x \leqslant y$, $x \leqslant z$ and the

(*) This terminology may be justified by noting that there exists a unique topology on E for which the open sets (resp. regular open sets) are those defined in Exercise 22 (a) (cf. *General Topology*, Chapter I, § 1, Exercise 2 and § 8, Exercise 20).

intervals $[y, \rightarrow[$ and $[z, \rightarrow[$ do not intersect. An antidirected set with no maximal elements (Exercise 23) is branched.

(b) Let E be the set of intervals in **R** of the form

$$[k.2^{-n}, (k+1)2^{-n}] \quad (0 \leqslant k < 2^n),$$

ordered by the relation $\supset$. Show that E is antidirected and has no maximal elements.

(c) Give an example of a branched set in which there exists no antidirected cofinal subset. (Take the product of the set E defined in (b) with a well-ordered set which contains no countable cofinal subset, and use Exercise 22.)

(d) Give an example of an ordered set E which is not antidirected, but which has an antidirected cofinal subset (note that an ordinal sum $\sum_{\xi \in E} F_\xi$ contains a cofinal subset isomorphic to E). *

§ 2

1. Show that, in the set of orderings on a set E, the minimal elements (with respect to the order relation "Γ is coarser than Γ′" between Γ and Γ′) are the total orderings on E, and that if Γ is any ordering on E, the graph of Γ is the intersection of the graphs of the total orderings on E which are coarser than Γ (apply Theorem 2 of no. 4). Deduce that every ordered set is isomorphic to a subset of a product of totally ordered sets.

2. Let E be an ordered set and let $\mathfrak{W}$ be the set of subsets of E which are well-ordered by the induced ordering. Show that the relation "X is a segment of Y" on $\mathfrak{W}$ is an order relation between X and Y and that $\mathfrak{W}$ is inductive with respect to this order relation. Deduce that there exist well-ordered subsets of E which have no strict upper bound in E.

3. Let E be an ordered set. Show that there exist two subsets A, B of E such that $A \cup B = E$ and $A \cap B = \emptyset$, and such that A is well-ordered and B has no least element (for example, take B to be the union of those subsets of E which have no least element). * Give an example in which there are several partitions of E into two subsets having these properties. *

¶ 4. An ordered set F is said to be *partially well-ordered* if every totally ordered subset of F is well-ordered. Show that in every ordered set E there exists a partially well-ordered subset which is cofinal in E. (Consider the set $\mathfrak{F}$ of partially well-ordered subsets of E, and the order relation "$X \subset Y$ and no element of $Y - X$ is bounded above by any element of X" between elements X and Y of $\mathfrak{F}$. Show that $\mathfrak{F}$ is inductive with respect to this order relation.)

5. Let E be an ordered set and let $\mathfrak{I}$ be the set of *free* subsets of E, ordered by the relation defined in § 1, Exercise 5. Show that, if E is inductive, then $\mathfrak{I}$ has a greatest element.

¶ 6. Let E be an ordered set and let f be a mapping of E into E such that $f(x) \geqslant x$ for all $x \in$ E.

(a) Let $\mathfrak{G}$ be the set of subsets M of E with the following properties : (1) the relation $x \in$ M implies $f(x) \in$ M; (2) if a non-empty subset of M has a least upper bound in E, then this least upper bound belongs to M. For each $a \in$ E, show that the intersection C_a of the sets of $\mathfrak{G}$ which contain a also belongs to $\mathfrak{G}$; that C_a is well-ordered; and that if C_a has an upper bound b in E, then $b \in C_a$ and $f(b) = b$. C_a is said to be the *chain* of a (with respect to the function f). (Consider the set $\mathfrak{M}$ whose elements are the empty set and the subsets X of E which contain a and have a least upper bound m in E such that either $m \notin$ X or $f(m) > m$, any apply Lemma 3 of no. 3 to the set $\mathfrak{M}$.)

(b) Deduce from (a) that if E is inductive, then there exists $b \in$ E such that $f(b) = b$.

¶ 7. Let E be an ordered set and let F be the set of all closures (§ 1, Exercise 13) in E. Order F by putting $u \leqslant v$ whenever $u(x) \leqslant v(x)$ for all $x \in$ E. Then F has a least element e, the identity mapping of E onto itself. For each $u \in$ F let I(u) denote the set of elements of E which are invariant under u.

(a) Show that $u \leqslant v$ in F if and only if I(v) $\subset$ I(u).

(b) Show that if every pair of elements of E has a greatest lower bound in E, then every pair of elements of F has a greatest lower bound in F. If E is a complete lattice, then so is F (§ 1, Exercise 11).

(c) Show that if E is inductive (with respect to the relation $\leqslant$), then every pair u, v of elements of F have a least upper bound in F. (Show that if $f(x) = v(u(x))$ and if $w(x)$ denotes the greatest element of the chain of x, relative to f (Exercise 6), then w is a closure in E and is the least upper bound of u and v.)

¶ 8. An ordered set E is said to be *ramified* (on the right) if, for each pair of elements x, y of E such that $x < y$, there exists $z > x$ such that y and z are not comparable. E is said to be *completely ramified* (on the right) if it is ramified and has no maximal elements. Every antidirected set (§ 1, Exercise 22) is ramified.

(a) Let E be an ordered set and let a be an element of E. Let $\mathfrak{R}_a$ denote the set of ramified subsets of E which have a as least element. Show that $\mathfrak{R}_a$, ordered by inclusion, has a maximal element.

(b) If E is branched (§ 1, Exercise 24), show that every maximal element of $\mathfrak{R}_a$ is completely ramified.

(c) Give an example of a branched set which is not ramified. The branched set defined in §1, Exercise 24 (c) is completely ramified.

(d) Let E be a set in which each interval $]\leftarrow, x]$ is totally ordered. Show that E has an antidirected cofinal subset (§1, Exercise 22) (use (b)).

9. An ordinal sum $\sum_{\iota \in I} E_\iota$ (§ 1, Exercise 3) is well-ordered if and only if I and each E_ι is well-ordered.

10. Let I be an ordered set and let $(E_\iota)_{\iota \in I}$ be a family of ordered sets, all equal to the same ordered set E. Show that the ordinal sum $\sum_{\iota \in I} E_\iota$ (§ 1, Exercise 3) is isomorphic to the lexicographical product of the sequence $(F_\lambda)_{\lambda \in \{\alpha, \beta\}}$, where the set $\{\alpha, \beta\}$ of two distinct elements is well-ordered by the relation whose graph is $\{(\alpha, \alpha), (\alpha, \beta), (\beta, \beta)\}$, and where $F_\alpha = I$ and $F_\beta = E$. This product is called the *lexicographic product* of E by I and is written E.I.

¶ * 11. Let I be a well-ordered set and let $(E_\iota)_{\iota \in I}$ be a family of ordered sets, each of which contains at least two distinct comparable elements. Then the lexicographic product of the E_ι is well-ordered if and only if each of the E_ι is well-ordered and I is *finite* (if I is infinite, construct a strictly decreasing infinite sequence in the lexicographic product of the E_ι). *

¶ 12. Let I be a totally ordered set and let $(E_\iota)_{\iota \in I}$ be a family of ordered sets indexed by I. Let $R\{x, y\}$ denote the following relation on $E = \prod_{\iota \in I} E_\iota$: "the set of indices $\iota \in I$ such that $pr_\iota x \neq pr_\iota y$ is well-ordered, and if $\varkappa$ is the least element of this subset of I, we have $pr_\varkappa x < pr_\varkappa y$". Show that $R\{x, y\}$ is an order relation between x and y on E. If the E_ι are totally ordered, show that the connected components of E with respect to the relation "x and y are comparable" (Chapter II, § 6, Exercise 10) are totally ordered sets. Suppose that each E_ι has at least two elements. Then E is totally ordered if and only if I is well-ordered and each E_ι is totally ordered (use Exercise 3); and E is then the lexicographic product of the E_ι.

13. (a) Let $Is(\Gamma, \Gamma')$ be the relation "Γ is an ordering (on E), and Γ' is an ordering (on E'), and there exists an isomorphism of E, ordered by Γ, onto E', ordered by Γ'". Show that $Is(\Gamma, \Gamma')$ is an equivalence relation on every set whose elements are orderings. The term $\tau_\Delta(Is(\Gamma, \Delta))$ is an ordering called the *order-type* of Γ and is denoted by Ord (Γ), or Ord(E) by abuse of notation. Two ordered sets are isomorphic if and only if their order-types are equal.

(b) Let $R\{\lambda, \mu\}$ be the relation: "λ is an order-type, and μ is an order-type, and there exists an isomorphism of the set ordered by λ onto a subset of the set ordered by μ". Show that $R\{\lambda, \mu\}$ is a preorder relation between λ and μ. It will be denoted by $\lambda \prec \mu$.

(c) Let I be an ordered set and let $(\lambda_\iota)_{\iota \in I}$ be a family of order-types indexed by I. The order-type of the ordinal sum (§ 1, Exercise 3) of the family of sets ordered by the λ_ι $(\iota \in I)$ is called the *ordinal sum* of the order-types λ_ι $(\iota \in I)$ and is denoted by $\sum_{\iota \in I} \lambda_\iota$. If $(E_\iota)_{\iota \in I}$ is a family of ordered sets, the order-type of $\sum_{\iota \in I} E_\iota$ is $\sum_{\iota \in I} \operatorname{Ord}(E_\iota)$. If I is the ordinal sum of a family $(J_\varkappa)_{\varkappa \in K}$, show that

$$\sum_{\varkappa \in K}\Big(\sum_{\iota \in J_\varkappa} \lambda_\iota\Big) = \sum_{\iota \in I} \lambda_\iota.$$

(d) Let I be a well-ordered set and $(\lambda_\iota)_{\iota \in I}$ a family of order-types indexed by I. The order-type of the lexicographic product of the family of sets ordered by the λ_ι $(\iota \in I)$ is called the *ordinal product* of the family $(\lambda_\iota)_{\iota \in I}$ and is denoted by $\mathbf{P}_{\iota \in I} \lambda_\iota$. If $(E_\iota)_{\iota \in I}$ is a family of ordered sets, the order-type of the lexicographic product of the family $(E_\iota)_{\iota \in I}$ is $\mathbf{P}_{\iota \in I} \operatorname{Ord}(E_\iota)$. If I is the ordinal sum of a family of well-ordered sets $(J_\varkappa)_{\varkappa \in K}$, indexed by a well-ordered set K, show that

$$\mathbf{P}_{\varkappa \in K}\Big(\mathbf{P}_{\iota \in J_\varkappa} \lambda_\iota\Big) = \mathbf{P}_{\iota \in I} \lambda_\iota.$$

(e) We denote by $\lambda + \mu$ (resp. $\mu\lambda$) the ordinal sum (resp. ordinal product) of the family $(\xi_\iota)_{\iota \in J}$, where $J = \{\alpha, \beta\}$ is a set with two distinct elements, ordered by the relation whose graph is $\{(\alpha, \alpha), (\alpha, \beta), (\beta, \beta)\}$, and where $\xi_\alpha = \lambda$, $\xi_\beta = \mu$. Show that if I is a well-ordered set of order-type λ, and if $(\mu_\iota)_{\iota \in I}$ is a family of order-types such that $\mu_\iota = \mu$ for each $\iota \in I$, then $\sum_{\iota \in I} \mu_\iota = \mu\lambda$. We have $(\lambda + \mu) + \nu = \lambda + (\mu + \nu)$, $(\lambda\mu)\nu = \lambda(\mu\nu)$, and $\lambda(\mu + \nu) = \lambda\mu + \lambda\nu$ (but in general $\lambda + \mu \neq \mu + \lambda$, $\lambda\mu \neq \mu\lambda$, $(\lambda + \mu)\nu \neq \lambda\nu + \mu\nu$).

(f) Let $(\lambda_\iota)_{\iota \in I}$ and $(\mu_\iota)_{\iota \in I}$ be two families of order-types indexed by the same ordered set I. Show that, if $\lambda_\iota \prec \mu_\iota$ for each $\iota \in I$, then $\sum_{\iota \in I} \lambda_\iota \prec \sum_{\iota \in I} \mu_\iota$ and (if I is well-ordered) $\mathbf{P}_{\iota \in I} \lambda_\iota \prec \mathbf{P}_{\iota \in I} \mu_\iota$. If J is a subset of I, show that $\sum_{\iota \in J} \lambda_\iota \prec \sum_{\iota \in I} \lambda_\iota$ and (if I is well-ordered and the λ_ι are non-empty) $\mathbf{P}_{\iota \in J} \lambda_\iota \prec \mathbf{P}_{\iota \in I} \lambda_\iota$.

(g) Let λ^* denote the order-type of the set ordered by the opposite of the ordering λ. Then we have

$$(\lambda^*)^* = \lambda \quad \text{and} \quad \Big(\sum_{\iota \in I} \lambda_\iota\Big)^* = \sum_{\iota \in I^*} \lambda_\iota^*,$$

where I^* denotes the set I endowed with the opposite of the ordering given on I.

¶ 14. An *ordinal* is the order-type of a well-ordered set (Exercise 13).

(a) Show that, if $(\lambda_\iota)_{\iota \in I}$ is a family of ordinals indexed by a well-ordered set I, then the ordinal sum $\sum_{\iota \in I} \lambda_\iota$ is an ordinal; * and that if, moreover, I is finite, the ordinal product $\mathrm{P}_{\iota \in I} \lambda_\iota$ is an ordinal (Exercise 11). *

The order-type of the empty set is denoted by 0, and that of a set with one element by 1 (by abuse of language, cf. § 3). Show that

$$\alpha + 0 = 0 + \alpha = \alpha \quad \text{and} \quad \alpha . 1 = 1 . \alpha = \alpha$$

for every ordinal α.

(b) Show that the relation "λ is an ordinal and μ is an ordinal and $\lambda \prec \mu$" is a *well-ordering relation*, denoted by $\lambda \leqslant \mu$. (Note that, if λ and μ are ordinals, the relation $\lambda \prec \mu$ is equivalent to "λ is equal to the order-type of a segment of μ" (no. 5, Theorem 3, Corollary 3): given a family $(\lambda_\iota)_{\iota \in I}$ of ordinals, consider a well-ordering on I and take the ordinal sum of the family of sets ordered by the λ_ι; finally, use Proposition 2 of no. 1.)

(c) Let α be an ordinal. Show that the relation "ξ is an ordinal and $\xi \leqslant \alpha$" is collectivizing in ξ, and that the set O_α of ordinals $< \alpha$ is a well-ordered set such that $\operatorname{Ord}(O_\alpha) = \alpha$. We shall often identify O_α with α.

(d) Show that for every family of ordinals $(\xi_\iota)_{\iota \in I}$ there exists a unique ordinal α such that the relation "λ is an ordinal and $\xi_\iota \leqslant \lambda$ for all $\iota \in I$" is equivalent to $\alpha \leqslant \lambda$. By abuse of language, α is called the *least upper bound* of the family of ordinals $(\xi_\iota)_{\iota \in I}$, and we write $\alpha = \sup_{\iota \in I} \xi_\iota$ (it is the greatest element of the union of $\{\alpha\}$ and the set of the ξ_ι). The least upper bound of the set of ordinals $\xi < \alpha$ is either α or an ordinal β such that $\alpha = \beta + 1$; in the latter case, β is said to be the *predecessor* of α.

15. (a) Let α and β be two ordinals. Show that the inequality $\alpha < \beta$ is equivalent to $\alpha + 1 \leqslant \beta$, and that it implies the inequalities $\xi + \alpha < \xi + \beta$, $\alpha + \xi \leqslant \beta + \xi$, $\alpha\xi \leqslant \beta\xi$ for all ordinals ξ, and $\xi\alpha < \xi\beta$ if $\xi > 0$.

(b) Deduce from (a) that there exists no set to which every ordinal belongs (use Exercise 14 (d)).

(c) Let α, β, μ be three ordinals. Show that each of the relations $\mu + \alpha < \mu + \beta$, $\alpha + \mu < \beta + \mu$ implies $\alpha < \beta$; and that each of the relations $\mu\alpha < \mu\beta$, $\alpha\mu < \beta\mu$ implies $\alpha < \beta$ provided that $\mu > 0$.

(d) Show that the relation $\mu + \alpha = \mu + \beta$ implies $\alpha = \beta$, and that $\mu\alpha = \mu\beta$ implies $\alpha = \beta$ provided that $\mu > 0$.

(e) Two ordinals α and β are such that $\alpha \leqslant \beta$ if and only if there exists an ordinal ξ such that $\beta = \alpha + \xi$. This ordinal ξ is then unique and is such that $\xi \leqslant \beta$; it is written $(-\alpha) + \beta$.

(f) Let α, β, ζ be three ordinals such that $\zeta < \alpha\beta$. Show that there exist two ordinals ξ, η such that $\zeta = \alpha\eta + \xi$ and $\xi < \alpha$, $\eta < \beta$ (cf. no. 5, Theorem 3, Corollary 3). Moreover, ξ and η are uniquely determined by these conditions.

¶ 16. An ordinal $\rho > 0$ is said to be *indecomposable* if there exists no pair of ordinals ξ, η such that $\xi < \rho$, $\eta < \rho$, and $\xi + \eta = \rho$.

(a) An ordinal ρ is indecomposable if and only if $\xi + \rho = \rho$ for every ordinal ξ such that $\xi < \rho$.

(b) If $\rho > 1$ is an indecomposable ordinal and if α is any ordinal > 0, then $\alpha\rho$ is indecomposable, and conversely (use Exercise 15 (f)).

(c) If ρ is indecomposable and if $0 < \alpha < \rho$, then $\rho = \alpha\xi$, where ξ is an indecomposable ordinal (use Exercise 15 (f)).

(d) Let α be an ordinal > 0. Show that there exists a greatest indecomposable ordinal among the indecomposable ordinals $\leqslant \alpha$ (consider the decompositions $\alpha = \rho + \xi$, where ρ is indecomposable).

(e) If E is any set of indecomposable ordinals, deduce from (d) that the least upper bound of E (Exercise 14 (d)) is an indecomposable ordinal.

¶ 17. Given an ordinal α_0, a term $f(\xi)$ is said to be an *ordinal functional symbol (with respect to* ξ*) defined for* $\xi \geqslant \alpha_0$ if the relation "ξ is an ordinal and $\xi \geqslant \alpha_0$" implies the relation "$f(\xi)$ is an ordinal"; $f(\xi)$ is said to be *normal* if the relation $\alpha_0 \leqslant \xi < \eta$ implies $f(\xi) < f(\eta)$ and if, for each family $(\xi_\iota)_{\iota\in I}$ of ordinals $\geqslant \alpha_0$, we have $\sup_{\iota\in I} f(\xi_\iota) = f(\sup_{\iota\in I} \xi_\iota)$ (cf. Exercise 14 (d)).

(a) Show that, for each ordinal $\alpha > 0$, $\alpha + \xi$ and $\alpha\xi$ are normal functional symbols defined for $\xi \geqslant 0$ (use Exercise 15 (f)).

(b) Let $w(\xi)$ be an ordinal functional symbol defined for $\xi \geqslant \alpha_0$ such that $w(\xi) \geqslant \xi$ and such that $\alpha_0 \leqslant \xi < \eta$ implies $w(\xi) < w(\eta)$. Also let $g(\xi, \eta)$ be a term such that the relation "ξ and η are ordinals $\geqslant \alpha_0$" implies the relation "$g(\xi, \eta)$ is an ordinal such that $g(\xi, \eta) > \xi$".

Define a term $f(\xi, \eta)$ with the following properties: (1) for each ordinal $\xi \geqslant \alpha_0$, $f(\xi, 1) = w(\xi)$; (2) for each ordinal $\xi \geqslant \alpha_0$ and each ordinal $\eta > 1$, $f(\xi, \eta) = \sup_{0<\zeta<\eta} g(f(\xi, \zeta), \xi)$ (use criterion C60 of no. 2). Show that if $f_1(\xi, \eta)$ is another term with these two properties, then $f(\xi, \eta) = f_1(\xi, \eta)$ for all $\xi \geqslant \alpha_0$ and all $\eta \geqslant 1$. Prove that, for each ordinal $\xi \geqslant \alpha_0$, $f(\xi, \eta)$ is a normal functional symbol with respect to η (defined for all $\eta \geqslant 1$). Show that $f(\xi, \eta) \geqslant \xi$ for all $\eta \geqslant 1$ and $\xi \geqslant \alpha_0$, and that $f(\xi, \eta) \geqslant \eta$ for all $\xi \geqslant \sup(\alpha_0, 1)$ and $\eta \geqslant 1$. Furthermore, for each pair (α, β) of ordinals such that $\alpha > 0$, $\alpha \geqslant \alpha_0$, and $\beta \geqslant w(\alpha)$ there exists a unique ordinal ξ such that

$$f(\alpha, \xi) \leqslant \beta < f(\alpha, \xi + 1),$$

and we have $\xi \leqslant \beta$.

(c) If we take $\alpha_0 = 0$, $w(\xi) = \xi + 1$, $g(\xi, \eta) = \xi + 1$, then $f(\xi, \eta) = \xi + \eta$. If we take $\alpha_0 = 1$, $w(\xi) = \xi$, $g(\xi, \eta) = \xi + \eta$, then $f(\xi, \eta) = \xi\eta$.

(d) Show that if the relations $\alpha_0 \leqslant \xi \leqslant \xi'$, $\alpha_0 \leqslant \eta \leqslant \eta'$ imply $g(\xi, \eta) \leqslant g(\xi', \eta')$, then the relations $\alpha_0 \leqslant \xi \leqslant \xi'$, $1 \leqslant \eta \leqslant \eta'$ imply $f(\xi, \eta) \leqslant f(\xi', \eta')$. If the relations $\alpha_0 \leqslant \xi \leqslant \xi'$, $\alpha_0 \leqslant \eta < \eta'$ imply $g(\xi, \eta) < g(\xi, \eta')$ and $g(\xi, \eta) \leqslant g(\xi', \eta)$, then the relations $\alpha_0 \leqslant \xi < \xi'$ and $\eta \geqslant 0$ imply $f(\xi, \eta + 1) < f(\xi', \eta + 1)$.

(e) Suppose that $w(\xi) = \xi$ and that the relations $\alpha_0 \leqslant \xi \leqslant \xi'$, $\alpha_0 \leqslant \eta < \eta'$ imply $g(\xi, \eta) < g(\xi, \eta')$ and $g(\xi, \eta) \leqslant g(\xi', \eta)$. Suppose, moreover, that for each $\xi \geqslant \alpha_0$, $g(\xi, \eta)$ is a normal functional symbol with respect to η (defined for $\eta \geqslant \alpha_0$), and that, whenever $\xi \geqslant \alpha_0$, $\eta \geqslant \alpha_0$, and $\zeta \geqslant \alpha_0$, we have the associativity relation

$$g(g(\xi, \eta), \zeta) = g(\xi, g(\eta, \zeta)).$$

Show that, if $\xi \geqslant \alpha_0$, $\eta \geqslant 1$, and $\zeta \geqslant 1$, then

$$g(f(\xi, \eta), f(\xi, \zeta)) = f(\xi, \eta + \zeta)$$

("distributivity" of g with respect to f) and

$$f(f(\xi, \eta), \zeta) = f(\xi, \eta\zeta)$$

("associativity" of f).

¶ 18. In the definition procedure defined in Exercise 17 (b), take $\alpha_0 = 1 + 1$ (denoted by 2 by abuse of language),

$$w(\xi) = \xi, \qquad g(\xi, \eta) = \xi\eta.$$

Denote $f(\xi, \eta)$ by ξ^η and define α^0 to be 1 for all ordinals α. Also define 0^β to be 0 and 1^β to be 1 for all ordinals $\beta \geqslant 1$.

(a) Show that if $\alpha > 1$ and $\beta < \beta'$, we have $\alpha^\beta < \alpha^{\beta'}$ and that, for each ordinal $\alpha > 1$, α^ξ is a normal functional symbol with respect to ξ. Moreover, if $0 < \alpha \leqslant \alpha'$, we have $\alpha^\beta \leqslant \alpha'^\beta$.

(b) Show that $\alpha^\xi . \alpha^\eta = \alpha^{\xi+\eta}$ and $(\alpha^\xi)^\eta = \alpha^{\xi\eta}$.

(c) Show that, if $\alpha \geqslant 2$ and $\beta \geqslant 1$, $\alpha^\beta \geqslant \alpha\beta$.

(d) For each pair of ordinals $\beta \geqslant 1$ and $\alpha \geqslant 2$, there exist three ordinals ξ, γ, δ such that $\beta = \alpha^\xi\gamma + \delta$, where $0 < \gamma < \alpha$ and $\delta < \alpha^\xi$, and these three ordinals are uniquely determined by these conditions.

* 19. Let α and β be two ordinals, and let E, F be two well-ordered sets such that $\mathrm{Ord}(E) = \alpha$ and $\mathrm{Ord}(F) = \beta$. In the set E^F of mappings of F into E, consider the subset G of mappings g such that $g(y)$ is equal to the least element of E for all but a *finite* number of elements $y \in F$. If F^* is the ordered set obtained by endowing F with the opposite order, show that G is a connected component with respect to the relation "x and y are comparable" (Chapter II, § 6, Exercise 10) in the product E^{F^*} endowed with the ordering defined in Exercise 12, and show that G is well-ordered. Furthermore, prove that $\mathrm{Ord}(G) = \alpha^\beta$ (use the uniqueness property of Exercise 17 (b)).

¶ 20. A set X is said to be *transitive* if the relation $x \in X$ implies $x \subset X$.

(a) If Y is a transitive set, then so is $Y \cup \{Y\}$. If $(Y_\iota)_{\iota \in I}$ is a family of transitive sets, then $\bigcup_{\iota \in I} Y_\iota$ and $\bigcap_{\iota \in I} Y_\iota$ are transitive.

(b) A set X is a *pseudo-ordinal* if every transitive set Y such that $Y \subset X$ and $Y \neq X$ is an element of X. A set S is said to be *decent* if the relation $x \in S$ implies $x \notin x$. Show that every pseudo-ordinal is transitive and decent (consider the union of the decent transitive subsets of X, and use (a)). If X is a pseudo-ordinal, then so is $X \cup \{X\}$.

(c) Let X be a transitive set, and suppose that each $x \in X$ is a pseudo-ordinal. Then X is a pseudo-ordinal (note that, for each $x \in X$, $x \cup \{x\}$ is a pseudo-ordinal contained in X).

(d) Show that $\varnothing$ is a pseudo-ordinal and that every element of a pseudo-ordinal X is a pseudo-ordinal. (Consider the union of the transitive subsets of X whose elements are pseudo-ordinals.)

(e) If $(X_\iota)_{\iota \in I}$ is a family of pseudo-ordinals, then $\bigcap_{\iota \in I} X_\iota$ is the least element of this family (with respect to the relation of inclusion). (Use (b).) Deduce that, if E is a pseudo-ordinal, the relation $x \subset y$ between elements x, y of E is a well-ordering relation.

(f) Show that for each ordinal α there exists a unique pseudo-ordinal E_α such that $\mathrm{Ord}(E_\alpha) = \alpha$ (use (e) and criterion C60). In particular, the pseudo-ordinals whose order-types are $0, 1, 2 = 1 + 1$, and $3 = 2 + 1$ are respectively

$$\varnothing, \quad \{\varnothing\}, \quad \{\varnothing, \{\varnothing\}\}, \quad \{\varnothing, \{\varnothing\}, \{\varnothing, \{\varnothing\}\}\}.$$

§ 3

¶ 1. Let E and F be two sets, let f be an injection of E into F, and let g be a mapping of F into E. Show that there exist two subsets A, B of E such that $B = E - A$, and two subsets A′, B′ of F such that $B' = F - A'$, for which $A' = f(A)$ and $B = g(B')$. (Let $R = E - g(F)$ and put $h = g \circ f$; take A to be the intersection of the subsets M of E such that $M \supset R \cup h(M)$.)

2. If E and F are distinct sets, show that $E^F \neq F^E$. Deduce that if E and F are the cardinals 2 and $4 (= 2 + 2)$, then at least one of the sets E^F, F^E is not a cardinal.

¶ 3. Let $(\mathfrak{a}_\iota)_{\iota \in I}$ and $(\mathfrak{b}_\iota)_{\iota \in I}$ be two families of cardinals such that $\mathfrak{b}_\iota \geqslant 2$ for each $\iota \in I$.

(a) Show that, if $\mathfrak{a}_\iota \leqslant \mathfrak{b}_\iota$ for each $\iota \in I$, then

$$\sum_{\iota \in I} \mathfrak{a}_\iota \leqslant \mathbf{P}_{\iota \in I} \mathfrak{b}_\iota.$$

(b) Show that, if $\mathfrak{a}_\iota < \mathfrak{b}_\iota$ for each $\iota \in I$, then

$$\sum_{\iota \in I} \mathfrak{a}_\iota < \mathbf{P}_{\iota \in I} \mathfrak{b}_\iota.$$

(Note that a product $\prod_{\iota \in I} E_\iota$ cannot be the union of a family $(A_\iota)_{\iota \in I}$ such that $\mathrm{Card}(A_\iota) < \mathrm{Card}(E_\iota)$ for all $\iota \in I$, by observing that $\mathrm{Card}(\mathrm{pr}_\iota(A_\iota)) < \mathrm{Card}(E_\iota)$.)

4. Let E be a set and let f be a mapping of $\mathfrak{P}(E) - \{\varnothing\}$ into E such that for each non-empty subset X of E we have $f(X) \in X$ ("choice function").

(a) Let $\mathfrak{b}$ be a cardinal and let A be the set of all $x \in E$ such that $\mathrm{Card}(\overset{-1}{f}(x)) \leqslant \mathfrak{b}$. Show that, if $\mathfrak{a} = \mathrm{Card}(A)$, then $2^{\mathfrak{a}} \leqslant 1 + \mathfrak{a}\mathfrak{b}$ (note that if $Y \subset A$ and $Y \neq \varnothing$, then $f(Y) \in A$).

(b) Let B be the set of all $x \in E$ such that, for each non-empty subset X of $\overset{-1}{f}(x)$, $\mathrm{Card}(X) \leqslant \mathfrak{b}$. Show that $\mathrm{Card}(B) \leqslant \mathfrak{b}$.

5. Let $(\lambda_\iota)_{\iota \in I}$ be a family of order-types (§ 2, Exercise 13), indexed by an ordered set I. Show that

$$\text{Card}\left(\sum_{\iota \in I} \lambda_\iota\right) = \sum_{\iota \in I} \text{Card}(\lambda_\iota)$$

and that, if I is well-ordered,

$$\text{Card}\left(\mathop{\mathsf{P}}_{\iota \in I} \lambda_\iota\right) = \mathop{\mathsf{P}}_{\iota \in I} \text{Card}(\lambda_\iota).$$

6. Show that for every set E there exists $X \subset E$ such that $X \notin E$ (use Theorem 2 of no. 6).

§ 4

1. (a) Let E be a set and let $\mathfrak{F}(E)$ be the set of finite subsets of E. Show that $\mathfrak{F}(E)$ is the smallest subset $\mathfrak{G}$ of $\mathfrak{P}(E)$ satisfying the following conditions: (i) $\varnothing \in \mathfrak{G}$; (ii) the relations $X \in \mathfrak{G}$ and $x \in E$ imply

$$X \cup \{x\} \in \mathfrak{G}.$$

(b) Deduce from (a) that the union of two finite subsets A, B of E is finite (consider the set of subsets X of E such that $X \cup A$ is finite; cf. § 5, no. 1, Proposition 1, Corollary 1).

(c) Deduce from (a) and (b) that for every finite set E the set $\mathfrak{P}(E)$ is finite (consider the set of subsets X of E such that $\mathfrak{P}(X)$ is finite; cf. § 5, no. 1, Proposition 1, Corollary 4).

2. Show that a set E is finite if and only if every non-empty subset of $\mathfrak{P}(E)$ has a maximal element (with respect to inclusion). (To show that the condition is sufficient, apply it to the set $\mathfrak{F}(E)$ of finite subsets of E.)

3. Show that if a well-ordered set E is such that the ordered set obtained by endowing E with the opposite ordering is also well-ordered, then E is finite (consider the greatest element x of E such that the segment S_x is finite).

4. Let E be a finite set with $n \geqslant 2$ elements, and let C be a subset of $E \times E$ such that, for each pair x, y of distinct elements of E, exactly one of the two elements (x, y), (y, x) of $E \times E$ belongs to C. Show that there exists a mapping f of the interval $[1, n]$ onto E such that $(f(i), f(i+1)) \in C$ for $1 \leqslant i \leqslant n-1$ (use induction on n).

¶ 5. Let E be an ordered set for which there exists an integer k such that k is the greatest number of elements in a free subset X of E (§ 1, Exercise 5). Show that E can be partitioned into k totally ordered subsets (with respect to the induced ordering). The proof is in two steps:

(a) If E is finite and has n elements, use induction on n; let a be a minimal element of E and let $E' = E - \{a\}$. If there exists a partition of E' into k totally ordered sets C_i ($1 \leqslant i \leqslant k$), let U_i be the set of all $x \in C_i$ which are $\geqslant a$. Show that there is at least one index i for which a free subset of $E' - U_i$ has at most $k - 1$ elements. The proof of this is by *reductio ad absurdum.* For each i, let S_i be a free subset of $E' - U_i$ which has k elements, let S be the union of the sets S_i, and let s_j be the least element of $S \cap C_j$ for each index $j \leqslant k$; show that the $k + 1$ elements $a, s_1, \ldots, s_k$ form a free subset of E.

(b) If E is arbitrary, the proof is by induction on k, as follows. A subset C of E is said to be *strongly related* in E if for each finite subset F of E there exists a partition of F into at most k totally ordered sets such that $C \cap F$ is contained in one of them. Show that there exists a maximal strongly related subset C_0, and that every free subset of $E - C_0$ has at most $k - 1$ elements. (Argue by contradiction, and suppose that there is a free subset $\{a_1, \ldots, a_k\}$ of k elements in $E - C_0$. Consider each set $C_0 \cup \{a_i\}$ ($1 \leqslant i \leqslant k$), and express the fact that it is not strongly related, thus introducing a finite subset F_i of E for each index i. Then consider the union F of the sets F_i and use the fact that C_0 is strongly related to obtain a contradiction.)

¶ 6. (a) Let A be a set and let $(X_i)_{1 \leqslant i \leqslant m}$, $(Y_j)_{m+1 \leqslant j \leqslant m+n}$ be two finite families of subsets of A. Let h be the least integer such that, for each integer $r \leqslant m - h$ and each subset $\{i_1, \ldots, i_{r+h}\}$ of $r + h$ elements of $[1, m]$, there exists a subset $\{j_1, \ldots, j_r\}$ of r elements of $[m + 1, m + n]$ for which the union of the sets X_{i_α} ($1 \leqslant \alpha \leqslant r + h$) meets each of the sets Y_{j_β} ($1 \leqslant \beta \leqslant r$) (which implies that $m \leqslant n + h$). Show that there exists a finite subset B of A with at most $n + h$ elements such that every X_i ($1 \leqslant i \leqslant m$) and every Y_j ($m + 1 \leqslant j \leqslant m + n$) meets B. (Consider the order relation on the interval $[1, m + n]$ whose graph is the union of the diagonal and the set of pairs (i, j) such that $1 \leqslant i \leqslant m$ and $m + 1 \leqslant j \leqslant m + n$ and $X_i \cap Y_j \neq \emptyset$, and apply Exercise 5 to this ordered set.)

(b) Let E and F be two finite sets and let $x \to A(x)$ be a mapping of E into $\mathfrak{P}(F)$. Then there exists an injection f of E into F such that $f(x) \in A(x)$ for each $x \in E$ if and only if for each subset H of E we have $\text{Card}\left(\bigcup_{x \in H} A(x)\right) \geqslant \text{Card}(H)$ (the method of proof is analogous to that of (a), with $h = 0$).

(c) With the hypotheses of (b), let G be a subset of F. Then there exists an injection f of E into F such that $f(x) \in A(x)$ for each $x \in E$

and such that $f(\mathrm{E}) \supset \mathrm{G}$ if and only if f satisfies the condition of (b) and for each subset L of G the cardinal of the set of all $x \in \mathrm{E}$ such that $\mathrm{A}(x) \cap \mathrm{L} \neq \emptyset$ is $\geqslant \mathrm{Card}(\mathrm{L})$. (Let $(a_i)_{1 \leqslant i \leqslant p}$ be the sequence of distinct elements of G, arranged in some order; let $(b_j)_{p+1 \leqslant j \leqslant p+m}$ be the sequence of distinct elements of F, arranged in some order; and let

$$(c_k)_{p+m+1 \leqslant k \leqslant p+m+n}$$

be the sequence of distinct elements of E, arranged in some order. Consider the order relation on the set $[1, p+m+n]$ whose graph is the union of the diagonal and the set of pairs (i, j) such that either

$$1 \leqslant i \leqslant p \quad \text{and} \quad p+1 \leqslant j \leqslant p+m \quad \text{and} \quad a_i = b_j,$$

$$\text{or} \quad 1 \leqslant i \leqslant p \quad \text{and} \quad p+m+1 \leqslant j \leqslant p+m+n \quad \text{and} \quad a_i \in \mathrm{A}(c_j),$$

$$\text{or } p+1 \leqslant i \leqslant p+m \text{ and } p+m+1 \leqslant j \leqslant p+m+n \text{ and } b_i \in \mathrm{A}(c_j);$$

then apply Exercise 5.)

7. An element a of a lattice E is said to be *irreducible* if the relation $\sup(x, y) = a$ implies either $x = a$ or $y = a$.

(a) Show that in a finite lattice E every element a can be written as $\sup(e_1, \ldots, e_k)$, where the e_i $(1 \leqslant i \leqslant k)$ are irreducible.

(b) Let E be a finite lattice and let J be the set of its irreducible elements. For each $x \in \mathrm{E}$ let $\mathrm{S}(x)$ be the set of all $y \in \mathrm{J}$ which are $\leqslant x$. Show that the mapping $x \to \mathrm{S}(x)$ is an isomorphism of E onto a subset of $\mathfrak{P}(\mathrm{J})$, ordered by inclusion, and that $\mathrm{S}(\inf(x, y)) = \mathrm{S}(x) \cap \mathrm{S}(y)$.

¶ 8. (a) Let E be a distributive lattice (§ 1, Exercise 16). If a is irreducible in E (Exercise 7), show that the relation $a \leqslant \sup(x, y)$ implies $a \leqslant x$ or $a \leqslant y$.

(b) Let E be a finite distributive lattice and let J be the set of its irreducible elements, ordered by the induced ordering. Show that the isomorphism $x \to \mathrm{S}(x)$ of E onto a subset of $\mathfrak{P}(\mathrm{J})$ defined in Exercise 7 (b) is such that $\mathrm{S}(\sup(x, y)) = \mathrm{S}(x) \cup \mathrm{S}(y)$. Deduce that if J^* is the ordered set obtained by endowing J with the opposite ordering, then E is isomorphic to the set $\mathcal{A}(\mathrm{J}^*, \mathrm{I})$ of increasing mappings of J^* into $\mathrm{I} = \{0, 1\}$ (§ 1, Exercise 6).

(c) With the hypotheses of (b), let P be the set of elements of J other then the least element of E. For each $x \in \mathrm{E}$, let $y_1, \ldots, y_k$ be the distinct minimal elements of the interval $]x, \rightarrow[$ in E; for each index i let q_i be an element of P such that $q_i \notin \mathrm{S}(x)$ and $q_i \in \mathrm{S}(y_i)$. Show that no two of the elements $q_1, \ldots, q_k$ are comparable.

(d) Conversely, let $q_1, \ldots, q_k$ be k elements of P, no two of which are comparable. Let $u = \sup(q_1, \ldots, q_k)$ and let

$$v_i = \sup_{1 \leqslant j \leqslant k,\, j \neq i} (q_j) \qquad (1 \leqslant i \leqslant k).$$

Show that $v_i < u$ for $1 \leqslant i \leqslant k$. Let $x = \inf(v_1, \ldots, v_k)$ and let

$$y_i = \inf_{1 \leqslant j \leqslant k,\, j \neq i} (v_j).$$

Show that $x < y_i$ for each index i, and deduce that the interval $]x, \rightarrow[$ has at least k distinct minimal elements.

¶ 9. A subset A of a lattice E is said to be a *sublattice* if for each pair (x, y) of elements of A, $\sup_E(x, y)$ and $\inf_E(x, y)$ belong to A.

(a) Let $(C_i)_{1 \leqslant i \leqslant n}$ be a finite family of totally ordered sets and let

$$E = \prod_{i=1}^{n} C_i$$

be their product. Let A be a sublattice of E. Show that A cannot have more than n irreducible elements (Exercise 7) no two of which are comparable. (The proof is by *reductio ad absurdum*. Suppose that there exist $r > n$ irreducible elements $a_1, \ldots, a_r$ in A, no two of which are comparable. Consider the elements $u = \sup(a_1, \ldots, a_r)$ and

$$v_j = \sup_{1 \leqslant j \leqslant r,\, j \neq i} (a_j)$$

of A. By projecting onto the factors, show that $u = v_i$ for some index i, and hence that two of the a_i are comparable.)

(b) Conversely, let F be a finite distributive lattice, let P be the set of irreducible elements of F other than the least element of F, and suppose that n is the greatest number of elements in a free subset of P (§ 1, Exercise 5). Show that F is isomorphic to a sublattice of a product of n totally ordered sets. (Apply Exercise 5, which shows that P is the union of n totally ordered sets P_i with no element in common. Let C_i be the totally ordered set obtained by adjoining a least element to P_i $(1 \leqslant i \leqslant n)$. With each $x \in F$ associate the family $(x_i)_{1 \leqslant i \leqslant n}$, where x_i is the least upper bound in C_i of the set of elements of P_i which are $\leqslant x$.)

¶ 10. (a) An ordered set E is isomorphic to a subset of a product of n totally ordered sets if and only if the graph of the ordering on E is the intersection of the graphs of n total orderings on E. (To show that the condition is necessary, show that if

$$F = \prod_{i=1}^{n} F_i$$

is a product of n totally ordered sets, then the graph of the product ordering on F is the intersection of n graphs of lexicographic orderings on F.)

(b) An ordered set E is isomorphic to a subset of a product of two totally ordered sets if and only if the ordering Γ on E is such that there exists another ordering Γ' on E with the property that any two distinct elements of E are comparable with respect to exactly one of the orderings Γ, Γ'.

(c) Let A be a finite set of n elements. Let E be the subset of $\mathfrak{P}(A)$ consisting of all subsets $\{x\}$ and $A - \{x\}$ as x runs through A. Show that n is the smallest integer m such that E, ordered by inclusion, is isomorphic to a subset of a product of m totally ordered sets (use (a)).

¶ 11. Let A be a set and let $\mathfrak{R}$ be a subset of the set $\mathfrak{F}(A)$ of finite subsets of A. $\mathfrak{R}$ is said to be *mobile* if it satisfies the following condition:

(MO) If X, Y are two distinct elements of $\mathfrak{R}$ and if $z \in X \cap Y$, then there exists $Z \subset X \cap Y$ belonging to $\mathfrak{R}$ such that $z \notin Z$.

A subset P of A is then said to be *pure* if it contains no set belonging to $\mathfrak{R}$.

(a) Show that every pure subset of A is contained in a maximal pure subset of A.

(b) Let M be a maximal pure subset of A. Show that for each $x \in \complement M$ there exists a unique finite subset $E_M(x)$ of M such that $E_M(x) \cup \{x\} \in \mathfrak{R}$. Moreover, if $y \in E_M(x)$, the set $(M \cup \{x\}) - \{y\}$ is a maximal pure subset of A.

(c) Let M, N be two maximal pure subsets of A, such that $N \cap \complement M$ is finite. Show that Card (M) = Card (N). (Proof by induction on the cardinal of $N \cap \complement M$, using (b).)

(d) Let M, N be two maximal pure subsets of A, and put $N' = N \cap \complement M$, $M' = M \cap \complement N$. Show that

$$M' \subset \bigcap_{x \in N'} E_M(x).$$

* Deduce that Card (M) = Card (N) (by virtue of (c), we are reduced to the case where N′ and M′ are infinite; show then that Card (M′) $\leqslant$ Card (N′)). *

EXERCISES

§ 5

1. Prove the formula

$$\sum_{k=q+1}^{n-p+q+1} \binom{n-k}{p-q-1}\binom{k-1}{q} = \binom{n}{p},$$

where $p \leqslant n$ and $q < p$ (generalize the argument of no. 8, Corollary to Proposition 14).

2. If $n \geqslant 1$, prove the relation

$$\binom{n}{0} - \binom{n}{1} + \binom{n}{2} - \cdots + (-1)^n\binom{n}{n} = 0.$$

(Define a one-to-one correspondence between the set of subsets of $[1, n]$ which have an even number of elements, and the set of subsets of $[1, n]$ which have an odd number of elements. Distinguish between the cases n even and n odd.)

3. Prove the relations

$$\binom{n}{0}\binom{n}{p} + \binom{n}{1}\binom{n-1}{p-1} + \binom{n}{2}\binom{n-2}{p-2} + \cdots + \binom{n}{p}\binom{n-p}{0} = 2^p\binom{n}{p},$$

$$\binom{n}{0}\binom{n}{p} - \binom{n}{1}\binom{n-1}{p-1} + \binom{n}{2}\binom{n-2}{p-2} - \cdots + (-1)^p\binom{n}{p}\binom{n-p}{0} = 0.$$

(Consider the subsets of p elements of $[1, n]$ which contain a given subset of k elements ($0 \leqslant k \leqslant p$), and use Exercise 2 for the second formula.)

4. Prove Proposition 15 of no. 8 by defining a bijection of the set of mappings u of $[1, h]$ into $[0, n]$ such that

$$\sum_{i=1}^{h} u(x) \leqslant n$$

onto the set of strictly increasing mappings of $[1, h]$ into $[1, n + h]$.

5. * (a) Let E be a distributive lattice and f be a mapping of E into a commutative semigroup M (written additively) such that

$$f(x) + f(y) = f(\sup(x, y)) + f(\inf(x, y))$$

for all x, y in E. Show that for each finite subset I of E we have

$$f(\sup(\mathrm{I})) + \sum_{2n \leqslant \mathrm{Card}(\mathrm{I})} \left(\sum_{\mathrm{H} \subset \mathrm{I},\ \mathrm{Card}(\mathrm{H}) = 2n} f(\inf(\mathrm{H})) \right) = \sum_{2n+1 \leqslant \mathrm{Card}(\mathrm{I})} \left(\sum_{\mathrm{H} \subset \mathrm{I},\ \mathrm{Card}(\mathrm{H}) = 2n+1} f(\inf(\mathrm{H})) \right)$$

(By induction on Card (I).) *

(b) In particular, let A be a set, let $(\mathrm{B}_\iota)_{\iota \in \mathrm{I}}$ be a finite family of finite subsets of A, and let B be the union of the B_ι. For each subset H of I, put $\mathrm{B}_\mathrm{H} = \bigcap_{\iota \in \mathrm{H}} \mathrm{B}_\iota$. Show that

$$\mathrm{Card}(\mathrm{B}) + \sum_{2n \leqslant \mathrm{Card}(\mathrm{I})} \left(\sum_{\mathrm{Card}(\mathrm{H}) = 2n} \mathrm{Card}(\mathrm{B}_\mathrm{H}) \right) = \sum_{2n+1 \leqslant \mathrm{Card}(\mathrm{I})} \left(\sum_{\mathrm{Card}(\mathrm{H}) = 2n+1} \mathrm{Card}(\mathrm{B}_\mathrm{H}) \right).$$

6. Prove the formula

$$\binom{n+h}{h} = 1 + \binom{h}{1}\binom{n+h-1}{h} - \binom{h}{2}\binom{n+h-2}{h} + \cdots + (-1)^h \binom{h}{h}\binom{n}{h}.$$

(If F denotes the set of mappings u of $[1, h]$ into $[0, n]$ such that

$$\sum_{x=1}^{h} u(x) \leqslant n,$$

consider for each subset H of $[1, h]$ the set of all $u \in \mathrm{F}$ such that $u(x) \geqslant 1$ for each $x \in \mathrm{H}$, and use Exercise 5.)

7. (a) Let $\mathrm{S}_{n,p}$ denote the number of mappings of $[1, n]$ *onto* $[1, p]$. Prove that

$$\mathrm{S}_{n,p} = p^n - \binom{p}{1}(p-1)^n + \binom{p}{2}(p-2)^n - \cdots + (-1)^{p-1}\binom{p}{p-1}.$$

(Note that $p^n = \mathrm{S}_{n,p} + \binom{p}{1}\mathrm{S}_{n,p-1} + \binom{p}{2}\mathrm{S}_{n,p-2} + \cdots + \binom{p}{p-1}$ and use Exercise 3.)

(b) Prove that $\mathrm{S}_{n,p} = p(\mathrm{S}_{n-1,p} + \mathrm{S}_{n-1,p-1})$ (method of no. 8, Proposition 13).

(c) Prove that

$$S_{n+1,n} = \frac{n}{2}(n+1)! \quad \text{and} \quad S_{n+2,n} = \frac{n(3n+1)}{24}(n+2)!$$

(consider the elements r of $[1, n]$ whose inverse image consists of more than one element).

(d) If $P_{n,p}$ is the number of partitions into p parts of a set of n elements, show that $S_{n,p} = p!P_{n,p}$.

8. Let p_n be the number of permutations u of a set E with n elements such that $u(x) \neq x$ for all $x \in E$. Show that

$$p_n = n! - \binom{n}{1}(n-1)! + \binom{n}{2}(n-2)! - \cdots + (-1)^n$$

* and hence that $p_n \sim n!/e$ as $n \to +\infty$ * (same method as in Exercise 7 (a)).

9. (a) Let E be a set with qn elements. Show that the number of partitions of E into n subsets each of q elements is equal to

$$(qn)!/(n!(q!)^n).$$

(b) Suppose that $E = [1, qn]$. Show that the number of partitions of E into n subsets of q elements, no one of which is an interval, is equal to

$$\frac{(qn)!}{n!(q!)^n} - \frac{(qn-q+1)!}{1!(n-1)!(q!)^{n-1}} + \frac{(qn-2q+2)!}{2!(n-2)!(q!)^{n-2}} - \cdots + (-1)^n$$

(same method as in Exercises 7 and 8).

10. Let $q_{n,k}$ be the number of strictly increasing mappings u of $[1, k]$ into $[1, n]$ such that for each even (resp. odd) x, $u(x)$ is even (resp. odd). Show that $q_{n,k} = q_{n-1,k-1} + q_{n-2,k}$ and deduce that

$$q_{n,k} = \binom{\left[\frac{n+k}{2}\right]}{k}.$$

¶ 11. Let E be a set with n elements and let S be a set of signs such that S is the disjoint union of E and a set consisting of a single element f. Suppose that f has weight 2 and that each element of E has weight 0 (Chapter I, Appendix, Exercise 3).

(a) Let M be the set of significant words in $L_0(S)$ which contain each element of E exactly once. Show that if u_n is the number of elements in M, then $u_{n+1} = (4n - 2)u_n$, and deduce that

$$u_n = 2.6 \dots (4n - 6) \quad (n \geqslant 2).$$

(This is the number of products of n different terms with respect to a non-associative law of composition.)

(b) Let x_i be the ith of the elements of E which appears in a word of M. Show that the number v_n of words of M, for which the sequence (x_i) is given, is equal to $\binom{2n-2}{n-1}/n$ and satisfies the relation

$$v_{n+1} = v_1 v_n + v_2 v_{n-1} + \cdots + v_{n-1} v_2 + v_n v_1.$$

¶ 12. (a) Let p and q be two integers $\geqslant 1$, let $n = 2p + q$, let E be a set with n elements, and let $N = \binom{n}{p} = \binom{n}{p+q}$. Let $(X_i)_{1 \leqslant i \leqslant N}$ (resp. $(Y_i)_{1 \leqslant i \leqslant N}$) be the sequence of all subsets of E which have p (resp. $p + q$) elements arranged in a certain order. Show that there exists a bijection φ of $[1, N]$ onto itself such that $X_{\varphi(i)} \subset Y_i$ for all i. (The method is analogous to that of Exercise 6 of § 4: observe that for each $r \leqslant N$ the number of sets Y_j which contain at least one of $X_1, \dots, X_r$ is $\geqslant r$.)

(b) Let h, k be two integers $\geqslant 1$, let n be an integer such that $2h + k < n$, let E be a set with n elements and let $(X_i)_{1 \leqslant i \leqslant r}$ be a sequence of distinct subsets of E, each having h elements. Show that there exists a sequence $(Y_j)_{1 \leqslant j \leqslant r+1}$ of distinct subsets of E, each having $h + k$ elements, such that each Y_j contains at least one X_i and each X_i is contained in at least one Y_j (by induction on n, using (a)).

¶ 13. Let E be a set with $2m$ elements, let q be an integer $< m$, and let $\mathcal{F}$ be the set of all subsets $\mathfrak{S}$ of $\mathfrak{P}(E)$ with the following property: if X and Y are two distinct elements of $\mathfrak{S}$ such that $X \subset Y$, then $Y - X$ has at most $2q$ elements.

(a) Let $\mathfrak{M} = (A_i)_{1 \leqslant i \leqslant p}$ be an element of $\mathcal{F}$ such that $p = \text{Card}(\mathfrak{M})$ is as large as possible. Show that $m - q \leqslant \text{Card}(A_i) \leqslant m + q$ for $1 \leqslant i \leqslant p$. (Argue by contradiction. Suppose, for example, that there exist indices i such that $\text{Card}(A_i) < m - q$, and consider those of the A_i for which $\text{Card}(A_i)$ has the least possible value $m - q - s$ (where $s \geqslant 1$). Let $A_1, \dots, A_r$, say, be these sets. Let $\mathfrak{G}$ be the set of subsets of E each of which is the union of some A_i $(1 \leqslant i \leqslant r)$ and a subset of $2q + 1$ elements contained in $E - A_i$. Show that $\mathfrak{G}$ contains at least $r + 1$ elements (cf. Exercise 12), and that if $B_1, \dots, B_{r+1}$ are $r + 1$ distinct elements of $\mathfrak{G}$, the set whose elements are $B_j (1 \leqslant j \leqslant r + 1)$ and $A_i (r + 1 \leqslant i \leqslant p)$ belongs to $\mathcal{F}$, contrary to hypothesis.)

(b) Deduce from (a) that the number of elements p of each $\mathfrak{S} \in \mathfrak{F}$ satisfies the inequality

$$p \leqslant \sum_{k=0}^{2q} \binom{2m}{m-q+k}.$$

(c) Establish results analogous to those of (a) and (b) when $2m$ or $2q$ is replaced by an uneven number.

¶ 14. Let E be a finite set with n elements, let $(a_j)_{1 \leqslant j \leqslant n}$ be the sequence of elements of E arranged in some order, and let $(\mathrm{A}_i)_{1 \leqslant i \leqslant m}$ be a sequence of subsets of E.

(a) For each index j, let k_j be the number of indices i such that $a_j \in \mathrm{A}_i$, and let $\mathrm{S}_i = \mathrm{Card}\,(\mathrm{A}_i)$. Show that

$$\sum_{j=1}^{n} k_j = \sum_{i=1}^{m} s_i.$$

(b) Suppose that, for each subset $\{x, y\}$ of two elements of E, there exists exactly one index i such that x and y are contained in A_i. Show that, if $a_j \notin \mathrm{A}_i$, then $\mathrm{S}_i \leqslant k_j$.

(c) With the hypotheses of (b), show that $m \geqslant n$. (Let k_n be the least of the numbers k_j. Show that we may suppose that, whenever $i \leqslant k_n$, $j \leqslant k_n$, and $i \neq j$, we have $a_j \notin \mathrm{A}_i$ and $a_n \notin \mathrm{A}_j$ for all $j \geqslant k_n$.)

(d) With the hypotheses of (b), show that $m = n$ if and only if one of the following two alternatives is true: (i) $\mathrm{A}_1 = \{a_1, a_2, \ldots, a_{n-1}\}$, $\mathrm{A}_i = \{a_{i-1}, a_n\}$ for $i = 2, \ldots, n$; (ii) $n = k(k-1) + 1$, each A_i has k elements, and each element of E belongs to exactly k sets A_i.

¶ 15. Let E be a finite set, let $\mathfrak{L}$ and $\mathfrak{C}$ be two disjoint non-empty subsets of $\mathfrak{P}(\mathrm{E})$, and let λ, h, k, l be four integers $\geqslant 1$ with the following properties: (i) for each $\mathrm{A} \in \mathfrak{L}$ and each $\mathrm{B} \in \mathfrak{C}$, $\mathrm{Card}\,(\mathrm{A} \cap \mathrm{B}) \geqslant \lambda$; (ii) for each $\mathrm{A} \in \mathfrak{L}$, $\mathrm{Card}\,(\mathrm{A}) \geqslant h$; (iii) for each $\mathrm{B} \in \mathfrak{C}$, $\mathrm{Card}\,(\mathrm{B}) \leqslant k$; (iv) for each $x \in \mathrm{E}$, the number of elements of $\mathfrak{L} \cup \mathfrak{C}$ which contain x is exactly l. Show that $\mathrm{Card}\,(\mathrm{E}) \leqslant hk/\lambda$. (Let $(a_i)_{1 \leqslant i \leqslant n}$ be the sequence of distinct elements of E arranged in some order, and for each i let r_i be the number of elements of $\mathfrak{L}$ to which a_i belongs. Show that, if $\mathrm{Card}\,(\mathfrak{L}) = s$ and $\mathrm{Card}\,(\mathfrak{C}) = t$, then we have

$$\sum_{i=1}^{n} r_i \leqslant sh, \qquad \sum_{i=1}^{n} (l - r_i) \leqslant tk, \qquad \text{and} \qquad \sum_{i=1}^{n} r_i(l - r_i) \geqslant \lambda st.)$$

For $\mathrm{Card}\,(\mathrm{E})$ to be equal to hk/λ it is necessary and sufficient that for each $\mathrm{A} \in \mathfrak{L}$ and each $\mathrm{B} \in \mathfrak{C}$ we have $\mathrm{Card}\,(\mathrm{A}) = h$, $\mathrm{Card}\,(\mathrm{B}) = k$,

Card $(A \cap B) = \lambda$, and that there exists an $r \leqslant l$ such that for each $x \in E$ the number of elements of $\mathfrak{L}$ to which x belongs is equal to r.

16. Let E be a finite set with n elements, let $\mathfrak{D}$ be a non-empty subset of $\mathfrak{P}(E)$, and let λ, k, l be three integers $\geqslant 1$ with the following properties : (i) if A, B are distinct elements of $\mathfrak{D}$, then Card $(A \cap B) = \lambda$; (ii) for each $A \in \mathfrak{D}$, Card $(A) \leqslant k$; (iii) for each $x \in E$ the number of elements of $\mathfrak{D}$ to which x belongs is equal to l. Show that

$$n(\lambda - 1) \leqslant k(k - 1),$$

and that if $n(\lambda - 1) = k(k - 1)$ then $\lambda = k$ and Card $(\mathfrak{D}) = n$. (Given $a \in E$, let $\mathfrak{L}$ be the set of all $A - \{a\}$ where $A \in \mathfrak{D}$ and $a \in A$, and let $\mathfrak{C}$ be the set of all $A \in \mathfrak{D}$ such that $a \notin A$. Apply the results of Exercise 15 to $\mathfrak{L}$ and $\mathfrak{C}$.)

¶ 17. Let i, h, k be three integers such that $i \geqslant 1$, $h \geqslant i$, $k \geqslant i$. Show that there exists an integer $m_i(h, k)$ with the following properties : for each finite set E with at least $m_i(h, k)$ elements, and each partition $(\mathfrak{X}, \mathfrak{Y})$ of the set $\mathfrak{F}_i(E)$ of subsets of i elements of E, it is impossible that every subset of h elements of E contains a subset $X \in \mathfrak{X}$ and that every subset of k elements of E contains a subset $Y \in \mathfrak{Y}$; in other words, if every subset of h elements of E contains some $X \in \mathfrak{X}$, there exists a subset A of k elements of E such that every subset of i elements of A belongs to $\mathfrak{X}$. (Proof by induction. Show that we may take

$$m_1(h, k) = h + k - 1, \qquad m_i(i, k) = k, \qquad \text{and} \qquad m_i(h, i) = h,$$

and finally $m_i(h, k) = m_{i-1}(m_i(h-1, k), m_i(h, k-1)) + 1$. If E is a set with $m_i(h, k)$ elements, if $a \in E$ and if $E' = E - \{a\}$, show that if the proposition were false, then every subset of $m_i(h - 1, k)$ elements of E' would contain a subset X' of $i - 1$ elements such that $X' \cup \{a\} \in \mathfrak{X}$, and that every subset of $m_i(h, k - 1)$ elements of E' would contain a subset Y' of $i - 1$ elements such that $Y' \cup \{a\} \in \mathfrak{Y}$.)

18. (a) Let E be a finite ordered set with p elements. If m,n are two integers such that $mn < p$, show that E has either a totally ordered subset of m elements or else a free subset (§ 1, Exercise 5) of n elements (use § 4, Exercise 5).

(b) Let h, k be two integers $\geqslant 1$ and let $r(h, k) = (h - 1)(k - 1) + 1$. Let I be a finite totally ordered set with at least $r(h, k)$ elements. Show that, for each finite sequence $(x_i)_{i \in I}$ of elements of a totally ordered set E, there exists either a subset H of h elements of I such that the sequence $(x_i)_{i \in H}$ is increasing, or else a subset K of k elements of I such that the sequence $(x_i)_{i \in K}$ is decreasing. (Use (a) applied to $I \times E$.)

§ 6

1. A set E is infinite if and only if for each mapping f of E into E there exists a non-empty subset S of E such that $S \neq E$ and $f(S) \subset S$.

2. Show that, if $\mathfrak{a}, \mathfrak{b}, \mathfrak{c}, \mathfrak{d}$ are four cardinals such that $\mathfrak{a} < \mathfrak{c}$ and $\mathfrak{b} < \mathfrak{d}$, then $\mathfrak{a} + \mathfrak{b} < \mathfrak{c} + \mathfrak{d}$ and $\mathfrak{a}\mathfrak{b} < \mathfrak{c}\mathfrak{d}$ (cf. Exercise 21 (c)).

3. If E is an infinite set, the set of subsets of E which are equipotent to E is equipotent to $\mathfrak{P}(E)$ (use Proposition 3 of no. 4).

4. If E is an infinite set, the set of all partitions of E is equipotent to $\mathfrak{P}(E)$ (associate a subset of $E \times E$ with each partition of E).

5. If E is an infinite set, the set of all permutations of E is equipotent to $\mathfrak{P}(E)$. (Use Proposition 3 of no. 4 to show that, for each subset A of E whose complement does not consist of a single element, there exists a permutation f of E such that A is the set of elements of E which are invariant under f.)

6. Let E, F be two infinite sets such that $\text{Card}(E) \leqslant \text{Card}(F)$. Show that (i) the set of all mappings of E onto F, (ii) the set of all mappings of E into F, and (iii) the set of all mappings of subsets of E into F are all equipotent to $\mathfrak{P}(F)$.

7. Let E, F be two infinite sets such that $\text{Card}(E) < \text{Card}(F)$. Show that the set of all subsets of F which are equipotent to E and the set of all injections of E into F are both equipotent to the set F^E of all mappings of E into F (for each mapping f of E into F, consider the injection $x \to (x, f(x))$ of E into $E \times F$).

8. Show that the set of well-orderings on an infinite set E (and *a fortiori* the set of orderings on E) is equipotent to $\mathfrak{P}(E)$ (use Exercise 5).

9. Let E be a non-empty well-ordered set in which every element x other than the least element of E has a predecessor (the greatest element of $]\leftarrow, x[$). Show that E is isomorphic to either **N** or an interval $[0, n]$ of **N** (remark that every segment $\neq E$ is finite by using Proposition 6 of no. 5; then use Theorem 3 of § 2, no. 5).

¶ 10. Let ω or ω_0 denote the ordinal Ord (**N**) (§ 2, Exercise 14). The set of all integers is then a well-ordered set isomorphic to the set of all ordinals $< \omega$. For each integer n we denote again by n (by abuse of language) the ordinal Ord $([0, n[)$.

(a) Show that for each cardinal $\mathfrak{a}$ the relation "ξ is an ordinal and $\text{Card}(\xi) < \mathfrak{a}$" is collectivizing (use Zermelo's theorem). Let $W(\mathfrak{a})$ denote the set of all ordinals ξ such that $\text{Card}(\xi) < \mathfrak{a}$.

(b) For each ordinal $\alpha > 0$ define a function f_α on the well-ordered set $O'(\alpha)$ of ordinals $\leqslant \alpha$ by transfinite induction as follows: $f_\alpha(0) = \omega_0 = \omega$, and for each ordinal ξ such that $0 < \xi \leqslant \alpha$, $f_\alpha(\xi)$ is the least upper bound (§ 2, Exercise 14 (d)) of the set of ordinals ζ such that $\mathrm{Card}(\zeta) \leqslant \mathrm{Card}(f_\alpha(\eta))$ for at least one ordinal $\eta < \xi$. Show that, if $0 \leqslant \eta < \xi \leqslant \alpha$, then $\mathrm{Card}(f_\alpha(\eta)) < \mathrm{Card}(f_\alpha(\xi))$ and that, if $\xi \leqslant \alpha \leqslant \beta$, then $f_\alpha(\xi) = f_\beta(\xi)$. Put $\omega_\alpha = f_\alpha(\alpha)$; ω_α is said to be the *initial ordinal* with index α. We have $\omega_\alpha \geqslant \alpha$. Put $\aleph_\alpha = \mathrm{Card}(\omega_\alpha)$; $\aleph_\alpha$ is said to be the *aleph of index* α. In particular, $\aleph_0 = \mathrm{Card}\,(\mathbf{N})$.

(c) Show that for each infinite cardinal $\mathfrak{a}$ the least upper bound λ of the set of ordinals $W(\mathfrak{a})$ is an initial ordinal ω_α, and that $\mathfrak{a} = \aleph_\alpha$ (consider the least ordinal μ such that $\omega_\mu \geqslant \lambda$); in other words, ω_α is the least ordinal ξ such that $\mathrm{Card}(\xi) = \aleph_\alpha$. For each ordinal α the mapping $\xi \to \aleph_\xi$, defined on $O'(\alpha)$, is an isomorphism of the well-ordered set $O'(\alpha)$ onto the well-ordered set of cardinals $\leqslant \aleph_\alpha$; in particular, $\aleph_{\alpha+1}$ is the least cardinal $> \aleph_\alpha$. Show that, if α has no predecessor, then for every strictly increasing mapping $\xi \to \sigma_\xi$ of an ordinal β into α such that $\alpha = \sup_{\xi<\beta} \sigma_\xi$, we have

$$\sum_{\xi<\beta} \aleph_{\sigma_\xi} = \aleph_\alpha.$$

(d) Deduce from (c) that ω_ξ is a normal ordinal functional symbol (§ 2, Exercise 17).

¶ 11. (a) Show that the ordinal ω is the least ordinal > 0 which has no predecessor, that ω is indecomposable (§ 2, Exercise 16), and that for each ordinal $\alpha > 0$, $\alpha\omega$ is the least indecomposable ordinal which is $> \alpha$ (note that $n\omega = \omega$ for each integer n). Deduce that

$$(\alpha + 1)\omega = \alpha\omega \quad \text{for each } \alpha > 0.$$

(b) Deduce from (a) that an ordinal is indecomposable if and only if it is of the form ω^β (use Exercise 18 (d) of § 2).

¶ 12. (a) Show that, for each ordinal α and each ordinal $\gamma > 1$, there exist two finite sequences of ordinals (λ_i) and (μ_i) $(1 \leqslant i \leqslant k)$ such that

$$\alpha = \gamma^{\lambda_1}\mu_1 + \gamma^{\lambda_2}\mu_2 + \cdots + \gamma^{\lambda_k}\mu_k,$$

where $0 < \mu_i < \gamma$ for each i, and $\lambda_i > \lambda_{i+1}$ for $1 \leqslant i \leqslant k - 1$ (use Exercise 18 (d) of § 2 and Exercise 3 of § 4). Moreover, the sequences (λ_i), (μ_i) are uniquely determined by these conditions. In particular, there exists a unique finite decreasing sequence $(\beta_j)_{1 \leqslant j \leqslant m}$ such that

$$\alpha = \omega^{\beta_1} + \omega^{\beta_2} + \cdots + \omega^{\beta_m}.$$

Let $\varphi(\alpha)$ denote the greatest ordinal ω^{β_1} in this decomposition.

(b) For each integer n let $f(n) \leqslant n!$ be the greatest number of elements in the set of ordinals of the form $\alpha_{\sigma(1)} + \alpha_{\sigma(2)} + \cdots + \alpha_{\sigma(n)}$, where $(\alpha_i)_{1 \leqslant i \leqslant n}$ is an arbitrary sequence of n ordinals, and σ runs through the set of permutations of the interval $[1, n]$. Show that

$$(1) \qquad f(n) = \sup_{1 \leqslant k \leqslant n-1} (k.2^{k-1} + 1) f(n-k).$$

(Consider first the case where all the $\varphi(\alpha_i)$ are equal and show that the largest possible number of distinct ordinals of the desired form is equal to n, by using Exercise 16 (a) of § 2. Then use induction on the number of ordinals α_i for which $\varphi(\alpha_i)$ takes the least possible value among the set of ordinals $\varphi(\alpha_j)$ $(1 \leqslant j \leqslant n)$.) Deduce from (1) that for $n \geqslant 20$ we have $f(n) = 81 f(n-5)$.

(c) Show that the $n!$ ordinals $(\omega + \sigma(1))(\omega + \sigma(2)) \ldots (\omega + \sigma(n))$, where σ runs through the set of all permutations of $[1, n]$, are all distinct.

¶ 13. (a) Let $w(\xi)$ be an ordinal functional symbol (§2, Exercise 17), defined for $\xi \geqslant \alpha_0$ and such that the relation $\alpha_0 \leqslant \xi < \xi'$ implies $w(\xi) < w(\xi')$. Show that, if $\xi \geqslant \alpha_0$, then $w(\xi + \eta) \geqslant w(\xi) + \eta$ for every ordinal η (argue by contradiction). Deduce that there exists α such that $w(\xi) \geqslant \xi$ for all $\xi \geqslant \alpha$ (take α to be the least indecomposable ordinal $\geqslant \alpha_0$; cf. Exercise 11 (a)).

(b) Let $f(\xi, \eta)$ be the ordinal functional symbol defined in §2, Exercise 17 (b). Suppose that the relations $\alpha_0 \leqslant \xi \leqslant \xi'$ and $\alpha_0 \leqslant \eta \leqslant \eta'$ imply $g(\xi, \eta) \leqslant g(\xi', \eta')$, so that the relations $\alpha_0 \leqslant \xi \leqslant \xi'$ and $1 \leqslant \eta \leqslant \eta'$ imply $f(\xi, \eta) \leqslant f(\xi', \eta')$ (§ 2, Exercise 17 (d)). Show that for each ordinal β there exists at most a finite number of ordinals η for which the equation $f(\xi, \eta) = \beta$ has at least one solution. (Note that if ξ_1 is the least solution of $f(\xi, \eta_1) = \beta$ and if ξ_2 is the least solution of $f(\xi, \eta_2) = \beta$, then the relation $\eta_1 < \eta_2$ implies $\xi_1 > \xi_2$.)

(c) A *critical ordinal* with respect to f is any infinite ordinal $\gamma > \alpha_0$ such that $f(\xi, \gamma) = \gamma$ for all ξ such that $\alpha_0 \leqslant \xi < \gamma$. Show that a critical ordinal (with respect to f) has no predecessor. If there exists a set A of ordinals such that $f(\xi, \gamma) = \gamma$ for all $\xi \in A$, and if γ is the least upper bound of A, show that γ is a critical ordinal.

(d) Let $h(\xi) = f(\xi, \xi)$ (defined for $\xi \geqslant \alpha_0$). Define inductively $\alpha_1 = \alpha_0 + 2$, $\alpha_{n+1} = h(\alpha_n)$ for $n \geqslant 1$. Show that the least upper bound of the sequence (α_n) is a critical ordinal with respect to f.

(e) Show that the least upper bound of every set of critical ordinals with respect to f is again a critical ordinal, and that every critical ordinal is indecomposable (note that $f(\xi, \eta + 1) \geqslant \omega(\xi) + \eta \geqslant \xi + \eta$ for all $\xi \geqslant \alpha_0$).

¶ 14. (a) Show that if $\alpha \geqslant 2$ and if β has no predecessor, then α^β is an indecomposable ordinal (cf. § 2, Exercise 16 (a)); if α is finite and if $\beta = \omega\gamma$, then $\alpha^\beta = \omega^\gamma$; if α is infinite and if π is the greatest indecomposable ordinal $\leqslant \alpha$, then $\alpha^\beta = \pi^\beta$ (use Exercise 11).

(b) An ordinal δ is critical with respect to the functional symbol $f(\xi, \eta) = \xi\eta$ if and only if, for each α such that $1 < \alpha \leqslant \delta$, the equation $\delta = \alpha^\xi$ has a solution; the unique solution ξ of this equation is then indecomposable (use Exercise 13 (e), together with Exercise 18 (d) of § 2). Conversely, for each $\alpha > 1$ and each indecomposable ordinal π, α^π is a critical ordinal with respect to $\xi\eta$ (use Exercise 13 (c)). Deduce that δ is a critical ordinal with respect to $\xi\eta$ if and only if δ is of the form ω^{ω^μ} (cf. Exercise 11 (b)).

(c) For an ordinal ε to be critical with respect to the functional symbol $f(\xi, \eta) = \xi^\eta$, i.e., such that $\gamma^\varepsilon = \varepsilon$ for each γ satisfying $2 \leqslant \gamma \leqslant \varepsilon$, it is sufficient that $2^\varepsilon = \varepsilon$. Show that the least critical ordinal ε_0 with respect to ξ^η is countable (cf. Exercise 13 (d)).

¶ 15. Let γ be an ordinal > 1, and for each ordinal α let $L(\alpha)$ denote the set of exponents λ_i in the expression for α given in Exercise 12 (a).

(a) Show that $\lambda_i \leqslant \alpha$ for each $\lambda_i \in L(\alpha)$, and that $\lambda_i = \alpha$ for one of these ordinals only if $\alpha = 0$ or if α is a critical ordinal with respect to ξ^η (Exercise 14 (c)).

(b) Define $L_n(\alpha)$ by induction on n as follows: $L_1(\alpha) = L(\alpha)$, and $L_n(\alpha)$ is the union of the sets $L(\beta)$ as β runs through $L_{n-1}(\alpha)$. Show that there exists an integer n_0 such that $L_{n+1}(\alpha) = L_n(\alpha)$ whenever $n \geqslant n_0$, and that the elements of $L_n(\alpha)$ are then either 0 or critical ordinals with respect to ξ^η. (Argue by contradiction: for each n, consider the set $M_n(\alpha)$ of elements $\beta \in L_n(\alpha)$ such that $\beta \notin L(\beta)$, and assume that $M_n(\alpha)$ is not empty for any n; use (a) to obtain a contradiction.)

16. Every totally ordered set has a well-ordered cofinal subset (§2, Exercise 2). The least of the ordinals Ord (M) of the well-ordered cofinal subsets M of E is called the *final character* of E.

(a) An ordinal ξ is said to be *regular* if it is equal to its final character, and *singular* otherwise. Show that every infinite regular ordinal is an initial ordinal ω_α (Exercise 10). Conversely, every initial ordinal ω_α, whose index α is either 0 or has a predecessor, is a regular ordinal. An initial ordinal ω_α whose index α has no predecessor is singular if $0 < \alpha < \omega_\alpha$; in particular, ω_ω is the least infinite singular initial ordinal.

(b) An initial ordinal ω_α is said to be *inaccessible* if it is regular and its index α has no predecessor. Show that, if $\alpha = 0$, then $\omega_\alpha = \alpha$; in other words, α is a critical ordinal with respect to the normal functional symbol ω_η (Exercise 10 (d) and 13 (c)). Let $\varkappa$ be the least critical ordinal with respect to this functional symbol. Show that $\omega_\varkappa$ is singular, with final character ω (cf. Exercise 13 (d)). In other words, there exists no inaccessible ordinal ω_α such that $0 < \alpha \leqslant \varkappa$ (*).

(c) Show that there exists only one regular ordinal which is cofinal in a given totally ordered set E; this ordinal is equal to the final character of E, and if E is not empty and has no greatest element, it is an initial ordinal. If $\omega_{\bar{\alpha}}$ is the final character of ω_α, then $\bar{\alpha} \leqslant \alpha$; and ω_α is regular if and only if $\alpha = \bar{\alpha}$.

(d) Let ω_α be a regular ordinal and let I be a well-ordered set such that Ord (I) $< \omega_\alpha$. Show that, for each family $(\xi_\iota)_{\iota \in I}$ of ordinals such that $\xi_\iota < \omega_\alpha$ for all $\iota \in I$, we have $\sum_{\iota \in I} \xi_\iota < \omega_\alpha$.

17. A cardinal $\aleph_\alpha$ is said to be *regular* (resp. *singular*) if the initial ordinal ω_α is regular (resp. singular). For $\aleph_\alpha$ to be regular it is necessary and sufficient that for every family $(\mathfrak{a}_\iota)_{\iota \in I}$ of cardinals such that Card (I) $< \aleph_\alpha$ and $\mathfrak{a}_\iota < \aleph_\alpha$ for all $\iota \in I$, we have

$$\sum_{\iota \in I} \mathfrak{a}_\iota < \aleph_\alpha.$$

$\aleph_\omega$ is the least singular cardinal.

¶ 18. (a) For each ordinal α and each cardinal $\mathfrak{m} \neq 0$ we have $\aleph_{\alpha+1}^{\mathfrak{m}} = \aleph_\alpha^{\mathfrak{m}} \cdot \aleph_{\alpha+1}$ (reduce to the case where $\mathfrak{m} < \aleph_{\alpha+1}$ and consider the mappings of the cardinal $\mathfrak{m}$ into the ordinal $\omega_{\alpha+1}$).

(b) Deduce from (a) that, for each ordinal γ such that Card $(\gamma) \leqslant \mathfrak{m}$, we have $\aleph_{\alpha+\gamma}^{\mathfrak{m}} = \aleph_\alpha^{\mathfrak{m}} \cdot \aleph_{\alpha+\gamma}^{\mathrm{Card}(\gamma)}$ (by transfinite induction on γ).

(c) Deduce from (b) that, for each ordinal α such that Card $(\alpha) \leqslant \mathfrak{m}$, we have $\aleph_\alpha^{\mathfrak{m}} = 2^{\mathfrak{m}} \cdot \aleph_\alpha^{\mathrm{Card}(\alpha)}$.

¶ 19. (a) Let α and β be two ordinals such that α has no predecessor, and let $\xi \to \sigma_\xi$ be a strictly increasing mapping of the ordinal ω_β into the ordinal α such that $\sup_{\xi < \omega_\beta} \sigma_\xi = \alpha$. Show that

$$\aleph_\alpha^{\aleph_\beta} = \prod_{\xi < \omega_\beta} \aleph_{\sigma_\xi}.$$

(With each mapping f of the ordinal ω_β into the ordinal ω_α associate an injective mapping $\bar{f}$ of ω_β into the set of all ω_{σ_ξ} $(\xi < \omega_\beta)$ such that

---

(*) At present it is not known whether or not there exist inaccessible ordinals other than ω.

$\bar{f}(\zeta) \leqslant f(\zeta)$ for all $\zeta < \omega_\beta$. Calculate the cardinal of the set of mappings f associated with the same $\bar{f}$ and observe that

$$\mathfrak{m} = \prod_{\xi < \omega_\beta} \aleph_{\sigma_\xi} \geqslant 2^{\mathrm{Card}\,(\omega_\beta)}$$

and $\mathfrak{m} \geqslant \aleph_\alpha$ (cf. § 3, Exercise 3).)

(b) Let $\bar{\alpha}$ be the ordinal such that $\omega_{\bar{\alpha}}$ is the final character of ω_α. Show that $\aleph_\alpha^{\aleph_{\bar{\alpha}}} > \aleph_\alpha$ and that if there exists $\mathfrak{n}$ such that $\aleph_\alpha = \mathfrak{n}^{\aleph_\gamma}$, then $\gamma < \bar{\alpha}$ (use (a) and Exercise 3 of § 3).

(c) Show that, if $\lambda < \bar{\alpha}$, then

$$\aleph_\alpha^{\aleph_\lambda} = \sum_{\xi < \alpha} \aleph_\xi^{\aleph_\lambda}$$

(argue as in Exercise 18 (a)).

¶ 20. (a) For a cardinal $\mathfrak{a}$ to be regular (Exercise 17) it is necessary that for every cardinal $\mathfrak{b} \neq 0$ we should have

$$\mathfrak{a}^{\mathfrak{b}} = \mathfrak{a} \,.\, \sum_{\mathfrak{m} < \mathfrak{a}} \mathfrak{m}^{\mathfrak{b}}.$$

(Use Exercise 19 and consider separately the cases (i) $\mathfrak{b}$ is finite, (ii) $\aleph_0 \leqslant \mathfrak{b} < \mathfrak{a}$, (iii) $\mathfrak{b} \geqslant \mathfrak{a}$; also use Exercise 3 of § 3.) The generalized continuum hypothesis implies that the above condition is also sufficient.

(b) Show that, if a cardinal $\mathfrak{a}$ is such that $\mathfrak{a}^{\mathfrak{m}} = \mathfrak{a}$ for every cardinal $\mathfrak{m}$ such that $0 < \mathfrak{m} < \mathfrak{a}$, then $\mathfrak{a}$ is regular (use Exercise 3 of § 3).

(c) Show that the proposition "for every regular cardinal $\mathfrak{a}$ and every cardinal $\mathfrak{m}$ such that $0 < \mathfrak{m} < \mathfrak{a}$, we have $\mathfrak{a} = \mathfrak{a}^{\mathfrak{m}}$" is equivalent to the generalized continuum hypothesis (use (a)).

¶ 21. An infinite cardinal $\mathfrak{a}$ is said to be *dominant* if, for each pair of cardinals $\mathfrak{m} < \mathfrak{a}$, $\mathfrak{n} < \mathfrak{a}$, we have $\mathfrak{m}^{\mathfrak{n}} < \mathfrak{a}$.

(a) For $\mathfrak{a}$ to be dominant it is sufficient that $2^{\mathfrak{m}} < \mathfrak{a}$ for every cardinal $\mathfrak{m} < \mathfrak{a}$.

(b) Define inductively a sequence $(\mathfrak{a}_n)$ of cardinals as follows: $\mathfrak{a}_0 = \aleph_0$, $\mathfrak{a}_{n+1} = 2^{\mathfrak{a}_n}$. Show that the sum $\mathfrak{d}$ of the sequence $(\mathfrak{a}_n)$ is a dominant cardinal. $\aleph_0$ and $\mathfrak{d}$ are the two smallest dominant cardinals.

(c) Show that $\mathfrak{d}^{\aleph_0} = \aleph_0^{\mathfrak{d}} = 2^{\mathfrak{d}}$ (note that $2^{\mathfrak{d}} \leqslant \mathfrak{d}^{\aleph_0}$). Deduce that $\mathfrak{d}^{\aleph_0} = (2^{\mathfrak{d}})^{\mathfrak{d}}$, although $\mathfrak{d} < 2^{\mathfrak{d}}$ and $\aleph_0 < \mathfrak{d}$.

¶ 22. A cardinal $\aleph_\alpha$ is said to be *inaccessible* if the ordinal ω_α is inaccessible (Exercise 16 (b)). We have then $\omega_\alpha = \alpha$ if $\omega_\alpha \neq \omega_0$. A cardinal $\mathfrak{a}$ is said to be *strongly inaccessible* if it is inaccessible and dominant (Exercise 21).

(a) The generalized continuum hypothesis implies that every inaccessible cardinal is strongly inaccessible.

(b) For a cardinal $\mathfrak{a} \geqslant 3$ to be strongly inaccessible it is necessary and sufficient that, for each family $(\mathfrak{a}_\iota)_{\iota \in I}$ of cardinals such that

$$\mathrm{Card}(I) < \mathfrak{a} \quad \text{and} \quad \mathfrak{a}_\iota < \mathfrak{a}$$

for all $\iota \in I$, we should have $\prod_{\iota \in I} \mathfrak{a}_\iota < \mathfrak{a}$.

(c) For an infinite cardinal $\mathfrak{a}$ to be strongly inaccessible it is necessary and sufficient that it should be dominant (Exercise 21) and that it should satisfy one of the following two conditions: (i) $\mathfrak{a}^{\mathfrak{b}} = \mathfrak{a}$ for every cardinal $\mathfrak{b}$ such that $0 < \mathfrak{b} < \mathfrak{a}$; (ii) $\mathfrak{a}^{\mathfrak{b}} = \mathfrak{a} . 2^{\mathfrak{b}}$ for every cardinal $\mathfrak{b} > 0$. (Use Exercises 20 and 21.)

¶ 23. Let α be an ordinal > 0. A mapping f of the ordinal α into itself is said to be *divergent* if for each ordinal $\lambda_0 < \alpha$ there exists an ordinal $\mu_0 < \alpha$ such that the relation $\mu_0 \leqslant \xi < \alpha$ implies $\lambda_0 \leqslant f(\xi) < \alpha$ (*).

(a) Let φ be a strictly increasing mapping of an ordinal β into α such that

$$\varphi(\sup_{\zeta<\gamma} \zeta) = \sup_{\zeta<\gamma} \varphi(\zeta) \quad \text{for all} \quad \gamma < \beta,$$

and such that

$$\sup_{\zeta<\beta} \varphi(\zeta) = \alpha (\dagger).$$

Then there exists a divergent mapping f of α into itself, such that $f(\xi) < \xi$ for all ξ satisfying $0 < \xi < \alpha$, if and only if there exists a divergent mapping of β into itself of the same type.

(b) Deduce from (a) that there exists a divergent mapping of ω_α into itself, such that $f(\xi) < \xi$ for all ξ satisfying $0 < \xi < \alpha$, if and only if the final character of ω_α is ω_0. (If ω_α is a regular ordinal $> \omega_0$, define inductively a strictly increasing sequence (η_n) as follows: $\eta_1 = 1$, and η_{n+1} is the least ordinal ζ such that $f(\xi) > \eta_n$ for all $\xi \geqslant \zeta$.)

(c) Let $\omega_{\bar{\alpha}}$ be the final character of ω_α (Exercise 16). Show that, if $\bar{\alpha} > 0$ and if f is a mapping of ω_α into itself such that $f(\xi) < \zeta$ for all ξ

(*) If the well-ordered set O'_α of ordinals $\leqslant \alpha$ is endowed with the topology $\mathcal{T}_-(O'_\alpha)$ (*General Topology*, Chapter I, § 2, Exercise 5), this condition may be written $\lim_{\xi \to \alpha, \xi < \alpha} f(\xi) = \alpha$.

(†) If we extend φ to O'_β by defining $\varphi(\beta) = \alpha$, the conditions above signify that φ is continuous with respect to the topologies $\mathcal{T}_-(O'_\alpha)$ and $\mathcal{T}_-(O'_\beta)$ on O'_α and O'_β respectively (*loc. cit.*).

such that $0 < \xi < \omega_\alpha$, then there exists an ordinal λ_0 such that the set of solutions of the equation $f(\xi) = \lambda_0$ has a cardinal $\geqslant \aleph_{\bar{\alpha}}$.

¶ 24. Let $\mathfrak{F}$ be a set of subsets of a set E such that for each $A \in \mathfrak{F}$ we have $\text{Card}(A) = \text{Card}(\mathfrak{F}) = \mathfrak{a} \geqslant \aleph_0$. Show that E has a subset P such that $\text{Card}(P) = \mathfrak{a}$ and such that no set of $\mathfrak{F}$ is contained in P. (If $\mathfrak{a} = \aleph_\alpha$, define by transfinite induction two injective mappings $\xi \to f(\xi)$, $\xi \to g(\xi)$ of ω_α into E such that the sets $P = f(\omega_\alpha)$ and $Q = g(\omega_\alpha)$ do not intersect and such that each of them meets every subset $A \in \mathfrak{F}$.)

(b) Suppose, moreover, that for each subset $\mathfrak{G}$ of $\mathfrak{F}$ such that $\text{Card}(\mathfrak{G}) < \mathfrak{a}$, the complement in E of the union of the sets $A \in \mathfrak{G}$ has cardinal $\geqslant \mathfrak{a}$. Show that E then has a subset P such that $\text{Card}(P) = \mathfrak{a}$ and such that, for each $A \in \mathfrak{G}$, $\text{Card}(P \cap A) < \mathfrak{a}$ (similar method).

¶ 25. (a) Let $\mathfrak{F}$ be a covering of an infinite set E. The *degree of disjointness* of $\mathfrak{F}$ is the least cardinal $\mathfrak{c}$ such that $\mathfrak{c}$ is *strictly greater* than the cardinals $\text{Card}(X \cap Y)$ for each pair of distinct sets $X, Y \in \mathfrak{F}$. If $\text{Card}(E) = \mathfrak{a}$ and $\text{Card}(\mathfrak{F}) = \mathfrak{b}$, show that $\mathfrak{b} \leqslant \mathfrak{a}^{\mathfrak{c}}$ (note that a subset of E of cardinal $\mathfrak{c}$ is contained in at most one set of $\mathfrak{F}$).

(b) Let ω_α be an initial ordinal and let F be a set such that $2 \leqslant \mathfrak{p} = \text{Card}(F) < \aleph_\alpha$. Let E be the set of mappings of segments of ω_α, other than ω_α itself, into F. Then we have $\text{Card}(E) \leqslant \mathfrak{p}^{\aleph_\alpha}$. For each mapping f of ω_α into F, let K_f be the subset of E consisting of the restrictions of f to the segments of ω_α (other than ω_α itself). Show that the set $\mathfrak{F}$ of subsets K_f is a covering of E such that $\text{Card}(\mathfrak{F}) = \mathfrak{p}^{\aleph_\alpha}$, and that its degree of disjointness is equal to $\aleph_\alpha$.

(c) Let E be an infinite set of cardinal $\mathfrak{a}$, and let $\mathfrak{c}$, $\mathfrak{p}$ be two cardinals > 1 such that $\mathfrak{p} < \mathfrak{c}$, $\mathfrak{p}^{\mathfrak{m}} < \mathfrak{a}$ for all $\mathfrak{m} < \mathfrak{c}$, and $\mathfrak{a} = \sum_{\mathfrak{m} < \mathfrak{c}} \mathfrak{p}^{\mathfrak{m}}$. Deduce from (b) that there exists a covering $\mathfrak{F}$ of E consisting of sets of cardinal $\mathfrak{c}$, with degree of disjunction equal to $\mathfrak{c}$, and such that $\text{Card}(\mathfrak{F}) = \mathfrak{p}^{\mathfrak{c}}$. In particular, if E is countably infinite, there exists a covering $\mathfrak{F}$ of E by infinite sets such that $\text{Card}(\mathfrak{F}) = 2^{\aleph_0}$ and such that the intersection of any two sets of $\mathfrak{F}$ is *finite*.

¶ 26. Let E be an infinite set and let $\mathfrak{F}$ be a set of subsets of E such that for each $A \in \mathfrak{F}$ we have

$$\text{Card}(A) = \text{Card}(\mathfrak{F}) = \text{Card}(E) = \mathfrak{a} \geqslant \aleph_0.$$

Show that there exists a partition $(B_\iota)_{\iota \in I}$ of E such that

$$\text{Card}(I) = \text{Card}(B_\iota) = \mathfrak{a}$$

for all $\iota \in I$, and such that $A \cap B_\iota \neq \emptyset$ for all $A \in \mathfrak{F}$ and all $\iota \in I$.

(With the notation of Exercise 24 (a), consider first a surjective mapping f of ω_α into $\mathfrak{F}$ such that for each $A \in \mathfrak{F}$ the set of all $\xi \in \omega_\alpha$ such that $f(\xi) = A$ has cardinal equal to $\mathfrak{a}$. Then, by transfinite induction, define a bijection g of ω_α onto E such that $g(\xi) \in f(\xi)$ for every $\xi \in \omega_\alpha$.)

¶ 27. Let L be an infinite set and let $(E_\lambda)_{\lambda \in L}$ be a family of sets indexed by L. Suppose that for each integer $n > 0$ the set of $\lambda \in L$ such that $\operatorname{Card}(E_\lambda) > n$ is equipotent to L. Show that there exists a subset F of the product $E = \prod_{\lambda \in L} E_\lambda$, such that $\operatorname{Card}(F) = 2^{\operatorname{Card}(L)}$, and such that F has the following property: for each finite sequence $(f_k)_{1 \leqslant k \leqslant n}$ of distinct elements of F there exists $\lambda \in L$ such that the elements

$$f_k(\lambda) \in E_\lambda \ (1 \leqslant k \leqslant n)$$

are all distinct. (Show first that there exists a partition $(L_j)_{j \in N}$ of L such that $\operatorname{Card}(L_j) = \operatorname{Card}(L)$ for all j, and such that $\operatorname{Card}(E_\lambda) \geqslant 2^j$ for each $\lambda \in L_j$. Hence reduce to the case where L is the sum of the countable family of sets X^j $(j \geqslant 1)$, where X is an infinite set, and $E_\lambda = 2^j$ for each $\lambda \in X^j$. With each mapping $g \in 2^X$ of X into 2, associate the element $f \in E$ such that $f(\lambda) = (g(x_1), \ldots, g(x_j))$ whenever $\lambda = (x_k)_{1 \leqslant k \leqslant j} \in X^j$; show that the set F of elements $f \in E$ so defined has the required property.)

¶ 28. Let E be an infinite set and let $(\mathfrak{X}_i)_{1 \leqslant i \leqslant m}$ be a finite partition of the set $\mathfrak{F}_n(E)$ of subsets of E having n elements. Show that there exists an index i and an infinite subset F of E such that every subset of F with n elements belongs to $\mathfrak{X}_i$. (Proof by induction on n. For each $a \in E$ show that there exists an index $j(a)$ and an infinite subset $M(a)$ of $E - \{a\}$ such that, for every subset A of $M(a)$ with $n-1$ elements, $\{a\} \cup A$ belongs to $\mathfrak{X}_{j(a)}$. Then define a sequence (a_i) of elements of E as follows: a_1 is an arbitrary element of E, a_2 is an arbitrary element of $M(a_1)$, a_3 is defined in terms of $M(a_1)$ and a_2 in the same way as a_2 was defined in terms of E and a_1, and so on. Show that the set F of elements of a suitable subsequence of the sequence (a_i) satisfies the required conditions.)

29. (a) In an ordered set E, every finite union of Noetherian subsets (with respect to the induced ordering) is Noetherian.

(b) An ordered set E is Noetherian if and only if for each $a \in E$ the interval $]a, \rightarrow[$ is Noetherian.

(c) Let E be an ordered set such that the ordered set obtained by endowing E with the opposite ordering is Noetherian. Let u be a letter and let $T\{u\}$ be a term. Show that there exists a set U and a mapping f of E onto U such that for each $x \in E$ we have $f(x) = T\{f^{(x)}\}$, where

$f^{(x)}$ denotes the mapping of $]\leftarrow, x[$ onto $f(]\leftarrow, x[)$ which coincides with f on this interval. Furthermore, U and f are determined uniquely by this condition.

(d) Let E be a Noetherian ordered set such that every finite subset of E has a least upper bound in E. Show that, if E has a least element, then E is a complete lattice (§ 1, Exercise 11); and that if E has no least element, the set E′ obtained by adjoining a least element to E (§ 1, no. 7, Proposition 3) is a complete lattice.

30. Let E be a lattice such that the set obtained by endowing E with the opposite ordering is Noetherian. Show that every element $a \in E$ can be written as $\sup(e_1, e_2, \ldots, e_n)$, where $e_1, \ldots, e_n$ are irreducible (§ 4, Exercise 7; show first that there exists an irreducible element e such that $a = \sup(e, b)$ if a is not irreducible). Generalize Exercise 7(b) of § 4 to E; also generalise Exercises 8(b) and 9(b) of § 4.

¶ 31. Let A be an infinite set and let E be the set of all infinite subsets of A, ordered by inclusion. Show that E is completely ramified (§ 2, Exercise 8) but not antidirected (§ 1, Exercise 23) and that E has an antidirected cofinal subset F. (Consider first the set $\mathfrak{D}(A)$ of countable infinite subsets of A (which is cofinal in E) and let $Z = R_0(\mathfrak{D}(A))$ (§ 1, Exercise 23). Write Z in the form $(z_\lambda)_{\lambda \in L}$, where L is a well-ordered set, and take F to be a set of countable subsets X_n^λ, where λ runs through a suitable subset of L, $n \in N$, $X_m^\lambda \supset X_n^\lambda$ whenever $m \leqslant n$, $X_n^\lambda - X_{n+1}^\lambda$ is infinite for all $n \geqslant 0$, and

$$\bigcap_{n \in N} X_n^\lambda = \emptyset;$$

the X_n^λ are to be defined by transfinite induction in such a way that the images of the sets X_n^λ under the canonical mapping $r : \mathfrak{D}(A) \to Z$ (§ 1, Exercise 23) are mutually disjoint and form a cofinal subset of Z.)

¶ *32. Let (M_n), (P_n) be two sequences of mutually disjoint finite sets (not all empty), indexed by the set **Z** of rational integers. Let $\alpha_n = \text{Card}(M_n)$, $\beta_n = \text{Card}(P_n)$. Suppose that there exists an integer $k > 0$ such that for each $n \in \mathbf{Z}$ and each integer $l \geqslant 1$ we have

$$\alpha_n + \alpha_{n+1} + \cdots + \alpha_{n+l} \leqslant \beta_{n-k} + \beta_{n-k+1} + \cdots + \beta_{n+l+k},$$
$$\beta_n + \beta_{n+1} + \cdots + \beta_{n+l} \leqslant \alpha_{n-k} + \alpha_{n-k+1} + \cdots + \alpha_{n+l+k}.$$

Let M be the union of the family (M_n) and let P be the union of the family (P_n). Show that there exists a bijection φ of M onto P such that

$$\varphi(M_n) \subset \bigcup_{i=n-k-1}^{n+k+1} P_i \quad \text{and} \quad \overset{-1}{\varphi}(P_n) \subset \bigcup_{i=n-k-1}^{n+k+1} M_i$$

for each $n \in \mathbf{Z}$. (Consider a total ordering on each M_n (resp. P_n) and take M (resp. P) to be the ordinal sum (§ 1, Exercise 3) of the family $(M_n)_{n \in \mathbf{Z}}$ (resp. $(P_n)_{n \in \mathbf{Z}}$). If n_0 is an index such that $M_{n_0} \neq \emptyset$, consider the isomorphisms of M onto P which transform the least element of M_{n_0} into one of the elements of $\bigcup_{j=n_0-k}^{n_0+k} P_j$, and show that one of these isomorphisms satisfies the required conditions. Let δ be the least of the numbers

$$\beta_{n-k} + \beta_{n-k+1} + \cdots + \beta_{n+l+k} - (\alpha_n + \alpha_{n+1} + \cdots + \alpha_{n+l}),$$
$$\alpha_{n-k} + \alpha_{n-k+1} + \cdots + \alpha_{n+l+k} - (\beta_n + \beta_{n+1} + \cdots + \beta_{n+l})$$

for all $n \in \mathbf{Z}$ and all $l \geqslant 1$. If $n \in \mathbf{Z}$ and $l \geqslant 1$ are such that, for example, $\beta_{n-k} + \beta_{n-k+1} + \cdots + \beta_{n+l+k} = \delta + \alpha_n + \alpha_{n+1} + \cdots + \alpha_{n+l}$, we may take φ to be such that the least element of P_{n-k} is the image under φ of the least element of M_n.)

§ 7

1. Let I be a directed set, let $(J_\lambda)_{\lambda \in L}$ be a family of subsets of I, indexed by a directed set L, such that (i) for each $\lambda \in L$, J_λ is directed with respect to the induced ordering; (ii) the relation $\lambda \leqslant \mu$ implies $J_\lambda \subset J_\mu$; (iii) I is the union of the family (J_λ). Let $(E_\alpha, f_{\alpha\beta})$ be an inverse system of sets relative to I, let E be its inverse limit, and for each $\lambda \in L$ let F_λ be the inverse limit of the system obtained from $(E_\alpha, f_{\alpha\beta})$ by restricting the index set to J_λ. For $\lambda \leqslant \mu$ let $g_{\lambda\mu}$ be the canonical mapping of F_μ into F_λ (no. 1). Show that $(F_\lambda, g_{\lambda\mu})$ is an inverse system of sets relative to L, and define a canonical bijection of $F = \varprojlim F_\lambda$ onto E.

2. Let $(E_\alpha, f_{\alpha\beta})$ be an inverse system of sets relative to a directed index set, let $E = \varprojlim E_\alpha$, and let $f_\alpha : E \to E_\alpha$ be the canonical mapping for each α. Show that, if all the $f_{\alpha\beta}$ are injective, then f_α is injective.

3. Let $(E_\alpha, f_{\alpha\beta})$ and $(F_\alpha, g_{\alpha\beta})$ be two inverse systems of sets relative to the same index set I. For each $\alpha \in I$, let u_α be a mapping of E_α into F_α, such that the u_α form an inverse system of mappings. Let $G_\alpha \subset E_\alpha \times F_\alpha$ be the graph of u_α. Show that (G_α) is an inverse system of subsets of $E_\alpha \times F_\alpha$ and that its inverse limit may be canonically identified with the graph of $u = \varprojlim u_\alpha$.

4. Let I be a non-empty directed set with no greatest element, and let F be the set of all sequences $x = (\alpha_1, \alpha_2, \ldots, \alpha_{2n-1}, \alpha_{2n})$ of an even number $\geqslant 2$ of elements of I with the following properties : (i) $\alpha_{2i-1} < \alpha_{2i}$ for $1 \leqslant i \leqslant n$; (ii) $\alpha_{2i-1} \not\leqslant \alpha_{2j-1}$ for $1 \leqslant j < i \leqslant n$. The set F is not empty. Put $r(x) = \alpha_{2n-1}$, $s(x) = \alpha_{2n}$. The integer n is called the *length* of x.

(a) For each $\alpha \in I$, let E_α be the set of all $x \in F$ such that $r(x) = \alpha$. Then E_α is not empty. For $\alpha \leqslant \beta$ in I, we define a mapping $f_{\alpha\beta}$ of E_β into the set of all finite sequences of elements of I, as follows : if

$$x = (\alpha_1, \ldots, \alpha_{2n-1}, \alpha_{2n}) \in E_\beta,$$

let j be the least index such that $\alpha \leqslant \alpha_{2j-1}$; then

$$f_{\alpha\beta}(x) = (\alpha_1, \ldots, \alpha_{2j-2}, \alpha, \alpha_{2j}).$$

Show that $f_{\alpha\beta}(E_\beta) = E_\alpha$ and that $(E_\alpha, f_{\alpha\beta})$ is an inverse system of sets relative to I.

(b) Show that if $x_\alpha \in E_\alpha$ and $x_\beta \in E_\beta$ are such that there exists an index γ for which $\gamma \geqslant \alpha$ and $\gamma \geqslant \beta$, and an element $x_\gamma \in E_\gamma$ for which $x_\alpha = f_{\alpha\gamma}(x_\gamma)$ and $x_\beta = f_{\beta\gamma}(x_\gamma)$, then, provided also x_α and x_β have the same length, we have $s(x_\alpha) = s(x_\beta)$.

(c) Deduce from (b) that, if $E = \varprojlim E_\alpha$ is not empty and if $y = (x_\alpha) \in E$, then the set of elements $s(x_\alpha)$ is countable and cofinal in I.

(d) Let I be the set of all finite subsets of an uncountable set A, ordered by inclusion. Show that I has no countable cofinal subset, and hence deduce from (c) an example of an inverse system of sets $(E_\alpha, f_{\alpha\beta})$ in which the E_α are non-empty and the $f_{\alpha\beta}$ are surjective, but for which $E = \varprojlim E_\alpha = \emptyset$.

(e) Deduce from (d) an example of an inverse system of mappings $u_\alpha : E_\alpha \to E'_\alpha$ such that each u_α is surjective but $\varprojlim u_\alpha$ is not surjective (let each E'_α consist of a single element).

¶ 5. Let I be a directed set and let $(E_\alpha)_{\alpha \in I}$ be a family of lattices such that each E_α, endowed with the opposite ordering, is Noetherian (§ 6, no. 5). For each pair (α, β) of indices in I such that $\alpha \leqslant \beta$ let

$$f_{\alpha\beta} : E_\beta \to E_\alpha$$

be an *increasing* mapping, and suppose that $(E_\alpha, f_{\alpha\beta})$ is an inverse system of sets relative to I. For each $\alpha \in I$ let G_α be a non-empty subset of E_α such that (i) no two distinct elements of G_α are comparable, (ii) $f_{\alpha\beta}(G_\beta) = G_\alpha$ whenever $\alpha \leqslant \beta$, (iii) for each $\alpha \leqslant \beta$ and each $x_\alpha \in G_\alpha$, $\overset{-1}{f_{\alpha\beta}}(x_\alpha)$ has a greatest element $M_{\alpha\beta}(x_\alpha)$ in E_β, (iv) whenever $\alpha \leqslant \beta$, if $h_\beta \in E_\beta$ is such that there exists $y_\beta \in G_\beta$ such that $y_\beta \leqslant h_\beta$, then for each $x_\alpha \in G_\alpha$ such that $x_\alpha \leqslant f_{\alpha\beta}(h_\beta)$ there exists $x_\beta \in G_\beta$ such that $x_\beta \leqslant h_\beta$ and $x_\alpha = f_{\alpha\beta}(x_\beta)$. Under these conditions the inverse limit of the inverse system of subsets (G_α) *is not empty*. The proof runs as follows:

(a) Let J be a *finite* subset of I. A family $(x_\alpha)_{\alpha \in J}$, where $x_\alpha \in G_\alpha$ for all $\alpha \in J$, is said to be *coherent* if it satisfies the following two conditions :

(i) if $\alpha \in J$, $\beta \in J$, $\alpha \leqslant \beta$, then $x_\alpha = f_{\alpha\beta}(x_\beta)$; (ii) for each upper bound γ of J in I there exists $x_\gamma \in G_\gamma$ such that $x_\alpha = f_{\alpha\gamma}(x_\gamma)$ for all $\alpha \in J$. Show that, for each upper bound γ of J in I, the set $\bigcap_{\alpha \in J} \overset{-1}{f}_{\alpha\gamma}(x_\alpha)$ has a greatest element equal to $\inf_{\alpha \in J} (M_{\alpha\gamma}(x_\alpha))$; furthermore, the intersection of G_γ and $\bigcap_{\alpha \in J} \overset{-1}{f}_{\alpha\gamma}(x_\alpha)$ is the set (non-empty by hypothesis) of all $y_\gamma \in G_\gamma$ such that

$$y_\gamma \leqslant \inf_{\alpha \in J} (M_{\alpha\gamma}(x_\alpha))$$

(use condition (i)).

(b) Let J be any subset of I. A family $x_J = (x_\alpha)_{\alpha \in J}$, where $x_\alpha \in G_\alpha$ for all $\alpha \in J$, is said to be *coherent* if every finite subfamily of x_J is coherent. If $J \neq I$ and if $\beta \in I - J$, show that there exists $x_\beta \in G_\beta$ such that the family $x_{J \cup \{\beta\}} = (x_\alpha)_{\alpha \in J \cup \{\beta\}}$ is coherent. (Using (a) and condition (iv), show that, for every finite subset F of J, if γ is an upper bound of $F \cup \{\beta\}$, then $f_{\beta\gamma}\Big(G_\gamma \cap \bigcap_{\alpha \in F} \overset{-1}{f}_{\alpha\gamma}(x_\alpha)\Big)$ is the (non-empty) set of all $y_\beta \in G_\beta$ which are $\leqslant f_{\beta\gamma}\Big(\inf_{\alpha \in F} (M_{\alpha\gamma}(x_\alpha))\Big)$. Using the fact that E_β endowed with the opposite ordering is Noetherian, show next that there exist a finite subset F_0 of J and an upper bound γ_0 of $F_0 \cup \{\beta\}$ such that for each finite subset F of J and each upper bound γ of $F \cup \{\beta\}$ we have

$$f_{\beta\gamma}\Big(\inf_{\alpha \in F} (M_{\alpha j}(x_\alpha))\Big) \geqslant f_{\beta\gamma_0}\Big(\inf_{\alpha \in F_0} (M_{\alpha\gamma_0}(x_\alpha))\Big).$$

Prove then that every element $x_\beta \in F_\beta$ which is $\leqslant f_{\beta\gamma_0}\Big(\inf_{\alpha \in F_0} (M_{\alpha\gamma_0}(x_\alpha))\Big)$ satisfies the required conditions.

(c) Finally, complete the proof by showing that there exists a coherent family whose index set is the whole of I. (Order the set of coherent families x_J by the relation "x_J is a subfamily of x_K", and apply (b) and Zorn's lemma.)

6. Let I be a directed set, and let $(J_\lambda)_{\lambda \in L}$ be a family of subsets of I satisfying the conditions of Exercise 1. Let $(E_\alpha, f_{\beta\alpha})$ be a direct system of sets indexed by I, let $E = \varinjlim E_\alpha$, and for each $\lambda \in L$ let F_λ be the direct limit of the direct system obtained from $(E_\alpha, f_{\beta\alpha})$ by restricting the index set to J_λ. Whenever $\lambda \leqslant \mu$, let $g_{\mu\lambda}$ be the canonical mapping of F_λ into F_μ (no. 6). Show that $(F_\lambda, g_{\mu\lambda})$ is a direct system of sets relative to L, and define a canonical bijection of E onto $F = \varinjlim F_\lambda$.

7. Let I be a directed set and let $(E_\alpha, f_{\beta\alpha})$ be a direct system of sets relative to I. For each $\alpha \in I$, let $f_\alpha \colon E_\alpha \to E = \varinjlim E_\alpha$ be the canonical mapping. In each E_α, let R_α be the equivalence relation $f_\alpha(x) = f_\alpha(y)$.

Show that, whenever $\alpha \leqslant \beta$, the mapping $f_{\beta\alpha}$ is compatible with the equivalence relations R_α and R_β. Let $E'_\alpha = E_\alpha/R_\alpha$, and let $f'_{\beta\alpha}$ be the mapping of E'_α into E'_β induced by $f_{\beta\alpha}$ on passing to the quotients. Show that $f'_{\beta\alpha}$ is injective and that $(E'_\alpha, f'_{\beta\alpha})$ is a direct system of sets, and define a canonical bijection of E onto $\varinjlim E'_\alpha$.

8. Let $(E_\alpha, f_{\beta\alpha})$ and $(F_\alpha, g_{\beta\alpha})$ be two direct systems of sets, both indexed by the same directed set I. For each $\alpha \in I$, let u_α be a mapping of E_α into F_α, such that the u_α form a direct system of mappings. Let $G_\alpha \subset E_\alpha \times F_\alpha$ be the graph of u_α. Show that (G_α) is a direct system of subsets of $E_\alpha \times F_\alpha$ and that its direct limit may be canonically identified with the graph of $u = \varinjlim u_\alpha$.

9. Let I be an *arbitrary* preordered set, and let $(E_\alpha)_{\alpha \in I}$ be a family of sets indexed by I. For each pair of indices (α, β) such that $\alpha \leqslant \beta$, let $f_{\beta\alpha}$ be a mapping of E_α into E_β, and suppose that these mappings satisfy conditions (LI_I) and (LI_{II}). Let G be the set which is the sum of the family (E_α) and (with the notation of no. 5) let $R\{x, y\}$ be the relation "$\lambda(x) = \alpha \leqslant \lambda(y) = \beta$ and $y = f_{\beta\alpha}(x)$" between two elements x, y of G. Let R′ be the equivalence relation on G whose graph is the smallest of the graphs of equivalence relations which contain the graph of R (Chapter II, § 6, Exercise 10). The set $E = G/R'$ is called the *direct limit* of the family (E_α) with respect to the family of mappings $(f_{\beta\alpha})$, and we write $E = \varinjlim E_\alpha$. When the index set I is *directed*, show that this definition agrees with that given in no. 5. In the general case, the restriction to E_α of the canonical mapping of G into E is called the canonical mapping of E_α into E and is denoted by f_α. Suppose we are given, for each $\alpha \in I$, a mapping u_α of E_α into F such that $u_\beta \circ f_{\beta\alpha} = u_\alpha$ whenever $\alpha \leqslant \beta$; show that there exists a unique mapping u of E into F such that $u = u_\alpha \circ f_\alpha$ for each $\alpha \in I$.

HISTORICAL NOTE

(Chapter III, § 5)

(Numbers in brackets refer to the bibliography at the end of this Note.)

The evolution of ideas relating to the notions of integer and cardinal number is inseparable from the history of the theory of sets and mathematical logic for which the reader is referred to the Historical Note following Chapter IV. The purpose of this Note is to indicate briefly some of the salient facts in the history of numeration and "combinatorial analysis".

History and archaeology have revealed to us a large number of "systems of numeration", the prime aim of which is to attach to each individual integer (up to some limit depending on the demands of practical use) a name and a written representation, formed from a restricted number of signs according to more or less regular laws. By far the most common procedure is to decompose the integers into sums of "successive units" $b_1, b_2, \ldots, b_n, \ldots$, each of which is an integral multiple of its predecessor; and although in general b_n/b_{n-1} is taken to be a fixed number b (the "base" of the system, usually 10), there are many known exceptions to this rule. For example, in the Babylonian system, b_n/b_{n-1} is sometimes 10 and sometimes 6 [1], and in the chronological system of the Mayas b_n/b_{n-1} is equal to 20 except for $n = 2$, and $b_2/b_1 = 18$ [2]. As to the corresponding written symbol, it must indicate the number of "units" b_i of each order i. In many systems (for example, the Egyptian, the Greek, and the Roman) the successive multiples $k.b_i$ (where k varies from 1 up to $(b_{i+1}/b_i) - 1$) are denoted by symbols which depend on both k and i. A first and important step foward was to denote all the numbers $k.b_i$ (for the same value of k) by the same sign: this is the principle of "numeration by position", where the index i is indicated by the fact that the symbol representing $k.b_i$ appears "in the ith place". The first system of this nature is that of the Babylonians, who, certainly as early as 2000 B.C., denoted by the same sign all the multiples $k.60^{\pm i}$ corresponding to various values of the exponent i ([1], pp. 93-109). The inconvenience of such a system is of course its ambiguity so long as there is nothing to indicate whether or not the units of a certain order are absent, i.e., so long as the system is not completed by the introduction of a "zero". Nevertheless, the Babylonians managed without such a sign for the greater part of their history, and did not make use of a "zero" except in the last two centuries B.C., and then only inside a number; up to that time, the context alone could clarify the meaning of the symbol under consideration. Only two other systems made systematic use of a "zero" : that of the Mayas (in use, apparently, since the beginning of the Christian era [2]) and our present decimal system, which comes (via the Arabs) from Hindu mathematics, where the use of zero is attested since the first centuries A.D. Moreover, the conception of zero as a number (and not merely as a separating sign) and its introduction into calculations are original contributions of the Hindus [3]. Of course, once the principle of "numeration by position" had been acquired, it was easy to extend it to an arbitrary base. A discussion of the merits of the different "bases" proposed since the 17th century depends on the techniques of numerical computation, and cannot be entered into here. We note only that the operation which lies at the root of these systems, the so-called "Euclidean division", did not appear before the time of the Greeks, and undoubtedly goes back to the early Pythagoreans, who made it the essential tool in their theoretical arithmetic.

The general problems of enumeration, grouped together under the name of "combinatorial analysis", seem not to have been attempted before the last centuries of classical antiquity; only the formula $\binom{n}{2} = \frac{1}{2}n(n-1)$ is attested, in the third century A.D. The Hindu mathematician Bhaskara (12th century) knew the general formula for $\binom{n}{p}$. A more systematic study is found in a manuscript of Levi ben Gerson (the beginning of the 13th century): he obtained the inductive formula for the number V_n^p of arrangements of n objects p at a time, and in particular for the number of permutations of n objects, and he stated rules which are equivalent to the relations $\binom{n}{p} = V_n^p/p!$ and $\binom{n}{p} = \binom{n}{n-p}$ ([4], pp. 64-65). But this manuscript seems to have remained unknown to his contemporaries, and the results were only gradually rediscovered by mathematicians in the subsequent centuries. As regards later progress, let us record that Cardan proved that the number of non-empty subsets of a set of n elements is $2^n - 1$. Pascal and Fermat, in founding the calculus of probability, rediscovered the expression for $\binom{n}{p}$, and Pascal was the first to observe the relation between these numbers and the binomial theorem, which seems to have been known to the Arabs since the 13th century, to the Chinese in the 14th century, and was rediscovered in the West at the beginning of the 16th century, together with the inductive method of calculation of the coefficients known as "Pascal's triangle" ([4], pp. 35-38). Finally, about 1676, Leibniz obtained (but did not publish) the general formula for "multinomial coefficients", which was rediscovered independently and published 20 years later by de Moivre.

Bibliography

1. O. Neugebauer, *Vorlesungen über die Geschichte der antiken Mathematik*, Bd. I : Vorgriechische Mathematik, Berlin (Springer), 1934.
2. S. G. Morley, *The Ancient Maya*, Stanford University Press, 1946.
3. B. Datta and A. N. Singh, *History of Hindu Mathematics*, vol. I, Lahore (Motilal Banarsi Das), 1935.
4. J. Tropfke, *Geschichte der Elementar-Mathematik*, vol. VI : Analysis, Analytische Geometrie, Berlin-Leipzig (de Gruyter), 1924.

CHAPTER IV

Structures

1. STRUCTURES AND ISOMORPHISMS

The purpose of this chapter is to describe once and for all a certain number of formative constructions and proofs (cf. Chapter I, § 1, no. 3 and § 2, no. 2) which arise very frequently in mathematics.

1. ECHELONS

An *echelon construction scheme* is a sequence $c_1, c_2, \ldots, c_m$ of ordered pairs of natural integers (*) $c_i = (a_i, b_i)$, satisfying the following conditions :

(a) If $b_i = 0$, then $1 \leqslant a_i \leqslant i-1$.
(b) If $a_i \neq 0$ and $b_i \neq 0$, then $1 \leqslant a_i \leqslant i-1$ and $1 \leqslant b_i \leqslant i-1$.

These conditions imply that $c_1 = (0, b_1)$, with $b_1 > 0$. If n is the largest of the integers b_i which appear in the pairs $(0, b_i)$, then $c_1, c_2, \ldots, c_m$ is said to be an echelon construction scheme *on n terms*.

Given an echelon construction scheme $S = (c_1, c_2, \ldots, c_m)$ on n terms, and given n terms $E_1, E_2, \ldots, E_n$ in a theory $\mathcal{T}$ which is stronger than the theory of sets, an *echelon construction of scheme* S *on* $E_1, \ldots, E_n$ is defined to be a sequence $A_1, A_2, \ldots, A_m$ of m terms in the theory $\mathcal{T}$, defined step by step by the following conditions :

(a) If $c_i = (0, b_i)$, then A_i is the term E_{b_i}.
(b) If $c_i = (a_i, 0)$, then A_i is the term $\mathfrak{P}(A_{a_i})$.
(c) If $c_i = (a_i, b_i)$, where $a_i \neq 0$ and $b_i \neq 0$, then A_i is the term $A_{a_i} \times A_{b_i}$.

(*) We use the notion of integer in the same manner as in Chapter I, that is to say, in the metamathematical sense of marks arranged in a certain order; this use has nothing to do with the mathematical theory of integers which was developed in Chapter III.

The last term A_m of the echelon construction of scheme S on $E_1, \ldots, E_n$ is called the *echelon of scheme* S *on the base sets* $E_1, \ldots, E_n$; in the general arguments which follow, it will be denoted by the notation $S(E_1, \ldots, E_n)$.

Example. Given two sets E, F, the set $\mathfrak{P}(\mathfrak{P}(E)) \times \mathfrak{P}(F)$ is an echelon on E, F, with scheme

$$(0, 1), (0, 2), (1, 0), (3, 0), (2, 0), (4, 5).$$

It is also the echelon on E, F with scheme

$$(0, 2), (0, 1), (1, 0), (2, 0), (4, 0), (5, 3).$$

Distinct schemes may therefore give rise to the same echelon on the same terms.

2. CANONICAL EXTENSIONS OF MAPPINGS

Let $S = (c_1, c_2, \ldots, c_m)$ be an echelon construction scheme on n terms. Let $E_1, \ldots, E_n, E'_1, \ldots, E'_n$ be sets (terms in $\mathcal{T}$) and let $f_1, \ldots, f_n$ be terms in $\mathcal{T}$ such that the relations "f_i is a mapping of E_i into E'_i" are theorems in $\mathcal{T}$ for $1 \leqslant i \leqslant n$. Let $A_1, \ldots, A_m$ (resp. $A'_1, \ldots, A'_m$) be the echelon construction of scheme S on $E_1, \ldots, E_n$ (resp. $E'_1, \ldots, E'_n$). We define step by step a sequence of m terms $g_1, \ldots, g_m$ such that g_i is a *mapping of* A_i *into* A'_i (for $1 \leqslant i \leqslant m$) by the following conditions:

(a) If $c_i = (0, b_i)$, so that $A_i = E_{b_i}$ and $A'_i = E'_{b_i}$, then g_i is the mapping f_{b_i}.

(b) If $c_i = (a_i, 0)$, so that $A_i = \mathfrak{P}(A_{a_i})$ and $A'_i = \mathfrak{P}(A'_{a_i})$, then g_i is the *canonical extension* $\hat{g}_{a_i}$ of g_{a_i} to sets of subsets (Chapter II, § 5, no. 1).

(c) If $c_i = (a_i, b_i)$, where $a_i \neq 0$ and $b_i \neq 0$, so that

$$A_i = A_{a_i} \times A_{b_i} \quad \text{and} \quad A'_i = A'_{a_i} \times A'_{b_i},$$

then g_i is the *canonical extension* $g_{a_i} \times g_{b_i}$ of g_{a_i} and g_{b_i} to $A_{a_i} \times A_b$ (Chapter II, § 3, no. 9).

The last term g_m of this sequence is called the *canonical extension, with scheme* S, *of the mappings* $f_1, \ldots, f_n$, and will be denoted by $\langle f_1, \ldots, f_n \rangle^S$.

¶ The following criteria can be verified step by step:

CST1. *If* f_i *is a mapping of* E_i *into* E'_i, *and if* f'_i *is a mapping of* E'_i *into* E''_i $(1 \leqslant i \leqslant n)$, *then for every echelon construction scheme* S *on* n *terms we have*

$$\langle f'_1 \circ f_1, f'_2 \circ f_2, \ldots, f'_n \circ f_n \rangle^S = \langle f'_1, f'_2, \ldots, f'_n \rangle^S \circ \langle f_1, f_2, \ldots, f_n \rangle^S.$$

CST2. *If f_i is injective* (resp. *surjective*) *for* $1 \leqslant i \leqslant n$, *then* $\langle f_1, \ldots, f_n\rangle^S$ *is injective* (resp. *surjective*).
This criterion follows from the corresponding properties of the extension $\hat{g}$ (Chapter II, § 5, no. 1, Proposition 1) and the extension $g \times h$ (Chapter II, § 3, no. 9).

CST3. *If f_i is a bijection of* E_i *onto* E'_i, *and if* f_i^{-1} *is the inverse bijection* (*), *then* $\langle f_1, \ldots, f_n\rangle^S$ *is a bijection and* $\langle f_1^{-1}, \ldots, f_n^{-1}\rangle^S$ *is its inverse*; *in other words*,

$$(\langle f_1, \ldots, f_n\rangle^S)^{-1} = \langle f_1^{-1}, \ldots, f_n^{-1}\rangle^S.$$

This follows immediately from CST1 and CST2.

3. TRANSPORTABLE RELATIONS

Let $\mathcal{T}$ be a theory which is stronger than the theory of sets, let $x_1, \ldots, x_n$, $s_1, \ldots, s_p$ be distinct letters which are distinct from the constants of $\mathcal{T}$, and let $A_1, \ldots, A_m$ be terms in $\mathcal{T}$ in which none of the letters x_i $(1 \leqslant i \leqslant n)$ and s_j $(1 \leqslant j \leqslant p)$ appears. Let $S_1, \ldots, S_p$ be echelon construction schemes on $n + m$ terms. Then the relation $T\{x_1, \ldots, x_n, s_1, \ldots, s_p\}$:

"$s_1 \in S_1(x_1, \ldots, x_n, A_1, \ldots, A_m)$ and $s_2 \in S_2(x_1, \ldots, x_n, A_1, \ldots, A_m)$ and $\ldots$ and $s_p \in S_p(x_1, \ldots, x_n, A_1, \ldots, A_m)$"

is called a *typification* of the letters $s_1, \ldots, s_p$.

¶ Let $R\{x_1, \ldots, x_n, s_1, \ldots, s_p\}$ be a relation in $\mathcal{T}$ which contains certain of the letters x_i, s_j (and possibly other letters as well). Then R is said to be *transportable* (*in* $\mathcal{T}$) *with respect to the typification* T, *the* x_i $(1 \leqslant i \leqslant n)$ *being considered as principal base sets and the* A_h $(1 \leqslant h \leqslant m)$ *as auxiliary base sets*, if the following condition is satisfied: let $y_1, \ldots, y_n, f_1, \ldots, f_n$ be distinct letters which are distinct from the x_i $(1 \leqslant i \leqslant n)$, the s_j $(1 \leqslant j \leqslant p)$, the constants of $\mathcal{T}$, and all the letters which appear in R or in the terms A_h $(1 \leqslant h \leqslant m)$, and let Id_h $(1 \leqslant h \leqslant m)$ denote the identity mapping of A_h on to itself. Then the relation

(1) "$T\{x_1, \ldots, x_n, s_1, \ldots, s_p\}$ and (f_1 is a bijection of x_1 onto y_1) and $\ldots$ and (f_n is a bijection of x_n onto y_n)"

implies, *in* $\mathcal{T}$, the relation

(2) $$R\{x_1, \ldots, x_n, s_1, \ldots, s_p\} \Longleftrightarrow R\{y_1, \ldots, y_n, s'_1, \ldots, s'_p\},$$

(*) For typographical reasons we write here f^{-1} instead of $\overset{-1}{f}$.

where

(3) $$s'_j = \langle f_1, \ldots, f_n, \mathrm{Id}_1, \ldots, \mathrm{Id}_m\rangle^{\wedge}{}_j(s_j) \qquad (1 \leqslant j \leqslant p).$$

There is an analogous but simpler definition in the case where there is no auxiliary set.

For example, if $n = p = 2$ and if the typification T is "$s_1 \in x_1$ and $s_2 \in x_1$", the relation $s_1 = s_2$ is transportable. On the other hand, the relation $x_1 = x_2$ is not transportable.

4. SPECIES OF STRUCTURES

Let $\mathcal{T}$ be a theory which is stronger than the theory of sets. *A species of structures* in $\mathcal{T}$ is a text Σ formed of the following assemblies:

(1) a certain number of letters $x_1, \ldots, x_n, s$, distinct from each other and from the constants of $\mathcal{T}$; $x_1, \ldots, x_n$ are called the *principal base sets* of the species of structures Σ;

(2) a certain number of terms $A_1, \ldots, A_m$ in $\mathcal{T}$ in which none of the letters $x_1, \ldots, x_n, s$ appears, and which are called the *auxiliary base sets* of Σ; Σ possibly contains no auxiliary base sets (but it must contain at least one principal base set);

(3) a typification $\mathrm{T}\{x_1, \ldots, x_n, s\}$:

$$s \in \mathrm{S}(x_1, \ldots, x_n, A_1, \ldots, A_m),$$

where S is an echelon construction scheme on $n + m$ terms (no. 1); $\mathrm{T}\{x_1, \ldots, x_n, s\}$ is called the *typical characterization* of the species of structures Σ;

(4) a relation $\mathrm{R}\{x_1, \ldots, x_n, s\}$ which is *transportable* (in $\mathcal{T}$) with respect to the typification T, the x_i being the principal base sets and the A_h the auxiliary base sets (no. 3); R is called the *axiom* of the species of structures Σ.

The theory $\mathcal{T}_\Sigma$ which has the same axiom schemes as $\mathcal{T}$ and whose explicit axioms are those of $\mathcal{T}$, together with the axiom "T and R", is called the *theory of the species of structures* Σ. The constants of $\mathcal{T}_\Sigma$ are therefore the constants of $\mathcal{T}$ and the letters which appear in T or in R.

¶ Let $\mathcal{T}'$ be a theory which is stronger than $\mathcal{T}$, and let $E_1, \ldots, E_n$, U be terms in $\mathcal{T}'$. In the theory $\mathcal{T}'$, U is said to be a *structure of species* Σ *on the principal base sets* $E_1, \ldots, E_n$, *with* $A_1, \ldots, A_m$ *as auxiliary base sets*, if the relation

$$\text{"}\mathrm{T}\{E_1, \ldots, E_n, U\} \text{ and } \mathrm{R}\{E_1, \ldots, E_n, U\}\text{"}$$

is a *theorem in* $\mathcal{T}'$. When this is so, then for each theorem $B\{x_1, \ldots, x_n, s\}$ in the theory $\mathcal{T}_\Sigma$ the relation $B\{E_1, \ldots, E_n, U\}$ is a *theorem in* $\mathcal{T}'$ (Chapter I, § 2, no. 3). In $\mathcal{T}_\Sigma$, the constant s is called the *generic structure of the species* Σ.

¶ In the theory $\mathcal{T}'$, the principal base sets $E_1, \ldots, E_n$ are said to be *endowed with the structure* U. Clearly, U is an element of the set

$$S(E_1, \ldots, E_n, A_1, \ldots, A_m).$$

The set of elements V of $S(E_1, \ldots, E_n, A_1, \ldots, A_m)$ which satisfy the relation $R\{E_1, \ldots, E_n, V\}$ is therefore the *set of structures of the species* Σ *on* $E_1, \ldots, E_n$ (and it may be empty).

Examples

(1) Take $\mathcal{T}$ to be the theory of sets, and consider the species of structures which has no auxiliary base set, one principal base set A, the typical characterization $s \in \mathfrak{P}(A \times A)$, and the axiom

$$s \circ s = s \quad \text{and} \quad s \cap \overset{-1}{s} = \Delta_A$$

(Δ_A being the diagonal of $A \times A$), which is a transportable relation with respect to the typification $s \in \mathfrak{P}(A \times A)$, as is easily verified. It is clear that the theory of this species of structures is just the theory of *ordered sets* (Chapter III, §1, no. 3); and therefore the species of structure so defined is also called the *species of order structures* on A. In Chapter III we saw many examples of sets endowed with structures of this species.

(2) Take $\mathcal{T}$ to be the theory of sets, and consider the species of structures which has no auxiliary base set, one principal base set A, the typical characterization $F \in \mathfrak{P}((A \times A) \times A)$, and as axiom the transportable relation "F is a functional graph whose domain is $A \times A$". The structures of this species are particular cases of what are called *algebraic structures*, and the function whose graph is F (a mapping of $A \times A$ into A) is called *the* (*everywhere defined*) *internal law of composition* of such a structure.

(3) As before, let $\mathcal{T}$ be the theory of sets, and consider the species of structures which has no auxiliary base set, one principal base set A, the typical characterization $V \in \mathfrak{P}(\mathfrak{P}(A))$, and as axiom the transportable relation

$$(\forall V)\,((V' \subset V) \Longrightarrow ((\bigcup_{X \in V'} X) \in V))$$
$$\text{and} \quad (\forall X)(\forall Y)((X \in V \text{ and } Y \in V) \Longrightarrow ((X \cap Y) \in V)).$$

This species of structures is called the *species of topological structures*. A structure of this species is also called a *topology*, and the relation $X \in V$ is expressed by saying that X is an *open set* in the topology V (*General Topology*, Chapter I, § 1).

* (4) Take $\mathcal{T}$ to be the theory of the species of division ring structures, which has (among other things) a constant K as unique (principal) base set. The species of structure of a *left vector space over* K has K as auxiliary base set, a principal base set E, and as typical characterization the relation

$$V \in \mathfrak{P}((E \times E) \times E) \times \mathfrak{P}((K \times E) \times E)$$

(pr_1V being the graph of addition in E, and pr_2V the graph of scalar multiplication); we shall not state here the axiom for this species of structures.

(5) Again let $\mathcal{T}$ be the theory of sets; in this theory, the field **C** of complex numbers is a term which contains no letters. The species of structure of a *complex analytic manifold of dimension n* has **C** as auxiliary base set, and one principal base set V. We shall not indicate here the typical characterization or the axiom of this species of structure. *

Remarks

(1) In applications it is often the case (as in Example 4 above) that the echelon $S(E_1, \ldots, E_n, A_1, \ldots, A_m)$ is a product of echelons

$$S_1(E_1, \ldots, A_m) \times \cdots \times S_p(E_1, \ldots, A_m).$$

If so, the letter s in the definition of Σ is often replaced by a "p-tuple" $(s_1, \ldots, s_p)$ (cf. Chapter II, § 2, no. 1).

Moreover, the axiom of a species of structures Σ is most frequently written as a conjunction of several transportable relations (as in Example 3 above). These relations are called *the axioms* of the species Σ.

(2) Names are given to the species of structures most frequently used in mathematics, and to sets endowed with structures of these species. Thus an *ordered set* (Chapter III, § 1) is a set endowed with an order structure (Example 1); * in the later Books of this series, we shall define the notions of *group, field, topological space, differentiable manifold*, etc., all of which denote sets endowed with certain structures. *

(3) By abuse of language, in the theory of sets $\mathcal{T}$, the giving of n distinct letters $x_1, \ldots, x_n$ (with no typical characterization and no axiom) is considered as a species of structure Σ_0, called the *structure of a set* on the n principal base sets $x_1, \ldots, x_n$.

5. ISOMORPHISMS AND TRANSPORT OF STRUCTURES

Let Σ be a species of structures in a theory $\mathcal{T}$, on n principal base sets $x_1, \ldots, x_n$, with m auxiliary base sets $A_1, \ldots, A_m$. Let S be the echelon construction scheme on $n + m$ letters which features in the typical characterization of Σ, and let R be the axiom of Σ. In a theory $\mathcal{T}'$ which is stronger than $\mathcal{T}$, let U be a structure of species Σ on sets $E_1, \ldots, E_n$ (as principal base sets) and let U′ be a structure *of the same species* on sets $E'_1, \ldots, E'_n$. Finally, let f_i (in $\mathcal{T}'$) be a *bijection* of E_i

onto E'_i $(1 \leqslant i \leqslant n)$. Then $(f_1, \ldots, f_n)$ is said to be an *isomorphism* of the sets $E_1, \ldots, E_n$, endowed with the structure U, onto the sets $E'_1, \ldots, E'_n$, endowed with the structure U′, if we have (in $\mathcal{T}'$)

$$(4) \qquad \langle f_1, \ldots, f_n, \mathrm{Id}_1, \ldots, \mathrm{Id}_m \rangle^S(U) = U'$$

where Id_h denotes the identity mapping of A_h onto itself $(1 \leqslant h \leqslant m)$.

¶ Let f'_i be the inverse of the bijection f_i $(1 \leqslant i \leqslant n)$. It follows immediately from (4) and the criterion CST3 (no. 2) that we have

$$\langle f'_1, \ldots, f'_n, \mathrm{Id}_1, \ldots, \mathrm{Id}_m \rangle^S(U') = U$$

and consequently that $(f'_1, \ldots, f'_n)$ is an *isomorphism* of $E'_1, \ldots, E'_n$, endowed with U′, onto $E_1, \ldots, E_n$, endowed with U. The isomorphisms $(f_1, \ldots, f_n)$ and $(f'_1, \ldots, f'_n)$ are said to be *inverses* of each other.

$E'_1, \ldots, E'_n$, endowed with U′, are said to be *isomorphic* to $E_1, \ldots, E_n$, endowed with U, if there exists an isomorphism of $E_1, \ldots, E_n$ onto $E'_1, \ldots, E'_n$; in this case the structures U and U′ are said to be *isomorphic*.

¶ The above definitions, together with CST1, imply the following criterion:

CST4. *Let* U, U′, U″ *be three structures of the same species* Σ *on the principal base sets* $E_1, \ldots, E_n$, $E'_1, \ldots, E'_n$, $E''_1, \ldots, E''_n$, *respectively. Let* f_i *be a bijection of* E_i *onto* E'_i, *and let* g_i *be a bijection of* E'_i *onto* E''_i $(1 \leqslant i \leqslant n)$. *If* $(f_1, \ldots, f_n)$ *and* $(g_1, \ldots, g_n)$ *are isomorphisms, then so is* $(g_1 \circ f_1, \ldots, g_n \circ f_n)$.

An isomorphism of $E_1, \ldots, E_n$ onto $E_1, \ldots, E_n$ (with respect to the *same* structure) is called an *automorphism* of $E_1, \ldots, E_n$. The composition of two automorphisms of $E_1, \ldots, E_n$ is an automorphism, and so is the inverse of an automorphism, * so that the automorphisms of $E_1, \ldots, E_n$ form a *group*. *

Remark. By abuse of language, if f_i is any bijection of E_i onto E'_i $(1 \leqslant i \leqslant n)$, $(f_1, \ldots, f_n)$ is said to be an isomorphism of $E_1, \ldots, E_n$ onto $E'_1, \ldots, E'_n$ with respect to the species of structure of a set (no. 4, Remark 3).

CST5. *In a theory* $\mathcal{T}'$ *which is stronger than* $\mathcal{T}$, *let* U *be a structure of species* Σ *on* $E_1, \ldots, E_n$, *and let* f_i *be a bijection of* E_i *onto a set* E'_i $(1 \leqslant i \leqslant n)$. *Then there exists a unique structure of species* Σ *on* $E'_1, \ldots, E'_n$ *such that* $(f_1, \ldots, f_n)$ *is an isomorphism of* $E_1, \ldots, E_n$ *onto* $E'_1, \ldots, E'_n$.

For this structure, if it exists, can only be the term U′ defined by the relation (4); it remains to be verified that this term is indeed a structure of species Σ, i.e., that the relation $R\{E'_1, \ldots, E'_n, U'\}$ is true in $\mathcal{T}'$. But this follows from the fact that $R\{x_1, \ldots, x_n, s\}$ is *transportable*, for

$R\{E'_1, \ldots, E'_n, U'\}$ is equivalent in $\mathcal{T}'$ to the relation $R\{E_1, \ldots, E_n, U\}$ (no. 3), which is true in $\mathcal{T}'$ by hypothesis.

¶ The structure U' is said to be obtained by *transporting the structure* U *to the sets* $E'_1, \ldots, E'_n$ *by means of the bijective mappings* $f_1, \ldots, f_n$. Thus two structures of the same species are isomorphic if and only if each is obtained from the other by transport of structure.

It may happen that *any* two structures of species Σ are *necessarily isomorphic*; the species of structure Σ is then said be *univalent*. * This is the case for the structure of an infinite cyclic group (isomorphic to **Z**), the structure of a prime field of characteristic zero (isomorphic to **Q**), the structure of a complete Archimedean ordered field (isomorphic to **R**), the structure of an algebraically closed, connected, locally compact topological field (isomorphic to **C**), and of the structure of a non-commutative, connected, locally compact topological division ring (isomorphic to **H**, the division ring of quaternions). For some of these species of structure, for example that of a prime field of characteristic zero, or that of a complete Archimedean ordered field, there is not even any automorphism other than the identity mapping; but there do exist such automorphisms for the other examples given above (for example, the symmetry $x \to -x$ in **Z**). *

It will be observed that the above species of structure are essentially those which are the basis of classical mathematics. On the other hand, * the species of group structures, the species of structures of an ordered set, and the species of topological structures, are not univalent. *

6. DEDUCTION OF STRUCTURES

Let Σ be a species of structures in a theory $\mathcal{T}$, on n principal base sets $x_1, \ldots, x_n$, with m auxiliary base sets $A_1, \ldots, A_m$. Let s be the generic structure of Σ, and let T be an echelon construction scheme on $n + m$ terms. A term $V\{x_1, \ldots, x_n, s\}$ which contains no letter other than the constants of $\mathcal{T}_\Sigma$ is said to be *intrinsic* for s, of type $T(x_1, \ldots, x_n, A_1, \ldots, A_m)$, if it satisfies the following conditions:

(1) the relation $V\{x_1, \ldots, x_n, s\} \in T(x_1, \ldots, x_n, A_1, \ldots, A_m)$ is a theorem in $\mathcal{T}_\Sigma$;

(2) let $\mathcal{T}'_\Sigma$ be the theory obtained by adjoining to the axioms of $\mathcal{T}_\Sigma$ the axioms "f_i is a bijection of x_i onto y_i" ($1 \leqslant i \leqslant n$) (where the letters y_i and f_i are distinct from each other and from the constants of $\mathcal{T}_\Sigma$, for $1 \leqslant i \leqslant n$); if s' is the structure obtained by transporting s by means of $(f_1, \ldots, f_n)$ (no. 5), then

$$V\{y_1, \ldots, y_n, s'\} = \langle f_1, \ldots, f_n, \mathrm{Id}_1, \ldots, \mathrm{Id}_m\rangle^T(V\{x_1, \ldots, x_n, s\})$$

is a theorem in $\mathcal{T}'_\Sigma$.

Most of the terms which one is led to define in the theory of a species of structures are intrinsic terms.

Let Θ be another species of structures in the theory $\mathcal{T}$, on r principal base sets $u_1, \ldots, u_r$, with p auxiliary base sets $B_1, \ldots, B_p$, and let $t \in T(u_1, \ldots, u_r, B_1, \ldots, B_p)$ be the typical characterization of Θ (no. 4). Then a system of $r+1$ terms $P, U_1, \ldots, U_r$, *intrinsic for* s, and such that P is a structure of species Θ on $U_1, \ldots, U_r$, *in the theory* $\mathcal{T}_\Sigma$, is called a *procedure of deduction of a structure of species* Θ *from a structure of species* Σ. By abuse of language, the term P alone is often called a procedure of deduction.

¶ Let $\mathcal{T}'$ be a theory stronger than $\mathcal{T}$. If $\mathcal{S}$ is a structure in $\mathcal{T}'$ of species Σ on $E_1, \ldots, E_n$, then $P\{E_1, \ldots, E_n, \mathcal{S}\}$ is a structure of species Θ on the r sets $F_j = U_j\{E_1, \ldots, E_n, \mathcal{S}\}$ $(1 \leqslant j \leqslant r)$, said to be *deduced from* $\mathcal{S}$ *by the procedure* P, *or subordinate to* $\mathcal{S}$. The hypothesis that the terms $P, U_1, \ldots, U_r$ are intrinsic for s moreover implies the following criterion:

CST6. *Let* $(g_1, \ldots, g_n)$ *be an isomorphism of* $E_1, \ldots, E_n$, *endowed with a structure* $\mathcal{S}$ *of species* Σ, *onto* $E'_1, \ldots, E'_n$, *endowed with a structure* $\mathcal{S}'$ *of the same species. If* U_j *is of type* $\mathfrak{P}(T_j)$, *put*

$$h_j = \langle g_1, \ldots, g_n, \mathrm{Id}_1, \ldots, \mathrm{Id}_m \rangle^{T_j} \quad (1 \leqslant j \leqslant r),$$

and let $F'_j = U_j\{E'_1, \ldots, E'_n, \mathcal{S}'\}$ $(1 \leqslant j \leqslant r)$. *Then* $(h_1, \ldots, h_r)$ *is an isomorphism of* $F_1, \ldots, F_r$ *onto* $F'_1, \ldots, F'_r$ *when these systems of sets are endowed with the structures of species* Θ *deduced from* $\mathcal{S}$ *and* $\mathcal{S}'$ *respectively by the procedure* P.

It is clear that the terms $x_1, \ldots, x_n$ are intrinsic for s. In many cases, the terms $U_1, \ldots, U_r$ are certain of the letters $x_1, \ldots, x_n$; the structure of species Θ deduced from s by the procedure P is then said to be a structure *underlying* s.

Examples

* (1) The species of *topological group* structures has a single principal base set A, no auxiliary base set, and the corresponding generic structure is a pair (s_1, s_2) (s_1 being the graph of the law of composition on A, and s_2 the set of open sets in the topology of A; cf. *General Topology*, Chapter III, § 1). Each of the terms s_1, s_2 is a procedure of deduction and provides respectively the *group structure* and the *topology* underlying the topological group structure (s_1, s_2).

Likewise, from a vector space structure can be deduced an underlying commutative group structure. From a ring structure can be deduced an underlying commutative group structure and a (multiplicative) semigroup structure. From the structure of a differentiable manifold can be deduced an underlying topology, etc.

(2) The species of vector space structures over **C** (resp. **R**) has a principal base set E, an auxiliary base set equal to **C** (resp. **R**), and typical characterization

$$s_1 \in \mathfrak{P}((E \times E) \times E) \quad \text{and} \quad s_2 \in \mathfrak{P}((\mathbf{C} \times E) \times E)$$
$$(\text{resp.} \quad s_1 \in \mathfrak{P}((E \times E) \times E) \quad \text{and} \quad s_2 \in \mathfrak{P}((\mathbf{R} \times E) \times E)).$$

The pair $(s_1, s_2 \cap ((\mathbf{R} \times E) \times E))$ is a procedure of deduction of a vector space structure over **R** from a vector space structure over **C** ("restriction of the field of scalars to **R**"). *

(3) Suppose that Θ has the *same* (principal and auxiliary) base sets as Σ, and the *same* typical characterization. If, moreover, the axiom of Σ *implies* (in $\mathcal{T}$) the axiom of Θ, it is clear that the term s is a procedure of deduction of a structure of species Θ from a structure of species Σ. Θ is then said to be *poorer* than Σ, and Σ is *richer* than Θ. Every structure of species Σ, in a theory $\mathcal{T}'$ which is stronger than $\mathcal{T}$, is then also a structure of species Θ. For example, the species of structures of *totally ordered* sets (obtained by taking as axiom the conjunction of the axiom of order structures (no. 4, Example 1) and the relation $s \cup \overset{-1}{s} = A \times A$) is richer than the species of order structures. *The species of commutative group structures is richer than the species of group structures. The species of compact topological space structures is richer than the species of topological structures, etc. *

* (4) When each of Σ and Θ is the species of group structures (resp. ring structures), there is defined in algebra a procedure of deduction which associates with each group structure (resp. ring structure) the group structure (resp. ring structure) on its *centre*. When Σ is the species of vector space structures over a field K, and when Θ is the species of algebraic structures over K, there are defined procedures of deduction which associate with every vector space over K its *tensor algebra* or its *exterior algebra*. We shall meet many other examples later in this series. *

Remark. When P is a "q-tuple" $(P_1, \ldots, P_q)$, it is also said that the terms $P_1, \ldots, P_q$ constitute a procedure of deduction of a structure of species Θ from a structure of species Σ.

7. EQUIVALENT SPECIES OF STRUCTURES

Let Σ and Θ be two species of structures, in the same theory $\mathcal{T}$, having the *same* principal base sets $x_1, \ldots, x_n$. Let s, t be the generic structures of the species Σ, Θ respectively. Suppose that the following conditions are satisfied :

(1) We have a procedure of deduction $P\{x_1, \ldots, x_n, s\}$ of a structure of species Θ on $x_1, \ldots, x_n$ from a structure of species Σ on $x_1, \ldots, x_n$.

(2) We have a procedure of deduction $Q\{x_1, \dots, x_n, t\}$ of a structure of species Σ on $x_1, \dots, x_n$ from a structure of species Θ on $x_1, \dots, x_n$.

(3) The relation $Q\{x_1, \dots, x_n, P\{x_1, \dots, x_n, s\}\} = s$ is a theorem in $\mathcal{T}_\Sigma$, and the relation $P\{x_1, \dots, x_n, Q\{x_1, \dots, x_n, t\}\} = t$ is a theorem in $\mathcal{T}_\Theta$.

The species of structures Σ and Θ are then said to be *equivalent by means of the procedures of deduction* P and Q. In this case, for every theorem $B\{x_1, \dots, x_n, s\}$ in the theory $\mathcal{T}_\Sigma$, the relation $B\{x_1, \dots, x_n, Q\}$ is a theorem in $\mathcal{T}_\Theta$; and conversely, for every theorem $C\{x_1, \dots, x_n, t\}$ in the theory $\mathcal{T}_\Theta$, the relation $C\{x_1, \dots, x_n, P\}$ is a theorem in $\mathcal{T}_\Sigma$.

¶ If U is a structure of species Σ, the structure deduced from U by the procedure P is said to be *equivalent* to U. Criterion CST6 implies the following:

CST7. *Let $\mathcal{S}$, $\mathcal{S}'$ be two structures of species Σ on the principal base sets $(E_1, \dots, E_n)$, $(E'_1, \dots, E'_n)$, respectively. Let $\mathcal{S}_0$, $\mathcal{S}'_0$ be structures of species Θ which are equivalent respectively to $\mathcal{S}$ and $\mathcal{S}'$. In order that $(g_1, \dots, g_n)$ should be an isomorphism with respect to the structures $\mathcal{S}_0$ and $\mathcal{S}'_0$, it is necessary and sufficient that $(g_1, \dots, g_n)$ should be an isomorphism with respect to the structures $\mathcal{S}$ and $\mathcal{S}'$.*

In practice, we make no distinction between the theories $\mathcal{T}_\Sigma$ and $\mathcal{T}_\Theta$ of two equivalent species of structures.

Examples

* (1) Let Σ be the species of commutative group structures; Σ has a single (principal) base set A, and its generic structure consists of a single letter F; the typical characterization of Σ is $F \in \mathfrak{P}((A \times A) \times A)$, and we denote the axiom of Σ by $R\{A, F\}$. This axiom implies in particular that F is the graph of a function (the "law of composition" of the group; cf. no. 4, Example 2). In the theory $\mathcal{T}_\Sigma$ (where $\mathcal{T}$ denotes the theory of sets) we define a term $M\{A, F\}$ which is a functional graph in $\mathfrak{P}((\mathbf{Z} \times A) \times A)$ and satisfies the following relation $B\{M, A, F\}$:

$$(\forall x)(\forall y)(\forall n)((x \in A \text{ and } y \in A \text{ and } n \in \mathbf{Z}) \Longrightarrow (M(n, F(x, y)) = F(M(n, x), M(n, y))))$$
$$\text{and } (\forall x)(\forall m)(\forall n)((x \in A \text{ and } m \in \mathbf{Z} \text{ and } n \in \mathbf{Z}) \Longrightarrow (M(m + n, x) = F(M(m, x), M(n, x))))$$
$$\text{and } (\forall x)(\forall m)(\forall n)((x \in A \text{ and } m \in \mathbf{Z} \text{ and } n \in \mathbf{Z}) \Longrightarrow (M(m, M(n, x)) = M(mn, x)))$$
$$\text{and } (\forall x)((x \in A) \Longrightarrow (M(1, x) = x)).$$

("multiplication of an element of A by an integer").

Consider the species Θ of **Z**-*module* structures, which has a single principal base set A, with **Z** as auxiliary set, and whose generic structure contains two letters G, L, with the typical characterization

$$G \in \mathfrak{P}((A \times A) \times A) \quad \text{and} \quad L \in \mathfrak{P}((\mathbf{Z} \times A) \times A)$$

and the axiom

$$\text{“}R\}A, G\{ \text{ and (L is a functional graph) and } B\}L, A, G\{\text{”}.$$

It is immediately verified that the terms F, M constitute a procedure of deduction of a structure of species Θ from a structure of species Σ, and that the term G is a procedure of deduction of a structure of species Σ from a structure of species Θ. Furthermore, the condition (3) above is trivially satisfied. We may therefore say that the species of commutative group structures and the species of **Z**-module structures are equivalent.

(2) Let Σ be the species of topological structures (no. 4, Example 3), let A be the (principal) base set, and let V be the generic structure of Σ. Consider the relation

$$x \in A \text{ and } X \subset A \text{ and } (\forall U)((U \in V \text{ and } x \in U) \Longrightarrow (X \cap U \neq \emptyset)).$$

This relation has a graph $P \subset \mathfrak{P}(A) \times A$ with respect to the pair (X, x); $P\{A, V\}$ is a term called "the set of all pairs (X, x) such that x lies in the *closure* of X with respect to the topology V". We can then prove (cf. *General Topology*, Chapter I, § 1) that the following relations are theorems in $\mathcal{T}_\Sigma$:

$$P(\emptyset) = \emptyset,$$
$$(\forall Y)((Y \subset A) \Longrightarrow (Y \subset P(Y))),$$
$$(\forall Y)((Y \subset A) \Longrightarrow (P(P(Y)) = P(Y))),$$
$$(\forall Y)(\forall Z)((Y \subset A \text{ and } Z \subset A) \Longrightarrow (P(Y \cup Z) = P(Y) \cup P(Z))).$$

Consider the species of structures Θ, having a single (principal) base set A, whose generic structure consists of a single letter W, which has as typical characterization $W \in \mathfrak{P}(\mathfrak{P}(A) \times A)$ and as axiom

$$W(\emptyset) = \emptyset \quad \text{and} \quad (\forall Y)((Y \subset A) \Longrightarrow (Y \subset W(Y)))$$
$$\text{and} \quad (\forall Y)((Y \subset A) \Longrightarrow (W(W(Y)) = W(Y)))$$
$$\text{and} \quad (\forall Y)(\forall Z)((Y \subset A \text{ and } Z \subset A) \Longrightarrow (W(Y \cup Z) = W(Y) \cup W(Z))).$$

Consider also the relation

$$U \subset A \quad \text{and} \quad (\forall x)((x \in U) \Longrightarrow x \notin W(A - U)).$$

The set of all $U \in \mathfrak{P}(A)$ which satisfy this relation is a subset $Q\{A, W\}$ of $\mathfrak{P}(A)$. We can then prove (*General Topology*, Chapter I, § 1, Exercise 10) that the following relations are theorems in $\mathcal{T}_\Theta$:

$$A \in Q,$$
$$(\forall M)((M \subset Q) \Longrightarrow \Big(\Big(\bigcup_{X \in M} X\Big) \in Q\Big),$$
$$(\forall X)(\forall Y)((X \in Q \text{ and } Y \in Q) \Longrightarrow ((X \cap Y) \in Q)).$$

Thus the terms P$\{$A, V$\{$ and Q$\}$A, W$\{$ satisfy conditions (1) and (2) above, and it is easily seen that they also satisfy condition (3). The species of structures Σ and Θ are therefore equivalent, and we therefore consider every structure of species Θ as a topology, namely that which corresponds to it under the procedure of deduction Q$\}$A, W$\{$. *

2. MORPHISMS AND DERIVED STRUCTURES

1. MORPHISMS

In this section and the next we shall assume for the sake of simplicity that the species of structures under consideration has only *one* base set (which is therefore a principal base set). The reader will have no difficulty in extending the definitions and results to the general case.

Let Σ be a species of structures in a theory $\mathcal{T}$ which is stronger than the theory of sets, and let x, y, s, t be four distinct letters which are different from the constants of $\mathcal{T}$. We recall that the notation $\mathcal{F}(x, y)$ denotes the set of mappings of x into y (Chapter II, § 5, no. 2). Suppose that we are given a term $\sigma\{x, y, s, t\}$ in $\mathcal{T}$ which satisfies the following conditions:

(MO_I) *The relation "s is a structure of species Σ on x, and t is a structure of species Σ on y" implies, in* E, *the relation* $\sigma\{x, y, s, t\} \subset \mathcal{F}(x, y)$.

(MO_{II}) *If, in a theory $\mathcal{T}'$ stronger than $\mathcal{T}$, we have three sets* E, E′, E″ *endowed respectively with structures $\mathcal{S}$, $\mathcal{S}'$, $\mathcal{S}''$ of species Σ, then the relations* $f \in \sigma\{E, E', \mathcal{S}, \mathcal{S}'\}$ *and* $g \in \sigma\{E', E'', \mathcal{S}', \mathcal{S}''\}$ *imply the relation*

$$g \circ f \in \sigma\{E, E'', \mathcal{S}, \mathcal{S}''\}.$$

(MO_{III}) *If, in a theory $\mathcal{T}'$ stronger than $\mathcal{T}$, we have two sets* E, E′ *endowed respectively with structures $\mathcal{S}$, $\mathcal{S}'$ of species Σ, then a bijection f of* E *onto* E′ *is an isomorphism if and only if* $f \in \sigma\{E, E', \mathcal{S}, \mathcal{S}'\}$ *and* $\overset{-1}{f} \in \sigma\{E', E, \mathcal{S}', \mathcal{S}\}$.

If Σ and σ are given, the relation $f \in \sigma\{x, y, s, t\}$ is expressed by saying that f is a *morphism* (or a *σ-morphism*) *of x, endowed with s, into y, endowed with t*. If (in a theory $\mathcal{T}'$ stronger than $\mathcal{T}$) E and E′ are two sets endowed with structures $\mathcal{S}$, $\mathcal{S}'$ of species Σ, then the term $\sigma\{E, E', \mathcal{S}, \mathcal{S}'\}$ is the *set of σ-morphisms of* E *into* E′.

Examples

(1) Take Σ to be the species of order structures and let $\sigma\{x, y, s, t\}$ denote the set of all mappings f of x into y such that the relation $(u, v) \in s$ implies $(f(u), f(v)) \in t$. With the notation of Chapter III, § 1,

this means that $u \leqslant v$ implies $f(u) \leqslant f(v)$, i.e., that f is *increasing*. The verification of axioms (MO_I), (MO_{II}), and (MO_{III}) is obvious.

(2) Take Σ to be a species of algebraic structures which has a single (internal) law of composition, everywhere defined (§1, no. 4, Example 2). Let A, A′ be two sets endowed with structures of species Σ, and let p, p' be the composition laws of these two structures. Consider the mappings f of A into A′ such that $p'(f(x), f(y)) = f(p(x, y))$ for all $x \in \mathrm{A}$ and all $y \in \mathrm{A}$. These mappings satisfy (MO_I), (MO_{II}), and (MO_{III}), and are called *homomorphisms* of A into A′.

(3) Take Σ to be the species of topological strcutures (§1, no. 4, Example 3). Let A, A′ be two sets endowed with topologies V, V′, respectively. Consider the mappings f of A into A′ such that the relation $\mathrm{X}' \in \mathrm{V}'$ implies $\overset{-1}{f}(\mathrm{X}') \in \mathrm{V}$ (in other words, such that the inverse image of every open set in the topology V′ is an open set in the topology V). These mappings, which satisfy (MO_I), (MO_{II}), and (MO_{III}), are the *continuous* mappings of A into A′ (with respect to the topologies V and V′) (cf. *General Topology*, Chapter I, § 2).

Remark. For given species of structures Σ we may have occasion to define various terms $\sigma\{x, y, s, t\}$ which satisfy the conditions (MO_I), (MO_{II}), and (MO_{III}). For example, if Σ is the species of topological structures, with the notation of Example 3 above, a mapping f of A into A′ is said to be *open* if the relation $\mathrm{X} \in \mathrm{V}$ implies $f(\mathrm{X}) \in \mathrm{V}'$ (in other words, if the image under f of every open set is an open set). It is easily checked that the open mappings also satisfy conditions (MO_I), (MO_{II}), and (MO_{III}) for the species Σ. * Moreover, it can be shown that a continuous mapping is not necessarily open, and that an open mapping is not necessarily continuous. * A given species of structures therefore *does not imply* a well-defined notion of morphisms.

Where order structures, algebraic structures, and topological structures are concerned, it is always to be understood that the morphisms are those which have been defined in the Examples above, unless the contrary is expressly stated.

The condition (MO_{III}) and the characterization of bijections (Chapter II, § 3, no. 8, Corollary to Proposition 8) imply the following criterion :

CST8. *Let* E, E′ *be two sets, each endowed with a structure of species* Σ. *Let* f *be a σ-morphism of* E *into* E′ *and let* g *be a σ-morhpism of* E′ *into* E. *If* $g \circ f$ *is the identity mapping of* E *onto itself, and if* $f \circ g$ *is the identity mapping of* E′ *onto itself, then* f *is an isomorphism of* E *onto* E′, *and* g *is the inverse isomorphism.*

It should be noted that a bijection of E onto E′ may be a σ-morphism without the inverse bijection necessarily being a σ-morphism. * For example, a bijective mapping of a topological space A onto a topological space A′ can be continuous without the inverse bijection being continuous (*General Topology*, Chapter I, § 2, no. 1, Remark 1). *

Remark. When a species of structures Σ has several principal base sets $x_1, \ldots, x_n$, and auxiliary base sets $A_1, \ldots, A_m$, a σ-morphism is a system $(f_1, \ldots, f_n)$, where f_i is a mapping of x_i into y_i $(1 \leqslant i \leqslant n)$, and these systems of mappings satisfy conditions analogous to (MO_{II}) and (MO_{III}), which the reader may easily state for himself.

2. FINER STRUCTURES

For the rest of this section we shall suppose that we are given a species of structures Σ and a notion of σ-morphism relative to this species of structures; *all the notions which will be introduced will depend not only on Σ but also on the notion of σ-morphism envisaged.* Usually we shall say "morphism" in place of "σ-morphism".

¶ Let E be a set and let $\mathcal{S}_1$, $\mathcal{S}_2$ be two structures of species Σ on E. The structure $\mathcal{S}_1$ is said to be *finer* than $\mathcal{S}_2$ (and $\mathcal{S}_2$ *coarser* than $\mathcal{S}_1$) if the identity mapping of E, endowed with $\mathcal{S}_1$, onto E, endowed with $\mathcal{S}_2$, is a morphism.

When necessary to avoid ambiguity, we shall say that $\mathcal{S}_1$ is finer than $\mathcal{S}_2$ *relative to the notion of σ-morphism under consideration*; and similarly for all the other notions to be defined in this section.

Suppose that $\mathcal{S}_1$ is finer than $\mathcal{S}_2$. If E′ is a set endowed with a structure $\mathcal{S}'$ of species Σ, and if f is a morphism of E, endowed with $\mathcal{S}_2$, into E′, endowed with $\mathcal{S}'$, then f is also a morphism of E, endowed with $\mathcal{S}_1$, into E′, endowed with $\mathcal{S}'$; this follows from the preceding definition and from (MO_{II}). Likewise, if g is a morphism of E′, endowed with $\mathcal{S}'$, into E, endowed with $\mathcal{S}_1$, then g is also a morphism of E′, endowed with $\mathcal{S}'$, into E, endowed with $\mathcal{S}_2$.

Thus the *finer* the structure (of species Σ) on E, the *more* morphisms there are with E as source, and the *fewer* morphisms with E as target.

The relation "$\mathcal{S}_1$ is coarser than $\mathcal{S}_2$" is an *order relation* between $\mathcal{S}_1$ and $\mathcal{S}_2$ on the set of all structures of the species Σ on E; for it is reflexive by (MO_{III}), transitive by (MO_{II}), and if a structure of species Σ is both finer and coarser than another, then the two structures are identical by virtue of (MO_{III}). In conformity with the general definitions (Chapter III, § 1, no. 14), two structures of species Σ on E are said to be *comparable* if one is finer than the other; a structure is said to be *strictly finer* (resp. *strictly coarser*) than another if it is *finer* (resp. *coarser*) than the other and is distinct from it.

Examples

(1) An order structure with graph s on a set A is finer than an order structure with graph s' if and only if $s \subset s'$. In other words, the relation $x \leqslant y$ with respect to s implies $x \leqslant y$ with respect to s'; this is in accordance with the definition given in Chapter III, § 1, no. 4, Example 3.

(2) Consider two algebraic structures F, F′ of the same species Σ on a set A, where F and F′ are the graphs of the (everywhere defined) laws of composition of these two structures. From the definition of morphisms in this case (no. 1, Example 2), F is finer than F′ if and only if $F \subset F'$. But since F and F′ are both functional graphs with the same domain $A \times A$, we must have $F = F'$. In other words, two *comparable* structures of species Σ are necessarily *identical*.

(3) Let V, V′ be two topologies on the same set A. To say that V is finer than V′ means, by virtue of the definition of morphisms in this case (no. 1, Example 3), that $V' \subset V$; in other words, every subset of A which is an open set in the topology V′ is also an open set in the topology V (and thus, the finer the topology, the more open sets there are).

Remark. We have just met an example (Example 2) in which two comparable structures of the same species Σ are necessarily identical. There are many other such examples : linear order structures, * compact topologies, Fréchet space structures (the morphisms being continuous linear mappings), topologies defined by an absolute value (or a valuation) on a division ring, etc. *

For such a species of structures Σ, a morphism f of E into E′ which is *bijective* is an *isomorphism*; for if we transport to E′ the structure $\mathcal{S}$ on E by means of f, we obtain a structure of species Σ which is finer than the structure $\mathcal{S}'$ on E′ and therefore coincides with $\mathcal{S}'$.

3. INITIAL STRUCTURES

Consider a family $(A_\iota)_{\iota \in I}$ of sets, each of which is endowed with a structure $\mathcal{S}_\iota$ of species Σ. Let E be a set, and for each $\iota \in I$ let f_ι be a mapping *of* E *into* A_ι. A structure $\mathfrak{J}$ of species Σ on E is said to be an *initial structure with respect to the family* $(A_\iota, \mathcal{S}_\iota, f_\iota)_{\iota \in I}$ if it has the following property :

(IN) Given any set E′, any structure of species $\mathcal{S}'$ on E′, and any mapping g *of* E′ *into* E, the relation

"g is a morphism of E′ into E"

is *equivalent* to the relation

"for each $\iota \in I$, $f_\iota \circ g$ is a morphism of E′ into A_ι".

CST9. *If there exists an initial structure on* E *with respect to the family* $(A_\iota, \mathcal{S}_\iota, f_\iota)_{\iota \in I}$, *it is the coarsest of all structures of species* Σ *on* E *for which each of the mappings* f_ι *is a morphism, and consequently is unique.*

Let $\mathfrak{J}$ be an initial structure on E, and let $\mathcal{S}$ be a structure of species Σ on E for which each of the f_ι is a morphism. If i denotes the identity mapping of E, endowed with $\mathcal{S}$, into E, endowed with $\mathfrak{J}$, then $f_\iota \circ i$ is a morphism for all $\iota \in I$, and the condition (IN) shows that i is a morphism, which means (no. 2) that $\mathcal{S}$ is *finer* than $\mathfrak{J}$. On the other hand, applying (IN) to the case in which g is the identity mapping of E (endowed with $\mathfrak{J}$) onto itself, we see (by (MO_{III})) that each f_ι is a morphism of E into A_ι, which completes the proof.

> It may happen that there exists a structure of species Σ on E which is the coarsest of all the structures of species Σ for which the f_ι are morphisms, but that this structure is not the initial structure with respect to $(A_\iota, \mathcal{S}_\iota, f_\iota)$ (Exercise 6).

We have the following *transitivity criterion* :

CST10. *Let* E *be a set, let* $(A_\iota)_{\iota \in I}$ *be a family of sets, and for each* $\iota \in I$ *let* $\mathcal{S}_\iota$ *be a structure of species* Σ *on* A_ι. *Let* $(J_\lambda)_{\lambda \in L}$ *be a partition of* I, *and let* $(B_\lambda)_{\lambda \in L}$ *be a family of sets indexed by* L. *For each* $\lambda \in L$ *let* h_λ *be a mapping of* E *into* B_λ, *and for each* $\lambda \in L$ *and each* $\iota \in J_\lambda$ *let* $g_{\lambda\iota}$ *be a mapping of* B_λ *into* A_ι, *and let* $f_\iota = g_{\lambda\iota} \circ h_\lambda$. *Suppose that, for each* $\lambda \in L$, *there exists an initial structure* $\mathcal{S}'_\lambda$ *on* B_λ *with respect to the family* $(A_\iota, \mathcal{S}_\iota, g_{\lambda\iota})_{\iota \in J_\lambda}$. *Then the following statements are equivalent* :

(a) *there exists an initial structure* $\mathfrak{J}$ *on* E *with respect to the family* $(A_\iota, \mathcal{S}_\iota, f_\iota)_{\iota \in I}$;

(b) *there exists an initial structure* $\mathfrak{J}'$ *on* E *with respect to the family* $(B_\lambda, \mathcal{S}'_\lambda, h_\lambda)_{\lambda \in L}$.

Furthermore, these statements imply that $\mathfrak{J} = \mathfrak{J}'$.

Let F be a set endowed with a structure of species Σ, and let u be a mapping of F into E. Observe that by definition the relation "$h_\lambda \circ u$ is a morphism of F into B_λ" is equivalent to the relation "for all $\iota \in J_\lambda$, $g_{\lambda\iota} \circ h_\lambda \circ u = f_\iota \circ u$ is a morphism of F into A_ι". The relation

(1) "for all $\lambda \in L$, $h_\lambda \circ u$ is a morphism of F into B_λ"

is therefore equivalent to the relation

(2) "for all $\iota \in I$, $f_\iota \circ u$ is a morphism of F into A_ι".

Now, to say that $\mathfrak{J}'$ is the initial structure with respect to the family $(B_\lambda, \mathcal{S}'_\lambda, h_\lambda)_{\lambda \in L}$ means that relation (1) is equivalent to the relation "u is a morphism of F into E endowed with $\mathfrak{J}'$"; and to say that $\mathfrak{J}$ is the initial structure with respect to the family $(A_\iota, \mathcal{S}_\iota, f_\iota)_{\iota \in I}$ means that relation (2) is equivalent to the relation "u is a morphism of F into E endowed with I". Hence the result, in view of the property of uniqueness of initial structure.

4. EXAMPLES OF INITIAL STRUCTURES

I. *Inverse image of a structure.* When I is a set consisting of a single element, the initial structure with respect to $(A, \mathcal{S}, f)$ is called the *inverse image under f of the structure $\mathcal{S}$* (when it exists).

> * A topology always has an inverse image under any mapping f; but this is not the case for an order structure or an algebraic structure. *

II. *Induced structure.* Let A be a set endowed with a structure $\mathcal{S}$ of species Σ, let B be a subset of A, and let j be the canonical injection of B into A. Then the inverse image under j of the structure B (if it exists) is called the *structure induced* by $\mathcal{S}$ on B.

> An order structure induces a structure of the same species on every subset of the set on which it is defined; but this is not the case for the structure of a directed set. * A topology induces a topology on every subset of the set on which it is defined, but a compact topology does not in general induce a compact topology. An algebraic structure on a set A does not in general induce a structure of the same species on an arbitrary subset B; if the given structure on A consists of laws of composition which are everywhere defined, then it is necessary that B should be stable with respect to each of these laws, but this necessary condition is not always sufficient. *

The general criterion CST10 gives us the following *transitivity criterion* for induced structures:

CST11. *Let* B *be a subset of* A, *let* C *be a subset of* B, *and let* $\mathcal{S}$ *be a structure of species* Σ *on* A *which induces a structure* $\mathcal{S}'$ *of the same species on* B. *Then* $\mathcal{S}$ *induces a structure of species* Σ *on* C *if and only if* $\mathcal{S}'$ *induces a structure of species* Σ *on* C, *and the structures induced on* C *by* $\mathcal{S}$ *and* $\mathcal{S}'$ *are then identical.*

CST12. *Let* A, A' *be two sets endowed with structures* $\mathcal{S}$, $\mathcal{S}'$ *of species* Σ. *Let* B *be a subset of* A, *and* B' *a subset of* A'. *Suppose that* $\mathcal{S}$ (resp. $\mathcal{S}'$) *induces a structure of species* Σ *on* B (resp. B'). *If* f *is a morphism of* A *into* A' *such that* $f(B) \subset B'$, *then the mapping* g *of* B *into* B' *which coincides with* f *on* B *is a morphism (with respect to the structures induced by* $\mathcal{S}$ *and* $\mathcal{S}'$*).*

Let j (resp. j') be the canonical injection of B (resp. B') into A (resp. A'). By definition we have $f \circ j = j' \circ g$. Since f and j are morphisms, so is $f \circ j$ by (MO_{II}); but then, $j' \circ g$ being a morphism, the mapping g is a morphism by the definition of initial structure.

III. *Product structure.* Let $(A_\iota)_{\iota \in I}$ be a family of sets, and on each set A_ι let $\mathcal{S}_\iota$ be a structure of species Σ. Let $E = \prod_{\iota \in I} A_\iota$ be the *product* of the family $(A_\iota)_{\iota \in I}$ (Chapter II, § 5), and let pr_ι denote the projection of E onto A_ι. The initial structure (if it exists) with respect to the family $(A_\iota, \mathcal{S}_\iota, pr_\iota)_{\iota \in I}$ is called the *product* of the structures $\mathcal{S}_\iota$.

A family of order structures always admits a product structure, but the same is not always true of a family of total order structures. * A family of group structures always admits a product structure, but the same need not be true of a family of division ring structures. A family of topologies always admits a product structure, but this is not always true of a family of structures of locally compact spaces; in this case, there is a product structure of the same species if the family is *finite*, but there need not be one if the family is *infinite* (cf. *General Topology*, Chapter I, § 9, no. 7, Prop. 14). *

Criterion CST10 gives rise to the following *associativity criterion* for product structures :

CST13. *Let* $(A_\iota)_{\iota \in I}$ *be a family of sets, and for each index* $\iota \in I$ *let* $\mathcal{S}_\iota$ *be a structure of species* Σ *on* A_ι. *Let* $(J_\lambda)_{\lambda \in L}$ *be a partition of* I. *Suppose that on each partial product* $B_\lambda = \prod_{\iota \in J_\lambda} A_\iota$ *the family* $(\mathcal{S}_\iota)_{\iota \in J_\lambda}$ *admits a product structure* $\mathcal{S}'_\lambda$. *Then the family* $(\mathcal{S}_\iota)_{\iota \in I}$ *admits a product structure* $\mathcal{S}$ *if and only if the family* $(\mathcal{S}'_\lambda)_{\lambda \in L}$ *admits a product structure* $\mathcal{S}'$, *and the canonical mapping of* $E = \prod_{\iota \in I} A_\iota$, *endowed with* $\mathcal{S}$, *onto* $F = \prod_{\lambda \in L} B_\lambda$, *endowed with* $\mathcal{S}'$ (Chapter II, § 5, no. 5), *is then an isomorphism.*

Another application of CST10 gives the following criterion concerning structures induced by a product structure :

CST14. *Let* $(A_\iota)_{\iota \in I}$ *be a family of sets, and for each* $\iota \in I$ *let* $\mathcal{S}_\iota$ *be a structure of species* Σ *on* A_ι. *For each* $\iota \in I$, *let* B_ι *be a subset of* A_ι. *Suppose that each* $\mathcal{S}_\iota$ *induces a structure* $\mathcal{S}'_\iota$ *on* B_ι, *and that on the product* $E = \prod_{\iota \in I} A_\iota$ *there exists a structure* $\mathcal{S}_0$ *which is the product of the family* $(\mathcal{S}_\iota)$. *Then the following statements are equivalent* :

(a) *on the set* $B = \prod_{\iota \in I} B_\iota \subset E$ *there exists a structure* $\mathcal{S}$ *induced by* $\mathcal{S}_0$;

(b) *on the set* B *there exists a structure* $\mathcal{S}'$ *which is the product of the family of structures* $(\mathcal{S}'_\iota)$.

Furthermore, these statements imply that $\mathcal{S} = \mathcal{S}'$.

Let j_ι be the canonical injection of B_ι into A_ι, let j be the canonical injection of B into E, let p_ι be the projection of E onto A_ι, and let p'_ι be the projection of B onto B_ι. Then we have $p_\iota \circ j = j_\iota \circ p'_\iota$ for all $\iota \in I$. By CST10, $\mathcal{S}$ is the initial structure with respect to the family $(A_\iota, \mathcal{S}_\iota, p_\iota \circ j)_{\iota \in I}$, and $\mathcal{S}'$ is the initial structure with respect to the family $(A_\iota, \mathcal{S}_\iota, j_\iota \circ p'_\iota)_{\iota \in I}$. Hence the result.

¶ The notions of inverse image and product are related by the following criterion:

CST15. *Let $(A_\iota)_{\iota \in I}$ be a family of sets, and for each $\iota \in I$ let $\mathcal{S}_\iota$ be a structure of species Σ on A_ι, and let f_ι be a mapping of a set E into A_ι. Suppose that on the product set $A = \prod_{\iota \in J} A_\iota$ there exists a product structure $\mathcal{S}$ of the family $(\mathcal{S}_\iota)_{\iota \in I}$. Then there exists an initial structure with respect to the family $(A_\iota, \mathcal{S}_\iota, f_\iota)_{\iota \in I}$ if and only if the structure $\mathcal{S}$ has an inverse image under the mapping $x \to f(x) = (f_\iota(x))$ of E into A, and these two structures are then identical.*

Since $f_\iota = \mathrm{pr}_\iota \circ f$, this criterion is a particular case of CST10.

> *Remark.* Let $(\mathcal{S}_\lambda)_{\lambda \in L}$ be a family of structures of species Σ on the *same* set A; let A_λ denote the set A endowed with the structure $\mathcal{S}_\lambda$, and let I_λ denote the identity mapping of A into A_λ. Let B be the product set $A^L = \prod_{\lambda \in L} A_\lambda$, and let Δ be the diagonal of this product (Chapter II, § 5, no. 3). Let h be the diagonal mapping of A onto Δ, so that $h(x)$ is the element $(x_\lambda)_{\lambda \in L}$ such that $x_\lambda = x$ for all $\lambda \in L$. Suppose that there exists on B a product structure $\mathcal{S}'$ of the family $(\mathcal{S}_\lambda)$. Since h is injective, criterion CST 15 shows that there exists an initial structure $\mathcal{S}$ with respect to the family $(A_\lambda, \mathcal{S}_\lambda, I_\lambda)_{\lambda \in L}$ if and only if there exists a structure $\mathcal{S}''$ on Δ induced by $\mathcal{S}'$; $\mathcal{S}''$ is then identical with the structure obtained by transporting $\mathcal{S}$ to Δ by means of h. In particular, when all the structures $\mathcal{S}_\lambda$ are identical, h is an *isomorphism* of A (endowed with this structure) onto Δ.

We have also the following criterion:

CST16. *Let $(A_\iota)_{\iota \in I}$, $(B_\iota)_{\iota \in I}$ be two families of sets indexed by the same set. For each $\iota \in I$, let $\mathcal{S}_\iota$ be a structure of species Σ on A_ι and let $\mathcal{S}'_\iota$ be a structure of species Σ on B_ι. Suppose that there exists on $A = \prod_{\iota \in I} A_\iota$* (resp. *on $B = \prod_{\iota \in I} B_\iota$*) *a product structure $\mathcal{S}$* (resp. $\mathcal{S}'$) *of the family $(\mathcal{S}_\iota)$* (resp. $(\mathcal{S}'_\iota)$). *For each $\iota \in I$, let f_ι be a morphism of A_ι into B_ι. Then the mapping $f = (f_\iota)_{\iota \in I}$ is a morphism of A into B.*

Let p_ι (resp. q_ι) be the projection of A onto A_ι (resp. of B onto B_ι). Then we have $q_\iota \circ f = f_\iota \circ p_\iota$. Since f_ι and p_ι are morphisms (criterion

CST9), $f_\iota \circ p_\iota$ is a morphism by $(\mathrm{MO_{II}})$; hence f is a morphism by (IN).

Remark. For most of the usual structures, the condition given in CST 16 is not only sufficient but also necessary for f to be a morphism (cf. Exercise 7). In particular, this is so in the following circumstances (which occur, for example, if Σ is the species of order structures, * or the species of group structures, or the species of topological structures $_*$, etc.; cf. Exercise 8) :

¶ There exists a family $(a_\iota)_{\iota \in \mathrm{I}}$ such that $a_\iota \in \mathrm{A}_\iota$ for all $\iota \in \mathrm{I}$ and such that, if we put $r_\iota(x_\iota) = (y_\varkappa)$, where $y_\iota = x_\iota$ and $y_\varkappa = a_\varkappa$ whenever $\varkappa \neq \iota$, each of the mappings r_ι is a *morphism* of A_ι into A. For if $f = (f_\iota)$ is a morphism of A into B, we may write $f_\iota = q_\iota \circ f \circ r_\iota$ for all $\iota \in \mathrm{I}$, and it is enough to apply $(\mathrm{MO_{II}})$.

¶ Note that r_ι is a morphism if the following condition is satisfied :

(a) For every set E endowed with a structure of species Σ, the constant mapping $z \to a_\iota$ is a morphism of E into A_ι; namely, for each $\varkappa \in \mathrm{I}$, $p_\varkappa \circ r_\iota$ is a morphism of A_ι into $\mathrm{A}_\varkappa$, since it is the identity mapping when $\varkappa = \iota$, and a constant mapping $z \to a_\varkappa$ when $\varkappa \neq \iota$; by the definition of product structure, r_ι is therefore a morphism of A_ι into A.

¶ The examples listed above satisfy not only (a), but also the following condition :

(b) On every set $\mathrm{A}'_\iota = \mathrm{A}_\iota \times \prod_{\varkappa \neq \iota} \{a_\varkappa\}$, the structure $\mathcal{S}$ induces a structure of species Σ.

¶ Let p'_ι denote the restriction of p_ι to A'_ι. If both conditions (a) and (b) are satisfied, then p'_ι is an *isomorphism* of A'_ι onto A_ι. For since $p'_\iota = p_\iota \circ j_\iota$, where j_ι is the canonical injection of A'_ι into A, p'_ι is a morphism by $(\mathrm{MO_{II}})$. Also we have $r_\iota = j_\iota \circ \overset{-1}{p'_\iota}$; hence $\overset{-1}{p'_\iota}$ is a morphism of A_ι into A'_ι by virtue of the definition of induced structure.

Finally, we have the following criterion, which characterizes the morphisms in many cases :

CST17. *Let* A *and* B *be two sets, endowed with structures* $\mathcal{S}_\mathrm{A}$, $\mathcal{S}_\mathrm{B}$ *of the same species* Σ. *Suppose that there exists on* $\mathrm{A} \times \mathrm{B}$ *the structure* $\mathcal{S}_{\mathrm{A}\times\mathrm{B}}$, *the product of* $\mathcal{S}_\mathrm{A}$ *and* $\mathcal{S}_\mathrm{B}$. *Let f be a mapping of* A *into* B, *let* F *be its graph, and let* π *be the bijection* $x \to (x, f(x))$ *of* A *onto* F. *Then, for f to be a morphism of* A *into* B, *it is necessary and sufficient that there should exist on* F *a structure of species* Σ *induced by* $\mathcal{S}_{\mathrm{A}\times\mathrm{B}}$ *and that, when* F *is endowed with this structure,* π *should be an isomorphism of* A *onto* F.

To prove sufficiency, let j be the canonical injection of F into $\mathrm{A} \times \mathrm{B}$. We may write $f = \mathrm{pr}_2 \circ j \circ \pi$, and f is then by hypothesis the composition of three morphisms.

¶ To prove necessity, let $\mathcal{S}_F$ be the structure of species Σ obtained by transporting the structure $\mathcal{S}_A$ to F by means of the bijection π (§ 1, no. 5). Then we must show that $\mathcal{S}_F$ is induced by $\mathcal{S}_{A\times B}$ on F. We remark first that j is a morphism of F into $A \times B$; for $j \circ \pi$ is the mapping $x \to (x, f(x))$ of A into $A \times B$ and is therefore a morphism by virtue of the hypothesis on f and the definition of the product structure; hence, by the definition of the structure $\mathcal{S}_F$, j is a morphism. It remains to be shown that if E is a set endowed with a structure of species Σ, and if g is a mapping of E into F such that $j \circ g$ is a morphism of E into $A \times B$, then g is a morphism; or, equivalently, that $g_1 = \overset{-1}{\pi} \circ g$ is a morphism of E into A. But since $g_1 = \mathrm{pr}_1 \circ (j \circ g)$, this follows from the hypothesis and the definition of the product structure.

5. FINAL STRUCTURES

Consider a family of sets $(A_\iota)_{\iota\in I}$, each endowed with a structure $\mathcal{S}_\iota$ of species Σ. Let E be a set, and for each $\iota \in I$ let g_ι be a mapping *of* A_ι *into* E. A structure $\mathcal{F}$ of species Σ on E is said to be a *final structure with respect to the family* $(A_\iota, \mathcal{S}_\iota, g_\iota)_{\iota\in I}$ if it has the following property :

(FI) Given any set E′, any structure $\mathcal{S}'$ of species Σ on E, and any mapping f *of* E *into* E′, the relation

"f is a morphism of E into E′"

is *equivalent* to the relation

"for all $\iota \in I$, $f \circ g_\iota$ is a morphism of A_ι into E′".

CST18. *If there exists a final structure on* E *with respect to the family* $(A_\iota, \mathcal{S}_\iota, g_\iota)_{\iota\in I}$, *then it is the finest structure of species* Σ *on* E *for which each of the mappings* g_ι *is a morphism, and is therefore unique.*

Let $\mathcal{F}$ be a final structure on E, and let $\mathcal{S}$ be a structure of species Σ on E for which each g_ι is a morphism. If i denotes the identity mapping of E, endowed with $\mathcal{F}$, onto E, endowed with $\mathcal{S}$, then $i \circ g_\iota$ is a morphism for each $\iota \in I$. The condition (FI) then shows that i is a morphism, which means (no. 2) that $\mathcal{S}$ is *coarser* than $\mathcal{F}$. Applying (FI) again to the case in which f is the identity mapping of E (endowed with $\mathcal{F}$) onto itself, we see (using (MO_{III})) that each g_ι is a morphism of A_ι into E. This completes the proof.

It may happen that there exists a structure of species Σ on E which is the finest of all structures of species Σ on E for which the g_ι are morphisms, but that this structure is not the final structure with respect to the family $(A_\iota, \mathcal{S}_\iota, g_\iota)$ (Exercise 6).

We have the following *transitivity criterion*:

CST19. *Let* E *be a set, let* $(A_\iota)_{\iota \in I}$ *be a family of sets, and for each* $\iota \in I$ *let* $\mathcal{S}_\iota$ *be a structure of species* Σ *on* A_ι. *Let* $(J_\lambda)_{\lambda \in L}$ *be a partition of* I, *and let* $(B_\lambda)_{\lambda \in L}$ *be a family of sets indexed by* L. *For each* $\lambda \in L$, *let* h_λ *be a mapping of* B_λ *into* E; *for each* $\lambda \in L$ *and each* $\iota \in J_\lambda$, *let* $g_{\iota\lambda}$ *be a mapping of* A_ι *into* B_λ, *and put* $f_\iota = h_\lambda \circ g_{\iota\lambda}$. *Suppose that, for each* $\lambda \in L$, *there exists a final structure* $\mathcal{S}'_\lambda$ *on* B_λ *with respect to the family* $(A_\iota, \mathcal{S}_\iota, g_{\iota\lambda})_{\iota \in J_\lambda}$. *Then the following statements are equivalent*:

(a) *There exists a final structure* $\mathcal{F}$ *on* E *with respect to the family* $(A_\iota, \mathcal{S}_\iota, f_\iota)_{\iota \in I}$.

(b) *There exists a final structure* $\mathcal{F}'$ *on* E *with respect to the family* $(B_\lambda, \mathcal{S}'_\lambda, h_\lambda)_{\lambda \in L}$.

Furthermore these statements imply that $\mathcal{F} = \mathcal{F}'$.

Let F be a set endowed with a structure of species Σ, and let u be a mapping of E into F. By definition, the relation "$u \circ h_\lambda$ is a morphism of B_λ into F" is equivalent to the relation

"for all $\iota \in J_\lambda$, $u \circ h_\lambda \circ g_{\iota\lambda} = u \circ f_\iota$ is a morphism of A_ι into F".

The relation

(3) "for all $\lambda \in L$, $u \circ h_\lambda$ is a morphism of B_λ into F"

is therefore equivalent to the relation

(4) "for all $\iota \in I$, $u \circ f_\iota$ is a morphism of A_ι into F".

To say that $\mathcal{F}'$ is the final structure with respect to the family $(B_\lambda, \mathcal{S}'_\lambda, h_\lambda)_{\lambda \in L}$ means that the relation (3) is equivalent to the relation "u is a morphism of E (endowed with $\mathcal{F}'$) into F"; and to say that E is the final structure with respect to the family $(A_\iota, \mathcal{S}_\iota, f_\iota)_{\iota \in I}$ means that the relation (4) is equivalent to the relation "u is a morphism of E (endowed with $\mathcal{F}$) into F"; hence the result, in view of the property of uniqueness of final structure.

6. EXAMPLES OF FINAL STRUCTURES

I. *Direct image of a structure.* When I is a set consisting of a single element, the final structure with respect to $(A, \mathcal{S}, f)$ is called the *direct image under* f *of the structure* $\mathcal{S}$ (when it exists).

II. *Quotient structure.* Let A be a set endowed with a structure $\mathcal{S}$ of species Σ, let R be an equivalence relation on A, and let φ be the canonical mapping of A onto the quotient set $E = A/R$ (Chapter II, §6, no. 2). The direct image of the structure $\mathscr{S}$ under the mapping φ is called (when it exists) the *quotient* of the structure $\mathscr{S}$ by the relation R.

* In general, an order structure or an algebraic structure does not admit quotient structures with respect to arbitrary equivalence relations (cf. Chapter III, § 1, Exercise 2). On the other hand, a topology always admits a quotient structure with respect to an arbitrary equivalence relation, but this is not necessarily the case for a Hausdorff topology. *

Let A, B be two sets endowed respectively with structures $\mathscr{S}$, $\mathscr{S}'$ of species Σ, and let f be a morphism of A into B. Let R be the equivalence relation $f(x) = f(y)$, let φ be the canonical mapping of A onto A/R, and let j be the canonical injection of $f(A)$ into B. Suppose that $\mathscr{S}$ admits a quotient structure $\mathscr{S}_0$ with respect to R, and that $\mathscr{S}'$ induces a structure $\mathscr{S}'_0$ on $f(A)$. Then, in the *canonical decomposition* $f = j \circ g \circ \varphi$ of f (Chapter II, §6, no. 5), the bijection g of A/R onto $f(A)$ which is associated with f is a *morphism* (but not necessarily an isomorphism) when A/R is endowed with $\mathscr{S}_0$ and $f(A)$ with $\mathscr{S}'_0$. For $j \circ g$ is a morphism of A/R into B by the definition of quotient structure, and g is therefore a morphism of A/R onto $f(A)$ by the definition of induced structure.

CST20. *Let* A, A′ *be two sets endowed with structures* $\mathscr{S}$, $\mathscr{S}'$ *of species* Σ, *and let* R (resp. R′) *be an equivalence relation on* A (resp. A′). *Suppose that there exists a quotient structure* $\mathscr{S}_0$ (resp. $\mathscr{S}'_0$) *of* $\mathscr{S}$ *by* R (resp. $\mathscr{S}'$ *by* R′). *If* f *is a morphism of* A *into* A′ *which is compatible with the equivalence relations* R *and* R′, *and if* g *is the mapping obtained from* f *by passing to the quotients, then* g *is a morphism of* A/R *into* A′/R′.

Let φ (resp. φ') be the canonical mapping of A onto A/R (resp. of A′ onto A′/R′); then we have $g \circ \varphi = \varphi' \circ f$. Since φ' and f are morphisms, so is $\varphi' \circ f$ by (MO_{II}). But then, $g \circ \varphi$ being a morphism, g is also a morphism by the definition of quotient structure.

¶ The transitivity criterion CST19 gives rise in particular to the following criterion:

CST21. *Let* A *be a set endowed with a structure* $\mathscr{S}$ *of species* Σ, *and let* R *be an equivalence relation on* A *such that there exists on* A/R *a quotient structure* $\mathscr{S}'$ *of* $\mathscr{S}$ *by* R. *Let* S *be an equivalence relation on* A *which is coarser than* R, *and let* S/R *denote the equivalence relation on* A/R *which is the quotient of* S *by* R (Chapter II, §6, no. 7). *Then there exists on* (A/R)/(S/R) *a quotient structure* $\mathscr{S}''$ *of* $\mathscr{S}'$ *by* S/R *if and only if there exists on* A/S *a quotient structure* $\mathscr{S}_0$ *of* $\mathscr{S}$ *by* S, *and the canonical mapping of* A/S (*endowed with* $\mathscr{S}_0$) *onto* (A/R)/(S/R) (*endowed with* $\mathscr{S}''$) *is an isomorphism.*

Let φ be the canonical mapping of A onto A/R, and let ψ be that of A/R onto (A/R)/(S/R). By virtue of CST19, $\mathcal{S}''$ is the quotient of $\mathcal{S}'$ by S/R if and only if $\mathcal{S}''$ is the final structure with respect to $(A, \mathcal{S}, \psi \circ \varphi)$. The criterion then follows from the fact that the relation $\psi(\varphi(x)) = \psi(\varphi(y))$ is equivalent to S.

Remark. Let A be a set endowed with a structure $\mathcal{S}$ of species Σ, and let R be an equivalence relation on A such that there exists on $E = A/R$ a quotient structure $\mathcal{S}'$ of $\mathcal{S}$ by R. Let φ be the canonical mapping of A onto E. In general, there exists no *section* s of φ (Chapter II, §3, no. 8) which is a *morphism* of E into A. Let us suppose that such a section s exists, and moreover that there exists a structure $\mathcal{S}''$ induced by $\mathcal{S}$ on $s(E)$. Then, if j denotes the canonical injection of $s(E)$ into A and if $s = j \circ f$, the bijection f is an *isomorphism* of E onto $s(E)$. For f is a morphism by the definition of induced structure, and $g = \varphi \circ j$ is a morphism of $s(E)$ onto E by reason of (MO_{II}). Since $g \circ f$ and $f \circ g$ are the identity mappings of E and $s(E)$, respectively, the assertion is a consequence of CST8.

3. UNIVERSAL MAPPINGS

1. UNIVERSAL SETS AND MAPPINGS

Let $\mathcal{T}$ be a theory which is stronger than the theory of sets, and let E be a term in $\mathcal{T}$. Let Σ be a species of structures in $\mathcal{T}$. For simplicity we shall suppose throughout that Σ is defined on a single (principal) base set, and for brevity we shall say "Σ-set" for "set endowed with a structure of species Σ". Furthermore, we shall suppose that the σ-morphisms have been defined for the species Σ (§2, no. 1; as in §2, we shall say "morphism" in place of "σ-morphism"). Finally, the species Σ being defined on the base set x and having s as generic structure (§1, no. 4), let us suppose that a term $\alpha\{x, s\}$ is defined in $\mathcal{T}_\Sigma$, satisfying the following conditions:

(QM_I) *The relation* $\alpha\{x, s\} \subset \mathcal{F}(E; x)$ *is true in* $\mathcal{T}_\Sigma$.

(QM_{II}) *If (in a theory* $\mathcal{T}'$ *which is stronger than* $\mathcal{T}$*)* F *and* F′ *are two sets endowed with structures* $\mathcal{S}$, $\mathcal{S}'$ *of species* Σ, *and if* f *is a morphism of* F *into* F′, *then the relation* $\varphi \in \alpha\{F, \mathcal{S}\}$ *implies* $f \circ \varphi \in \alpha\{F', \mathcal{S}'\}$.

We shall express the relation $\varphi \in \alpha\{x, s\}$ by saying that φ *is an α-mapping of* E *into* x *(endowed with* s*)*.

A Σ-set F_E and an α-mapping φ_E of E into F_E are said to be *universal* if the following condition is satisfied :

(AU) *For each α-mapping φ of* E *into a Σ-set* F *there exists a unique morphism* f *of* F_E *into* F *such that* $\varphi = f \circ \varphi_E$.

The pair (F_E, φ_E) is then also said to be a *solution of the universal mapping problem* for E (relative to Σ, σ, and α).

¶ Let (F'_E, φ'_E) and (F''_E, φ''_E) be two solutions of the universal mapping problem for E. The condition (AU) shows then that there exists a unique morphism f_1 of F'_E into F''_E and a unique morphism f_2 of F''_E into F'_E such that $\varphi''_E = f_1 \circ \varphi'_E$ and $\varphi'_E = f_2 \circ \varphi''_E$. We have therefore $\varphi'_E = f_2 \circ f_1 \circ \varphi'_E$ and $\varphi''_E = f_1 \circ f_2 \circ \varphi''_E$. Applying (AU) to the case where $F = F'_E$ and $\varphi = \varphi'_E$, we find that $f_2 \circ f_1$ is the identity mapping of F'_E onto itself. Similarly, $f_1 \circ f_2$ is the identity mapping of F''_E onto itself. Consequently (§ 2, no. 1, criterion CST8) f_1 is an *isomorphism* of F'_E onto F''_E, and f_2 is its inverse isomorphism. This result is expressed by saying that the solution of the universal mapping problem for E is *unique up to isomorphism.*

To verify that a pair (F_E, φ_E) is a solution of the universal mapping problem for E, it is often convenient to verify the following two conditions :

(AU'_I) *For every Σ-set* F *and every α-mapping φ of* E *into* F, *there exists a morphism* f *of* F_E *into* F *such that* $\varphi = f \circ \varphi_E$.

(AU'_{II}) *For every Σ-set* F, *two morphisms of* F_E *into* F *which agree on* $\varphi_E(E)$ *are equal.*

For if these two conditions are satisfied, the morphism f whose existence is ensured by (AU'_I) is unique by (AU'_{II}). Conversely, it is clear that (AU) implies (AU'_I); furthermore, if f and f' are two morphisms of F_E into F which agree on $\varphi_E(E)$, we have $f \circ \varphi_E = f' \circ \varphi_E$, whence $f = f'$ by applying (AU) to the α-mapping $f \circ \varphi_E$. Hence (AU) implies (AU'_{II}).

2. EXISTENCE OF UNIVERSAL MAPPINGS

A universal mapping problem does not necessarily have a solution (Exercise 1). However, we shall show that the following conditions imply the existence of a solution :

(CU_I) *On every product of a family of Σ-sets there exists a product structure of species* Σ (§ 2, no. 4).

(CU_{II}) *Let* $(F_\iota)_{\iota \in I}$ *be a family of Σ-sets, and for each* $\iota \in I$ *let* φ_ι *be an α-mapping of* E *into* F_ι. *Then the mapping* $(\varphi_\iota)_{\iota \in I}$ *of* E *into* $\prod_{\iota \in I} F_\iota$ *(endowed with the product structure) is an α-mapping.*

A subset G of a Σ-set F will be said to be *Σ-admissible* if the structure on F induces a structure of species Σ on G (§ 2, no. 4).

(CU_{III}) *There exists a cardinal* $\mathfrak{a}$ *with the following properties: for every* Σ*-set* F *and every* α*-mapping* φ *of* E *into* F *there exists a* Σ*-admissible subset* G *of* F *which contains* $\varphi(E)$, *has cardinal* $\leqslant \mathfrak{a}$, *is such that the mapping of* E *into* G *having the same graph as* φ *is an* α*-mapping, and such that any two morphisms of* G *into a* Σ*-set, which agree on* $\varphi(E)$, *are equal.*

CST22. *If the conditions* (CU_I) *to* (CU_{III}) *are satisfied, then the universal mapping problem for* E *has a solution.*

We shall first show that if there exists a pair (F_E, φ_E) which satisfies (AU'_I), then there also exists a solution of the universal mapping problem for E. For by (CU_{III}) there exists a Σ-admissible subset F'_E of F_E containing $\varphi_E(E)$, such that the mapping φ'_E of E into F'_E which has the same graph as φ_E is an α-mapping, and such that any two morphisms of F'_E into a Σ-set which agree on $\varphi_E(E)$ are equal. Let j be the canonical injection of F'_E into F_E, so that $\varphi_E = j \circ \varphi'_E$. For every morphism f of F_E into a Σ-set F, $f \circ j$ is a morphism of F'_E into F, and we have $f \circ \varphi_E = (f \circ j) \circ \varphi'_E$. It is therefore clear that (F'_E, φ'_E) satisfies (AU'_I) and (AU'_{II}).

It remains for us to establish the existence of a pair (F_E, φ_E) satisfying (AU'_I). Let $s \in S(x)$ be the typical characterization of the species of structures Σ, and consider the subset L of $\mathfrak{P}(\mathfrak{a}) \times S(\mathfrak{a}) \times \mathfrak{P}(E \times \mathfrak{a})$ consisting of all triples $\lambda = (X, V, P)$ having the following property: "V is a structure of species Σ on $X \subset \mathfrak{a}$, and P is the graph of an α-mapping of E into X (with respect to the structure V)" (observe that we have $S(X) \subset S(\mathfrak{a})$, as is easily seen by arguing step by step on the length of the echelon construction scheme S). For every $\lambda = (X, V, P) \in L$, we denote by X_λ the set X endowed with the structure V, and by φ_λ the mapping of E into X_λ whose graph is P.

Let F_E be the Σ-set which is the product of the X_λ (it exists by (CU_I)), and let φ_E be the mapping $x \to (\varphi_\lambda(x))$ of E into F_E, which is an α-mapping by virtue of (CU_{II}). Let us show that the pair (F_E, φ_E) satisfies (AU'_I). Given an α-mapping φ of E into a Σ-set F, let G be a subset of F which satisfies the conditions stated in (CU_{III}). Let j be the canonical injection of G into F, and let ψ be the mapping of E into G which has the same graph as φ, so that $\varphi = j \circ \psi$. It follows from (CU_{III}) that ψ is an α-mapping of E into G. Since Card (G) $\leqslant \mathfrak{a}$, there is a subset G' of $\mathfrak{a}$ equipotent to G. Let g be a bijection of G onto G'. If we transport by means of g the structure of species Σ on G, there exists by definition an element λ of L such that G' (endowed with the transported structure) is equal to X_λ and such that $g \circ \psi = \varphi_\lambda$. Then

$f = j \circ \overset{-1}{g} \circ \mathrm{pr}_\lambda$ is a morphism of F_E into F such that $\varphi = f \circ \varphi_E$, and the proof is complete.

CST23. *Let* (F_E, φ_E) *be a solution of the universal mapping problem for* E. *Then* φ_E *is an injection of* E *into* F_E *if and only if, for each pair of distinct elements* x, y *of* E, *there exists an* α*-mapping* φ *of* E *into a* Σ*-set* F *such that* $\varphi(x) \neq \varphi(y)$.

Since φ_E is an α-mapping, the criterion is an immediate consequence of the definitions.

¶ In this case the α-mappings are said to *separate* the elements of E, and we do not ordinarily make any distinction, in the terminology, between the elements of E and their images under φ_E. With this convention, if (F_E, φ_E) is a solution of a universal mapping problem, and if condition (CU_{III}) is satisfied, then every α-mapping of E into a Σ-set F *extends uniquely to a morphism of* F_E *into* F.

3. EXAMPLES OF UNIVERSAL MAPPINGS

* The examples which follow will, for the most part, be treated in detail elsewhere in this series.

I. *Free algebraic structures.* Let E be a set and let Σ be a species of algebraic structures, defined by one or more laws of composition. We take as morphisms the *homomorphisms* for the species Σ under consideration, and as α-mappings *arbitrary* mappings of E into a Σ-set (in other words, $\alpha\{x, s\} = \mathcal{F}(E, x)$). All the usual species of algebraic structures satisfy (CU_{III}); with the exception of division ring structures, they also satisfy (CU_I), and (CU_{II}) is here a trivial consequence of (CU_I).

Since in general there exist structures of species Σ defined on sets with at least two elements, the α-mappings separate the elements of E, and E may therefore be considered as embedded in F_E. F_E is said to be the *free* Σ-set generated by E. Thus in algebra we speak of *free semigroups, free groups, free modules,* and *free algebras.*

II. *Rings and fields of fractions.* Let E be a commutative ring with an identity element and let S be a multiplicatively closed subset of E which does not contain 0. We take Σ to be the species of structures of commutative rings with identity element, and the morphisms to be ring homomorphisms which transform identity element into identity element. The α-mappings will be the homomorphisms φ of E into a commutative ring A with an identity element, such that $\varphi(1) = 1$ and $\varphi(S)$ contains only *units* of A. The conditions (QM_{II}), (CU_I) throngh (CU_{III}) (with $\mathfrak{a} = \mathrm{Card}\,(E)\,\mathrm{Card}\,(\mathbf{N})$) are immediately verified. The universal mapping problem therefore always has a solution (F_E, φ_E), but in general φ_E

is not injective. The most frequently arising case is that in which E is an integral domain; in this case, φ_E is injective. If, moreover, we take $S = E - \{0\}$, then F_E is a field, called the *field of fractions* of E.

III. *Tensor product of two modules.* Let E be the product $A \times B$ of two modules over a commutative ring C which has an identity element. Take Σ to be the species of C-module structures, the morphisms to be linear mappings, and the α-mappings to be *bilinear* mappings of $A \times B$ into a C-module. The condition (QM_{II}) is evidently satisfied, and so are (CU_I) through (CU_{III}) (with $\mathfrak{a} = \text{Card}(E)\ \text{Card}(C)\ \text{Card}(\mathbf{N})$). The universal C-module F_E corresponding to the pair (A, B) is called the *tensor product* of A and B and is written $A \otimes B$. The universal mapping φ_E is written $(x, y) \to x \otimes y$; it is bilinear but not, in general, injective.

IV. *Extension of the ring of operators of a module.* Let A be a commutative ring with an identity element, let B be a subring of A containing the identity element of A, and let E be a B-module. The species Σ is the species of A-module structures, the morphisms are A-*linear* mappings, and the α-mappings are the B-*linear* mappings of E into an A-module. The universal A-module F_E corresponding to the B-module E is said to be obtained by *extending to* A the ring of operators B of E.

V. *Completion of a uniform space.* Let E be a uniform space. Take Σ to be the species of structures of complete Hausdorff uniform spaces, the morphisms to be uniformly continuous mappings, and the α-mappings to be uniformly continuous mappings of E into a complete Hausdorff uniform space. The Σ-admissible subsets of a complete Hausdorff uniform space are here the *closed* subsets (with respect to the topology of the space), and conditions (QM_{II}) and (CU_I) through (CU_{III}) are satisfied (with $\mathfrak{a} = 2^{2^{\text{Card}(E)}}$). The complete Hausdorff uniform space is (up to isomorphism) the *completion* of the Hausdorff uniform space associated with E (*General Topology*, Chapter II, § 3, no. 7).

VI. *Stone-Čech compactification.* Let E be a completely regular space. Σ is the species of structures of compact spaces, the morphisms being continuous mappings (of a compact space into a compact space), and the α-mappings are continuous mappings of E into a compact space. The Σ-admissible subsets are again the *closed* subsets, and conditions (QM_{II}), (CU_I) through (CU_{III}) are easily verified (with the same cardinal as in Example V). The compact space F_E is (up to isomorphism) the "Stone-Čech compactification" obtained by completing E with respect to the coarsest uniformity for which all the continuous mappings of E into the interval [0, 1] of $\mathbf{R}$ are uniformly continuous (*General Topology*, Chapter IX, § 1, Exercise 7); the mapping φ_E is injective, because any two distinct points of E can be separated by a continuous mapping of E into [0, 1].

VII. *Free topological groups.* Let E be a completely regular space, let Σ be the species of Hausdorff topological group structures, the morphisms being continuous homomorphisms; and take the α-mappings to be the continuous mappings of E into a Hausdorff topological group. Conditions (QM_{II}) and (CU_I) to (CU_{III}) are easily verified, with

$$\mathfrak{a} = \text{Card (E) Card (N)}.$$

The Hausdorff topological group F_E which is the solution of this universal mapping problem for E is called the *free topological group generated by the space* E. Since any two distinct points of E can be separated by a continuous mapping of E into the Hausdorff topological group R, the mapping φ_E is injective; it can be shown that φ_E is a homeomorphism of E onto the subspace $\varphi_E(E)$ of F_E (*). Instead of taking Σ to be the species of structures of Hausdorff topological groups, we could also take other species of structures, such as those of Hausdorff topological abelian groups, compact groups, Hausdorff topological rings, Hausdorff topological vector spaces (over a topological division ring, considered as an auxiliary base set), etc.

VIII. *Almost periodic functions on a topological group.* Let E be a topological group. Take Σ to be the species of compact group structures, the morphisms being continuous homomorphisms, and the α-mappings being continuous homomorphisms of E into a compact group. Conditions (QM_{II}), (CU_I) through (CU_{III}) are satisfied, with $\mathfrak{a} = 2^{2^{\text{Card (E)}}}$. The compact group F_E which is the solution of this universal mapping problem for E is called the *compact group associated* with E; the mapping φ_E is not necessarily injective. Every continuous real-valued function on E, of the form $g \circ \varphi_E$, where g is a continuous real-valued function on E, is called an *almost periodic function* on E.

IX. *Albanese variety.* Let E be an algebraic variety, let Σ be the species of structures of abelian varieties over the same base field as E (an abelian variety is a complete algebraic variety endowed with an algebraic group structure; it is necessarily commutative). The morphisms are rational mappings of one abelian variety into another (each morphism is necessarily the composition of a homomorphism and a translation). The α-mappings are rational mappings of E into an abelian variety. Condition (CU_I) is not satisfied, yet this universal mapping problem for E admits a solution F_E, called the *Albanese variety* of E. In general, the rational mapping φ_E is not injective. *

(*) See P. Samuel, "On universal mappings and free topological groups", *Bull. Amer. Math. Soc.*, **54** (1948), pp. 591-598.

EXERCISES

§ 1

1. Let S be the set of signs P, X, $x_1, \ldots, x_n$, the letters x_i being of weight 0, P of weight 1, and X of weight 2. Let T be a balanced word in L(S). Such a word will be called an *echelon type* on $x_1, \ldots, x_n$.

Let $E_1, \ldots, E_n$ be n terms in a theory stronger than the theory of sets. For each echelon type T on $x_1, \ldots, x_n$ define a term $T(E_1, \ldots, E_n)$ as follows:

(i) If T is a letter x_i, then $T(E_1, \ldots, E_n)$ is the set E_i.

(ii) If T is of the form PU, then $T(E_1, \ldots, E_n)$ is the set

$$\mathfrak{P}(U(E_1, \ldots, E_n)).$$

(iii) If T is of the form XUV, where U and V are assemblies antecedent to T, then $T(E_1, \ldots, E_n)$ is the set $U(E_1, \ldots, E_n) \times V(E_1, \ldots, E_n)$.

Show that, for each echelon type T on $x_1, \ldots, x_n$, $T(E_1, \ldots, E_n)$ is an echelon on the terms $E_1, \ldots, E_n$, and conversely (proof by induction on the length of the echelon type or of the echelon construction scheme). $T(E_1, \ldots, E_n)$ is called the *realization of the echelon type* T on the terms $E_1, \ldots, E_n$. More precisely, every echelon on n distinct letters can be written *in exactly one way* in the form $T(x_1, \ldots, x_n)$, where T is an echelon type.

Also show how one may associate with an echelon type T on n letters, and n mappings $f_1, \ldots, f_n$, a canonical extension of these mappings. Deduce that if two echelon construction schemes S, S′ on n letters are such that $S(x_1, \ldots, x_n) = S'(x_1, \ldots, x_n)$ ($x_1, \ldots, x_n$ being distinct letters), then $\langle f_1, \ldots, f_n \rangle^S = \langle f_1, \ldots, f_n \rangle^{S'}$.

§ 2

1. Let S be the set of signs P, P⁻, X, X⁻, $x_1, \dots, x_n$, the letters x_i being of weight 0, P and P⁻ of weight 1, and X, X⁻ of weight 2. For each word A in L(S) we define the *variance* of A as follows: each letter x_i, and the signs P, X, have variance 0; the signs P⁻, X⁻ have variance 1; the variance of A is the sum * (in the field $\mathbf{F}_2$) * of the variances of the signs which appear in A (in other words, it is 0 if A contains an *even* number of signs of variance 1, and is 1 otherwise).

A *signed echelon type* is a balanced word A in L(S) (Chapter I, Appendix, no. 3) which satisfies one of the following two conditions: (1) A is one of the letters x_i; (2) if A is of the form $fA_1A_2 \dots A_p$ (Chapter I, Appendix, no. 3, Corollary 2 of Proposition 2), where f is one of the signs P, P⁻, X, X⁻, p is equal to 1 or 2 and the A_i $(1 \leqslant i \leqslant p)$ are balanced words, then each of the A_i is a signed echelon type; further, if $f = X$, then A_1 and A_2 have variance 0, and if $f = X^-$, then A_1 and A_2 have variance 1.

A signed echelon type is said to be *covariant* if it has variance 0, *contravariant* if it has variance 1. If in a signed echelon type A we replace P⁻ and X⁻ by P and X, respectively, we obtain an echelon type A* (§ 1, Exercise 1). Every realization of the echelon type A* on n terms $E_1, \dots, E_n$ is called a *realization of the signed echelon type* A on $E_1, \dots, E_n$, and is written $A(E_1, \dots, E_n)$.

Let $E_1, \dots, E_n, E'_1, \dots, E'_n$ be sets, and let f_i be a mapping of E_i into E'_i $(1 \leqslant i \leqslant n)$. Show that to each signed echelon type S on $x_1, \dots, x_n$ we may associate a mapping $\{f_1, \dots, f_n\}^S$ having the following properties:

(i) If S is covariant (resp. contravariant), then $\{f_1, \dots, f_n\}^S$ is a mapping of $S(E_1, \dots, E_n)$ into $S(E'_1, \dots, E'_n)$ (resp. of $S(E'_1, \dots, E'_n)$ into $S(E_1, \dots, E_n)$).

(ii) If S is a letter x_i, then $\{f_1, \dots, f_n\}^S$ is f_i.

(iii) If S is PT (resp. P⁻T), and if $g = \{f_1, \dots, f_n\}^T$ is a mapping of F into F′, then $\{f_1, \dots, f_n\}^S$ is the extension of g to sets of subsets (resp. the extension of $\overset{-1}{g}$ to sets of subsets).

(iv) If S is XTU or X⁻TU, if $\{f_1, \dots, f_n\}^T$ is a mapping $g: F \to F'$ and if $\{f_1, \dots, f_n\}^U$ is a mapping $h: G \to G'$, then $\{f_1, \dots, f_n\}^S$ is the extension $g \times h: F \times G \to F' \times G'$.

The mapping $\{f_1, \dots, f_n\}^S$ is called the *signed canonical extension* of $f_1, \dots, f_n$ corresponding to the signed echelon type S. If S is an echelon type (i.e., if P⁻ and X⁻ do not appear in S), then the signed canonical extension $\{f_1, \dots, f_n\}^S$ is equal to $\langle f_1, \dots, f_n \rangle^S$.

Show that, if f_i is a mapping of E_i into E'_i, and if f'_i is a mapping of E'_i into E''_i $(1 \leqslant i \leqslant n)$, then for a covariant signed echelon type S we have

$$\{f'_1 \circ f_1, \dots, f'_n \circ f_n\}^S = \{f'_1, \dots, f'_n\}^S \circ \{f_1, \dots, f_n\}^S,$$

and for a contravariant signed echelon type we have

$$\{f'_1 \circ f_1, \ldots, f'_n \circ f_n\}^S = \{f_1, \ldots, f_n\}^S \circ \{f'_1, \ldots, f'_n\}^S.$$

Deduce, that if f_i is a bijection of E_i onto E'_i and if f'_i is its inverse bijection ($1 \leqslant i \leqslant n$), then $\{f_1, \ldots, f_n\}^S$ is a bijection and $\{f'_1, \ldots, f'_n\}^S$ its inverse. Moreover, if S^* is the (unsigned) echelon type corresponding to the signed echelon type S, then $\{f_1, \ldots, f_n\}^S$ is equal to $\langle f_1, \ldots, f_n\rangle^{S^*}$ or to $\langle f'_1, \ldots, f'_n\rangle^{S^*}$ according as S is covariant or contravariant.

2. Let S be a signed echelon type on $n + m$ letters (Exercise 1). Let Σ be a species of structures having $x_1, \ldots, x_n$ as principal base sets and $A_1, \ldots, A_m$ as auxiliary base sets, whose typical characterization is of the form $s \in \mathfrak{P}(S(x_1, \ldots, x_n, A_1, \ldots, A_m))$. Show that we may define a notion of σ-morphism for this species of structures as follows: given n sets $E_1, \ldots, E_n$ endowed with a structure U of species Σ, n sets $E'_1, \ldots, E'_n$ endowed with a structure U′ of species Σ, and mappings $f_i : E_i \to E'_i$ ($1 \leqslant i \leqslant n$), then $(f_1, \ldots, f_n)$ is said to be a σ-morphism if the mappings f_i satisfy the following conditions:

(i) If S is a covariant echelon type, then

$$\{f_1, \ldots, f_n, I_1, \ldots, I_m\}^S\langle U\rangle \subset U'.$$

(ii) If S is a contravariant echelon type, then

$$\{f_1, \ldots, f_n, I_1, \ldots, I_m\}^S\langle U'\rangle \subset U.$$

Show that by assigning the variances suitably we may obtain in this way the definition of the morphisms for order structures, algebraic structures, and topological structures.

3. Let A, B, C be three sets endowed with structures of the same species Σ, let f be a morphism of A *onto* B, and let g be a morphism of B into C. Show that if $g \circ f$ is an isomorphism of A onto C, then g and f are isomorphisms.

4. Let A, B, C, D be four sets endowed with structures of the same species Σ, let f be a morphism of A into B, g a morphism of B into C, and h a morphism of C into D. Show that if $g \circ f$ and $h \circ g$ are isomorphisms, then f, g, and h are isomorphisms (cf. Chapter II, §3, Exercise 9).

5. Let A, B be two sets endowed with structures $\mathcal{S}$, $\mathcal{S}'$ of the same species Σ. Let f be a morphism of A into B, and let g be a morphism of B into A. Let M (resp. N) be the set of all $x \in A$ (resp. $y \in B$) such that $g(f(x)) = x$ (resp. $f((g)y) = y$). Suppose that $\mathcal{S}$ (resp. $\mathcal{S}'$) induces on M (resp. N) a structure of species Σ. Show that M and N, endowed with these structures, are isomorphic.

6. Let Σ be the species of structure having a principal base set A and an auxiliary set k, whose generic structure (F, H) has as typical characterization

$$F \in \mathfrak{P}((A\times A)\times A)\times\mathfrak{P}((A\times A)\times A)\times\mathfrak{P}((k\times A)\times A) \quad \text{and} \quad H\in\mathfrak{P}(A)$$

and whose axiom (which can be written more explicitly) is the following: "F is a structure of commutative k-algebra with identity element, and H is an irreducible ideal of the algebra A" (we recall that an irreducible ideal in a k-algebra A is an ideal $\mathfrak{a}$ such that, for any ideals $\mathfrak{b}$, $\mathfrak{c}$ of A satisfying $\mathfrak{b}\cap\mathfrak{c}=\mathfrak{a}$, either $\mathfrak{b}=\mathfrak{a}$ or $\mathfrak{c}=\mathfrak{a}$). If A, A′ are two sets endowed respectively with structures (F, H), (F′, H′) of species Σ, we define the σ-morphisms of A into A′ to be homomorphisms f of k-algebras, which map identity element to identity element, and satisfy $f(H)\subset H'$. Give an example of a family $(\mathcal{S}_\lambda)$ of structures of species Σ on a set A, such that there exists a least upper bound for this family of structures (in the ordered set of structures of species Σ on A) but such that this least upper bound is not an initial structure for the family $(A_\lambda, \mathcal{S}_\lambda, \mathrm{Id}_\lambda)$, where A_λ is the set A endowed with the structure $\mathcal{S}_\lambda$, and Id_λ is the identity mapping $A\to A_\lambda$. (Consider a polynomial ring $A=k[T]$: the irreducible ideals in A are powers of maximal ideals. If F is the k-algebra structure of A, consider the two structures (F, $\mathfrak{p}_1$) and (F, $\mathfrak{p}_2$) of species Σ, where $\mathfrak{p}_1$, $\mathfrak{p}_2$ are distinct maximal ideals of A; show that the least upper bound of these two structures is (F, (0)), but that this is not the initial structure for the family in question. To prove this, consider the k-subalgebra $B=k+(\mathfrak{p}_1\cap\mathfrak{p}_2)$ of A, in which $\mathfrak{p}_1\cap\mathfrak{p}_2$ is a maximal ideal, and the injection $B\to A$.)
¶ Also give an example of a species of structure Σ', and a family $(\mathcal{S}'_\lambda)$ of structures of species Σ' on a set A′, such that the greatest lower bound of this family of structures exists but is not the final structure for the family. (On the dual of A (considered as a vector space) consider the structure of linearly compact k-coalgebra deduced from the k-algebra structure of A by transposition, and define the species of structures Σ' by transposition.)

* 7. Let Σ be the species of structures which has a principal base set A, and the set $\mathbf{R}$ as auxiliary base set, whose generic structure consists of two letters (V, φ) with the typical characterization

$$V\in\mathfrak{P}(\mathfrak{P}(A)) \quad \text{and} \quad \varphi\in\mathfrak{P}(\mathbf{R}\times A),$$

and whose axioms are (i) the axiom $R\{V\}$ of the species of topological structures (§ 1, no. 4, Example 3), (ii) the relation "there exists $a>0$ such that φ is the graph of a continuous injective mapping (with respect to the topology V) of the interval $[0, a]$ into A".

If A, A′ are two sets endowed respectively with structures (V, φ), (V′, φ') of the species Σ, the σ-morphisms of A into A′ are defined to be the mappings f of A into A′ which are continuous (with respect to the topologies V, V′) and whose graph F is such that $F \circ \varphi \subset \varphi'$. Show that this notion of σ-morphisms can be defined by the procedure of Exercise 2, and that there exists a product structure on the product of two sets A_1, A_2 endowed with arbitrary structures of species Σ. But give an example in which the direct image, under the first projection pr_1, of the product structure on $A_1 \times A_2$ is not the original structure given on A_1 (take A_1 to be a space homeomorphic to **R**). *

8. Let Σ be the species of structures which has a single (principal) base set, whose generic structure (V, a, b) consists of three letters, with the typical characterization

$$V \in \mathfrak{P}(\mathfrak{P}(A)) \quad \text{and} \quad a \in A \quad \text{and} \quad b \in A$$

and whose axiom is the relation

$$R\{V\} \quad \text{and} \quad a \neq b,$$

where $R\{V\}$ is the axiom of the species of topological structures (§ 1, no. 4, Example 3). If A, A′ are two sets endowed respectively with structures (V, a, b), (V', a', b') of species Σ, the σ-morphisms of A into A′ are defined to be the continuous mappings $f: A \to A'$ (with respect to the topologies V, V′) such that $f(a) = a'$ and $f(b) = b'$. Show that by replacing Σ by an equivalent species of structures, we can obtain this notion of σ-morphisms by the procedure of Exercise 2. Show that if A, B are two sets endowed with structures S, S′ of species Σ, there exists a product structure on A × B, and that the direct image of this structure under pr_1 (resp. pr_2) is S (resp. S′). * But give an example where there is no section of pr_1 which is a σ-morphism of A into A × B (take A to be connected and B discrete). *

* 9. Let Σ be the species of division ring structures. Show that a notion of σ-morphisms for this species of structures may be defined by taking the morphisms of a division ring K into a division ring K′ to be the homomorphisms f of K into K′ in the sense of Example 2 of no. 4, together with the mapping f_0 of K into K′ for which $f_0(0) = 0$ and $f_0(x) = 1$ for all $x \neq 0$ in K. Show that, with this notion of morphisms, we have the following properties: for every division ring K (of any characteristic) the division ring structure of K induces a division ring structure (isomorphic to that of $\mathbf{F}_2$) on the set $\{0, 1\}$; and that if R is the equivalence relation whose equivalence classes are $\{0\}$ and $K^* = K - \{0\}$, there exists a quotient structure (isomorphic to that of $\mathbf{F}_2$) of the structure of K by the relation R. *

*10. Let Σ be the species of structures of a complete Archimedean ordered field. For each set A endowed with a structure of species Σ, let φ_A be the unique isomorphism of A onto **R**. If A, B are two sets endowed with structures of species Σ, show that the morphisms of A into B may be taken to be the mappings f of A into B such that $\varphi_B(f(x)) \geqslant \varphi_A(x)$ for all $x \in A$. For this notion of morphisms show that there exist bijective morphisms which are not isomorphisms, although the species Σ is univalent. *

11. Let Σ be a species of structures (in a theory $\mathfrak{T}$) which has only one base set. Let $s \in F(x)$ be the typical characterization, and $R\{x, s\}$ the axiom of Σ. Let $A(x)$ denote the set of structures of species Σ on x. Let $\sigma\{x, y, s, t\}$ be a term which satisfies the conditions (MO_I), (MO_{II}), and the following condition :

(MO'_{III}) Given two sets E, E' (in a theory $\mathfrak{T}'$ which is stronger than $\mathfrak{T}$) endowed with structures $\mathcal{S}$, $\mathcal{S}'$ of species Σ, respectively, if f is any isomorphism of E onto E', then $f \in \sigma\{E, E', \mathcal{S}, \mathcal{S}'\}$.

(a) Let I_x be the identity mapping of x onto itself. Show that the relation $Q\{x, s, t\}$:

$$s \in A(x) \quad \text{and} \quad t \in A(x) \quad \text{and} \quad I_x \in \sigma\{x, x, s, t\} \cap \sigma\{x, x, t, s\}$$

is an equivalence relation between s and t on $A(x)$. Let $B(x)$ be the quotient set $A(x)/Q$, and let φ_x be the canonical mapping of $A(x)$ onto $B(x)$. Suppose that the relation $s' \in B(x)$ is transportable (see the Appendix for conditions which will ensure that this is so), and let Θ denote the species of structures with typical characterization $s' \in \mathfrak{P}(F(x))$ and axiom $s' \in B(x)$.

(b) Let $\overline{\sigma}\{x, y, s', t'\}$ be the set of mappings $f \in \mathcal{F}(x, y)$ which satisfy the following relation : "$s' \in B(x)$ and $t' \in B(y)$ and there exist $s \in A(x)$ and $t \in A(y)$ such that $s' = \varphi_x(s)$ and $t' = \varphi_y(t)$ and $f \in \sigma\{x, y, s, t\}$". Show that, for the species of structures Θ, the term $\overline{\sigma}$ satisfies (MO_I), (MO_{II}), and (MO_{III}) and that we have

$$\sigma\{x, y, s, t\} \subset \overline{\sigma}\{x, y, \varphi_x(s), \varphi_y(t)\}.$$

§ 3

* 1. Let E be a topological space and let Σ be one of the species of structures defined in Exercises 7 and 8 of § 2. Take the morphisms to be those defined in the same exercises, and the α-mappings to be the continuous mappings of E into a Σ-set. Show that the universal mapping problem for E (relative to the preceding definitions) has in general no solution. *

* 2. Let E be a field, Σ the species of algebraically closed field structures. Take the morphisms to be homomorphisms, and the α-map-

pings to be homomorphisms of E into an algebraically closed field. Show that the algebraic closure F_E of E and the canonical injection of E into F_E satisfy (AU'_I), but that there is in general no solution to the universal mapping problem for E. *

3. Let Σ be a species of structures, let $(A_\iota)_{\iota \in I}$ be a family of sets, and for each $\iota \in I$ let $\mathcal{S}_\iota$ be a structure of species Σ on A_ι. Let E be the set which is the sum of the family $(A_\iota)_{\iota \in I}$, each A_ι being considered as a subset of E. Suppose that we are given a notion of σ-morphisms for the species Σ, and define an α-mapping to be a mapping φ of E into a Σ-set F such that, for each $\iota \in I$, the restriction of φ to A_ι is a morphism of A_ι into F. Show that, if there exists a solution (F_E, φ_E) to the universal mapping problem for E, then the structure of species Σ on F_E is the *final structure* with respect to the family $(A_\iota, \mathcal{S}_\iota, \varphi_\iota)_{\iota \in I}$, where φ_ι denotes the restriction of φ_E to A_ι.

Furthermore, let F be a set, and for each $\iota \in I$ let f_ι be a mapping of A_ι into F. If there exists a final structure of species Σ on F with respect to the family $(A_\iota, B_\iota, f_\iota)_{\iota \in I}$, then we may write $f_\iota = f \circ \varphi_\iota$, where f is a morphism of F_E into F, and the structure on F is the *direct image* under f of the structure on F_E.

Consider the applications to the following cases :

* (i) Σ is a species of algebraic structures, the morphisms being homomorphisms, and the conditions (CU_I) through (CU_{III}) are satisfied. This is so for semi-group structures, group structures, module structures, algebra structures, etc. The Σ-set F_E is called the *free product* of the A_ι in the case of groups, *direct sum* in the case of modules, *direct product* in the case of algebras.

(ii) Σ is the species of topological group structures or the species of topological vector space structures. The conditions (CU_I) through (CU_{III}) are then satisfied. In the case of locally convex topological vector spaces, F_E is called the *topological direct sum* of the A_ι. *

HISTORICAL NOTE
(Chapters I through IV)

"Aliquot selectos homines rem intra quinquennium absolvere posse puto"
Leibniz ([12 b], vol. 7, p. 187).

(Numbers in brackets refer to the bibliography at the end of this Note.)

The study of what is usually called the "foundations of mathematics", which has been unflaggingly pursued since the beginning of the nineteenth century, could not have achieved success without a parallel effort of systematization of logic, or at least of those parts of logic which govern the relationships of mathematical propositions. Thus the history of the theory of sets and the formalization of mathematics cannot be separated from that of "mathematical logic". But traditional logic, like that of the modern philosophers, extends in principle over a field of applications far wider than mathematics. The reader will therefore not find a history of logic, even in a summary form, in this Note; we have limited ourselves, so far as possible, to retracing the evolution of logic only in so far as it has influenced the evolution of mathematics. Thus we shall say nothing about non-classical logics (many-valued logics, modal logics); *a fortiori*, we shall not go into the history of the controversies which, from the time of the Sophists to that of the Vienna school, have divided philosophers as to the possibility and manner of applying logic to the objects of the external world or the concepts of the human mind.

¶ Today it is well established that mathematics was already highly developed in pre-Hellenic times. Not only are the (extremely abstract) notions of integers and measurement of magnitudes freely used in the most ancient documents which have come down to us from Egypt and Chaldaea, but the algebra of the Babylonians, by the elegance and sureness of its methods, cannot be regarded as simply a collection of problems solved by empirical devices. And, although there is nothing in the texts resembling a "proof" in the formal sense of the word, we may justly consider that the discovery of such methods of solution, whose generality is apparent from the particular numerical applications treated, could not have been achieved without a minimum of logical reasoning (perhaps not entirely conscious, but rather of the type on which a modern algebraist relies when undertaking a calculation, before writing out all the details) ([1], pp. 203 ff.).

The originality of the Greeks consists precisely in a conscious effort of arranging mathematical proofs in a sequence so that the passage from one step to the next leaves nothing in doubt and compels universal assent. Of course, the Greek mathematicians, just like their present-day successors, made use of "heuristic" rather than rigorous arguments in the course of their researches; for example, Archimedes, in his *Treatise on Method* [4 *bis*], refers to results "found, but not proved" by earlier mathematicians (*). But, from the earliest detailed texts known to us (which date from the middle of the fifth century B.C.) the ideal "canon" of a mathematical text is fixed. It found its complete realization in the works of the great masters Euclid, Archimedes, and Apollonius. Their notion of proof differs in no respect from ours.

We do not possess any text which would allow us to trace the first steps of this "deductive method". At its first known appearance, it is already nearly perfected. We can only think that it was a natural development from the perpetual search for "explanations" of the world, which characterized Greek thought from the time of the Ionian philosophers of the seventh century B.C. onward; furthermore, tradition is unanimous in ascribing the development and perfection of the method to the Pythagorean school, at a period somewhere between the end of the sixth century and the middle of the fifth.

This "deductive" mathematics, precise in its aims and rigorous in its methods, was to occupy the philosophical and mathematical reflections of the succeeding ages. We see, on the one hand, the edifice of "formal" logic slowly being built up, on the model of mathematics, and culminating in the creation of formalized languages; and on the other hand, principally from the beginning of the nineteenth century onward, especially after the advent of the theory of sets, the concepts which lie at the root of mathematics being more and more closely questioned and examined in an effort to clarify their nature.

(*) Notably Democritus, to whom Archimedes attributes the discovery of the formula for the volume of a pyramid ([4 *bis*], p. 13). This allusion should be compared with a celebrated fragment ascribed to Democritus (but of doubtful authenticity) in which he declares : "*nobody has ever surpassed me in the construction of figures by means of proofs, not even the Egyptian so-called 'harpedonaptes'*" (H. Diels, *Die Fragmente der Vorsokratiker*, 2nd edition, vol. I, p. 439 and vol. II.1, pp. 727-728, Berlin (Weidmann), 1906-1907). Archimedes' remark, and the fact the Egyptian texts which have come down to us contain no proofs (in the classical sense), lead us to think that the " proofs " to which Democritus alludes were no longer considered to be proofs in the classical period, and would not be so considered today.

The formalization of logic

The general impression that we get from the (extremely incomplete) texts in our possession is that Greek philosophic thought of the fifth century B.C. was dominated by an increasingly conscious effort to extend to the whole field of human thought the procedures of articulate discourse which were put to use with such great success in rhetoric and contemporary mathematics — in other words, to create Logic, in the most general sense of the word. The tone of philosophical writings underwent an abrupt change in this epoch. Whereas in the seventh and sixth centuries the philosophers asserted and prophesied (or, at best, sketched vague arguments, founded on equally vague analogies), from Parmenides and especially Zeno onward they reasoned and sought to extract general principles to serve as a basis for their dialectic. The first statement of the principle of the excluded middle is to be found in the work of Parmenides, and the proofs "by contradiction" of Zeno of Elea have remained famous. But Zeno flourished in the middle of the fifth century and, in spite of the uncertainty of documentation (*), it is very likely that at this period the mathematicians also used these principles in their own sphere.

As we have already said, it is not within the scope of this book to relate the innumerable difficulties which arose at each stage in the development of this Logic, and the resulting polemics which occupied the Eleatic school, and the Sophists, as well as Plato and Aristotle. Let us merely note the part played in this evolution by the assiduous culture of the art of oratory and the analysis of language which naturally followed from it — developments which are attributed mainly to the Sophists of the fifth century. On the other hand, the influence of mathematics, if not always explicitly acknowledged, is no less clear, in particular in the writings of Plato and Aristotle. It could be said that Plato was almost obsessed by mathematics; although himself not an inventor in this domain, he kept himself informed of the discoveries of the contemporary mathematicians (many of whom were his friends or his pupils), and remained always most directly interested in the subject, to the extent of suggesting new directions for research. In his writings, it is mathematics which constantly serves as illustration or model (and on occasion, as with the Pythagoreans, feeds his penchant toward mysticism). Aristotle, as Plato's pupil, could hardly have failed to receive the minimum of mathematical education demanded of the pupils at the Academy, and a volume of selections from his writings which relate or

---

(*) The most beautiful classical example of *reductio ad absurdum* in mathematics is the proof that $\sqrt{2}$ is irrational, to which Aristotle alludes on several occasions. But the experts have not succeeded in assigning any precise date to this discovery; some put it at the beginning and others right at the end of the fifth century B.C. (see the Historical Note to *General Topology*, Chapter IV, and the references cited there).

allude to mathematics has been compiled [2 *bis*]. However, he seems never to have made much effort to keep in contact with the mathematical movements of his times, and in this domain he quotes only results which had long since become common knowledge. The later philosophers were, for the most part, even less well acquainted with mathematics. Many of them lacked the necessary technical knowledge, and their references to mathematics relate to stages long since passed in the evolution of the subject.

The culmination of this period, in so far as logic is concerned, is the monumental work of Aristotle [2], whose great merit is to have succeeded for the first time in systematizing and codifying methods of reasoning which had hitherto remained vague or unformulated (*). For our purpose we need to keep in mind the general thesis of this work, namely that it is possible to reduce every correct argument to the systematic application of a small number of immutable rules which are independent of the particular nature of the objects under consideration (this independence is clearly brought out by the use of letters to denote concepts or propositions — a usage which Aristotle very probably borrowed from the mathematicians). But Aristotle concentrated his attention almost exclusively on a particular type of relations and chains of reasoning, namely what he called "syllogisms"; essentially, these are relations of the form $A \subset B$ or $A \cap B \neq \emptyset$, in the language of set theory (†), and chains of such relations or their

(*) In spite of the simplicity and "obviousness" to our minds of the logical rules formulated by Aristotle, it is only necessary to place them in their historical context for us to appreciate the difficulties in the way of a precise conception of these rules, and to understand the effort which Aristotle must have required to surmount these difficulties. Plato, in his dialogues, which were addressed to an educated public, allows his characters to become entangled in questions as elementary as the relationship between the negation of $A \subset B$, and $A \cap B = \emptyset$ (in modern language) before producing the correct answer (cf. R. Robinson, "Plato's consciousness of fallacy", *Mind*, vol. 51 (1942), pp. 97-114).

(†) The corresponding statements in Aristotle are "every A is B" and "some A is B". In this notation, A (the "subject") and B (the "predicate") stand for concepts, and to say that "every A is B" means that we can attribute the concept B to every entity to which the concept A can be attributed (A is the concept "man" and B the concept "mortal" in the classical examples). The interpretation which we give in this series consists in considering the sets of entities to which the concepts A and B respectively apply; this is the "extensional" point of view and was already known to Aristotle. But Aristotle considers the relation "every A is a B" from another point of view, the "comprehensional", in which B is envisaged as one of the concepts which in some sense constitute the more complex concept A, or, as Aristotle says, "belong" to A. At first sight, these two points of view seem equally natural; but the "comprehensional" point of view has been a constant source of difficulties in the development of logic (it seems more remote from intuition than the former, and leads rather easily to errors, especially in schemes which involve negations; cf. [12 *bis*], pp. 21-32).

negations linked together by means of the scheme

$$(A \subset B \text{ and } B \subset C) \Longrightarrow (A \subset C).$$

Aristotle was too well acquainted with the mathematics of his time not to have realized that schemes of this nature were not sufficient to account for all the logical operations used by mathematicians, nor *a fortiori* for other applications of logic ([2], *Analytica Priora*, I, 35; [2 *bis*], pp. 25-26) (*). At any rate, he undertook a close study of the various forms of "syllogism" (almost entirely devoted to the elucidation of the perpetual difficulties which arise from the ambiguity or obscurity of the terms involved in reasoning), which led him to formulate, among other things, rules for taking the negation of a proposition ([2], *Analytica Priora*, I, 46). To Aristotle also belongs the merit of having distinguished with great clarity the role of "universal" propositions from that of "particular" propositions: a first step toward quantifiers (†). But it is only too well known how the influence of his writings — often narrowly and unintelligently interpreted — which remained considerable up to the nineteenth century, led philosophers to neglect the study of mathematics and impeded the progress of formal logic (**). Nevertheless, logic continued to make progress in antiquity, in the Megarian and Stoic schools, the rivals of the Peripatetics. Our knowledge of these doctrines is unfortunately all at second hand, often derived from their adversaries or mediocre commentators. It seems that the contribution of these logicians was the foundation of a "propositional calculus" in the modern sense of the word; instead of limiting themselves, like Aristotle, to propositions of the particular form $A \subset B$, they stated rules concerning entirely *indeterminate* propositions. Moreover, they analyzed so closely the logical relationships between these rules that they were able to deduce them all from five elementary rules, which were classed as "unprovable", by methods very similar to those we described in Chapter I, §§ 2 and 3 [5]. Unfortunately their influence was ephemeral, and their results lapsed into oblivion until rediscovered by the logicians of the nineteenth century. Aristotle remained the unchallenged master, in logic, until the seventeenth century. In particular, the scholastic philosophers were entirely under his influence, and although their contribution to formal logic is far from negligible, it contains no advance of the first order in comparison with that due to the philosophers of antiquity.

(*) For a critical discussion of the syllogism and its inadequacies, see, for example, [12 *bis*], pp. 432-441, or [32], pp. 44-50.

(†) The absence of genuine quantifiers (in the modern sense) up to the end of the nineteenth century was one of the causes of the stagnation of formal logic.

(**) We may mention that during a recent lecture at Princeton an eminent academic is reported to have said in the presence of Gödel that nothing new had been done in logic since Aristotle!

However, it does not appear that the works of Aristotle or his successors caused any notable repercussions in mathematics. The Greek mathematicians pursued their research along the paths opened by the Pythagoreans and their successors of the fourth century (Theodorus, Theaetetus, Eudoxus) without apparently concerning themselves with formal logic in the presentation of their results. This is hardly surprising when one compares the flexibility and precision of mathematical reasoning at this period with the extremely rudimentary state of Aristotelian logic. And, later, when logic passed beyond this stage, its evolution was guided by the new conquests of mathematics.

With the development of algebra, one could hardly fail to be struck by the similarity between the rules of formal logic and those of algebra, both being applicable to unspecified objects (propositions or numbers). And when, in the seventeenth century, algebraic notation took its definitive form in the hands of Vieta and Descartes, almost immediately there appeared various attempts to represent logical operations symbolically. But before Leibniz these attempts (as for example, that of Hérigone (1644) to write symbolically the proofs of elementary geometry, or that of Pell (1659) to do the same for arithmetic) were superficial and did not lead to any progress in the analysis of mathematical reasoning.

Leibniz was a philosopher who was also a mathematician of the first rank, and who was able to draw from his mathematical experience the germs of the ideas which were to release formal logic from the scholastic impasse (*). A universal genius if ever there was one, and an inexhaustible source of original and fertile ideas, Leibniz was all the more interested in logic because it lay at the heart of his grand projects of formalizing language and thought, at which he continued to work throughout his life. Having broken with scholastic logic already in his childhood, he was attracted by the idea (going back to Raymond Lulle) of a method of resolving all human concepts into primitive concepts, which thus constituted an "alpha-

(*) Although Descartes and (to a lesser extent) Pascal had devoted part of their philosophical works to the foundations of mathematics, their contribution to the progress of formal logic was negligible. Undoubtedly the reason for this limitation is to be found in the fundamental trend of their thoughts, which was to emancipate themselves from the scholastic yoke and which led them to reject everything connected with scholastic philosophy, and first of all formal logic. In his *Réflexions sur l'esprit géométrique,* Pascal essentially limited himself — as he himself realized — to fashioning into striking formulae the known principles of Euclid's proofs (for example, the famous precept "*Substituer toujours mentalement les définitions à la place des définis*" ([11], vol. IX, p. 280) had already been known to Aristotle ([2], *Topics*, VI, 4; [2 *bis*], p. 87)). As to Descartes, the rules of reasoning which he put forward were, above all, psychological precepts (and rather vague ones) rather than logical criteria; and as Leibniz says ([12 *bis*], pp. 94 and 202-203), their validity is therefore only subjective.

bet of human thoughts", and recombining them in a quasi-mechanical fashion so as to obtain all true propositions ([12 b], vol. VII, p. 185; cf. [12 *bis*], Chap. II). When still a very young man, he also conceived another, much more original idea, that of the utility of symbolic notation as a "thread of Ariadne" of thought (*). "*The true method*", he said "*should provide us with a* filum Ariadnes, *that is to say, a certain sensible material means which will lead the intellect, like the lines drawn in geometry and the forms of operations which are prescribed to novices in arithmetic. Without this, our minds will not be able to travel a long path without going astray*" ([12 b], vol. VII, p. 22; cf. [12 *bis*], p. 90). He had little knowledge of the mathematics of his time until the age of twenty-five, and first presented his projects in the form of a "universal language". But when he came into contact with algebra, he adopted it as the model for his "universal characteristic". By this phrase he meant a sort of symbolic language, capable of expressing all human thoughts unambiguously, of strengthening our powers of deduction, of avoiding errors by a purely mechanical effort of attention, and constructed in such a way that "*les chimeres, que celuy meme qui les avance n'entend pas, ne pourrount pas estre ecrites en ces caracteres*" ([12 a], vol. I, p. 187). In countless passages in his writings where Leibniz alludes to this grandiose project and the progress which its achievement would bring (cf. [12 *bis*], Chapters IV and VI), one sees how clearly he conceived the notion of a formalized language, as a pure combination of signs in which only the order of writing is important (†), so that a machine would be capable of producing all the theorems (**) and all controversies could be resolved by a simple calculation ([12 *bis*], vol. VII, pp. 198-203). Although these hopes may appear extravagant, it is nonetheless true that this constant theme in Leibniz' thought underlies a good part of his mathematical output, beginning with his work on the symbolism of infinitesimal calculus. He himself was perfectly aware of this fact and explicitly associated his ideas on indicial notation and determinants ([12 a], vol. II, p. 240; cf. [12 *bis*], pp. 481-487) and his sketch of "geometrical calculus" (cf. [12 *bis*], Chapter IX) with his "characteristic". In his mind, the essential part of his scheme was to be symbolic logic or, in his words, a "Calculus ratiocinator"; and if he did not succeed in creating this calculus, he made at least three attempts at it. In the first attempt he had the idea of associating with each "prim-

(*) Of course, the usefulness of such a symbolism had not escaped the predecessors of Leibniz so far as mathematics was concerned. Descartes, for example, recommended the replacement of whole figures "*by very short signs*" (XVIe Règle pour la direction de l'esprit; [10], vol. X, p. 454). But no one before Leibniz had insisted so strongly on the universal range of application of this principle.

(†) It is striking that Leibniz cites, as examples of "formal" reasoning, "*un compte de receveur*" or even a legal text ([12 b], vol. IV, p. 295).

(**) This conception of a "logical machine" is used nowadays in metamathematics with great effect ([48], Chapter VIII).

itive" term a prime number, every term formed of several primitive terms being represented by the product of the corresponding prime numbers (*); he sought to translate into this system the usual rules of syllogism, but met considerable complications caused by negation (which he attempted, naturally enough, to represent by a change of sign) and soon abandoned this approach ([12 c], pp. 42-96; cf. [12 *bis*], pp. 326-344). In later attemps, he sought to give Aristotelian logic a more algebraic form. Sometimes he retained the notation AB for the conjunction of two concepts, sometimes he used the notation A + B (†). He observed (in the multiplicative notation) the law of idempotence AA = A, noted that the proposition "every A is B" may be replaced by the equality A = AB, and that from this starting-point most of the rules of Aristotle may be obtained by purely algebraic calculation ([12 c], pp. 229-237 and 356-399; cf. [12 *bis*], pp. 345-364). Also he had the idea of the empty concept ("non Ens") and recognized, for example, the equivalence of the propositions "every A is B" and "A. (not B) is not true" (*loc. cit.*). Furthermore, he noted that his logical calculus applied not only to the logic of concepts but also to that of propositions ([12 c], p. 377). Thus he came very close to "Boolean calculus". Unfortunately, he seems not to have succeeded completely in casting off the scholastic influence; not only did he limit the aim of his calculus almost entirely to the transcription of the rules of syllogism in his notation (**), but he went as far as sacrificing his most felicitous ideas to the desire of recovering the Aristotelian rules in their entirety, even those which are incompatible with the notion of the empty set (††).

The work of Leibniz remained, for the most part, unpublished until the beginning of the twentieth century, and had little direct influence. Throughout the eighteenth century and the beginning of the nineteenth, various authors (de Segner, Lambert, Ploucquet, Holland, De Castillon, Gergonne) gave outlines of attempts which were similar to those of Leibniz but which never really passed beyond the point at which he had halted. Their works made very little impression, and most of them were unaware of

(*) This idea has been taken up successfully, in a slightly different form, by Gödel in his metamathematical works (cf. [44a] and [48], p. 254).

(†) Leibniz does not attempt to introduce disjunction into his calculus, except in one or two fragments (where it is denoted by A + B), and seems not to have succeeded in simultaneously handling this operation and conjunction in a satisfactory manner ([12 *bis*], p. 363).

(**) Leibniz knew perfectly well that Aristotelian logic was insufficient to translate mathematical texts into formal language, but in spite of various attempts he never succeeded in improving it in this respect ([12 *bis*], pp. 435 and 560).

(††) These are the so-called "rules of conversion", based on the postulate that "every A is a B" implies "some A is a B", which of course presupposes that A is not empty.

the results of their predecessors (*). Another who wrote under the same conditions was G. Boole, who can be considered as the true creator of modern symbolic logic [16]. His master idea was to take the "extensional" point of view systematically and thus to calculate directly with sets, denoting by xy the intersection of two sets, and by $x+y$ their union when x and y are disjoint. He also introduced a "universe", denoted by 1 (the set of all objects), and the empty set, denoted by 0, and he wrote $1-x$ for the complement of x. As Leibniz had done, he interpreted the relation of inclusion by the relation $xy = x$ (from which he easily deduced the justifications of the rules of the classical syllogism), and his notation for unions and complements gave his system a flexibility which its predecessors had lacked (†). Moreover, by associating with each proposition the set of "cases" in which it is satisfied, he interpreted the relation of implication as an inclusion, and his calculus of sets gave him in this way the rules of "propositional calculus".

In the second half of the nineteenth century, Boole's system became the basis of the work of an active school of logicians, who improved and completed it in various respects. Thus Jevons (1864) widened the sense of the operation of union $x+y$ by extending it to the case where x and y are arbitrary. A. De Morgan in 1858, and C. S. Peirce in 1867, proved the duality relations

$$(\complement A) \cap (\complement B) = \complement(A \cup B), \qquad (\complement A) \cup (\complement B) = \complement(A \cap B) \quad (**).$$

In 1860 De Morgan also began the study of relations, and defined inversion and composition of binary relations (i.e., the operations which correspond to the operations $\overset{-1}{G}$ and $G_1 \circ G_2$ on graphs) (‡). All this work was

(*) The influence of Kant, from the middle of the eighteenth century onward, is without doubt partly responsible for the lack of interest in formal logic at this period. Kant thought that "*we have no need of any new invention in logic*", the form given to it by Aristotle being sufficient for all possible applications (*Werke*, vol. VIII, Berlin (B. Cassirer), 1923, p. 340). For the dogmatic conceptions of Kant on mathematics and logic, consult the article of L. Couturat, *Revue de Métaph. et de Morale*, vol. 12 (1904) pp. 321-383.

(†) Note in particular that Boole uses the distributivity of intersection with respect to union, which seems to have been noticed for the first time by J. Lambert.

(**) It should be noted that statements equivalent to these rules are to be found in the writings of certain scholastic philosophers ([6], pp. 67-ff).

(‡) However, the notion of the "cartesian" product of two arbitrary sets seems to have been first introduced by G. Cantor ([25], p. 286), who was also the first to define exponentiation A^B (*loc. cit.*, p. 287). The general notion of infinite product is due to A. N. Whitehead (*Amer. Jour. of Math.* **24** (1902), p. 369). The use of graphs of relations is fairly recent; apart, of course, from the classical case of real-valued functions of real variables, they appeared first in the work of the Italian geometers, especially C. Segre, in the study of algebraic correspondences.

systematically expounded and developed in the massive and prolix volumes of Schröder [23]. But it is rather curious to observe that the logicians we have just mentioned seemed scarcely interested in the application of their results to mathematics and that, on the contrary, the principal aim of Boole, and Schröder in particular, was apparently to develop "Boolean" algebra by imitating the methods and duplicating the problems of classical algebra (often in a rather artificial fashion). No doubt the reason for this attitude lies in the fact that Boolean calculus was still not suitable for transcribing the majority of mathematical arguments (*) and therefore was only a very incomplete answer to the grand dream of Leibniz. The construction of formalisms better adapted to mathematics — the introduction of variables and quantifiers, due independently to Frege [24] and C. S. Peirce [22 b], the capital step forward in this direction — was the work of logicians and mathematicians who, in contrast to their predecessors, were primarily interested in applications to the foundations of mathematics.

Frege's project [24 b and c] was to base arithmetic on logic formalized by means of a symbolism to which he gave the name *Begriffsschrift* (concept-script); we shall return later to the method by which he defined the natural integers. His publications were characterized by extreme precision and minute care in the analysis of concepts, which led him to introduce many distinctions that have since shown themselves to be of great importance in modern logic. For example, he was the first to distinguish between the statement of a proposition and the assertion that the proposition is true; between the relation of membership and that of inclusion; between an object x and the set $\{x\}$ whose only member is x, etc. This formalized logic, which contains not only "variables" in the mathematical sense of the word but also "propositional variables" which represent undetermined relations and are susceptible of quantification, was later (through the work of Russell and Whitehead) to provide the fundamental tool of metamathematics. Unfortunately the notation adopted by Frege was unsuggestive, of appalling typographical complexity, and far removed from current mathematical usage; and the resulting unreadability had the effect of considerably diminishing his influence on his contemporaries.

Peano's aim was both more ambitions and more down-to-earth. He set about publishing a *Formulaire de mathematiques* written entirely in formalized language, and containing not merely mathematical logic but all the results of the most important branches of mathematics. The speed with which he succeeded in completing this ambitious project, assisted by a constellation of enthusiastic collaborators (Vailati, Pieri, Padoa, Vacca,

(*) For *each* relation obtained from one or more given relations by applying our quantifiers it would be necessary, in this calculus, to introduce an *ad hoc* notation of the type $\overset{-1}{\mathrm{G}}$ and $\mathrm{G}_1 \circ \mathrm{G}_2$ (cf. for example [22 b]).

Vivanti, Fano, Burali-Forti) bears witness to the excellence of the system he had adopted. Closely following current mathematical usage, and introducing many well-chosen abbreviating symbols, his language succeeded moreover in being fairly readable, thanks mainly to an ingenious system of replacing brackets by separating points [29 c]. Many of the notations due to Peano are now in common use by the majority of mathematicians : for example, $\in$, $\supset$ (but, contrary to the present usage, in the sense of "is contained" or "implies" (*)), $\cup$, $\cap$, $A - B$ (the set of differences $a - b$, where $a \in A$ and $b \in B$). Besides this, the *Formulaire* contains, for the first time, a close analysis of the general notion of a function, of direct (†) and inverse images, and the remark that a sequence is just a function defined on **N**. But in Peano's hands quantification is subjected to irksome restrictions (in his system, in principle, only relations of the form $A \Longrightarrow B$, $A \Longleftrightarrow B$, or $A = B$ can be quantified). Moreover, the almost fanatic zeal of some of his disciples laid him wide open to ridicule, and the often unjust criticism of H. Poincaré in particular was a serious blow to Peano's school and hindered the dissemination of his doctrines in the mathematical world.

With Frege and Peano we have the essential ingredients of the formalized languages in use today. The most famous of such languages is, without doubt, that created by Russell and Whitehead in their great work *Principia Mathematica*, which happily combines the precision of Frege with the convenience of Peano. Most of the formalized languages in current use differ from each other only in modifications of secondary importance whose object is to simplify their use. One of the most ingenious is the "functional" notation for relations (for example, $\in xy$ in place of $x \in y$), due to Lukasiewicz, which does away completely with the need for brackets. But the most interesting is certainly the introduction by Hilbert of the symbol τ, which allows one to consider the quantifiers $\exists$ and $\forall$ as abbreviating symbols, to avoid the "universal" functional symbol ι of Peano and Russell (which applies only to functional relations) and to dispense with the axiom of choice in the theory of sets ([31], p. 183 (**)).

The notion of truth in mathematics

Mathematicians have always been convinced that what they prove is "true". It is clear that such a conviction can be only of a sentimental

(*) This indicates how deeply rooted, even in Peano, was the old habit of thinking in terms of "comprehension" rather than "extension".

(†) The introduction of this appears to be due to Dedekind, in his work *Was sind und was sollen die Zahlen*, which we shall come to later ([26], vol. III, p. 348).

(**) It should be observed that what Hilbert denotes by $\tau_x(A)$ in this context is denoted by $\tau_x(\text{not } A)$ in Chapter I.

or metaphysical order, and cannot be justified, or even ascribed a meaning which is not tautological, within the domain of mathematics. The history of the concept of truth in mathematics therefore belongs to the history of philosophy and not of mathematics; but the evolution of this concept has had an undeniable influence on the development of mathematics, and for this reason we cannot pass over it in silence.

First of all, a mathematician who is well schooled in philosophy is as rare as a philosopher with a wide knowledge of mathematics. The opinions of mathematicians on philosophical questions, even questions which touch on their own science, are most often second- or third-hand and from sources of doubtful value. But, precisely for this reason, these "average" opinions are of as much interest to the historian of mathematics as the original views of thinkers such as Descartes or Leibniz (to cite two who were also mathematicians of the highest order), Plato (who was at least familiar with the mathematics of this time), and Aristotle or Kant (for whom one cannot say as much).

The traditional notion of mathematical truth goes back to the Renaissance. In this conception, there is no great difference between the objects studied by mathematics and those studied by the natural sciences. Both are knowable, and can be apprehended by intuition and by reason; neither intuition nor reason is fallible if properly employed. "*Only an entirely false intellect*", says Pascal, "*could reason wrongly from principles so obvious that it is almost impossible for them to escape notice*" ([11], vol. XII, p. 9). Descartes convinced himself that "*it is only mathematicians who have been able to find proofs, that is to say, certain and evident reasons*" ([10], vol. VI, p. 19), and this (if we accept his account) before he had constructed a metaphysic in which he says, "*this, which a little while ago I took as a rule, namely that the things we conceive very clearly and very distinctly are all true, is assured only because God is or exists and is a perfect being*" ([10], vol. VI, p. 38). Although Leibniz objects that it is not obvious how one is to recognize that an idea is "clear and distinct" (*), yet he too considers the axioms to be self-evident and ineluctable consequences of the definitions as soon as the terms are understood (†). Moreover, it should not be forgotten that at this period mathe-

(*) "*Those who have given us methods*", he says in this context, "*have given us without doubt good precepts, but not the means of observing them*" ([12 b], vol. VII, p. 21). Elsewhere, deriding the rules of Descartes, he compares them to the alchemists' formulae : "*Take what is necessary, do as you ought, and you will obtain what you desire!*" ([12 b], vol. IV, p. 329).

(†) On this point, Leibniz was still under the influence of the schoolmen; he thought always of propositions as establishing a relationship of "subject" to "predicate" between concepts. As soon as the concepts had been resolved into "primitive" concepts (which, as we have seen, was one of his fundamental ideas), everything was reduced, for Leibniz, to a problem of verifying relations of "inclusion" by means of what he called the "identical axioms" (which are essentially

matics embraced many sciences — sometimes even the art of the engineer which we would not nowadays regard as mathematical. The remarkable success of its applications to "natural philosophy", the "mechanical arts", and navigation was a principal cause of the confidence it inspired.

From this point of view the axioms were no more open to discussion or doubt than the rules of logical reasoning; at most, it was a matter of personal choice whether one reasoned "in the manner of the ancients" or gave free rein to one's intuition. The choice of starting point was also a matter of individual preference. Thus many editions of Euclid's work were published in which the solid logical framework of the *Elements* was strangely travestied, and "deductive" accounts of infinitesimal calculus or rational mechanics were given in which the foundations were singularly badly laid. Spinoza therefore perhaps acted in good faith when he claimed that his *Ethics* was "proved in the manner of the geometers" (*more geometrico demonstrata*). Although it is hard to find two mathematicians in the seventeenth century in agreement on any single question whatsoever, although the controversies were frequent, interminable, and acrimonious, yet the notion of truth was free from all imputation. "*Since there is only one truth for each thing*", said Descartes, "*whoever finds it knows as much as can be known*" ([10], vol. VI, p. 21).

Although no Greek mathematical text of the great period on these questions has been preserved, it is probable that the point of view of the Greek mathematicians on this subject was much less rigid. It is only by experience that the rules of reasoning have been developed to the point of inspiring complete confidence; before they could be considered as above all discussion they must have passed through many hesitations and tentative forms. Furthermore, it would have been out of keeping with the critical spirit of the Greeks and their taste for discussion and sophistry, if those "axioms" which Pascal considered to be self-evident (and which, according to a legend originated by his sister, he himself discovered, with an infallible instinct, as a child) had not been the object of long discussions. In a domain which is not, strictly speaking, that of geometry, the paradoxes of the Eleatic philosophers have preserved for us some traces of such controversies; and Archimedes, when he observes ([4], p. 265) that his predecessors have on many occasions made use of the axiom which bears

---

the propositions $A = A$ and $A \subset A$) and the principle of "substitution of equivalents" (if $A = B$, we may replace A everywhere by B, which is our scheme S6 of Chapter I, § 5) ([12 *bis*], pp. 184-206). It is interesting in this connection to note that, in accordance with his desire to reduce everything to logic and to "prove all that can be proved", Leibniz proved the symmetry and transitivity of the equality relation, starting from the axiom $A = A$ and the principle of substitution of equivalents; the proofs he obtained are essentially those we gave in Chapter I, § 5 ([12 a], vol. VII, pp. 77-78).

his name, adds that what is proved with the help of this axiom "*has been admitted no less than what is proved without it*", and that it is enough for him that his own results should be admitted on the same footing. Plato, in conformity with his metaphysical views, presented mathematics as a means of access to "truth in itself", and the objects with which it deals as existing in their own right in the world of ideas. He characterizes the mathematical method with accuracy in a famous passage of the *Republic*: "*Those who study geometry and arithmetic... assume the existence of odd and even numbers, and three kinds of angles; these things they take as known, and consider that there is no need to justify them either to themselves or to others, because they are self-evident to everyone; and, starting from them, they proceed consistently step by step to the propositions which they set out to examine*" (Book VI, 510c-e). Thus what constitutes a proof is first of all a starting point, which is to some extent arbitrary (although "self-evident to everyone") and from which, he says a little later, one does not attempt to go further back; next, a path which passes in order through a sequence of intermediate stations; finally, at each step the consent of the interlocutor guaranteeing the correctness of the reasoning. It should be added that, once the axioms had been stated, no new appeal to intuition was in principle admitted; Proclus, quoting Geminus, recalls that "*we have taught even the pioneers in this science to take no account of merely plausible conclusions when concerned with reasoning that is to form a part of our geometric doctrine*" ([3], vol. I, p. 203).

Thus the rules of mathematical reasoning were developed by experience and the fire of criticism. If it is true, as has been plausibly argued (*), that Book VIII of Euclid has preserved for us a part of the arithmetic of Archytas, then it is not surprising to find there the somewhat pedantic stiffness of reasoning which never fails to appear in any mathematical school which discovers or thinks it has discovered "rigour". But once these rules of reasoning had entered into the common practice of mathematicians, it does not appear that they were ever subjected to doubt until a very recent epoch; although Aristotle and the Stoics deduced certain of these rules from certain others by schemes of reasoning, the primitive rules were always admitted as self-evident. Likewise, having gone back as far as the "hypotheses", "axioms", and "postulates" which seemed to them to provide a solid foundation for the science of their period (such, for example, as must have appeared in the first *Elements*, which tradition ascribes to Hippocrates of Chio, about 450 B.C.), the Greek mathematicians of the classical period seem to have devoted their efforts to the discovery of new results rather than to a critique of the foundations, which at this period could only have been sterile; and, metaphysical preoccupations apart, the text from Plato quoted earlier bears witness to this general

(*) Cf. B. L. van der Waerden, "Die Arithmetik der Pythagoreer", *Math. Ann.* **120** (1947) pp. 127-153.

agreement among mathematicians concerning the bases of their science.

Furthermore, the Greek mathematicians seem not to have believed that the "primary notions" which served for them as a starting point — straight line, surface, ratio of magnitudes — were capable of further elucidation. If they gave "definitions", it was visibly to clear their consciences and without any illusions about the validity of these definitions. It goes without saying that, by contrast, the Greek mathematicians and philosophers had perfectly clear ideas on definitions other than those of the "primary notions" (often called "nominal" definitions, and playing the same role as our "abbreviating symbols"). In this connection there intervenes explicitly, without doubt for the first time, the question of "existence" in mathematics. Aristotle did not omit to remark that a definition does not imply the existence of the thing defined, and that either a postulate or a proof of existence is necessary. Undoubtedly his observation was derived from the practice of mathematicians. In every case Euclid is careful to postulate the existence of the circle, to prove that of the equilateral triangle, parallel lines, the square, etc. as he finds it necessary to introduce them into his arguments ([3], Book I); these proofs are "constructions" — in other words, he exhibits, with the aid of his axioms, mathematical objects which he then proves satisfy the definitions in question.

Thus we see Greek mathematics of the classical period come to a sort of empirical certainty (whatever may have been the metaphysical basis, according to this or that philosopher). If the rules of mathematical reasoning are considered to be above question, then the success of Greek science, and the feeling that critical revision would have been inopportune, are largely due to the confidence inspired by the axioms : a confidence of the same order as that accorded in the last century to the principles of theoretical physics. It is what is suggested by the proverb "*nihil est in intellectu quod non prius fuerit in sensu*" with which Descartes rightly took issue as not providing a firm enough basis for what he proposed to achieve by the use of reason.

It was not until the beginning of the nineteenth century that mathematicians retreated from the arrogant position of Descartes (not to speak of Kant or of Hegel; the latter was a little behind the time, as was to be expected, in relation to the science of his age (*)) to a position as flexible as that of the Greeks. The first blow to the classical conceptions was the construction of hyperbolic non-Euclidean geometry by Gauss, Lobatschevsky, and Bolyai at the beginning of the century. We shall not undertake here a detailed account of the genesis of this discovery, the culmination of numerous fruitless attempts to prove the parallel postulate. At the time,

(*) In his inaugural dissertation, he "proved" that there could exist only seven planets, in the same year that an eighth was discovered.

its effect on the principles of mathematics was perhaps not so deep as is sometimes asserted. It simply demanded the abandonment of the claims of the preceding century to the "absolute truth" of Euclidean geometry and, *a fortiori*, of the Leibnizian view that the definitions imply the axioms; the latter no longer appear at all as "self-evident" but rather as hypotheses which are to be judged according to whether they are adapted to the mathematical representation of the physical world. Gauss and Lobatschevsky believed that the debate between the various possible geometries could be settled by experiment ([15], p. 76). This was also the point of view of Riemann, the aim of whose famous inaugural lecture "*On the hypotheses which lie at the foundations of geometry*" was to provide a general mathematical framework for various natural phenomena. "*The question remains to be settled*", he said, "*as to how far and to what extent these hypotheses will be confirmed by experiment*" ([19], p. 284). But clearly this is a problem which no longer has anything to do with mathematics; and none of the authors just mentioned seems to have had any doubts that, even if a "geometry" did not correspond to experimental reality, its theorems were any less "mathematically true" (*).

However, if this is so, such a conviction can no longer be attributed to an unlimited confidence in classical "geometrical intuition". The description which Riemann sought to give of "manifolds *n* times extended", the object of his work, relies on "intuitive" considerations (†) only to the point of justifying the introduction of "local coordinates"; from then on, he seems to have felt that he was on firm ground, namely that of Analysis. But the latter is ultimately based on the concept of real number, which up to that time had remained on a very intuitive basis; and progress in the theory of functions was leading to most disturbing results in this respect. With the work of Riemann himself on integration, and even more with the examples of curves with no tangents, constructed by Bolzano and Weierstrass, the whole pathology of mathematics was beginning to emerge. A century later, we have seen so many monsters of this sort that we have become a little blasé. But the effect produced on the majority of nineteenth century mathematicians ranged from disgust to consternation. "*How*", asked H. Poincaré, "*can intuition deceive us at this point?*" ([33 b], p. 19); and Hermite (not without a turn of humour, which not all the commentators on this celebrated phrase seem to have perceived) declared that he

(*) Cf. the arguments of Poincaré in favour of the "simplicity" and "convenience" of Euclidean geometry ([33 *a*], p. 67), and the analysis by which, a little later, he arrives at the conclusion that experience provides no absolute criterion for the choice of one geometry over another as the framework for natural phenomena.

(†) Again, this word is justified only for $n \leqslant 3$; for larger values of n it is in fact argument by analogy which is involved.

"*turned away with fright and horror from this lamentable plague of continuous functions with no derivatives*" ([27], vol. II, p. 318). The most serious feature of the malaise was that it was no longer possible to attribute these phenomena, so contrary to common sense, to badly elucidated notions, as in the days of the "indivisibles", because they arose after the reforms of Bolzano, Abel, and Cauchy, which had succeeded in founding the notion of limit as rigorously as the theory of proportions. They were therefore to be ascribed to the gross and incomplete character of our geometrical intuition, which since that time has been discredited as a method of proof.

This conclusion inevitably affected classical mathematics, beginning with geometry. However much respect was paid to the axiomatic construction of Euclid, the imperfections in it had been noticed, from antiquity onward. The parallel postulate had been the object of the largest number of criticisms and attempts at proof; but Euclid's successors and commentators had also sought to prove other postulates (notably that of the equality of right angles) or had recognized the inadequacy of certain definitions, such as those of a straight line or a plane. In the 16th century Clavius, an editor of the *Elements*, noticed the absence of a postulate guaranteeing the existence of the fourth proportional; Leibniz, for his part, remarked that Euclid makes use of geometrical intuition without mentioning it explicitly, for example when he assumes (*Elements*, Book I, Proposition 1) that two circles, each of which passes though the centre of the other, have a common point ([12 b], vol. VII, p. 166). Gauss (who did not deny himself the use of such topological considerations) drew attention to the part played in Euclid's constructions by the notion of a point (or a line) situated "between" two others, a notion which nevertheless had not been defined ([14], vol. VIII, p. 222). Finally, the use of displacements — notably in the case of "congruent triangles" — long accepted as standing to reason (*), soon appeared to the critics of the nineteenth century as a concept which rested on unformulated axioms. Thus, in the period from 1860 to 1885, many partial revisions of the foundations of geometry appeared (Helmholtz, Méray, Houël) with the purpose of filling some of these gaps. But it was not until M. Pasch [28] that the abandonment of all appeal to intuition was clearly formulated as a programme and carried out with perfect rigour. The success of his enterprise soon produced numerous emulators who, principally between 1890 and 1910, gave varied presentations of the axioms of Euclidean geometry. The most famous of these works were Peano's, written in his symbolic language [29 b], and above all Hilbert's *Grundlagen der Geometrie*, which appeared in 1899, and by the lucidity and profundity of its exposition immediately and justly became the model of

(*) It should however be noted that, already in the sixteenth century, a commentator on Euclid, J. Peletier, protested against this method of proof, in terms close to these of modern critics ([3], vol. I, p. 249).

modern axiomatics, to the extent that its predecessors were forgotten. Hilbert was not content with giving a complete system of axioms for Euclidean geometry, but classed them in various groups according to their nature, and set about determining the exact range of each group of axioms, not only by developing the logical consequences of each group in isolation, but also by discussing the various "geometries" obtained by suppressing or modifying certain axioms (geometries among which those of Lobatschevsky and Riemann appear as special cases (*)). In this way he brought clearly into view, in a domain which until that time had been considered as one of the closest to external reality, the liberty which a mathematician has in the choice of his postulates. In spite of the disarray caused among some philosophers by these "metageometries" with their weird properties, the thesis of the *Grundlagen* was rapidly and almost unanimously adopted by mathematicians. Thus H. Poincaré, who could hardly be suspected of partiality towards formalism, asserted in 1902 that the axioms of geometry are conventions for which the notion of "truth", in the everyday sense of the word, has no meaning ([33 a], pp. 66-67). "Mathematical truth" therefore consists entirely in logical deduction from premises posed arbitrarily as axioms. As we shall see later, the validity of the rules of reasoning according to which the deductions operate was soon to be called in question, leading thus to a complete overhauling of the conceptions at the base of mathematics.

Objects, models, structures

A. *Mathematical objects and structures.* From antiquity to the nineteenth century there was common agreement on the principal objects of study of a mathematician, namely those mentioned by Plato in the passage quoted earlier: numbers, magnitudes, and figures. Although, for the Greeks, this list should be enlarged to include the objects and phenomena proper to mechanics, astronomy, optics, and music, they nevertheless always made a clear distinction between these "mathematical" disciplines and arithmetic and geometry; and from the time of the Renaissance the former soon acquired the status of independent sciences.

Whatever the philosophical nuances which coloured the conception of mathematical objects in the mind of this or that mathematician or philosopher, there was at least one point of unanimous agreement: that these objects are *given* and that it does not lie in our power to ascribe arbitrary properties to them, any more than a physicist can alter a natural

(*) What seems to have struck Hilbert's contemporaries most is "non-Archimedean" geometry, i.e., geometry over a non-Archimedean ordered base field (or division ring), the commutative case of which had been introduced some years before by Veronese (*Fondamenti di geometria*, Padova, 1891).

phenomenon. In truth, these views are partly the product of psychological reactions, which it is not for us to go into, but which every mathematician is well aware of when he exhausts himself in vain efforts to obtain a proof which seems perpetually to escape him. From there it is only one step to assimilating this resistance to external obstacles; and even today more than one mathematician, who parades an intransigent formalism, would voluntarily subscribe in his conscience to this avowal by Hermite : "*I believe that the numbers and functions of Analysis are not the arbitrary product of our minds; that they exist outside of us, with the same character of necessity as the objects of the material world, and we meet or discover them, and study them, just like the physicists, chemists, and zoologists* ([27], vol. II, p. 398).

In the classical conception, mathematics was the study of only numbers and figures. But this official doctrine, to which every mathematician felt bound to give his verbal adherence, gradually became more and more of an intolerable constraint under the pressure of new ideas. The embarrassment of algebraists in the presence of negative numbers vanished only when analytical geometry provided them with a convenient "interpretation". But even in the eighteenth century d'Alembert (although a convinced "positivist"), when discussing the question in the *Encyclopédie* [13], suddenly lost courage after a column of somewhat confused explanations, and contented himself with concluding that "*the rules of algebraic operations on negative quantities are generally admitted by everyone and are generally received as correct, whatever interpretation is to be attached to these quantities*". As regards imaginary numbers, the scandal was far worse still; for if they were "impossible" roots and if (until about 1800) there was no known way of "interpreting" them, how could these undefinable objects be talked about without contradiction, and above all how could they be introduced? On this topic d'Alembert prudently kept his silence, and did not even state the questions, no doubt because he realized that he could not answer in terms other than those naïvely used by A. Girard a century earlier [9] : "*On pourroit dire : à quoy sert ces solutions qui sont impossibles? Je réponds : pour trois choses, pour la certitude de la reigle générale, et qu'il n'y a point d'autres solutions, et pour son utilité*".

In analysis, the situation in the eighteenth century was hardly better. It was a happy coincidence that analytical geometry appeared, as if on cue, to provide an "interpretation", in the form of a geometrical figure, of the great creation of the 17th century, the notion of function, and thus to assist powerfully (in the hands of Fermat, Pascal, and Barrow) at the birth of infinitesimal calculus. But the philosophico-mathematical controversies to which the notions of infinitely small and indivisible quantities gave rise are only too well known. And although d'Alembert was on firmer ground here, and realized that in the "metaphysics" of the infinitesimal calculus there is nothing but the concept of limit, he was no more able than his contemporaries to comprehend the true meaning of expansions in divergent series, or to explain the paradox of correct results obtained by calculations with expressions which are devoid of all numerical interpretation. Finally, even in the domain of "geometrical certainty", the Euclidean façade was cracking. When Stirling, in 1717, did not hesitate to say that a certain curve has an "imaginary double point at infinity" (*), he would have been hard put to relate such an "object" to commonly received notions; and Poncelet, who, at the beginning of the nineteenth century, gave a considerable impetus to such ideas by founding projective geometry, was content to invoke as justification an entirely metaphysical "principle of continuity".

In these circumstances (and, paradoxically, at the very time when the "absolute truth" of mathematics was being proclaimed with the greatest insistence), the notion of proof seemed to become more and more blurred as the eighteenth century ran its course, because the notions employed, and their fundamental properties, could not be definitively fixed in the manner of the Greeks. The return to rigour, which was set in motion at the beginning of the nineteenth century, brought some improvements to this state of affairs, but did not correspondingly stem the flood of new ideas. Thus in algebra appeared the imaginaries of Galois, the ideal numbers of Kummer, followed by vectors and quaternions, n-dimensional spaces, multivectors, and tensors, not to mention Boolean algebra. Undoubtedly one of the great advances (which made possible a return to rigour without abandonment of any part of the conquests of the past) was the possibility of constructing "models" of these new notions in more classical terms. Thus ideal numbers and the imaginary numbers of Galois were interpreted in terms of the theory of congruences, n-dimensional geometry could be regarded (if one so wished) purely as a language for expressing the results of algebra "of n variables"; and, as regards the classical imaginary numbers — whose geometrical representation by the points of a plane marked the beginning of this efflorescence of algebra — there was soon the choice between this geometrical "model" and an interpretation in terms of congruences. But mathematicians at last began to feel that this makeshift approach was contrary to the natural development of their science and that it should be legitimate, in mathematics, to work with objects which have no concrete "interpretation". "*It is not of the essence of mathematics*", said Boole in 1854, "*to be conversant with the ideas of number and quantity*" ([16 b], p. 13) (**). The same considerations led Grassmann,

---

(*) J. Stirling, *Lineae tertii ordinis Newtonianae...* (1717).

(**) In this respect again, Leibniz appeared as a forerunner. "*The universal Mathematics*", he says, "*is, so to speak, the logic of imagination*", and should concern itself with "*everything in the domain of imagination which is capable of exact determination*" ([12 c], p. 348; cf. [12 *bis*], pp. 290-291). In his view the keystone of mathematics so conceived is what he calls the "art of formulae", i.e., essentially the

in his "*Ausdehnungslehre*" of 1844, to present his calculus in a form in which the notions of number and geometrical object were completely excluded from the start (*). And a little later, Riemann, in his inaugural lecture, was careful not to speak of "points" but of "determinations" (*Bestimmungsweise*) in his description of "manifolds *n* times extended", and emphasized that in such a manifold the "metrical relations" (*Massverhältnisse*) "*can be studied only for abstract quantities and can be represented only by formulae; under certain assumptions they can however be decomposed into relations, each of which separately is capable of a geometrical representation, and thereby the results of calculations may be expressed in geometrical terms*" ([19], p. 276).

From this moment onward, the enlargement of the axiomatic method was an accepted fact. Although for some time yet it was felt to be useful to check "abstract" results, whenever possible, against geometrical intuition, at least it was agreed that the "classical" objects were not the only legitimate objects of study for a mathematician. Precisely because of the multiplicity of possible "interpretations" or "models", it came to be recognized that the "nature" of mathematical objects is ultimately of secondary importance, and that it matters little, for example, whether a result is presented as a theorem of "pure" geometry or as a theorem of algebra *via* analytical geometry. In other words, the essence of mathematics — that fleeting and insubstantial notion which hitherto could

---

science of abstract relations between mathematical objects. But whereas at that time practically the only relations considered in mathematics were relations of magnitude (equality, inequality, proportion), Leibniz conceived of many other types of relations which, in his opinion, ought to be systematically investigated by mathematicians, such as inclusion, or what he called the relation of (single-valued or many-valued)" determination" (i.e., the notions of mapping and correspondence) ([12 *bis*], pp. 307-310). Many other modern ideas made their appearance under his pen in this context; he observed that the various equivalence relations in classical geometry have in common the properties of symmetry and transitivity; he conceived the notion of a relation which is compatible with an equivalence relation, and expressly observed that an arbitrary relation does not necessarily have this property ([12 *bis*], pp. 313-315). Of course, here as everywhere, he advocated the use of a formalized language and even introduced a sign designed to denote an undetermined relation ([12 *bis*], p. 301).

(*) It should be noted that his quasi-philosophical language was hardly likely to attract the majority of mathematicians, who felt ill at ease in the presence of a statement such as the following: "*Pure mathematics is the science of the particular entity in so far as it is created by thought*" (die Wissenschaft des *besonderen* Seins als eines durch das Denken *gewordenen*). But it is apparent from the context that what Grassmann had quite clearly in mind was axiomatic mathematics in the modern sense (except that, rather curiously, he followed Leibniz in considering that the bases of this "formal science", as he called it, were the definitions and not the axioms). In any case, like Boole, he emphasized that "*the name of science of magnitudes is not a suitable one for the whole of mathematics*" ([17], vol. I_1, pp. 22-23).

only be expressed by vague names such as "*reigle generale*" or "*metaphysic*" — appeared as the study of *relations* between objects which do not of themselves intrude on our consciousness, but are known to us by means of *some* of their properties, namely those which serve as the axioms at the basis of their theory. Boole had understood this clearly in 1847 when he wrote that mathematics is concerned with "*operations considered in themselves, independently of the various ways in which they may be applied*" ([16 a], p. 3). Hankel, in 1867, inaugurating the axiomatization of algebra, defended mathematics as being "*purely intellectual, a pure theory of forms whose purpose is the study, not of the combination of magnitudes or of their images, namely numbers, but objects of thought* (Gedankendinge), *which may correspond to concrete objects or relations, although such a correspondence is not necessary*" ([20], p. 10). Cantor, in 1883, echoes this claim for a "free mathematics", proclaiming that "*mathematics is entirely free in its development, and its concepts are bound only by the necessity that they should be non-contradictory and coordinated with concepts previously introduced by precise definitions*" ([25], p. 182). Finally, the revision of Euclidean geometry helped to disseminate and popularize these ideas. Pasch himself, although still attached to a certain "reality" of geometrical objects, realized that geometry is a fact independent of their signification, and consists purely in the study of their relations ([28], p. 90): a conception which Hilbert pushed to its logical conclusion by emphasizing that even the names of the basic notions of a mathematical theory may be chosen at random (*), and which Poincaré expressed by saying that axioms are "disguised definitions", thereby completely reversing the point of view of the scholastic philosophers.

It is therefore tempting to assert that the modern notion of "structure" was substantially in existence by 1900, but in fact another thirty years of preparation were required before it made its full-fledged appearance. Certainly it is not difficult to recognize structures of the same species when they are of a sufficiently simple nature; with group-structures, for example, this point was attained in the middle of the nineteenth century. But at the same period Hankel was still struggling, without complete success, to extract the general ideas of a field and an extension field, which he managed to express only in the semi-metaphysical form of a "principle of permanence" [20], and which were definitively formulated only by Steinitz forty years later. It has been especially difficult to escape from the feeling that

(*) According to a well-known anecdote, Hilbert was wont to express this idea by saying that the words "point", "line", and "plane" could be replaced by "table", "chair", and "beer-mug" without changing geometry in the least. It is curious to find an anticipation of this conceit (in the older sense of the word) in d'Alembert: "*We may assign to words any meanings we like*", he writes in the *Encyclopédie* ([13], s.v. DEFINITION); "*(we could) build up the elements of Geometry in full rigour by calling a triangle what is usually called a circle*".

mathematical objects are "given" *together with their structure*, and it has taken many years of functional analysis to make modern mathematicians familiar with the idea that, for example, there are several "natural" topologies on the set of rational numbers, and several measures on the real line. With this dissociation the passage to the general definition of structures such as has been presented in this book, was finally achieved.

B. *Models and isomorphisms.* In several places we have had occasion to refer to the notion of a "model" or an "interpretation" of a mathematical theory constructed with the help of another theory. This is not a recent idea, and it should undoubtedly be seen as a continually recurring manifestation of a deep-lying feeling of the unity of the various "mathematical sciences". If the traditional maxim "*All things are numbers*" of the early Pythagoreans is taken as authentic, it may be considered as the vestige of a first attempt to reduce the geometry and algebra of the time to arithmetic. Although the discovery of irrationals appeared to close this path for good, the reaction it provoked in Greek mathematics was a second attempt at synthesis, this time taking geometry as the basis and embracing, among other things, the methods of solution of algebraic equations inherited from the Babylonians (*). This conception reigned supreme until the fundamental reform of R. Bombelli and Descartes, which assimilated every measurement of magnitudes to a measurement of length (in other words, to a real number). But with the creation of analytical geometry by Descartes and Fermat, the trend was again reversed, and a far closer fusion of geometry and algebra was obtained, this time to the benefit of algebra. Descartes went further and at one stroke conceived of the essential unity of "*all the sciences which are commonly called mathematical... Although their objects are different*", he wrote, "*they are all concordant with one another in that they are concerned only with the various ratios or proportions which are present*" ([10], vol. VI, pp. 19-20) (†). However, this point of view tended only to make algebra the fundamental mathematical science: a conclusion against which vigorous protests were uttered by Leibniz, who, as we have seen, had also conceived of a "universal mathematics", but on a far vaster scale

(*) Arithmetic, however, remained outside this synthesis. Thus Euclid, having developed the general theory of proportions between arbitrary magnitudes, then developed the theory of rational numbers independently of the notion of ratios of magnitudes.

(†) It is rather curious, in this connection, to see Descartes compare arithmetic with "*combinations of numbers*", and the "*arts... in which order is dominant, such as those of the weavers and the tapestry-makers, the women who embroider and make lace*" ([10], vol. X, p. 403), as if in anticipation of modern studies of symmetry and its relationship with the notion of a group (cf. H. Weyl, *Symmetry*, Princeton University Press, 1952).

and already in close accord with modern ideas. Making quite explicit the "concordance" of which Descartes spoke, he perceived for the first time what is effectively the general notion of isomorphism (which he called "similitude") and the possibility of "identifying" isomorphic relations or operations; as an example he cites addition and multiplication ([12 *bis*], pp. 301-303). But these bold conceptions found no echo among his contemporaries, and the realization of his dreams had to await the expansion of algebra in the middle of the nineteenth century. We have already observed that it was at this point that the "models" multiplied and the habit of passing from one theory to another by simple changes of language was formed; the most striking example of the latter is perhaps that of duality in projective geometry, where the practice of printing theorems "dual" to each other side by side in two columns undoubtedly had a large share in the acceptance of the notion of isomorphism. On a more technical level, it is certain that the notion of isomorphic groups was known to Gauss for commutative groups, and to Galois for groups of permutations; it was in defined for general arbitrary groups about the middle of the nineteenth century (*). Thereafter, with each new axiomatic theory it became natural to define a notion of isomorphism; but it is only with the modern notion of structure that it has been finally recognized that every structure contains in itself a notion of isomorphism, and that it is superfluous to give a special definition for each species of structures.

C. *The arithmetization of classical mathematics.* The increasingly widespread use of the notion of "model" was also to permit the nineteenth century mathematicians to achieve the unification of mathematics dreamed of by the Pythagoreans. At the beginning of the century, whole numbers and continuous quantities appeared to be as irreconcilable as they were in antiquity; the real numbers were related to the notion of geometric magnitude (especially that of length), and the "models" of negative numbers and imaginary numbers were constructed on this basis. Even the rational numbers were traditionally associated with the idea of "subdivision" of a magnitude into equal parts. Only the integers remained apart, as "*exclusive products of our intellect*", as Gauss said in 1832, when contrasting them with the notion of space ([14], vol. VIII, p. 201). The first attempts, by Martin Ohm (1822), to bring together arithmetic and analysis were concerned with the rational numbers (positive and negative); they were taken up again around 1860 by several authors, notably Grassmann,

(*) The word "isomorphism" was introduced into group theory at the same period. But at first it was used also to denote surjective homomorphisms, which were called "meriedric isomorphisms", in contrast to "holoedric isomorphisms", which were isomorphisms in the modern sense of the word. This terminology remained in use up to the time of E. Noether.

Hankel, and Weierstrass (in unpublished lectures). To Weierstrass is apparently due the idea of obtaining a "model" of the positive or negative rational numbers by considering classes of ordered pairs of natural integers. But the most important step still remained to be taken, namely to construct a "model" of the irrational numbers from within the theory of rational numbers. By 1870 this had become an urgent problem because of the necessity, after the discovery of "pathological" phenomena in analysis, to purge every trace of geometrical intuition and the vague notion of "magnitude" from the definition of real numbers. The problem was in fact solved, at about this time, almost simultaneously by Cantor, Dedekind, Méray, and Weierstrass, using quite different methods from one another.

From then on the integers became the foundation of all classical mathematics. Furthermore, the "models" founded on arithmetic acquired still greater importance with the extension of the axiomatic method and the conception of mathematical objects as free creations of the human intellect. For there remained a limitation to the freedom claimed by Cantor, namely the limitation raised by the question of "existence", which had already occupied the Greeks, and which arose now so much more urgently precisely because all appeal to intuitive representation was now abandoned. We shall see later what a philosophico-mathematical maelström was to be generated by the notion of "existence" in the early years of the twentieth century. But in the nineteenth century this stage had not yet been reached, and to prove the existence of a mathematical object having a given set of properties meant simply, just as for Euclid, to "construct" an object with the required properties. This was precisely the purpose of the arithmetical "models"; once the real numbers had been "interpreted" in terms of integers, then the same was done for complex numbers and Euclidean geometry, thanks to analytic geometry, and likewise for all the new algebraic objects introduced since the beginning of the century. Finally, in a discovery which achieved great fame, Beltrami and Klein obtained Euclidean "models" of the non-Euclidean geometries of Lobatschevsky and Riemann and thereby "arithmetized" (and completely justified) these theories which at first had aroused so much distrust.

D. *The axiomatization of arithmetic.* It was natural, in this line of development, that attention should next be directed toward the foundations of arithmetic itself, and in fact this is what happened around 1880. Apparently, before the nineteenth century, no one had attempted to define addition and multiplication in any other way than by a direct appeal to intuition. Leibniz alone, faithful to his principles, pointed out explicitly that "truths" as "obvious" as $2 + 2 = 4$ are no less capable of proof, if one reflects on the definitions of the numbers which appear therein ([12 b], vol. IV, p. 403; cf. [12 *bis*], p. 203); and he did not by any means regard the commutativity of addition and multiplication as self-evident (*). He did not, however, push his reflections on this subject any further, and in the middle of the nineteenth century progress had still not been made in this direction. Even Weierstrass, whose lectures contributed greatly to the the spread of "arithmetizing" point of view, seems not to have realized the necessity of a logical clarification of the theory of integers. The first steps in this direction are apparently due to Grassmann who, in 1861 ([17], vol. II_2, p. 295), gave a definition of addition and multiplication of integers and proved their fundamental properties (commutativity, associativity, distributivity), using nothing but the operation $x \to x + 1$ and the principle of induction. The latter had been clearly conceived and used for the first time in the seventeenth century by B. Pascal ([11], vol. III, p. 456) (†) — though more or less explicit applications of it are to be found in the mathematics of antiquity — and it was in current use by mathematicians from the second half of the seventeenth century. But it was not until 1888 that Dedekind ([26], vol. III, pp. 359-361) enunciated a complete system of axioms for arithmetic (reproduced three years later by Peano and usually known under his name [29 a]), which contained in particular a precise formulation of the principle of induction (which Grassmann had used without enunciating it explicitly).

With this axiomatization it seemed that the definitive foundations of mathematics had been attained. But, in fact, at the very moment when the axioms of arithmetic were being clearly formulated, arithmetic itself was being dethroned from this role of primordial science in the eyes of many mathematicians (beginning with Dedekind and Peano), in favour of the most recent of mathematical theories, namely the theory of sets; and the controversies which were to surround the notion of integers cannot be isolated from the great "crisis of foundations" of the years 1900-1930.

The theory of sets

It may be said that at every period of history mathematicians and philosophers have used arguments of the theory of sets more or less consciously. But in the history of their conceptions on this subject, it is necessary to separate clearly all questions relating to the idea of cardinal numbers (and, in particular, relating to the notion of infinity) from those which involve only the notions of membership and inclusion. The latter are

(*) As examples of non-commutative operations, he cited subtraction, division, and exponentiation ([12 b], vol. VII, p. 31). At one time he even attempted to introduce such operations into his logical calculus ([12 *bis*], p. 353).

(†) See H. Freudenthal, Zur Geschichte der vollständigen Induktion, *Arch. Int. Hist. Sci.*, **33** (1953), p. 17.

intuitively clear and seem never to have given rise to controversies. They serve as the easiest foundation for a theory of the syllogisms (as Leibniz and Euler were to show), for axioms such as "the whole is greater than the part", and for that part of geometry which is concerned with intersections of curves and surfaces. Up to the end of the nineteenth century, there was equally no objection raised to speaking of the set (or "class" according to some writers) of all objects having one or another given property (*); and the celebrated "definition" given by Cantor ("*By a set we mean a grouping into one entity of distinct objects of our intuition or our thought*") ([25], p. 282) provoked hardly any objections at the time of its publication (†). But the situation was quite different where the notions of number or magnitude became entangled with the notion of set. The question of the indefinite divisibility of length (raised, certainly, by the early Pythagoreans) led, as is well known, to considerable philosophical difficulties; from the Eleatic school to Bolzano and Cantor, mathematicians and philosophers unsuccessfully came up against the paradox of the finite magnitude composed of an infinite number of points without magnitude. For us it is irrelevant to retrace, even in summary, the interminable and passionate polemics aroused by this problem, which constituted a particularly favourable terrain for metaphysical or theological aberrations. Let us only take note of the point of view held since antiquity by the majority of mathematicians. It consisted essentially in refusing to discuss the question since it could not be irrefutably decided, an attitude which we shall find again among the modern formalists: just as the latter contrive to eliminate all appearances of "paradoxical" sets (see later, page 328), so the classical mathematicians carefully avoided introducing into their arguments the "actual infinite" (that is to say, sets containing an infinity of objects conceived as existing simultaneously, at least in thought) and contented themselves with the "potential infinite", i.e., the possibility of increasing any given magnitude (or of diminishing it, if a "continuous" magnitude was under consideration) (**). If this point of view contained a certain measure of hypocrisy (††),

---

(*) We said earlier that Boole did not hesitate to include in his logical calculus the "Universe" (denoted by 1), the set of all objects. It does not appear that this conception was criticized at the time, although it had been rejected by Aristotle, who gave a rather obscure proof which purported to demonstrate its absurdity ([2], Met. B, 3, 998 b).

(†) Frege seems to have been one of the few contemporary mathematicians who, not without reason, protested at the spate of similar "definitions" ([24 c], vol. I, p. 2).

(**) A typical example of this conception is Euclid's statement: "*For every given quantity of prime numbers, there is one which is greater than all of them*", which we would express nowadays by saying that the set of prime numbers is infinite.

(††) Classically it would, of course, be correct to say that a pointbelongs to a line, but to draw the conclusion that a line is "composed of points" would violate

it nevertheless allowed the development of the greater part of classical mathematics (including the theory of proportion and, later, the infinitesimal calculus) (*); it appeared to be an excellent lifeline, especially after the quarrels aroused by infinitesimals, and had become an almost universally accepted dogma until well into the nineteenth century.

A first glimmer of the general notion of equipotence appeared in a remark of Galileo ([8], vol. VIII, pp. 78-80). He observed that the mapping $n \to n^2$ established a one-to-one correspondence between the natural integers and their squares, and consequently that the axiom "the whole is greater than the part" could not be applied to infinite sets. But, far from inaugurating a rational study of infinite sets, this remark seems to have had no other effect than to reinforce distrust of the "actual infinite"; this was the moral that Galileo himself drew, and Cauchy quotes him in 1833 only to endorse his attitude.

The requirements of analysis — and especially the deeper study of functions of real variables, which was pursued throughout the nineteenth century — are at the origin of what was to become the modern theory of sets. When Bolzano in 1817 proved the existence of the greatest lower bound of a set bounded below in $\mathbf{R}$, he argued, as most of his contemporaries did, in terms of "comprehension" and considered, not an arbitrary set

---

the taboo of the "actual infinite", and Aristotle devoted long passages to justifying this interdict. Probably in order to escape any objection of this sort many mathematicians in the nineteenth century avoided speaking of sets, and reasoned systematically in terms of "comprehension"; for example, Galois did not speak of a number field, but only of the properties common to all the elements of such a field. Even Pasch and Hilbert, in their axiomatic presentations of Euclidean geometry, still abstained from saying that lines and planes were sets of points; Peano was the only one who freely used the language of set theory in elementary geometry.

(*) The reason for this fact is undoubtedly the circumstance that the sets envisaged in classical mathematics belong to a small number of single types, and can for the most part be completely described by a finite number of numerical "parameters", so that consideration of them reduces to that of a finite set of numbers (this is the case, for example, with algebraic curves and surfaces, which for a long time constituted almost exclusively the "figures" of classical geometry). Before the progress of analysis led, in the nineteenth century, to the consideration of arbitrary subsets of the line and $\mathbf{R}^n$, it was rare to encounter sets which did not conform to the above types. For example, Leibniz, original as always, considered the closed disc minus its centre as a "geometrical locus", or (by a curious presentiment of the theory of ideals) considered that in arithmetic an integer is the "type" of the set of its multiples, and remarked that the set of multiples of 6 is the intersection of the set of multiples of 2 and the set of multiples of 3 ([12 b], vol. VII, p. 292). From the beginning of the nineteenth century sets of this latter type became familiar, in algebra and in number theory: for example, the classes of quadratic forms, introduced by Gauss, and fields and ideals, defined by Dedekind before the Cantorian revolution.

of real numbers, but an arbitrary property of them. But when, thirty years later, he wrote his *Paradoxien des Unendlichen* [18] (published in 1851, three years after his death), he did not hesitate to claim the possibility of existence for the "actual infinite" and to speak of arbitrary sets. In this work he defined the general notion of equipotence of two sets and proved that any two compact intervals in **R** are equipotent (provided each contains more than a single point). He observed also that the characteristic difference between finite and infinite sets lies in the fact that an infinite set E is equipotent to some proper subset of E, but he gave no convincing proof of this assertion. The general tone of the work is more philosophical than mathematical; and since he did not distinguish precisely enough between the notion of the power of a set and that of magnitude or order of infinity, Bolzano failed in his attempts to construct infinite sets of greater and greater powers, and his discussion in this connection was entangled with considerations of divergent series which are entirely devoid of meaning.

The creation of the theory of sets, as it is understood today, is due to the genius of G. Cantor. He too began as an analyst, and his work on trigonometric series, inspired by that of Riemann, led him naturally in 1872 to a first essay in classifying the "exceptional" sets which arise in this theory (*) through the notion of successive "derived sets", which he introduced in this connection. Undoubtedly it was these researches, and also his method for defining real numbers, which led Cantor to interest himself in problems of equipotence; for in 1873 he noted that the set of rational numbers (or the set of algebraic numbers) is countable. In his correspondence with Dedekind, which began at about this time [25 *bis*], he raised the question whether the set of integers and the set of real numbers are equipotent, and succeeded a few weeks later in showing that they are not. Next, from 1874 onward, the problem of dimension preoccupied him, and for three years he sought in vain to establish the impossibility of a one-to-one correspondence between **R** and $\mathbf{R}^n$ ($n > 1$), before he came to construct such a correspondence, to his own stupefaction (†). Having obtained these new and surprising results, he devoted himself entirely to the theory of sets. In a series of six memoirs published in the *Mathematische Annalen* between 1878 and 1884, he considered problems of equipotence, the theory of totally ordered sets, topological properties of **R** and $\mathbf{R}^n$, and the problem of measure; and it is admirable to see

(*) These are sets $E \subset \mathbf{R}$ with the property that if a trigonometric series $\sum_{-\infty}^{+\infty} c_n e^{nix}$ converges to 0 except at points of E, then $c_n = 0$ for all n ([25], p. 99).

(†) "*Je le vois, mais je ne le crois pas*", he wrote to Dedekind ([25 *bis*], p. 34; in French in the text).

how in his hands the notions which appeared to be so inextricably rooted in the classical conception of the "continuum" were gradually brought out. In 1880 he had the idea of iterating "transfinitely" the formation of "derived sets"; but this idea did not take substance until two years later, with the introduction of well-ordered sets — one of the most original of Cantor's discoveries, thanks to which he was able to initiate a detailed study of cardinal numbers and to formulate the "problem of the continuum" [25].

It was not to be expected that such bold conceptions, which ran counter to a tradition two thousand years old and led to such unlikely results of so paradoxical an appearance, should have been accepted without resistance. In fact, among the influential mathematicians in Germany, Weierstrass was the only one to follow with some favour the work of Cantor (his former pupil); but Cantor met with the irreconcilable opposition of Schwarz, and above all of Kronecker (*). It seems to have been as much the constant tension generated by opposition to his ideas as his fruitless efforts to prove the continuum hypothesis which produced in Cantor the first symptoms of a nervous illness which was to affect his mathematical productivity (†). In fact he did not resume his interest in the theory of sets until about 1887, and his last publications, which are concerned mainly with the theory of totally ordered sets and the calculus of ordinals, date from the period 1895-97. He had also proved in 1890 the inequality $\mathfrak{m} < 2^{\mathfrak{m}}$. However, not only did the problem of the continuum remain unanswered, but there was a more serious gap in the theory of cardinals, for Cantor had not been able to prove the existence of a well-ordering relation between arbitrary cardinals. This gap was to be filled partly by the theorem of F. Bernstein (1897) showing that the relations $\mathfrak{a} \leqslant \mathfrak{b}$ and $\mathfrak{b} \leqslant \mathfrak{a}$ imply $\mathfrak{a} = \mathfrak{b}$ (**), and above all by the theorem of Zermelo [36 a] which proved the existence of a well-ordering on any set: a theorem that Cantor had conjectured already in 1883 ([25], p. 169).

Dedekind, however, had followed the work of Cantor with sustained interest since the beginning. But whereas the latter concentrated his

---

(*) Kronecker's contemporaries made frequent allusions to his doctrinal position on the foundations of mathematics; it is thus to be presumed that he expressed himself more explicitly in personal contacts than in his publications (which, so far as the role of the natural integers is concerned, contain nothing more than comments on "arithmetization", which by 1880 had become rather banal) (cf. H. Weber, "Leopold Kronecker", *Math. Ann.*, **43** (1893), pp. 1-25, in particular pp. 14-15).

(†) For this period of Cantor's life, see A. Schoenflies, *Acta Math.* **50** (1928), pp. 1-23.

(**) This theorem had already been obtained in 1887 by Dedekind, but he did not publish his proof ([26], vol. III, p. 447).

attention on infinite sets and their classification, Dedekind pursued his own thoughts on the notion of number (which had already led to his definition of irrational numbers by means of "cuts"). In his pamphlet *Was sind und was sollen die Zahlen*, which was published in 1888, but whose essential content dates from the period 1872-8 ([26], vol. III, p. 335), he showed how the notion of natural integer (which, as we have seen, was now the basis for the whole of classical mathematics) could itself be derived from the fundamental notions of the theory of sets. He was certainly the first to develop explicitly the elementary properties of arbitrary mappings of one set into another (hitherto neglected by Cantor, who was interested only in injective correspondences) and introduced, for any mapping f of a set E into itself, the notion of the "chain" of an element $a \in E$ relative to f, namely the intersection of all sets $K \subset E$ such that $a \in K$ and $f(K) \subset K$ (*). Next, he took as the *definition* of an infinite set E the fact that there exists a one-to-one mapping φ of E into E such that $\varphi(E) \neq E$ (†). If, furthermore, there exists such a mapping φ and an element $a \notin \varphi(E)$ for which the whole set E is the chain of a, Dedekind said that E is "simply infinite"; moreover, he noted that "Peano's axioms" are then satisfied and showed (before Peano did) how, from that starting-point, all the elementary theorems of arithmetic may be obtained. The only thing lacking in his presentation is the axiom of infinity, which Dedekind (following Bolzano) believed he could prove by considering the "world of thoughts" (*Gedankenwelt*) as a set (**).

(*) A notion closely related to this forms the basis of Zermelo's second proof of his theorem ([36 b]; see Chapter III, § 2, Exercise 6).

(†) As we have seen, Bolzano had already noticed this characterization of infinite sets, but his work (which was apparently not well known in mathematical circles) was not known to Dedekind at the time of writing *Was sind und was sollen die Zahlen*.

(**) Another method of defining the notion of natural integer and deducing its fundamental properties had been proposed by Frege in 1884 [24 b]. First of all, he sought to give the notion of the cardinal of a set a more precise meaning than Cantor; at this date the latter had defined only the notions of equipotent sets and of a set having a power at most equal to that of another set, and the definition of "cardinal number" which he gave later ([25], p. 282) was as obscure and unusable as Euclid's definition of a straight line. Frege, with his usual care for accuracy, had the idea of taking as the definition of the cardinal of a set A the set of all sets equipotent to A ([24 b], § 68); then, having defined $\varphi(\mathfrak{a}) = \mathfrak{a} + 1$ for every cardinal $\mathfrak{a}$ (§ 76), he considered the set C of all cardinals, and defined the relation "$\mathfrak{b}$ is a φ-successor of $\mathfrak{a}$" to mean that $\mathfrak{b}$ belongs to the intersection of all sets $X \subset C$ such that $\varphi(\mathfrak{a}) \in X$ and $\varphi(X) \subset X$ (§ 79). Finally, he defined a natural integer as a φ-successor of 0 (§ 83; all these definitions are of course expressed by Frege in his *Begriffsschrift*). Unfortunately, this construction was later seen to be defective, since the set C and the set of all sets equipotent to a set A are "paradoxical" (see below).

On another front, Dedekind had been led by his arithmetical work (and especially the theory of ideals) to envisage the notion of ordered set in a more general aspect than Cantor had done. Whereas the latter restricted himself entirely to totally ordered sets (*), Dedekind investigated the general case and in particular made a close study of lattices ([26], vol. II, pp. 236-271). This work attracted hardly any attention at the time; and although his results, rediscovered by various authors, have been the object of numerous publications since 1935, their historical importance lies far less in the possible applications of this theory (which indeed are rather scanty) than in the fact it was one of the first examples of careful axiomatic construction. On the other hand, Cantor's first results on countable sets soon produced many important applications, even to the most classical questions of analysis (†) (quite apart from those parts of his work which inaugurated general topology and measure theory; for these the reader is referred to the Historical Notes in *General Topology*). Moreover, the last years of the nineteenth century saw the first applications of the principle of transfinite induction which, especially since the proof of Zermelo's theorem, has become an indispensable instrument in all parts of modern mathematics. Kuratowski in 1922 gave a version of this principle, subsequently rediscovered by Zorn [45] and customarily named after him, which is often more convenient to apply since it avoids the use of well-ordered sets ([40], p. 89); it is this form of the principle that is generally used nowadays (**).

By the end of the nineteenth century, therefore, the essential conceptions of Cantor had carried the day (††). As we have seen, the formalization of mathematics was achieved at the same period, and the use of the axio-

---

(*) It is curious to note that Cantor would never admit the existence of "non-Archimedean" ordered groups, because they brought in the notion of "actual infinitesimal" ([25], pp. 156 and 172). Such order relations presented themselves naturally in the researches of Du Bois-Reymond on orders of infinity and were studied systematically by Veronese (*Fundamenta di Geometria*, Padova, 1891).

(†) In 1874 Weierstrass had announced, in a letter to Du Bois-Reymond, an application to functions of a real variable of Cantor's theorem on the possibility of arranging the rational numbers in a sequence (*Acta Math.* **30** (1924), pp. 206).

(**) For this reason, the importance attached to Cantor's ordinals has declined considerably; and, generally speaking, many of the results of Cantor and his successors on the arithmetic of uncountable ordinals and cardinals have so far remained somewhat isolated.

(††) The first International Congress of Mathematicians (Zurich, 1897) may be taken as the official consecration of the theory of sets; on this occasion Hadamard and Hurwitz announced important applications of set theory to analysis. The increasing influence of Hilbert at this period contributed greatly to the dissemination of Cantor's ideas, especially in Germany.

matic method had become almost universally accepted. In other words, the intense labours of the years 1875-1895 had acquired for mathematics the essential content of the material presented in this book. But the same period saw the beginning of a "crisis of the foundations" of rare violence, which was to shake the mathematical world for more than thirty years and seemed at times to put in danger not only all the recent acquisitions but even the most classical parts of mathematics.

The paradoxes of the theory of sets and the crisis of the foundations

The first "paradoxical" sets appeared in the theory of cardinals and ordinals. In 1897, Burali-Forti noted that there cannot be a set formed of *all the ordinals*, for such a set would be well-ordered and therefore isomorphic to a proper segment of itself, which is absurd (*). In 1899, Cantor observed (in a letter to Dedekind) that we can no more say that the cardinals form a set, nor speak of "the set of all sets" without contradiction (the set of all subsets of the latter "set" Ω would be equipotent to a subset of Ω, which is contrary to the inequality $\mathfrak{m} < 2^{\mathfrak{m}}$) ([25], pp. 444-448). Finally, in 1905, by analyzing the proof of this inequality Russell showed that the reasoning which establishes it also proves (without any appeal to the theory of cardinals) that the notion of "the set of all sets which are not elements of themselves" is self-contradictory [37] (†).

It might be thought that such "antinomies" would appear only in the peripheral regions of mathematics characterized by the consideration of sets so "large" as to be inaccessible to intuition. But other paradoxes were soon to threaten the most classical parts of mathematics. Berry and Russell [37], simplifying an argument due to J. Richard [34], observed that the set of integers which can be named in less than nineteen syllables is finite, but that it is nevertheless contradictory to define an integer as "the least integer not nameable in fewer than nineteen syllables", since this definition contains only eighteen syllables.

Although such arguments, so remote from the current usage of mathematicians, must have appeared to many as a sort of play upon words, yet they indicated the need for a revision of the basis of mathematics in order to eliminate "paradoxes" of this nature. But although there was

---

(*) C. Burali-Forti, "Sopra un teorema del Sig. G. Cantor", *Atti Accad. Torino*, **32** (1896-7), pp. 229-237. This remark had already been made by Cantor in 1896 (in an unpublished letter to Hilbert).

(†) Russell's argument is to be compared with the ancient paradoxes of the "liar" type, which were the subject of innumerable commentaries in classical formal logic; the question is whether a man who says "I am lying" is telling the truth or not when he speaks these words (cf. A. Rüstow, *Der Lügner*, Diss. Erlangen, 1910).

unanimous agreement on the urgency of such a revision, radical divergences of opinion soon arose as to the manner of achieving it. For one group of mathematicians, the "idealists" and the "formalists" (*), the situation created by the "paradoxes" of set theory was very similar to that caused in geometry by the discovery of non-Euclidean geometries or "pathological" curves (such as curves with no tangents); it should therefore lead to a similar, but more general conclusion, namely that it is futile to attempt to found *any* mathematical theory on an appeal (explicit or otherwise) to "intuition". This position may be summarized in the words of the principal adversary of the formalist school : "... *the view of formalism*", said Brouwer ([38 a], p. 83), "*which maintains that human reason does not have at its disposal exact images either of straight lines or of numbers larger than ten, for example... It is true that from certain relations among mathematical entities, which we assume as axioms, we deduce other relations according to fixed laws, in the conviction that in this way we derive truths from truths by logical reasoning... For the formalist, therefore, mathematical exactness consists merely in the method of developing the series of relations, and is independent of the significance one might want to give to the relations or the entities which they relate*".

For the formalist, therefore, what is required is to provide an axiomatic basis for the theory of sets quite analogous to that for elementary geometry, in which it is not necessary to know what the "things" are that are called "sets", nor what the relation $x \in y$ means, but in which the conditions imposed on this relation are enumerated; naturally, this should be done in such a way as to include, so far as possible, all the results of Cantor's theory, while rendering impossible the existence of "paradoxical" sets. The first example of such an axiomatization was given by Zermelo in 1908 [36 c]; he avoided sets which are "too big" by introducing a "selection axiom" (*Aussonderungsaxiom*), according to which a property $P\{x\}$ determines a set formed by the elements which possess this property only if $P\{x\}$ already implies a relation of the form $x \in A$ (†). But the elimination of paradoxes analogous to "Richard's paradox" could be achieved only by restricting the sense attached to the notion of "property".

(*) The differences between these two schools are mainly of a philosophical order, and we cannot here enter into more details on this subject. The essential point for us is that they were in agreement on strictly mathematical questions. For example, Hadamard (a typical representative of the "idealists") adopted a point of view very close to that of the formalists, so far as the validity of the arguments of the theory of sets is concerned, but without expressing himself in axiomatic terms ([35], p. 271).

(†) For example, Russell's paradox would be valid in Zermelo's system only if the relation $(\exists z)((x \notin x) \implies (x \in z))$ were proved. Of course, such a proof, if it were to be found, would entail as an immediate consequence the necessity of substantial modification of the system in question.

In this connection, Zermelo was content to describe in extremely vague terms a type of properties which he called "definite" and to indicate that the application of the selection axiom must be limited to properties of this type. This point was made precise by Skolem [39] and Fraenkel [41] (*); as they observed, its elucidation demands a completely formalized system (such as that described in Chapters I and II of this book), in which the notions of "property" and "relation" have lost all "meaning" and have become simply designations for assemblies formed in accordance with explicit laws. This, of course, necessitates that the rules of logic used be incorporated in the system, which was not yet the case in the Zermelo-Fraenkel system; apart from this, it is essentially their system which we have described in Chapters I and II.

Other axiomatizations of the theory of sets were subsequently proposed. We shall consider principally that due to von Neumann [43 a and b], which comes closer than the Zermelo-Fraenkel system to Cantor's primitive conception; the latter, in order to avoid "paradoxical" sets, had already proposed (in his correspondence with Dedekind; [25], pp. 445-448) to distinguish two kinds of sets, "multiplicities" (*Vielheiten*) and "sets" (*Mengen*) in the strict sense, the latter being characterized by the fact that they could be thought of as single objects. It is this idea which von Neumann made precise by distinguishing between two types of objects, "sets" and "classes". In his system (almost completely formalized) the classes are distinguished from sets by the fact that they cannot be placed on the left of the sign $\in$. One of the advantages of such a system is that it rehabilitates the notion of "universal class" (which, of course, is not a set) used by the logicians of the nineteenth century. We should add that von Neumann's system avoids (for set theory) the introduction of axiom schemes, which are replaced by suitable axioms, thereby making the logical study of the system easier. Variants of von Neumann's system have been given by Bernays and Gödel [44 b].

These systems seem to have succeeded in eliminating the paradoxes, but at the cost of restrictions which cannot but appear highly arbitrary. In favour of the Zermelo-Fraenkel system it can be said that it limits itself to formulating prohibitions which do no more than sanction current practice in the applications of the notion of set to various mathematical theories. The systems of von Neumann and Gödel are more remote from usual conceptions. On the other hand, we cannot exclude the possibility that

---

(*) Skolem [39] and Fraenkel [41] also noted that Zermelo's axioms are insufficient to prove, for example, the existence of uncountable cardinals $\mathfrak{m}$ such that $2^{\mathfrak{n}} < \mathfrak{m}$ for every cardinal $\mathfrak{n} < \mathfrak{m}$. As we have seen (Chapter III, § 6, Exercise 21), the existence of such cardinals can be proved by reinforcing the selection axiom (scheme S8, Chapter II, § 1, no. 6); the axiom thus introduced is a variant of those proposed by Skolem and Fraenkel.

it may be easier to insert the basis of some mathematical theories into the framework of such systems than into the more rigid framework of the Zermelo-Fraenkel system.

In any case, it cannot be said that any of these solutions gives the impression of finality. They satisfy the formalists, because the formalists refuse to take into consideration the individual psychological reactions of every mathematician; they consider that a formalized language has done its duty if it is capable of transcribing mathematical arguments in a form which contains no ambiguities, and hence of serving as a vehicle for mathematical thought. It is everyone's right, they would say, to think what they like about the "nature" of mathematical entities or the "truth" of the theorems they use, provided that their reasoning can be transcribed into the common language (*).

In other words, from a philosophical point of view, the attitude of the formalists is characterized by an indifference to the problem posed by the "paradoxes", resulting from the abandonment of the Platonic position which claimed to attribute to mathematical notions an intellectual "content" common to all mathematicians. Many mathematicians rebelled against this break with tradition. Russell, for example, sought to avoid the paradoxes by analyzing their structure more deeply. Taking up an idea first advanced by J. Richard (in the article [34] in which he presented his "paradox") and later developed by H. Poincaré [33 c], Russell and Whitehead observed that the definitions of the paradoxical sets all violate the following principle, called the "vicious circle principle": "whatever involves *all* of a collection must not be one of the collection" ([37], vol. I, p. 40). This statement served as a basis for the *Principia*, and the "theory of types" was developed in order to accommodate it. Like that of Frege which inspired it, the logic of Russell and Whitehead (much vaster than the mathematical logic used in Chapter I of this Book) contains "propositional variables". The theory of types proceeds to a classification of these variables, in broad outline as follows.

Starting from an undefined "domain of individuals", which may be qualified as "objects of order 0", relations in which the variables (free or bouud) are individuals are called "first-order objects"; and in general, relations in which the variables are objects of order $\leqslant n$ (and at least one of them is of order n) are called "objects of order $n + 1$" (*). A set of objects

(*) Hilbert, at any rate, apparently always believed in an objective mathematical "truth" ([30], pp. 315 and 323). Even some formalists, such as H. Curry, took a position very close to that which we have just summarized, rejecting with a sort of indignation the idea that mathematics can be regarded simply as a game, and insisted that it is an "objective science" (H. Curry, *Outlines of a Formalist Philosophy of Mathematics*, Amsterdam (North Holland Publ. Co.), 1951, p. 57).

of order n can therefore be defined only by a relation of order $n+1$, and this condition allows the elimination of paradoxical sets without any difficulty (†). But the principle of the "hierarchy of types" is so restrictive that, if strictly adhered to, it leads to a mathematics of inextricable complexity (**). To escape this consequence, Russell and Whitehead were obliged to introduce an "axiom of reducibility" which asserts the existence, for every relation between "individuals", of an equivalent relation of the *first* order. This condition is as arbitrary as the axioms of the formalists, and reduces considerably the interest of the construction of the *Principia.* The system of Russell and Whitehead has been more popular with logicians than with mathematicians. Moreover, it is not completely formalized (††), so that there are numerous obscurities of detail. Various efforts have been made to simplify and clarify it (Ramsey, Chwistek, Quine, Rosser); these authors tend to use more and more completely formalized languages, and replace the rules of the *Principia* (which still contain a certain intuitive substratum) by restrictions which take account only of the writing down of the assemblies considered; not only do these rules thus appear as gratuitous as the prohibitions formulated in the systems of Zermelo-Fraenkel or von Neumann, but also, being more remote from mathematical usage, they have in many cases led to unacceptable consequences which the author had not foreseen (such as Burali-Forti's paradox, or the negation of the axiom of choice).

For the mathematicians of these schools, the essential aim was to avoid renouncing any part of the heritage of the past. *"No one shall expel us"*,

(*) In fact, this is only the beginning of the classification of "types", which cannot be accurately described without going into very long explanations. The reader who wishes to have a more detailed account may refer especially to the introduction to Volume II of *Principia Mathematica* [37].

(†) In the system of Russell and Whitehead, the relation $x \in x$ therefore cannot be legitimately written, in contrast to the Zermelo-Fraenkel system for example (cf. Chapter II, § 1, no. 4).

(**) For example, equality is not a primitive notion in the system of *Principia*; two objects a, b are equal if, for *every* property $P\{x\}$, $P\{a\}$ and $P\{b\}$ are equivalent propositions. But this definition is meaningless in the theory of types. To give it meaning, it would at any rate be necessary to specify the "order" of P, and one would thus be led to distinguish an infinite number of equality relations! Zermelo had noted in 1908 [36 b] that many definitions of classical mathematics (for example, that of the greatest lower bound of a set in $\mathbf{R}$) do not respect the "vicious circle principle", and the adoption of this principle therefore risked imposing an interdict on important parts of the most traditional mathematical theories.

(††) Russell and Whitehead (like Frege) adhered to the classical position touching mathematical formulae, which for them always possessed a "meaning" related to an underlying activity of the mind.

said Hilbert ([30], p. 274), *"from the paradise Cantor has created for us"*. To achieve this aim, they were prepared to accept limitations on mathematical reasoning which, though not of vital importance because they conformed with mathematical usage, did not appear to be imposed by our mental habits and the intuitive notion of a set. For them, anything was preferable to the intrusion of psychology in the criteria of validity in mathematics; and rather than *"take into account the properties of our brains"*, as Hadamard said ([35], p. 270), they resigned themselves to the imposition of largely arbitrary boundaries, provided that all of classical mathematics was contained in them and that they did not threaten to impede later progress.

Quite different was the attitude of the mathematicians who adhered to the following doctrine. Whereas the formalists agreed to abandon control by the "eyes of the mind" so far as mathematical reasoning was concerned, the mathematicians who have been labeled as "empiricists", "realists", or "intuitionists" refused to consent to this abdication; they insisted on a sort of inner certainty guaranteeing the "existence" of the mathematical objects they studied. They had no serious objection to the renunciation of spatial intuition, because arithmetical "models" allowed them to shelter behind the intuitive notion of integers. But irreconcilable opposition arose over the question of reducing the notion of integer to that of set (which is intuitively far less clear) and then of erecting barriers, devoid of intuitive foundations, to the manipulation of sets. The first person to voice this opposition (and the most influential, by reason of the authority of his genius) was H. Poincaré. Having accepted not only the axiomatic point of view toward geometry and the arithmetization of analysis but also a good part of Cantor's theory, he refused to consider that arithmetic, too, might be amenable to axiomatic treatment. The principle of mathematical induction, in particular, was in his eyes a fundamental intuition of our intellects and could not be regarded purely as a convention (*) [33 c]. He was in principle against formalized languages, whose usefulness he contested, and was prone to confuse the notion of integers in formalized mathematics with the use of integers in the theory of proof (which was then emerging and of which we shall speak later). No doubt it was not easy at that time to make this distinction as precisely as we do today — after more than fifty years of study and discussion — although the distinction had been grasped clearly by men such as Hilbert and Russell.

(*) Poincaré goes so far as to say, in substance, that it is impossible to define a structure which satisfies all of Peano's axiom except for the principle of mathematical induction ([33 a], p. 65). The example (due to Padoa) of the integers with the mapping $x \to x + 2$ replacing $x \to x + 1$ shows that this assertion is incorrect. Curiously, this example is already to be found in Frege, in almost the same terms ([24 b], p. 21 e).

Criticisms of this nature multiplied after the introduction of the axiom of choice by Zermelo in 1904 [36 a]. Its use in many previous proofs in analysis and set theory had hitherto gone almost unnoticed (*), and it was by following up an idea suggested by Erhard Schmidt that Zermelo, explicitly stating this axiom (and in the two forms given in the Summary of Results, §4, no. 10), deduced by an ingenious method (which we reproduced in essence in Chapter III, §2) a satisfactory proof of the existence of a well-ordering on any set. It appears that, coming as it did at the same time as the "paradoxes", this new method of reasoning, with its unfamiliar appearance, caused confusion among many mathematicians; one need only read the strange misconceptions which arose in this connection, in the following volume of the *Mathematische Annalen*, by authors as familiar with Cantor's methods as Schoenflies and F. Bernstein. The criticisms of E. Borel, published in the same volume, have more substance and clearly reflect the point of view of Poincaré on integers; they were developed and discussed in an exchange of letters between E. Borel, Baire, Hadamard, and Lebesgue, which has become a classic of the French mathematical tradition [35]. Borel began by denying the validity of the axiom of choice because in general it involves an uncountable infinity of choices, which cannot be intuitively imagined. But Hadamard and Lebesgue observed that a countable infinity of successive *arbitrary* choices is no more intuitive, because it involves an infinite number of operations, which it is impossible to conceive of being actually carried out. In the view of Lebesgue, who enlarged the debate, everything depended on what was meant by the assertion that a mathematical object "exists"; for him, it is necessary to "name" explicitly a property (a "functional" property, we should say) which defines the object uniquely. A function such as that used by Zermelo in his proof is what Lebesgue called a "law" of choice. If, he continued, this requirement is not satisfied, and we merely "consider" this function instead of "naming" it, can we be sure that throughout our reasoning we are always considering the same function? ([35], p. 267) This consideration led Lebesgue to new doubts: even the choice of a single element in a set appears to raise difficulties; it is necessary to be certain that such an element "exists", i.e., that at least one element of the set can be "named" (†). Can we therefore speak of the "existence" of a set if

(*) In 1890, Peano, in the course of proving his theorem on the existence of integrals of differential equations, remarked that he had been led naturally to "*applying an infinite number of times an arbitrary law which associates with a class an individual member of the class*"; but he added immediately that such an argument was not admissible, in his opinion (*Math. Ann.*, **37** (1890), p. 210). In 1902, B. Levi noted that the same line of reasoning had been implicitly employed by F. Bernstein in a proof in the theory of cardinals (*Ist. Lombardo Sci. Lett., Rend.* (2) **35** (1902), p. 863).

(†) The so-called "choice" of an element in a set has in fact nothing to do with the axiom of choice; it is simply a manner of speaking, and whenever we use

we cannot "name" every element of it? Baire did not hesitate to deny the "existence" of the set of subsets of a given infinite set (*loc. cit.*, pp. 263-264); in vain Hadamard observed that these restrictions would lead to renouncing even the right to speak of the set of real numbers, and E. Borel in the end came to this conclusion. Apart from the fact that countable sets seem to have been granted the franchise, we are almost back to the classical position of the opponents of the "actual infinite".

None of these objections was very systematic; it was reserved to Brouwer and his school to undertake a complete recasting of mathematics on similar but even more radical lines. We shall not here presume to summarize a doctrine so complex as intuitionism, which is as much of philosophy as mathematics, and we shall do no more than indicate some of its most striking features, referring for more details to the works of Brouwer himself [38] and Heyting's exposition [46]. In Brouwer's view, mathematics is identical with the "exact" part of our thought, founded on the primary intuition of the sequence of natural numbers, and cannot be translated into a formal system without mutilation. Moreover, it is "exact" only in the minds of mathematicians, and it is illusory to hope to construct an instrument of communication between mathematicians which is free of all the imperfections and ambiguities of languages; the most that can be hoped for is to arouse in the interlocutor a favourable state of mind for more or less vague descriptions ([46], pp. 11-13). Intuitionist mathematics attaches hardly more importance to logic than to language; a proof is conclusive not by virtue of the rules of logic, fixed once and for all, but by reason of the "immediate self-evidence" of each of its steps. This "self-evidence" is moreover to be interpreted in an even more restrictive sense than that of E. Borel and his supporters; thus in intuitionist mathematics we may not assert that a relation of the form "R or (not R)" is true (the principle of the excluded middle) unless, for every system of values given to the variables appearing in R, we can prove that one of the two propositions R, "not R" is true. For example, from the equation $ab = 0$ between two real numbers, we may not infer "$a = 0$ or $b = 0$", because it is easy to construct explicit examples of real numbers a, b such that we have $ab = 0$ without at present being able to prove either one of the two propositions $a = 0, b = 0$ ([36], p. 21).

---

this expression we are in fact using the method of the auxiliary constant (Chapter I, §3, no. 3) which depends only on the most elementary logical laws (not involving the sign τ). Of course, the application of this method to a set A requires that $A \neq \emptyset$ should have been proved; this is the point of Lebesgue's argument, for such a proof is not valid for him unless an element of A has been "named". For example, Lebesgue does not accept as valid the argument of Cantor proving the existence of transcendental numbers; for him their existence is proved only because it is possible to "name" particular transcendental numbers, such as Liouville's numbers or the numbers e or π.

It is hardly surprising that the intuitionist mathematicians, starting from such principles, were led to results quite different from the classical theorems. A large number of the latter disappeared, for example most of the "existence" theorems of analysis (such as the theorems of Bolzano and Weierstrass for real functions); if a function of a real variable "exists" in the intuitionist sense, it is *ipso facto* continuous, and a monotone bounded sequence of real numbers does not necessarily have a limit. Furthermore, many of the classical notions ramify into several fundamentally distinct notions in intuitionist mathematics. Thus there are two notions of convergence (for a sequence of real numbers) and eight notions of countability. It goes without saying that transfinite induction and its applications to modern analysis are, like most of Cantor's theory, condemned without appeal.

According to Brouwer, it is only in this way that mathematical propositions can acquire a "content"; formalist arguments which go beyond what is admitted by intuitionism are judged to be worthless, because it is impossible to give them a "meaning" to which the intuitive notion of "truth" could be applied. Clearly, judgments of this sort can only rest on a previous notion of "truth", which is of a psychological or metaphysical nature; that is to say, in practice, that they are beyond discussion.

Certainly, the vigorous attacks from the intuitionist camp have from time to time obliged not only the avant-garde mathematical schools, but even the partisans of traditional mathematics, to be on the defensive. A well-known mathematician has admitted to being impressed by these attacks to the point that he voluntarily restricted his work to those branches of mathematics considered to be "certain". But such cases must have been rather uncommon. The intuitionist school, whose memory will undoubtedly survive only as a historical curiosity, has at least rendered the service of having obliged its opponents, that is to say the vast majority of mathematicians, to clarify their own positions and to become more consciously aware of the reasons (whether logical or sentimental) for their confidence in mathematics.

Metamathematics

Absence from contradiction has always been considered to be a *sine qua non* of all mathematics, and from the time of Aristotle logic was sufficiently developed for it to be realized that anything could be deduced from a contradictory theory. Proofs of "existence", which have been regarded as indispensable ever since antiquity, clearly served only to guarantee that the introduction of a new concept did not risk entailing a contradiction, particularly if the concept was too complicated to fall immediately under "intuition". We have seen how this demand for a proof of non-contradiction became more imperative with the advent of the axiomatic point of view in the nineteenth century, and how the construction of arithmetic "models" answered it. But might arithmetic itself not be contradictory? Such a question could not have been asked before the end of the nineteenth century; so much did the integers appear to belong to what is surest in our intuition. But after the "paradoxes" everything seemed open to doubt, and the feeling of insecurity they created understandably led mathematicians, round about 1900, to look more closely at the problem of the consistency of arithmetic, in order to save at least classical mathematics from shipwreck. Thus this problem was the second of those listed by Hilbert in his famous address to the International Congress of 1900 ([31], pp. 229-301). In doing so, he put forward a new principle which was to make a deep impression. Whereas, in traditional logic, the non-contradiction of a concept only made the concept "possible", for Hilbert it was equivalent to the *existence* of the concept (at any rate where axiomatically defined mathematical concepts were concerned). This point of view apparently entailed the necessity of proving *a priori* the consistency of a mathematical theory before even being able to develop the theory legitimately; certainly this was how the principle was understood by H. Poincaré, who took up Hilbert's idea as a convenient stick with which to beat the formalists, by pointing out with malicious satisfaction how far the formalists in this period were from being able to satisfy this condition ([33 c], p. 163). We shall see later how Hilbert took up this challenge. But first, it should be noted that, under his influence and that of Poincaré, the demands put forward by the latter were accepted without reserve for a long time, as much by the formalists as by their opponents. One consequence was the belief, which was very widespread even among the formalists, that Hilbert's theory of proof was an integral part of mathematics, and constituted an indispensable prolegomenon to mathematics proper. We explained in the Introduction why this dogma appears unjustified to us (*), and we hold that the role of metamathematics in an account of logic and mathematics can and should be confined to the very elementary part which deals with the manipulation of abbreviating symbols and deductive criteria. Contrary to what Poincaré asserted, it is thus not a question of "claiming freedom from contradiction", but rather of considering, as Hadamard did, that consistency, even when it cannot be proved, can be ascertained ([35], p. 270).

It remains for us to give a brief historical sketch of the efforts of Hilbert and his school. Although the theory of proof is not touched on in this series, it is not without interest to retrace in outline not only the evolution of ideas which finally led to Gödel's negative result and justified *a posteriori*

(*) According to the pure formalist doctrine, the words "there exists" in a formalized text have no more "meaning" than any others, and there is no other type of "existence" to be considered in formalized proofs.

Hadamard's scepticism, but also all the progress which has resulted in an understanding of the mechanism of mathematical reasoning, and which has elevated modern metamathematics to the position of an autonomous science of undeniable importance.

In 1904, in an address to the International Congress ([30], pp. 247-261), Hilbert attacked the problem of the consistency of arithmetic. First, he established that a proof of consistency could not be obtained by recourse to a "model" (*), and he indicated in broad outline the principle of another method. He proposed to consider the true propositions of formalized arithmetic as assemblies of signs without any meaning and to show that, by using the rules governing the formation and juxtaposition of these assemblies, an assembly could never be obtained which was a true proposition and whose negation was also a true proposition. He sketched a proof of this nature for a formalism less elaborate than that of arithmetic; but, as H. Poincaré observed soon afterward ([33 c], p. 185), this proof made essential use of the principle of induction, and therefore appeared to be founded on a vicious circle. Hilbert did not reply to this criticism immediately, and fifteen years went by before anyone tried to develop his ideas. It was only in 1917 that (moved by the desire to answer the attacks of the intuitionists) he devoted himself again to the problem of the foundations of mathematics, which from then on occupied him continuously until the end of his scientific career. In his work on this subject, which stretches from 1920 to about 1930, and in which a whole school of young mathematicians (Ackermann, Bernays, Herbrand, von Neumann) were active participants, Hilbert gradually extracted the principles of his "theory of proof" in a more precise fashion. Realizing implicitly the justice of Poincaré's criticism, he admitted that in metamathematics the arithmetical arguments used could be based only on our intuition of the integers (and not on formalized arithmetic). Thus it appeared essential to restrict these arguments to "finite procedures" (*finite Prozesse*) of a type allowed by the intuitionists. For example, a proof by *reductio ad absurdum* cannot establish the metamathematical existence of an assembly or a sequence of assemblies; it is necessary to give an explicit law of construction (†). Furthermore, Hilbert enlarged his initial program in two directions; not only did he attack the problem of the consistency of arithmetic, but he also hoped to prove the consistency of the theory of

(*) The "models" provided by the definitions of Dedekind and Frege serve only to shift the question to the consistency of the theory of sets, which is certainly a more difficult problem than the consistency of arithmetic, and must have appeared even more so at this period, when no serious attempt to get round the "paradoxes" had yet been proposed.

(†) For a detailed and precise description of the finite procedures allowed in metamathematics, the reader may consult, for example, Herbrand's thesis [42].

real numbers and even of the theory of sets (*). To the problems of consistency were added those of independence of the axioms, of categoricity, and the decision problem. We shall give a brief review of these various questions and mention the principal researches which they have occasioned.

A proof of the *independence* of a system of propositions $A_1, A_2, \ldots, A_n$ consists in showing that, for each index i, A_i is not a theorem in the theory $\mathcal{T}_i$ obtained by taking as axioms the A_j where $j \neq i$. All we need is to find a non-contradictory theory $\mathcal{T}'_i$ in which the A_j $(j \neq i)$ are theorems, together with "not A_i"; we may therefore consider the problem in two aspects according as we do or do not assume that certain theories (such as arithmetic or set theory) are consistent. In the latter case, we have a problem of "absolute" consistency. On the other hand, the first type of problem is to be solved, like problems of "relative" consistency, by the construction of appropriate "models", and many proofs of this nature were devised well before mathematics had taken on a completely formalized aspect. We may cite as examples the models of non-Euclidean geometry, the questions of the independence of the axioms of elementary geometry treated by Hilbert in the *Grundlagen der Geometrie* [30], the work of Steinitz on the axiomatization of algebra, and the work of Hausdorff and his successors on the axiomatization of topology.

A theory $\mathcal{T}$ is said to be *categorical* if, for every proposition A in $\mathcal{T}$ which contains no letters other than the constants of $\mathcal{T}$, one of the two propositions A, (not A) is a theorem in $\mathcal{T}$ (†). Apart from some very rudimentary formalism whose categoricity is easily proved ([32], p. 35; cf. Chapter I, Appendix, Exercise 7), the results obtained in this area are essentially negative. The most important is due to K. Gödel who showed that if $\mathcal{T}$ is non-contradictory and if the axioms of formalized arithmetic are theorems in $\mathcal{T}$, then $\mathcal{T}$ is not categorical. The fundamental idea of his ingenious method consists in establishing a one-to-one correspondence (of course, by means of "finite procedures") between metamathematical statements and certain propositions of formalized arithmetic. We give here an outline sketch of the argument (**). With each assembly A which is a term or a relation in $\mathcal{T}$, we associate (by an explicit constructive procedure which can be applied quasi-mechanically) an integer $g(\mathrm{A})$ in a one-to-one manner. Likewise, with every proof D in $\mathcal{T}$ (considered as a succession of assemblies; cf. Chapter I, § 2, no. 2) we may associate

(*) When we speak of the consistency of the theory of real numbers, we suppose that this theory has been axiomatically defined without recourse to the theory of sets (or at least without recourse to certain axioms of the latter, such as the axiom of choice or the axiom of the set of subsets).

(†) This is often expressed by saying that if A is not a theorem in $\mathcal{T}$, then the theory $\mathcal{T}'$ obtained by adjoining A to the axioms of $\mathcal{T}$ is contradictory.

(**) For more details, see [44 a] or [48], pp. 181-258.

an integer $h(\mathrm{D})$ in a one-to-one manner. Finally, we can give an explicit procedure for constructing a relation $\mathrm{P}\{x, y, z\}$ in $\mathcal{T}$ (*) such that, in $\mathcal{T}$, $\mathrm{P}\{x, y, z\}$ implies that x, y, z are integers, and satisfies the following two conditions :

(1) If D is a proof of $\mathrm{A}\{\lambda\}$, where $\mathrm{A}\{x\}$ is a relation in $\mathcal{T}$, and λ is a definite integer (that is, a term in $\mathcal{T}$ which is an integer), then

$$\mathrm{P}\{\lambda, g(\mathrm{A}\{x\}), h(\mathrm{D})\}$$

is a theorem in $\mathcal{T}$.

(2) If the definite integer μ is not of the form $h(\mathrm{D})$, or if $\mu = h(\mathrm{D})$ and D is not a proof $\mathrm{A}\{\lambda\}$, then (not $\mathrm{P}\{\lambda, g(\mathrm{A}\{x\}), \mu\}$) is a theorem in $\mathcal{T}$.

Now let $\mathrm{S}\{x\}$ be the relation (not $(\exists z)\mathrm{P}\{x, x, z\}$), and let $\gamma = g(\mathrm{S}\{x\})$, which is a term in $\mathcal{T}$. If $\mathcal{T}$ is not contradictory, there is *no proof* of the proposition $\mathrm{S}\{\gamma\}$ in $\mathcal{T}$. For if D were such a proof, then

$$\mathrm{P}\{\gamma, g(\mathrm{S}\{x\}), h(\mathrm{D})\}$$

would be a theorem in $\mathcal{T}$. This relation is just $\mathrm{P}\{\gamma, \gamma, h(\mathrm{D})\}$, and consequently $(\exists z)\mathrm{P}\{\gamma, \gamma, z\}$ would also be a theorem in $\mathcal{T}$; but since the last relation is equivalent to (not $\mathrm{S}\{\gamma\}$), $\mathcal{T}$ would be contradictory. Furthermore, what has just been said shows that, for each definite integer μ, (not $\mathrm{P}\{\gamma, \gamma, \mu\}$) is a theorem in $\mathcal{T}$. It follows that there is *no proof* in $\mathcal{T}$ of the proposition (not $\mathrm{S}\{\gamma\}$), for this relation is equivalent to $(\exists z)\,\mathrm{P}\{\gamma, \gamma, z\}$, and the existence of an integer μ such that $\mathrm{P}\{\gamma, \gamma, \mu\}$ would imply that $\mathcal{T}$ was contradictory, by virtue of the preceding argument (†). This metamathematical theorem of Gödel has been subsequently generalized in various directions ([48], Chapter XI) (**).

(*) The detailed description of $g(\mathrm{A})$, $h(\mathrm{D})$, and $\mathrm{P}\{x, y, z\}$ is extremely long and tedious, and to write out $\mathrm{P}\{x, y, z\}$ would require so large a number of signs as to be impossible in practice. It is this type of difficulty that we discussed in the Introduction, but no mathematician would consider that this diminishes in any way the validity of these constructions.

(†) In fact, the last part of this argument presupposes a little more than the consistency of $\mathcal{T}$, namely what is called the "ω-consistency" of $\mathcal{T}$. This means that there is no relation $\mathrm{R}\{x\}$ in $\mathcal{T}$ such that $\mathrm{R}\{x\}$ implies $x \in \mathrm{N}$ and such that, for each *definite* integer μ, $\mathrm{R}\{\mu\}$ is a theorem in $\mathcal{T}$, although $(\exists x)(x \in \mathrm{N}$ and (not $\mathrm{R}\{x\}))$ is also a theorem in $\mathcal{T}$. However, Rosser has shown that Gödel's argument can be modified so as to require only the consistency of $\mathcal{T}$ ([48], p. 208).

(**) The reader may have noticed the analogy between Gödel's argument and the sophism of the liar : the proposition $\mathrm{S}\{\gamma\}$ implies its own falsity when interpreted in metamathematical terms! It should also be noted that the proposition $(\forall z)((z \in \mathrm{N}) \Longrightarrow (\text{not } \mathrm{P}\{\gamma, \gamma, z\}))$ is intuitively true, because we have a proof

¶ The relation $S\{\gamma\}$ in $\mathfrak{T}$, which is thus shown to have the property that there exists no proof in $\mathfrak{T}$ either of it or of its negation, has obviously been manufactured to fit the requirements of the argument and is not related in any natural way to any mathematical problem. What is much more interesting is the fact that, if $\mathfrak{T}$ denotes the theory of sets (with the system of axioms of von Neumann-Bernays), *neither the continuum hypothesis nor its negation are provable in* $\mathfrak{T}$. This remarkable result has been established in two steps : in 1940, Gödel proved that the theory obtained by adjoining to $\mathfrak{T}$ the hypothesis $2^{\aleph_0} = \aleph_1$ is not contradictory [44b]; and quite recently. P. Cohen has proved that the same is met when the relation $2^{\aleph_0} = \aleph_2$ (or $2^{\aleph_0} = \aleph_n$ for any integer $n > 1$) is adjoined to $\mathfrak{T}$ [49].

The decision problem (*Entscheidungsproblem*) is certainly the most ambitious of all in metamathematics. The problem is whether, for a given formalized language, a quasi-mechanical "universal procedure" can be conceived which, when applied to any relation whatever in the formalism under consideration, will indicate in a finite number of operations whether the relation is true or not. The solution of this problem formed a substantial part of the grand designs of Leibniz, and it seems that at one point the Hilbert school believed they were very close to a solution. It is a fact that such procedures can be described for formalisms which contain few primitive signs and axioms ([48], pp. 136-141; cf. Chapter I, Appendix, Exercise 7). But the efforts to make precise the decision problem by defining exactly what is to be meant by "universal procedure" have led so far only to negative results ([48], pp. 432-439). Moreover, the solution of the decision problem for a theory $\mathfrak{T}$ would tell immediately whether or not $\mathfrak{T}$ is contradictory, because the "universal procedure" could be

---

of "not $P\{\gamma, \gamma, \mu\}$" in $\mathfrak{T}$ for each *definite* integer μ. Nevertheless this proposition is unprovable in $\mathfrak{T}$. The above situation should be compared with a result obtained earlier by Lowenheim and Skolem (see [39]). The latter defines metamathematically a relation between two *natural integers* x, y which, when written as $x \in y$, satisfies von Neumann's axioms of set theory. Hence, at first sight, we have a new "paradox", because in this "model" all the infinite sets are countable, contrary to Cantor's inequality $\mathfrak{m} < 2^{\mathfrak{m}}$. But, in fact, the relation defined by Skolem cannot be written in formalized set theory any more than the "theorem" asserting that the set of subsets of an infinite set has only a countable infinity of "elements". At bottom, this "paradox" is only a more subtle form of the banal remark that one can write down only a finite number of assemblies in a formalized theory, and that it is therefore absurd to conceive of an uncountable set of terms of the theory — a remark parallel to that which had already led to "Richard's paradox". Similar arguments show that the formalization of set theory is indispensable if we wish to preserve the essentials of Cantor's edifice. Mathematicians are in agreement that there is hardly more than a superficial concordance between our "intuitive" conceptions of the notion of set or of integer, and the formalisms which are supposed to account for them; the disagreement begins with the question of choosing between them.

applied to a relation in $\mathfrak{T}$ and its negation; and, as we shall see, the possibility of resolving the question in this way is excluded, so far as the usual mathematical theories are concerned (*).

It is in fact in the question of consistency of mathematical theories — the origin and the very heart of mathematics — that the results have turned out to be most deceptive. During the years 1920-1930, Hilbert and his school developed new methods for attacking these problems. Having proved the consistency of some partial formalisms, covering a part of arithmetic (cf. [42], [43 c]), they thought they were on the verge of victory and that they would succeed in proving the consistency not only of arithmetic, but also of set theory, when Gödel used the non-categoricity of arithmetic to deduce that it is impossible to prove, by means of Hilbert's "finite procedures", the consistency of any theory E which contains arithmetic (†).

Nevertheless, Gödel's theorem does not close the door altogether to attempts to prove consistency, provided that Hilbert's restrictions concerning "finite procedures" are (at least partially) abandoned. Thus, in 1936, Gentzen [47] succeeded in proving the consistency of formalized arithmetic by "intuitive" use of transfinite induction as far as the countable ordinal ε_0 (Chapter III, § 6, Exercise 14) (**). The degree of "certainty" which can

---

(*) The decision problem should be carefully distinguished from the belief, held by many mathematicians and often forcefully expressed by Hilbert in particular, that whether a given mathematical proposition is true, false, or undecidable will always ultimately be settled. This is a pure act of faith, and lies outside our discussion.

(†) With the notation introduced above, Gödel's precise result is as follows. To say that $\mathfrak{T}$ is consistent means that there is no proof, in $\mathfrak{T}$, of the relation $0 \neq 0$. This implies, for each definite integer μ, that (not $\mathrm{P}\}0, g(x \neq x), \mu\{$) is a theorem in $\mathfrak{T}$. Consider the proposition

$$(\forall z)((z \in \mathrm{N}) \Longrightarrow \text{not } \mathrm{P}\}0, g(x \neq x), z\{),$$

which we denote by C. By "translating" into formalized arithmetic the argument (reproduced above) by which it is proved metamathematically that "if $\mathfrak{T}$ is consistent, then there is no proof of $\mathrm{S}\{\gamma\}$ in $\mathfrak{T}$", we can show that $\mathrm{C} \Longrightarrow (\text{not } (\exists z)(\mathrm{P}\}\gamma, \gamma, z\{))$ is a theorem in $\mathfrak{T}$, i.e., that $\mathrm{C} \Longrightarrow \mathrm{S}\{\gamma\}$ is a theorem in $\mathfrak{T}$. It follows that, if $\mathfrak{T}$ is consistent, C *is not a theorem in* $\mathfrak{T}$, because in these conditions $\mathrm{S}\{\gamma\}$ is not a theorem in $\mathfrak{T}$. This is the exact statement of Gödel's theorem.

(**) Gentzen associates with each proof D in formalized arithmetic an ordinal $\alpha(\mathrm{D}) < \varepsilon_0$. He also describes a procedure which, starting from any proof D which leads to a contradiction, produces a proof D′ which also leads to a contradiction and is such that $\alpha(\mathrm{D}') < \alpha(\mathrm{D})$. The theory of well-ordered sets then allows him to conclude that such a proof D does not exist (a type of reasoning which extends the classical "infinite descent" of number theory).

be attributed to such an argument is undoubtedly less than for those which satisfy Hilbert's original restrictions, and is essentially a matter depending on the personal psychology of the individual mathematician. But it is no less true that similar "proofs" using "intuitive" transfinite induction as far as some given ordinal would be considered as an important step forward if they could be applied, for example, to the theory of real numbers or to an important part of set theory.

Furthermore, within set theory itself there are many problems of "relative" consistency connected with the many "hypotheses" at large in the theory. The most remarkable result in this area is again due to Gödel. In 1940 he proved that, if the theory whose axioms are those of the von Neumann-Bernays system, except for the axiom of choice, is consistent, then the theory obtained by adjoining to these axioms a very strong form of the axiom of choice (of the type which the use of the symbol τ gives us) and the generalized continuum hypothesis is also consistent [44 b]. More recently, I. Novak-Gal has shown that the consistency of the Zermelo-Fraenkel system implies that of the von Neumann-Bernays-Gödel system (*).

BIBLIOGRAPHY

1. O. Neugebauer, *Vorlesungen über die Geschichte der antiken Mathematik*, Vol. I, Vorgriechische Mathematik, Berlin (Springer), 1934.

2. *The Works of Aristotle*, translated under the editorship of J. A. Smith and W. D. Ross (12 volumes, Oxford, 1908-1952).

2 (*bis*). T. L. Heath, *Mathematics in Aristotle*, Oxford (Clarendon Press), 1949.

3. T. L. Heath, *The Thirteen Books of Euclid's Elements*..., 3 volumes, Cambridge, 1908.

4. *Archimedis Opera Omnia*, 3 volumes, edited by J. L. Heiberg, 2nd edition, 1913-1915.

4 (*bis*). T. L. Heath, *The Method of Archimedes*, Cambridge, 1912.

5. J. M. Bochenski, *Ancient Formal Logic*, Studies in Logic, Amsterdam (North Holland Publ. Co.), 1951.

6. P. Böhner, *Medieval Logic, an Outline of its Development from* 1250 *to ca.* 1400, Chicago, 1952.

7. D. Francisci Maurolyci, *Abbatis Messamensis, Mathematici Celeberrimi, Arithmeticorum Libri Duo*, Venice, 1575.

8. Galileo Galilei, *Opere*, Ristampa della Edizione Nazionale, 20 volumes, Florence (Barbara), 1929-39.

(*) I. Novak-Gal, "A construction for models of consistent systems", *Fund. Math.*, **37** (1950), pp. 87-110.

9. A. Girard, *Invention nouvelle en l'Algèbre*, 1629 (new edition, Bierens de Haan, 1884).
10. R. Descartes, *Œuvres*, edited by C. Adam and P. Tannery, 11 volumes, Paris (L. Cerf), 1897-1909.
11. B. Pascal, *Œuvres*, edited by Brunschvicg, 14 volumes, Paris (Hachette), 1904-1914.
12. G. W. Leibniz : (a) *Mathematische Schriften*, edited by C. I. Gerhardt, 7 volumes, Berlin-Halle (Asher-Schmidt), 1849-1863; (b) *Philosophische Schriften*, edited by C. I. Gerhardt, 7 volumes, Berlin, 1840-1890; (c) *Opuscules et fragments inédits*, edited by L. Couturat, Paris (Alcan), 1903.
12 (*bis*). L. Couturat, *La logique de Leibniz d'après des documents inédits*, Paris (Alcan), 1901.
13. D'Alembert, *Encyclopédie*, Paris, 1751-1765, articles "Négatif", "Imaginaire", "Définition".
14. C. F. Gauss, *Werke*, 12 volumes, Göttingen, 1870-1927.
15. N. Lobatschevsky, *Pangeometrie*, Ostwald's Klassiker, no. 130, Leipzig (Engelmann), 1902.
16. G. Boole : (a) *The Mathematical Analysis of Logic*, Cambridge-London, 1847 (= *Collected Logical Works*, edited by P. Jourdain, Chicago-London, 1916, vol. I); (b) *An Investigation of the Laws of Thought*, Cambridge-London, 1854 (= *Collected Logical Works*, vol. II).
17. H. Grassmann, *Gesammelte Werke*, 6 volumes, Leipzig (Teubner), 1894.
18. B. Bolzano, *Paradoxien des Unendlichen*, Leipzig, 1851.
19. B. Riemann, *Gesammelte mathematische Werke*, 2nd edition, Leipzig (Teubner), 1892.
20. H. Hankel, *Theorie der complexen Zahlensysteme*, Leipzig (Voss), 1867.
21. A. De Morgan : (a) "On the syllogism (III)", *Trans. Camb. Phil. Soc.*, **10** (1858), pp. 173-230; (b) "On the syllogism (IV) and on the logic of relations", *Trans. Camb. Phil. Soc.*, **10** (1860) pp. 331-358.
22. C. S. Peirce, (a) "Upon the logic of mathematics", *Proc. Amer. Acad. Arts and Sci.*, **7** (1865-1868), pp. 402-412; (b) "On the algebra of logic", *Amer. Journ. of Math.*, **3** (1880), pp. 49-57; (c) "On the algebra of logic", *Amer. Journ. of Math.*, **7** (1884), pp. 190-202.
23. E. Schröder, *Vorlesungen über die Algebra der Logik*, 3 volumes, Leipzig (Teubner), 1890.
24. G. Frege, (a) *Begriffsschrift eine der arithmetischen nachgebildete Formelsprache des reinen Denkens*, Halle, 1879; (b) *Die Grundlagen der Arithmetik*, 2nd edition with an English translation by J. L. Austin, New York, 1950; (c) *Grundgesetze der Arithmetik, begriffsschriftlich abgeleitet*, 2 volumes, Jena, 1893-1903.
25. G. Cantor, *Gesammelte Abhandlungen*, Berlin (Springer), 1932.
25 (*bis*). G. Cantor, R. Dedekind, *Briefwechsel*, edited by J. Cavaillès and E. Noether, *Act. Sci. et Ind.*, no. 518, Paris (Hermann), 1937.

26. R. Dedekind, *Gesammelte mathematische Werke*, 3 volumes, Braunschweig (Vieweg), 1932.
27. C. Hermite, T. Stieltjes, *Correspondance*, 2 volumes, Paris (Gauthier-Villars), 1905.
28. M. Pasch und M. Dehn, *Vorlesungen über neuere Geometrie*, 2nd edition, Berlin, (Springer), 1926.
29. G. Peano, (a) *Arithmeticae principia, novo modo exposita*, Turin, 1889; (b) *I principii di Geometria, logicamente espositi*, Turin, 1889; (c) *Formulaire de Mathematiques*, 5 volumes, Turin, 1895-1905.
30. D. Hilbert, *Grundlagen der Geometrie*, 7th edition, Leipzig-Berlin (Teubner), 1930.
31. D. Hilbert, *Gesammelte Abhandlungen*, vol. III, Berlin (Springer), 1935.
32. D. Hilbert, W. Ackermann, *Grundzüge der theoretischen Logik*, 3rd edition, Berlin (Springer), 1949.
33. H. Poincaré : (a) *Science et hypothèse*, Paris (Flammarion), 1906; (b) *La valeur de la science*, Paris (Flammarion), 1905; (c) *Science et méthode*, Paris (Flammarion), 1920.
34. J. Richard, "Les principes des mathématiques et le problème des ensembles", *Rev. Gen. des Sci. Pures et Appl.*, **16** (1905), pp. 541-543.
35. R. Baire, E. Borel, J. Hadamard, H. Lebesgue, "Cinq lettres sur la théorie des ensembles", *Bull. Soc. Math. de France*, **33** (1905), pp. 261-273.
36. E. Zermelo, (a) "Beweis dass jede Menge wohlgeordnet werden kann", *Math. Ann.*, **59** (1904), pp. 514-516; (b) "Neuer Beweis fur die Möglichkeit einer Wohlordnung", *Math. Ann.*, **65** (1908), pp. 107-128; (c) "Untersuchungen über die Grundlagen der Mengenlehre", *Math. Ann.*, **65** (1908), pp. 261-281.
37. B. Russell and A. N. Whitehead, *Principia Mathematica*, 3 volumes, Cambridge, 1910-1913.
38. L. E. J. Brouwer, (a) "Intuitionism and formalism", *Bull. Amer. Math. Soc.*, **20** (1913), pp. 81-96; (b) "Zur Begründung des intuitionistischen Mathematik", *Math. Ann.*, **93** (1925), pp. 244-257; **95** (1926), pp. 453-473; **96** (1926), pp. 451-458.
39. T. Skolem, "Einige Bemerkungen zur axiomatischen Begründung der Mengenlehre", *Wiss. Vorträge, 5 Kongress Skand. Math.*, Helsingfors, 1922.
40. K. Kuratowski, "Une méthode d'élimination des nombres transfinis des raisonnements mathématiques", *Fund. Math.*, **5** (1922), pp. 76-108.
41. A. Fraenkel : (a) "Zu den Grundlagen der Cantor-Zermeloschen Mengenlehre", *Math. Ann.*, **86** (1922), pp. 230-237; (b) *Zehn Vorlesungen über die Grundlgaung der Mengenlehre*, Wiss. und Hypothese vol. 31, Leipzig-Berlin, 1927; (c) *Einleitung in die Mengenlehre*, 3rd edition, Berlin (Springer), 1928.

42. J. Herbrand, "Recherches sur la théorie de la démonstration", *Trav. Soc. Sci. et Lett. Varsovie*, cl. II (1930), pp. 33-160.
43. J. von Neumann : (a) "Eine Axiomatisierung der Mengenlehre", *Crelle*, **154** (1925), pp. 219-240; (b) "Die Axiomatisierung der Mengenlehre", *Math. Zeitschr.*, **27** (1928), pp. 669-752; (c) "Zur Hilbertschen Beweistheorie", *Math. Zeitschr.*, **26** (1927), pp. 1-46.
44. K. Gödel : (a) "Über formal unentscheidbare Sätze der Principia Mathematica und verwandter Systeme", *Monatsh. für Math. u. Phys.*, **38** (1931), pp. 173-198; (b) *The consistency of the axiom of choice and of the generalized continuum hypothesis*, Ann. of Math. Studies, no. 3, Princeton, 1940.
45. M. Zorn, "A remark on method in transfinite algebra", *Bull. Amer. Math. Soc.*, **41** (1935), pp. 667-670.
46. A. Heyting, *Mathematische Grundlagenforschung. Intuitionismus. Beweistheorie*, Ergebnisse der Math., vol. 3, Berlin (Springer), 1934.
47. G. Gentzen, *Die gegenwartige Lage in der mathematischen Grundlagenforschung. Neue Fassung des Widerspruchfreiheitsbeweises für die reine Zahlentheorie*, Forschungen der Logik..., Heft 4, Leipzig (Hirzel), 1938.
48. S. Kleene, *Introduction to Metamathematics*, New York, 1952.
49. P. J. Cohen, "The independence of the continuum hypothesis", *Proc. Nat. Acad. Sci.*, **50** (1963), pp. 1143-1148 and **51** (1964), pp. 105-110.

Summary of Results

INTRODUCTION

This Summary contains all the definitions and all the results, but none of the proofs, from the theory of sets which will be used in the remainder of this series. As for the notions and terms introduced below without definitions, the reader may safely take them with their usual meanings; this will not cause any difficulties as far as the remainder of the series is concerned, and renders almost trivial the majority of the propositions (*).

1. ELEMENTS AND SUBSETS OF A SET

1. A *set* consists of *elements* which are capable of possessing certain *properties* and of having certain *relations* between themselves or with elements of other sets.

2. Sets and elements are denoted in mathematical arguments by graphical symbols, which in general are letters (from various alphabets) or combinations of letters and other signs. Relations between elements of one or more sets are denoted by inserting the symbols which denote these elements into a scheme characteristic of the relation considered (†); and similarly for properties.

A letter may denote either a *fixed* element or an *arbitrary* element (also called a *variable*, an *argument*, or a *generic* element) of a set. When an arbitrary element is replaced by a fixed element in a relation (or property), the arbitrary element is said to be given this fixed element as *value*.

(*) The reader will not fail to observe that the "naïve" point of view taken here is in direct opposition to the "formalist" point of view taken in Chapters I to IV. Of course, this contrast is deliberate, and corresponds to the different purposes of this Summary and the rest of the volume.

(†) When the symbol which denotes an element is a combination of several signs and is to be inserted into a relation in the place of a single letter, it is customary to put it in brackets in order to avoid possible confusion.

In order to indicate the elements which appear in a relation which is not explicitly written down, we represent the relation by a notation such as $R\{x, y, z\}$ (if x, y, z are the elements in question).

3. A relation or a property in which arbitrary elements feature (*) is said to be an *identity* if it becomes a true proposition whatever values we give to these arbitrary elements. If R and S denote two relations (or properties), R is said to *imply* S if S is true whenever the arbitrary elements which enter in these relations are fixed in such a way that R is true. The relations (or properties) R and S are said to be *equivalent* if each implies the other.

4. Let $R\{x, y, z\}$ be a relation between the variables x, y, z. The phrase "for all x, $R\{x, y, z\}$" is a relation *between y and z*, which will be considered to be true for a system of given values of these latter variables if R is true for these values of y and z and *every* value of x. The phrase "there exists x such that $R\{x, y, z\}$" (or "for some x, $R\{x, y, z\}$") is again a relation between y and z, which will be considered to be true for a system of given values of y and z if, these variables being thus fixed, there is *at least one* value of x for which R is true. Similarly for a relation between any number of variables.

If $\overline{R}$ denotes the *negation* of R, then the negation of "for all x, R" is "there exists x such that $\overline{R}$"; the negation of "there exists x such that R" is "for all x, $\overline{R}$".

5. If R, S denote two relations, we regard "R and S" as a *single* relation, which is considered as true whenever *both* R *and* S are true. Likewise, "R or S" is a relation which is considered to be true whenever *at least one* of the relations R, S is true (and, in particular, whenever they are both true. The word "or" thus does not have the disjunctive sense here which it sometimes has in ordinary speech). Let $\overline{R}$, $\overline{S}$ denote the negations of R, S, respectively. Then the negation of "R and S" is "$\overline{R}$ or $\overline{S}$", and the negation of "R or S" is "$\overline{R}$ and $\overline{S}$".

6. By writing two symbols one on each side of the sign "=" (read "equals"), we have a relation called the relation of *equality*, which means that the two symbols represent the *same* element. The negation of this relation is obtained by writing the same symbols one on each side of the sign "≠" (read "is not equal to" or "is different from").

(*) It should be emphasized that, when we speak of a *property of a generic* element of a set E, this in no way implies that the property is true for *every* element of E, but simply that it *has a meaning* for *every* element of E; it may be true for some of these elements and false for others. Similarly for relations.

7. Given a set E and a *property* of a generic element of E, those of the elements of E which have this property form a new set, called a *subset* of E. Two *equivalent* properties therefore define the *same* subset of E, and conversely.

Let A be a subset of E. When x is a generic element of E, the property "x belongs to A" (i.e., "x is an element of A") is written "$x \in A$"; the set of elements which have this property is clearly just A.

The negation of this property is written "$x \notin A$" (read "x does not belong to A"); the set of elements of E which have this property is called the *complement* of A and is written $\complement A$ or $E - A$.

8. Some properties, for example $x = x$, are true for *all* elements of E. Any two such properties are equivalent, and the subset they define is the set E itself.

On the other hand, some properties, for example $x \neq x$, are not true for *any* element of E. Again, any two such properties are equivalent, and the subset they define is called the *empty subset* of E, which is denoted by $\emptyset$.

Note that E and $\emptyset$ are *complements* of each other.

9. Let a be a determinate element of E. Some properties, for example $x = a$, are true only for the *single* element a. Any two such properties are equivalent; the subset they define is denoted by $\{a\}$, and is called the subset *consisting of a alone.*

10. The set whose elements are all the *subsets* of a set E is called the *set of subsets* of E, and is denoted by $\mathfrak{P}(E)$. We have $\emptyset \in \mathfrak{P}(E)$, $E \in \mathfrak{P}(E)$, and $\{x\} \in \mathfrak{P}(E)$ for all $x \in E$. If x denotes a generic element of E, and X a generic element of $\mathfrak{P}(E)$, the relation "$x \in X$" between x and X is called the *relation of membership.*

11. Let x and y be two elements of E and let X be a generic element of $\mathfrak{P}(E)$. Then the relation of equality "$x = y$" is *equivalent* to the relation "for all X such that $x \in X$, we have $y \in X$".

12. Let X, Y be two subsets of a set E. If the property $x \in X$ implies the property $x \in Y$, in other words if every element of X belongs to Y, then we say that X is *contained in* Y, or that Y *contains* X, or that X *is a subset of* Y. This relation between X and Y is called the relation of *inclusion* (of X in Y), and is denoted by "$X \subset Y$" or "$Y \supset X$". Its negation is denoted by "$X \not\subset Y$" or "$Y \not\supset X$".

For all subsets X of E we have $\emptyset \subset X$ and $X \subset E$. The relation of membership "$x \in X$" is equivalent to "$\{x\} \subset X$".

The relation "$X \subset Y$ and $Y \subset Z$" implies "$X \subset Z$".

The relation "$X \subset Y$" does not exclude the possibility of "$X = Y$". The relation "$X \subset Y$ and $Y \subset X$" is equivalent to "$X = Y$".

13. Let X and Y be any two subsets of E. The set of all elements which have the property "$x \in X$ or $x \in Y$" is denoted by $X \cup Y$ and is called the *union* of X and Y. The set of all elements which have the property "$x \in X$ and $x \in Y$" is denoted by $X \cap Y$ and is called the *intersection* of X and Y.

The union and intersection of several subsets of E are defined in the same way.

If x, y, z are three elements of E, the union $\{x\} \cup \{y\} \cup \{z\}$ is denoted by $\{x, y, z\}$. Similarly for any number of (individually named) elements.

Let X and Y be two subsets of E. According as $X \cap Y \neq \emptyset$ or $X \cap Y = \emptyset$, we say that X and Y *intersect* or are *disjoint*.

14. In the statements of the following propositions, X, Y, Z denote any subsets of the same set E.

(a) We have $\emptyset = \complement E$, $E = \complement \emptyset$.

(b) For all X we have

$$\complement(\complement X) = X; \tag{1}$$

$$X \cup X = X, \quad X \cap X = X; \tag{2}$$

$$X \cup (\complement X) = E, \quad X \cap (\complement X) = \emptyset; \tag{3}$$

$$X \cup \emptyset = X, \quad X \cap E = X; \tag{4}$$

$$X \cup E = E, \quad X \cap \emptyset = \emptyset. \tag{5}$$

(c) For all X, Y we have

$$X \cup Y = Y \cup X, \quad X \cap Y = Y \cap X \quad \text{(commutativity)}; \tag{6}$$

$$X \subset X \cup Y, \quad X \cap Y \subset X; \tag{7}$$

$$\complement(X \cup Y) = (\complement X) \cap (\complement Y), \quad \complement(X \cap Y) = (\complement X) \cup (\complement Y). \tag{8}$$

(d) The relations $X \subset Y$, $\complement X \supset \complement Y$, $X \cup Y = Y$, $X \cap Y = X$ are *equivalent*.

(e) The relations $X \cap Y = \emptyset$, $X \subset \complement Y$, $Y \subset \complement X$ are *equivalent*.

(f) The relations $X \cup Y = E$, $\complement X \subset Y$, $\complement Y \subset X$ are *equivalent*.

(g) For all X, Y, Z we have

$$\left.\begin{aligned} X \cup (Y \cup Z) &= (X \cup Y) \cup Z = X \cup Y \cup Z \\ X \cap (Y \cap Z) &= (X \cap Y) \cap Z = X \cap Y \cap Z \end{aligned}\right\} \text{(associativity)}; \tag{9}$$

$$\left.\begin{aligned} X \cup (Y \cap Z) &= (X \cup Y) \cap (X \cup Z) \\ X \cap (Y \cup Z) &= (X \cap Y) \cup (X \cap Z) \end{aligned}\right\} \text{(distributivity)}. \tag{10}$$

(h) The relation "$X \subset Y$" implies the relations "$X \cup Z \subset Y \cup Z$" and "$X \cap Z \subset Y \cap Z$".

(i) The relation "$Z \subset X$ and $Z \subset Y$" is equivalent to "$Z \subset X \cap Y$". The relation "$X \subset Z$ and $Y \subset Z$" is equivalent to "$X \cup Y \subset Z$".

15. From the identities (8) we conclude that if a subset A of E is obtained from other subsets X, Y, Z of E by applying *only* the operations $\complement$, $\cup$, $\cap$ (in any order), then the complement $\complement A$ can be obtained by replacing the subsets X, Y, Z by their respective complements, and the operations $\cup$, $\cap$ by $\cap$, $\cup$, respectively, while preserving the order of the operations. This is the *duality rule*. Given an equality $A = B$ between subsets of the above form, consider the equivalent equality $\complement A = \complement B$. If we replace $\complement A$ and $\complement B$ by the expressions obtained by applying the duality rule, and if then we replace $\complement X$, $\complement Y$, $\complement Z$ by X, Y, Z, respectively, and vice versa, we obtain an equality called the *dual* of $A = B$. We can do the same for an inclusion relation $A \subset B$, but then we must take care to replace the sign "$\subset$" by "$\supset$".

The identities above which carry the same number are duals of each other.

16. In certain questions we have to consider a fixed subset A of a set E. If X is any arbitrary subset of E, the set $A \cap X$ is called the *trace* of X on A, and is sometimes denoted by X_A; it is considered, in this case, as a subset of A. For all subsets X, Y of E we have

$$(X \cup Y)_A = X_A \cup Y_A, \qquad (X \cap Y)_A = X_A \cap Y_A,$$

and

$$\complement_A X_A = (\complement_E X)_A,$$

where $\complement_E X$ denotes the complement of X in E and $\complement_A X_A$ denotes the complement of X_A in A.

If $\mathfrak{E}$ denotes a set of subsets of E, then the set $\mathfrak{E}_A$ of the traces on A of the sets of $\mathfrak{E}$ is called the *trace* of $\mathfrak{E}$ on A.

2. FUNCTIONS

1. Let E and F be two sets, which may or may not be distinct. A relation between a variable element x of E and a variable element y of F is called a *functional relation in* y if, *for all* $x \in E$, *there exists a unique* $y \in F$ *which is in the given relation with* x.

We give the name of *function* to the operation which in this way associates with every element $x \in E$ the element $y \in F$ which is in the given relation with x; y is said to be the *value* of the function at the element x, and the function is said to be *determined* by the given functional relation. Two *equivalent* functional relations determine the *same* function. Such a function is said to "take its values in F" and to be "defined on E". More briefly, it is also said to be a *mapping of* E *into* F.

2. The mappings of a set E into a set F are the *elements* of a new set, the *set of mappings of* E *into* F. If f is any element of this set, the value of f at the element x of E is often denoted by $f(x)$. In some situations, the notation f_x (called the *indicial* notation; the set E is then called the *index* set) is preferable. The relation "$y = f(x)$" is a functional relation in y, which determines f.

When a relation of the form $y = \langle x \rangle$ (where $\langle x \rangle$ denotes a combination of signs in which x may appear) is a functional relation in y, the function it determines is often denoted by the notation $x \rightarrow \langle x \rangle$, or even simply by $\langle x \rangle$; this is a very common abuse of language. For example, if X and Y are two generic subsets of a set E, the relation $Y = \complement X$ is functional in Y; the mapping of $\mathfrak{P}(E)$ into $\mathfrak{P}(E)$ it determines is denoted by $X \rightarrow \complement X$, or simply by $\complement X$.

Instead of saying "let f be a mapping of E into F", we shall often say, more simply, "let $f: E \rightarrow F$".

To describe a situation in which several mappings are involved, we shall also make use of *diagrams* such as

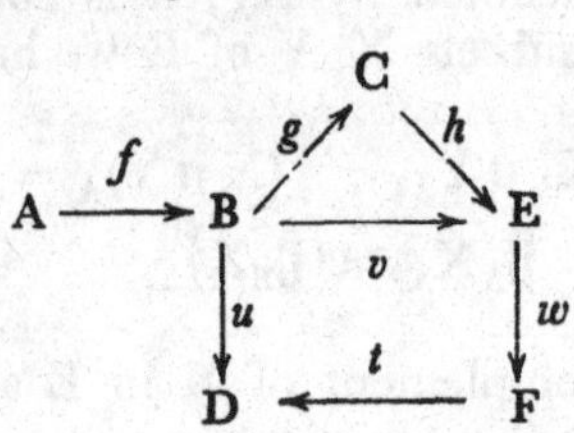

in which the letter attached to an arrow denotes a mapping of the set at the tail of the arrow into the set at its head.

The relation of equality "$f = g$" between mappings of E into F is equivalent to the relation "for all $x \in E, f(x) = g(x)$".

3. A function, defined on a set E, which takes the same value a for every element x of E, is called a *constant* function on E; it is determined by the functional relation $y = a$.

The mapping of E into E which associates with each element x of E this same element is called the *identity* mapping. It is determined by the functional relation $y = x$.

If A is any subset of E, the mapping of A into E which associates with each element x of A the same element x considered as an element of E is called the *canonical* mapping of A into E.

Let f be a mapping of a set E into itself. An element x of E is said to be *fixed* (or *invariant*) *under* f if $f(x) = x$.

An element x of E is said to be *fixed* (or *invariant*) under a set of mappings of E into E if it is fixed under each of them.

4. Let f be a mapping of E into F and let X be any subset of E. Then the *image of* X *under* f is the subset Y of F consisting of all elements y which have the property "there exists $x \in E$ such that $x \in X$ and $f(x) = y$".

This defines a relation between X and Y which is functional in Y and therefore determines a mapping of $\mathfrak{P}(E)$ into $\mathfrak{P}(F)$. This mapping is called the *canonical extension of f to sets of subsets;* by abuse of language it, too, is denoted by f, and we write $Y = f(X)$.

For all f and x we have

$$f(\emptyset) = \emptyset \quad \text{and} \quad f(\{x\}) = \{f(x)\}.$$

By abuse of language, the *value* $f(x)$ of f at x is also called the *image of* x *under* f.

If y is a generic element of F, the property "$y \in f(E)$" may be expressed by saying that "*y is of the form $f(x)$*".

By abuse of language, the image $f(E)$ of E under f is sometimes called the *image of f*.

If we have $f(E) = F$, that is if for all $y \in F$ there exists $x \in E$ such that $y = f(x)$, then f is said to be a mapping of E *onto* F, or a *surjective* mapping, or a *surjection*.

Let x be any element and X any subset of E. Instead of saying that $f(x)$ is the value of f at x, and $f(X)$ the image of X under f, it is sometimes said that f *transforms* (or *maps*) x into $f(x)$ and X into $f(X)$; $f(x)$ and $f(X)$ are then called the *transforms* by f of x and X, respectively.

If f is a mapping of E into itself, a subset X of E is said to be *stable under* f if $f(X) \subset X$. The subset X is said to be *stable* under a set of mappings of E into E if X is stable under each of these mappings.

5. Let f be a mapping of E into F. Then we have the following propositions, in which X and Y denote arbitrary subsets of E:

(a) The relation $X \subset Y$ implies $f(X) \subset f(Y)$.
(b) The property $X \neq \emptyset$ is equivalent to $f(X) \neq \emptyset$.
(c) For all X, Y we have

$$f(X \cup Y) = f(X) \cup f(Y), \tag{11}$$
$$f(X \cap Y) \subset f(X) \cap f(Y). \tag{12}$$

6. Let f be a mapping of E into F, and let Y be any subset of F. The *inverse image of* Y *under* f is the subset X of E consisting of all elements x which have the property $f(x) \in Y$.

This defines a relation between X and Y which is functional in X and therefore determines a mapping of $\mathfrak{P}(F)$ into $\mathfrak{P}(E)$, called the *inverse extension of f to sets of subsets*, and denoted by $\overset{-1}{f}$; thus we write $X = \overset{-1}{f}(Y)$.

In particular, if y is an element of F, then $\overset{-1}{f}(\{y\})$ will be the set of all $x \in E$ such that $f(x) = y$. The relations "$f(x) = y$" and "$x \in \overset{-1}{f}(\{y\})$" are equivalent. By abuse of language we often write $\overset{-1}{f}(y)$ in place of $\overset{-1}{f}(\{y\})$.

The *trace* X_A of a subset X of E on a given subset A is the inverse image of X under the canonical mapping of A into E (no. 3).

7. Let f be a mapping of E into F. Then we have the following propositions, in which X and Y denote arbitrary subsets of F:

(a) The relation $X \subset Y$ implies $\overset{-1}{f}(X) \subset \overset{-1}{f}(Y)$.

(b) For all X, Y we have

$$\overset{-1}{f}(X \cup Y) = \overset{-1}{f}(X) \cup \overset{-1}{f}(Y), \tag{13}$$
$$\overset{-1}{f}(X \cap Y) = \overset{-1}{f}(X) \cap \overset{-1}{f}(Y), \tag{14}$$
$$\overset{-1}{f}(\complement X) = \complement \overset{-1}{f}(X). \tag{15}$$

Note the difference between formulae (12) and (14); (14) would not be true for all X and Y if we replaced $\overset{-1}{f}$ by an arbitrary mapping of F into E. Again, there is no analogue of (15) for the extension of an arbitrary mapping.

Furthermore, we have $\overset{-1}{f}(\emptyset) = \emptyset$; but we can also have $\overset{-1}{f}(X) = \emptyset$ for a non-empty subset X of F. In order that $X \neq \emptyset$ should imply $\overset{-1}{f}(X) \neq \emptyset$, it is necessary and sufficient that f should be a mapping of E *onto* F.

8. If a mapping f of E into F is such that for all $y \in F$ there exists *at most one* $x \in E$ such that $y = f(x)$ (in other words, the set $\overset{-1}{f}(y)$ is either empty or consists of a single element), then f is said to be a *one-to-one* mapping of E into F, or an *injective* mapping, or an *injection*. We have then, for all subsets X, Y of E,

$$f(X \cap Y) = f(X) \cap f(Y). \tag{16}$$

9. If a mapping f of E into F is such that for all $y \in F$ *there exists exactly one* $x \in E$ such that $y = f(x)$ (in other words, $\overset{-1}{f}(y)$ consists of a single element), then f is said to be a *one-to-one* mapping of E *onto* F,

or a *bijective* mapping, or a *bijection*. Such a mapping may be characterized as being both a mapping of E *onto* F and a *one-to-one* mapping of E into F.

If f is a one-to-one mapping of E onto F, the relation $y = f(x)$ is not only functional in y, but also *functional in* x. As a functional relation in x, it determines a one-to-one mapping of F onto E, called the *inverse* of the mapping f.

Note that the *extension of the inverse of* f is the same as the *inverse of the extension of* f.

Let g be the inverse of f. Then the relations "$y = f(x)$" and "$x = g(y)$" are *equivalent*. The inverse of g is f. If f is a one-to-one mapping of E onto F, we have not only the relation (16) but also, for all subsets X of E,

$$f(\complement X) = \complement f(X).$$

Moreover, the extension of f is a one-to-one mapping of $\mathfrak{P}(E)$ onto $\mathfrak{P}(F)$.

A one-to-one mapping of E onto F, together with its inverse mapping, are said to *realize a one-to-one correspondence between* E *and* F; alternatively, we say that E and F *are put in one-to-one correspondence by these mappings*.

A one-to-one mapping of a set E onto itself is called a *permutation* of E. The identity mapping is a permutation. If a permutation is identical with its inverse, it is said to be *involutory*; for example, the mapping $X \to \complement X$ of $\mathfrak{P}(E)$ onto itself is involutory.

10. In the following propositions, X denotes an arbitrary subset of E, and Y an arbitrary subset of F:

(a) If f is a mapping of E *into* F, we have

$$\overset{-1}{f}(Y) = \overset{-1}{f}(Y \cap f(E)), \tag{17}$$

$$X \subset \overset{-1}{f}(f(X)), \tag{18}$$

$$f(\overset{-1}{f}(Y)) \subset Y. \tag{19}$$

(b) The properties "for all Y, $f(\overset{-1}{f}(Y)) = Y$" and "$f$ is a mapping of E *onto* F" are *equivalent*.

(c) The properties "for all X, $\overset{-1}{f}(f(X)) = X$" and "$f$ is a *one-to-one* mapping of E *into* F" are *equivalent*.

(d) The properties "for all X and Y,

$$\overset{-1}{f}(f(X)) = X \quad \text{and} \quad f(\overset{-1}{f}(Y)) = Y\text{"}$$

and "f is a *one-to-one* mapping of E *onto* F" are *equivalent*.

11. Let E, F, G be three sets, which may or may not be distinct. Let f be a mapping of E into F, and let g be a mapping of F into G. The mapping of E into G whose value at any element x of E is $g(f(x))$ is called the *composition* of g and f, and is denoted by $g \circ f$, or simply gf if there is no risk of ambiguity.

The equality $h = g \circ f$ is called a *factorization* of h.

> Note that, if G is distinct from E, we may not speak of the composition of f and g (in that order), and the notation $f \circ g$ has no meaning. If G is identical with E, then $f \circ g$ and $g \circ f$ are not elements of the same set unless F is also identical with E; and even when this is so, it is usually the case that $f \circ g \neq g \circ f$. Thus the *order* of composition of f and g is essential.

Let φ be the composition of g and f, let X be any subset of E, and let Z be any subset of G. Then we have

$$\varphi(\mathrm{X}) = g(f(\mathrm{X})), \tag{20}$$

$$\overset{-1}{\varphi}(\mathrm{Z}) = \overset{-1}{f}(\overset{-1}{g}(\mathrm{Z})). \tag{21}$$

If f is a *one-to-one* mapping of E *onto* F, and if g is a *one-to-one* mapping of F *onto* G, then $g \circ f$ is a *one-to-one* mapping of E *onto* G.

Let h be a mapping of G into a set H. Then we have

$$h \circ (g \circ f) = (h \circ g) \circ f;$$

this mapping of E into H is written $h \circ g \circ f$, and is called the *composition* of the three mappings h, g, f in this order. The composition of more than three mappings is defined similarly.

If f is a mapping of E into itself, the *iterates* of f are defined to be the mappings f^n (n an integer $\geqslant 1$) of E into itself, defined by induction on n by means of the relations $f^1 = f$, $f^n = f^{n-1} \circ f$; f^n is called the *nth iterate* of f. We have $f^{m+n} = f^m \circ f^n$.

12. In general, the composition $\overset{-1}{f} \circ f$ of the inverse extension and the extension of a mapping f is not the identity mapping of $\mathfrak{P}(\mathrm{E})$ onto itself. Likewise, $f \circ \overset{-1}{f}$ is not usually the identity mapping of $\mathfrak{P}(\mathrm{E})$ onto itself. These two conditions are satisfied simultaneously only if f is a *bijection* of E onto F.

If f is a bijection of E onto F, and if g denotes the inverse of f, then the compositions $g \circ f$ and $f \circ g$ are respectively the identity mapping of E onto E and the identity mapping of F onto F.

Conversely, if f is a mapping of E into F, and g a mapping of F into E, such that $g \circ f$ is a permutation of E and $f \circ g$ is a permutation of F, then f is a bijection of E onto F and g is a bijection of F onto E.

If, moreover, $g \circ f$ is the identity mapping of E onto itself, then g is the inverse of f.

13. Let f be a mapping of E into F, and let A be any subset of E. The mapping f_A of A into F whose value at any element x of A is $f(x)$ is called the *restriction of f to the subset* A; it is just the composition of f and the canonical mapping of A into E. If two mappings f, g of E into F have the same restriction to A, they are said to *agree* (or *coincide*) on A. Conversely, f is said to be an *extension* of f_A to E.

14. A mapping of a set E *onto* a set F is also called a *parametric representation of* F *by means of* E; E is then called the *parameter set* of this representation, and the elements of E take the name of *parameters*.

A *family of elements* of a set F is by definition a subset of F endowed with a parametric representation; in other words, to be given a family of elements of F is equivalent to being given a mapping of some set E into F. The image of E under this mapping is called the *set of elements of the family*. Note that two distinct families of elements of F may have the same subset of F as the set of their elements.

With each subset A of a set F we may always associate a family of elements whose set of elements is A. It is enough to consider the family defined by the *canonical* mapping of A into F.

A family of elements of F, defined by a mapping $\iota \to x_\iota$ of a set I into F, is denoted by $(x_\iota)_{\iota \in I}$, or simply (x_ι) if there is no possible ambiguity about the index set.

If J is a subset of I, the family $(x_\iota)_{\iota \in J}$ is called the *subfamily* corresponding to J of the family $(x_\iota)_{\iota \in I}$; it is defined by the restriction to J of the mapping $\iota \to x_\iota$.

3. PRODUCTS OF SETS

1. Let E, F be two sets, which may or may not be distinct. The *ordered pairs* (x, y), whose first element x is any element of E and whose second element y is any element of F, are the elements of a new set, called the *product of* E *by* F, and denoted by $E \times F$; E and F are called the *factors* of $E \times F$. Two ordered pairs are considered to be identical only if they have the same first element and the same second element; in other words, the relation "$(x, y) = (x', y')$" is *equivalent* to the relation "$x = x'$ and $y = y'$". If z is any element of $E \times F$, the relation "x is the first element of the ordered pair z" is a functional relation in x; it determines a mapping of $E \times F$ *onto* E, which is called the *first coordinate function*, or the *first projection*, and is denoted by pr_1. Instead of saying "x is the first element of the ordered pair z", we also say "x is the first coordinate

(or projection) of z" or "$x = \mathrm{pr}_1 z$". Similarly we define the *second coordinate function*, or *second projection*, which is a mapping of $E \times F$ *onto* F, denoted by pr_2.

The relation "$x = \mathrm{pr}_1 z$ and $y = \mathrm{pr}_2 z$" is equivalent to "$z = (x, y)$".

The extension of the function pr_1 to sets of subsets is denoted by the same symbol, in accordance with the general conventions, and is again called the *first projection* (here we do not use the term "coordinate"); similarly for the extension of the second projection.

2. A relation R between a generic element x of E and a generic element y of F is a property of the pair (x, y), and consequently defines a subset of the product $E \times F$, called the *graph* of R. Conversely, every subset A of $E \times F$ is the graph of the relation $(x, y) \in A$ between x and y.

Let A be a subset of E and let B be a subset of F. We denote by $A \times B$ the subset of $E \times F$ defined by the relation "$x \in A$ and $y \in B$" between x and y.

3. In the following propositions, X and X′ denote arbitrary subsets of E, Y and Y′ arbitrary subsets of F, and Z an arbitrary subset of $E \times F$.

(a) The relation "$X \times Y = \varnothing$" is *equivalent* to "$X = \varnothing$ or $Y = \varnothing$".

(b) If $X \times Y \neq \varnothing$, the relation "$X \times Y \subset X' \times Y'$" is *equivalent* to "$X \subset X'$ and $Y \subset Y'$".

(c) For all X, X′, Y we have

$$(X \times Y) \cup (X' \times Y) = (X \cup X') \times Y. \tag{22}$$

(d) For all X, X′, Y, Y′ we have

$$(X \times Y) \cap (X' \times Y') = (X \cap X') \times (Y \cap Y'). \tag{23}$$

(e) For all X, Y we have

$$\overset{-1}{\mathrm{pr}_1}(X) = X \times F, \qquad \overset{-1}{\mathrm{pr}_2}(Y) = E \times Y. \tag{24}$$

(f) If $Y \neq \varnothing$, then for all X we have

$$\mathrm{pr}_1(X \times Y) = X. \tag{25}$$

(g) For all Z we have

$$Z \subset \mathrm{pr}_1(Z) \times \mathrm{pr}_2(Z). \tag{26}$$

(h) Let a be an element of E. Then the mapping $(a, y) \to y$ of the set $\{a\} \times F$ *onto* F (i.e., the restriction of pr_2 to the subset $\{a\} \times F$) is *one-to-one*.

4. The mapping

$$(x, y) \to (y, x) \tag{27}$$

is a *one-to-one* mapping of $E \times F$ *onto* $F \times E$, and is called *canonical.* When E and F are the same set, the mapping (27) is called the *canonical symmetry*; it is then *involutory.* The elements (x, y) of $E \times E$ which are fixed under this symmetry are those which have the property $x = y$; the set Δ of these elements is called the *diagonal* of $E \times E$. The mapping $x \to (x, x)$ is a *bijection* of E onto Δ, called the *diagonal mapping* of E into $E \times E$.

If Z is any subset of $E \times F$, the image of Z under the canonical mapping of $E \times F$ onto $F \times E$ is denoted by $\overset{-1}{Z}$. If X is any subset of E and Y any subset of F, then

$$\overset{-1}{\overbrace{X \times Y}} = Y \times X.$$

If a relation R between x and y, considered as a property of the ordered pair (x, y), defines a subset A of $E \times F$, then the *same* relation, considered as a property of the pair (y, x), defines the subset $\overset{-1}{A}$ of $F \times E$. R is equivalent to each of the relations $(x, y) \in A$, $(y, x) \in \overset{-1}{A}$. If E and F are the same set, the relation R and the corresponding subset A are said to be *symmetric* when $A = \overset{-1}{A}$. The diagonal Δ (defined by the relation of equality) is symmetric. If Z is any subset of $E \times E$, then $Z \cup \overset{-1}{Z}$ and $Z \cap \overset{-1}{Z}$ are symmetric.

5. Let A be a subset of a set E, and let f be a mapping of A into a set F. The relation "$x \in A$ and $y = f(x)$" between a generic element x of E and a generic element y of F defines a subset of $E \times F$ called the *graph* of the function f. If B is a subset of E which contains A, and if g is an *extension* (§ 2, no. 13) of f to B, then the graph of f is *contained* in the graph of g.

Conversely, let C be a subset of $E \times F$ such that, for each $x \in E$, there exists *at most one* $y \in F$ such that $(x, y) \in C$. Then the relation $(x, y) \in C$ between a generic element x *of the set* $\mathrm{pr}_1(C)$ and a generic element y of F is a functional relation in y and determines a mapping of $\mathrm{pr}_1(C)$ into F whose graph is C.

The set of subsets C of $E \times F$ which have the property "for all $x \in E$ there is at most one $y \in F$ such that $(x, y) \in C$" (a set which is a subset of $\mathfrak{P}(E \times F)$) can therefore be put into one-to-one correspondence with *the set of mappings of subsets of* E *into* F.

Let f be an injective mapping of E into F, and let g be the inverse of f considered as a bijection of E onto $f(\mathrm{E})$. If C is the graph of f, then the graph of g is $\overset{-1}{\mathrm{C}}$.

6. If f is a mapping of E into F, and C is its graph in E × F, the relation "$y=f(x)$" is equivalent to "$(x, y) \in \mathrm{C}$". The relation "$y \in f(\mathrm{X})$" is equivalent to "there exists x such that $x \in \mathrm{X}$ and $(x, y) \in \mathrm{C}$".

Now let K be *any* subset of E × F, and let X be any subset of E. Let K(X) denote the subset of F consisting of all elements y which satisfy the relation "there exists x such that $x \in \mathrm{X}$ and $(x, y) \in \mathrm{K}$"; this relation is thus equivalent to "$y \in \mathrm{K}(\mathrm{X})$". The mapping $\mathrm{X} \to \mathrm{K}(\mathrm{X})$ of $\mathfrak{P}(\mathrm{E})$ into $\mathfrak{P}(\mathrm{F})$ is said to be *defined by the subset* K of E × F. Note that K(X) is the second projection of the set $\mathrm{K} \cap (\mathrm{X} \times \mathrm{F})$. If K is the graph of a mapping f of E into F, then the mapping $\mathrm{X} \to \mathrm{K}(\mathrm{X})$ is the canonical extension of f to sets of subsets.

7. If x is a generic element of E, then $x \to \mathrm{K}(\{x\})$ is a mapping of E into $\mathfrak{P}(\mathrm{F})$ whose value $\mathrm{K}(\{x\})$ (denoted also by $\mathrm{K}(x)$, by abuse of language) is called the *section of* K *at* x. The relation $(x, y) \in \mathrm{K}$ is equivalent to $y \in \mathrm{K}(x)$.

Conversely, *every* mapping $x \to \Phi(x)$ of E into $\mathfrak{P}(\mathrm{F})$ can be obtained in this way; for the relation $y \in \Phi(x)$ defines a subset K of E × F, and $\Phi(x)$ is precisely the section of K at x. The set $\mathfrak{P}(\mathrm{E} \times \mathrm{F})$ and the set of mappings of E into $\mathfrak{P}(\mathrm{F})$ are thus in one-to-one correspondence.

8. Every mapping $\mathrm{X} \to \mathrm{K}(\mathrm{X})$ defined by a subset K of E × F has the following properties, which generalize those of the canonical extension of a mapping of E into F (§2, nos. 4 and 5):

(a) $\mathrm{K}(\emptyset) = \emptyset$.

(b) "$\mathrm{X} \subset \mathrm{Y}$" implies "$\mathrm{K}(\mathrm{X}) \subset \mathrm{K}(\mathrm{Y})$".

(c) For all X, Y we have

$$\mathrm{K}(\mathrm{X} \cup \mathrm{Y}) = \mathrm{K}(\mathrm{X}) \cup \mathrm{K}(\mathrm{Y}), \tag{28}$$

$$\mathrm{K}(\mathrm{X} \cap \mathrm{Y}) \subset \mathrm{K}(\mathrm{X}) \cap \mathrm{K}(\mathrm{Y}). \tag{29}$$

If K and K′ are two subsets of E × F such that $\mathrm{K} \subset \mathrm{K}'$, then we have $\mathrm{K}(\mathrm{X}) \subset \mathrm{K}'(\mathrm{X})$ for all $\mathrm{X} \subset \mathrm{E}$; in particular, $\mathrm{K}(x) \subset \mathrm{K}'(x)$ for all $x \in \mathrm{E}$. Conversely, if $\mathrm{K}(x) \subset \mathrm{K}'(x)$ for all $x \in \mathrm{E}$, then $\mathrm{K} \subset \mathrm{K}'$.

9. A relation between a generic element of E and a generic element of F defines a subset K of E × F and a subset $\overset{-1}{\mathrm{K}}$ of F × E, and hence a mapping $\mathrm{X} \to \mathrm{K}(\mathrm{X})$ of $\mathfrak{P}(\mathrm{E})$ into $\mathfrak{P}(\mathrm{F})$ and a mapping $\mathrm{Y} \to \overset{-1}{\mathrm{K}}(\mathrm{Y})$ of $\mathfrak{P}(\mathrm{F})$ into $\mathfrak{P}(\mathrm{E})$.

If K is the graph of a mapping f of E into F, the mapping $\mathrm{Y} \to \overset{-1}{\mathrm{K}}(\mathrm{Y})$ is the inverse extension of f.

It should be noted that the relations (18) and (19) do not generalize to the mappings $X \to K(X)$ and $Y \to \overset{-1}{K}(Y)$, when K is an arbitrary subset of $E \times F$.

10. Let E, F, G be three sets, which may or may not be distinct, let A be a subset of $E \times F$ and let B be a subset of $F \times G$. Then the elements (x, z) of $E \times G$ which have the property "there exists $y \in F$ such that $(x, y) \in A$ and $(y, z) \in B$" form a subset of $E \times G$, called the *composition of* B *and* A, and denoted by $B \circ A$, or simply BA when there is no risk of confusion. Here again, the order of composition is essential. ¶ The mapping $X \to BA(X)$ of $\mathfrak{P}(E)$ into $\mathfrak{P}(G)$ is the composition of $Y \to B(Y)$ and $X \to A(X)$; in other words, for all $X \subset E$ we have

$$BA(X) = B(A(X)). \tag{30}$$

Let H be another set, not necessarily distinct from E, F, G, and let C be a subset of $G \times H$. Then we have $C \circ (B \circ A) = (C \circ B) \circ A$; this set is also denoted by $C \circ B \circ A$ (or simply CBA) and is called the *composition* of C, B, A taken in this order.

Let f be a mapping of E into F, and let g be a mapping of F into G. If A (resp. B) is the graph of f (resp. g), then the composition BA is the graph of the composite mapping $g \circ f$.

11. We have

$$\overset{-1}{\widehat{B \circ A}} = \overset{-1}{A} \circ \overset{-1}{B}. \tag{31}$$

Let A, A′ be two subsets of $E \times F$, and let B, B′ be two subsets of $F \times G$. Then the relation

$$\text{"}A \subset A' \text{ and } B \subset B'\text{" implies "}B \circ A \subset B' \circ A'\text{".}$$

Let A be a subset of $E \times F$, Δ the diagonal of $E \times E$, and Δ' the diagonal of $F \times F$. Then we have

$$A \circ \Delta = \Delta' \circ A = A. \tag{32}$$

12. Let E, F, G be three sets, which may or may not be distinct. Their *product* $E \times F \times G$ is the set of ordered *triples* (x, y, z), where $x \in E$, $y \in F$, and $z \in G$, the relation "$(x, y, z) = (x', y', z')$" being equivalent to "$x = x'$ and $y = y'$ and $z = z'$". The three mappings $(x, y, z) \to x$, $(x, y, z) \to y$, $(x, y, z) \to z$ of $E \times F \times G$ onto E, F, G respectively are called the *first, second,* and *third coordinate functions* (or *projections*); similarly,

for example, the *projection with indices* 1, 2 is the mapping

$$(x, y, z) \to (x, y)$$

of $E \times F \times G$ onto $E \times F$, and is denoted by $pr_{1,2}$.

The definitions and propositions of nos. 2, 3, and 4 generalize easily to the product of three sets.

Furthermore, instead of considering the product $E \times F \times G$ of three sets, we may equivalently consider the product $(E \times F) \times G$, obtained by a double application of the operation of forming the product of two sets. In fact, $(x, y, z) \to ((x, y), z)$ is a *one-to-one* mapping of $E \times F \times G$ *onto* $(E \times F) \times G$, called *canonical*. Similarly we define one-to-one mappings of $E \times F \times G$ onto the set $E \times (F \times G)$, and onto all the sets obtained from $E \times F \times G$, $(E \times F) \times G$, and $E \times (F \times G)$ by permuting the three letters E, F, G.

There are analogous definitions and properties for the product of more than three sets.

13. If a function f, which takes its values in any set E', is defined on a product of three sets E, F, G, it is said to be a function of *three variables*, each of which runs through one of the sets E, F, G. The value of f at the element (x, y, z) of $E \times F \times G$ is denoted by $f(x, y, z)$.

Let a be any element of E. Then $(y, z) \to f(a, y, z)$ is a mapping of $F \times G$ into E', called a *partial mapping* (or *function*) *determined by* f corresponding to the value a of x; it is also the composition of f and the mapping $(y, z) \to (a, y, z)$ of $F \times G$ into $E \times F \times G$.

Likewise, if b is an element of F, then $z \to f(a, b, z)$ is a mapping of G into E', called the partial mapping determined by f corresponding to the values a, b of x, y.

Inversely, let g be a mapping of E into E'. Then $(x, y, z) \to g(x)$ is a mapping h of $E \times F \times G$ into E' such that every partial mapping of E into E' determined by h, corresponding to any values of y and z, is identical with g. This fact is often expressed by saying that a function of an argument x can always be envisaged as a function of all the arguments which need to be considered at a given moment and which will, of course, include x.

14. Let f, g, h be three mappings of E into E', F into F', G into G', respectively. The mapping $(x, y, z) \to (f(x), g(y), h(z))$ of $E \times F \times G$ into $E' \times F' \times G'$ is denoted by (f, g, h) or $f \times g \times h$ and is called the *extension of* f, g, h *to products*. If *all three* of f, g, h are injective (resp. surjective, bijective), then $f \times g \times h$ is injective (resp. surjective, bijective).

In this and the previous subsection we have considered only the case of *three* sets merely to fix the ideas; analogous considerations hold for any finite number of sets.

4. UNION, INTERSECTION, PRODUCT OF A FAMILY OF SETS

1. In this section we consider a family $(X_\iota)_{\iota \in I}$ of *subsets* of a set E, in which the index set I is arbitrary; the *set of subsets of* E *belonging to the family* will be denoted by $\mathfrak{F}$ (which is therefore a subset of $\mathfrak{P}(E)$).

If I is finite, the consideration of the family (X_ι) reduces to that of several subsets of E, which may or may not be distinct, and in number equal to the number of elements of I. For example, any three subsets X_1, X_2, X_3 of E form a family of subsets of E, the index set here consisting of the numbers 1, 2, 3.

2. Let J be any subset of I, and consider the set of all elements x which have the property "there exists $\iota \in J$ such that $x \in X_\iota$". This set is called the *union of the family of sets* $(X_\iota)_{\iota \in J}$, and is denoted by $\bigcup_{\iota \in J} X_\iota$.

We may also formulate this definition as follows : to the mapping $\iota \to X_\iota$ of I into $\mathfrak{P}(E)$ there corresponds a well-defined subset C of $I \times E$ such that $X_\iota = C(\iota)$ (§ 3, no. 7), and we have $\bigcup_{\iota \in J} X_\iota = C(J)$.

In particular,

$$\bigcup_{\iota \in \emptyset} X_\iota = C(\emptyset) = \emptyset. \tag{33}$$

When $J = I$, we often write $\bigcup_\iota X_\iota$, or simply $\bigcup X_\iota$, in place of $\bigcup_{\iota \in I} X_\iota$.

The union $\bigcup_\iota X_\iota$ *depends only on the set* $\mathfrak{F}$; in other words, it is the same for two families corresponding to the same subset $\mathfrak{F}$ of $\mathfrak{P}(E)$. In particular, it is equal to the union of the family defined by the canonical mapping of $\mathfrak{F}$ into $\mathfrak{P}(E)$, and we may therefore write $\bigcup_{X \in \mathfrak{F}} X_\iota$, which is also called the *union of the sets belonging to* $\mathfrak{F}$.

When I is a set whose elements are explicitly designated, for example the numbers 1, 2, 3, we have

$$\bigcup_{\iota \in I} X_\iota = X_1 \cup X_2 \cup X_3,$$

which justifies the name "union" given in the general case to the set $\bigcup_\iota X_\iota$.

3. For any $J \subset I$ we have

$$\bigcup_{\iota \in J} X_\iota \subset \bigcup_{\iota \in I} X_\iota.$$

In particular, for all $\varkappa \in I$, we have

$$X_\varkappa \subset \bigcup_{\iota \in I} X_\iota.$$

Conversely, if Y is a subset of E such that $X_\iota \subset Y$ for all $\iota \in I$, then

$$\bigcup_{\iota} X_\iota \subset Y.$$

More generally, if (Y_ι) is another family of subsets of E, indexed by the same set I, and if $X_\iota \subset Y_\iota$ for all $\iota \in I$, then

$$\bigcup_{\iota} X_\iota \subset \bigcup_{\iota} Y_\iota.$$

Let F be another set, and let $X \to K(X)$ be the mapping of $\mathfrak{P}(E)$ into $\mathfrak{P}(F)$ defined by a subset K of $E \times F$. Then we have

$$K\Big(\bigcup_{\iota \in J} X_\iota\Big) = \bigcup_{\iota \in J} K(X_\iota). \tag{34}$$

Now let L be another index set, and $(J_\lambda)_{\lambda \in L}$ a family of subsets of I. Then

$$\bigcup_{\iota \in \bigcup_{\lambda \in L} J_\lambda} X_\iota = \bigcup_{\lambda \in L}\Big(\bigcup_{\iota \in J_\lambda} X_\iota\Big). \tag{35}$$

This is the general *associativity* formula for unions. When I and L are sets whose elements are explicitly designated, the relations we obtain have already been given (see § 1, no. 14). If L alone satisfies this condition and consists, say, of the numbers 1, 2, we have

$$\bigcup_{\iota \in J_1 \cup J_2} X_\iota = \Big(\bigcup_{\iota \in J_1} X_\iota\Big) \cup \Big(\bigcup_{\iota \in J_2} X_\iota\Big). \tag{36}$$

Let $(X_\iota)_{\iota \in I}$ and $(Y_\varkappa)_{\varkappa \in K}$ be any two families of subsets of E. Then we have

$$\Big(\bigcup_{\iota \in I} X_\iota\Big) \cap \Big(\bigcup_{\varkappa \in K} Y_\varkappa\Big) = \bigcup_{(\iota, \varkappa) \in I \times K} (X_\iota \cap Y_\varkappa), \tag{37}$$

the *distributivity* formula, which includes the second formula of (10) as a particular case.

If $(X_\iota)_{\iota \in I}$ is a family of subsets of E, and if $(Y_\varkappa)_{\varkappa \in K}$ is a family of subsets of F, then

$$(38) \qquad \left(\bigcup_{\iota \in I} X_\iota\right) \times \left(\bigcup_{\varkappa \in K} Y_\varkappa\right) = \bigcup_{(\iota, \varkappa) \in I \times K} (X_\iota \times Y_\varkappa).$$

4. A family $(X_\iota)_{\iota \in I}$ of subsets of E is a *covering* of a subset A of E, or *covers* A, if

$$A \subset \bigcup_{\iota \in I} X_\iota.$$

In particular, if (X_ι) is a covering of E, we have

$$\bigcup_{\iota \in I} X_\iota = E.$$

A *partition* of E is a covering (X_ι) of E such that

(a) $X_\iota \neq \emptyset$ for all $\iota \in I$;

(b) $X_\iota \cap X_\varkappa = \emptyset$ for each pair of *different* indices ι, $\varkappa$ in I. (This second condition may be expressed by saying that the X_ι are *pairwise disjoint*.)

These conditions imply that $\iota \to X_\iota$ is a *bijection* of I *onto* the set $\mathfrak{F}$ of subsets of the partition. Hence, if $\mathfrak{F}$ is given, the family is determined to within a one-to-one correspondence of index sets. In particular, a partition may be considered indifferently as a *set* of subsets or as a *family* of subsets.

5. Let $(X_\iota)_{\iota \in I}$ be any family of non-empty subsets of a set E. In the product $I \times E$, let X'_ι denote the subset $\{\iota\} \times X_\iota$ for each $\iota \in I$. The set

$$S = \bigcup_{\iota \in I} X'_\iota$$

is called the *sum* of the family $(X_\iota)_{\iota \in I}$. It is clear that the family $(X'_\iota)_{\iota \in I}$ is a partition of S and that for each $\iota \in I$ the mapping $x_\iota \to (\iota, x_\iota)$ is a bijection of X_ι onto X'_ι. By abuse of language, any set in one-to-one correspondence with S is often called the sum of the family $(X_\iota)_{\iota \in I}$, and the X_ι are usually identified with the subsets of this set to which they correspond.

The sum of two non-empty sets E and F is often said to be obtained by *adjoining* the set F to E.

6. With the notation of no. 2, the set of elements x of E which have the property "for all $\iota \in J$, $x \in X_\iota$" is called the *intersection of the family of sets*

$(X_\iota)_{\iota\in J}$, and is denoted by $\bigcap_{\iota\in J} X_\iota$; when $J = I$, we often write $\bigcap_\iota X_\iota$, or simply $\bigcap X_\iota$, instead of $\bigcap_{\iota\in I} X_\iota$.

We have

$$\complement\Big(\bigcup_{\iota\in J} X_\iota\Big) = \bigcap_{\iota\in J} (\complement X_\iota). \tag{39}$$

In particular, if $J = \emptyset$,

$$\bigcap_{\iota\in\emptyset} X_\iota = E. \tag{40}$$

The intersection $\bigcap_\iota X_\iota$ depends only on the set $\mathfrak{F}$, and may be written $\bigcap_{X\in\mathfrak{F}} X$. For example, if I consists of the numbers 1, 2, 3, we have

$$\bigcap_\iota X_\iota = X_1 \cap X_2 \cap X_3.$$

7. Formula (39) allows us to generalize the *duality rule*. If a subset A of E is obtained from other subsets X, Y, Z and families (X_ι), $(Y_\varkappa)$, (Z_λ) of subsets of E by applying (in any order) *only* the operations $\complement$, $\cup$, $\cap$, $\bigcup$, $\bigcap$, then we shall obtain the complement $\complement A$ by replacing the subsets X, Y, Z, X_ι, $Y_\varkappa$, Z_λ by their complements, and the operations $\cup$, $\cap$, $\bigcup$, $\bigcap$ by $\cap$, $\cup$, $\bigcap$, $\bigcup$, respectively, the order of operations being preserved; of course, the operations of intersection and union are not to be altered where they apply to *index* sets written under the signs $\bigcup$ and $\bigcap$.

As in § 1, no. 15, we define the *dual* of a relation $A = B$ or $A \subset B$ where A and B are subsets of E of the above form.

8. For all $J \subset I$ we have

$$\bigcap_{\iota\in I} X_\iota \subset \bigcap_{\iota\in J} X_\iota.$$

In particular, for all $\varkappa \in I$ we have

$$\bigcap_\iota X_\iota \subset X_\varkappa.$$

Conversely, if $Y \subset X_\iota$ for all $\iota \in I$, then

$$Y \subset \bigcap_\iota X_\iota.$$

More generally, if (Y_ι) is another family of subsets of E, indexed by the same set I, and if $X_\iota \subset Y_\iota$ for all $\iota \in I$, then

$$\bigcap_\iota X_\iota \subset \bigcap_\iota Y_\iota.$$

The *union* of the sets X_ι is the *intersection* of all sets Y such that $X_\iota \subset Y$ for all $\iota \in I$. The *intersection* of the X_ι is the *union* of all Z such that $Z \subset X_\iota$ for all $\iota \in I$.

The following formulae are the duals of (35) and (37), respectively:

(41) $$\bigcap_{\iota \in \bigcup_{\lambda \in L} J_\lambda} X_\iota = \bigcap_{\lambda \in L}\Big(\bigcap_{\iota \in J_\lambda} X_\iota\Big) \quad \text{(associativity);}$$

(42) $$\Big(\bigcap_{\iota \in I} X_\iota\Big) \cup \Big(\bigcap_{\varkappa \in K} Y_\varkappa\Big) = \bigcap_{(\iota,\varkappa) \in I \times K} (X_\iota \cup Y_\varkappa) \quad \text{(distributivity).}$$

If $(X_\iota)_{\iota \in I}$ is a family of subsets of E, and $(Y_\varkappa)_{\varkappa \in K}$ a family of subset of F, then

(43) $$\Big(\bigcap_{\iota \in I} X_\iota\Big) \times \Big(\bigcap_{\varkappa \in K} Y_\varkappa\Big) = \bigcap_{(\iota,\varkappa) \in I \times K} (X_\iota \times Y_\varkappa).$$

Moreover, if (X_ι) and (Y_ι) are families of subsets of E and F, respectively, indexed by the *same* set I, then

(44) $$\Big(\bigcap_{\iota \in I} X_\iota\Big) \times \Big(\bigcap_{\iota \in I} Y_\iota\Big) = \bigcap_{\iota \in I} (X_\iota \times Y_\iota).$$

The formula (34) has no dual; in general all we can say is

(45) $$K\Big(\bigcap_{\iota \in J} X_\iota\Big) \subset \bigcap_{\iota \in J} K(X_\iota).$$

We have equality in (45) for all families (X_ι) only if $X \to K(X)$ is the *inverse extension* of a mapping of a subset of F into E. Consequently, if f is a *mapping of* F *into* E, we have (generalizing (14))

(46) $$\overset{-1}{f}\Big(\bigcap_{\iota \in J} X_\iota\Big) = \bigcap_{\iota \in J} \overset{-1}{f}(X_\iota).$$

9. Let E be any set and let I be any index set. The set of all *families* $(x_\iota)_{\iota \in I}$ *of elements* of E, indexed by I, is denoted by E^I, and the operation of passing from E to E^I is called *exponentiation*. E^I is thus in one-to-one correspondence with the set of all mappings of I into E (which for this reason is often denoted by E^I, by abuse of language), as well as with a

subset of $\mathfrak{P}(I \times E)$, by considering the graphs of these mappings. The sets E^J corresponding to subsets J of the set I may therefore be considered all as subsets of the same set, which is in one-to-one correspondence with a subset of $\mathfrak{P}(I \times E)$.

Now let $(X_\iota)_{\iota \in I}$ be a *family of subsets* of E, indexed by the same set I, and let J be any subset of I. The property "for all $\iota \in J$, $x_\iota \in X_\iota$" of the family $(x_\iota)_{\iota \in J}$ defines a subset of E^J, called the *product of the family of sets* $(X_\iota)_{\iota \in J}$, and denoted by $\prod_{\iota \in J} X_\iota$ $\left(\text{or simply } \prod_\iota X_\iota \text{ when } J = I\right)$. The X_ι are called the *factors* of the product. Note that $\prod_{\iota \in \emptyset} X_\iota$ is a set consisting of one element (corresponding to the empty subset of $I \times E$). If $X_\iota = E$ for all $\iota \in J$, then we have

$$\prod_{\iota \in J} X_\iota = E^J.$$

If, for example, I consists of the three numbers 1, 2, 3, then $\prod_{\iota \in J} X_\iota$ is in one-to-one correspondence with the set $X_1 \times X_2 \times X_3$.

10. If $R\{x, y\}$ is a relation between a generic element x of a set E and a generic element y of a set F, then the following propositions are *equivalent*:

"for each x there exists y such that $R\{x, y\}$"

and

"there exists a mapping f of E into F such that $R\{x, f(x)\}$ for all x".

The assertion of this equivalence is known as the *axiom of choice* (or *Zermelo's* axiom). We shall sometimes indicate whether the proof of a theorem depends on it.

The axiom of choice is *equivalent* to the following proposition: "if, for each $\iota \in I$, we have $X_\iota \neq \emptyset$, then $\prod_{\iota \in I} X_\iota \neq \emptyset$".

11. In this and the following subsection, we shall consider a non-empty product $\prod_{\iota \in I} A_\iota$, where (A_ι) is any family of (non-empty) subsets of E.

Let J be a subset of I. The mapping $(x_\iota)_{\iota \in I} \rightarrow (x_\iota)_{\iota \in J}$ of $\prod_{\iota \in J} A_\iota$ *onto* $\prod_{\iota \in J} A_\iota$ is called the *projection* of $\prod_{\iota \in I} A_\iota$ onto $\prod_{\iota \in J} A_\iota$, and is denoted by pr_J. In particular, the mapping $(x_\iota)_{\iota \in I} \rightarrow x_\varkappa$ of $\prod_{\iota \in I} A_\iota$ onto $A_\varkappa$ is called the *coordinate function* (or *projection*) *of index* $\varkappa$, and is denoted by $\mathrm{pr}_\varkappa$.

If z is an element of $\prod_\iota A_\iota$, we have $z = (\mathrm{pr}_\iota z)_{\iota \in I}$.

Let J_1, J_2 be two sets forming a *partition* of I. Then $z \to (\mathrm{pr}_{J_1} z, \mathrm{pr}_{J_2} z)$ is a *one-to-one* mapping of $\prod_{\iota \in I} A_\iota$ *onto* $\prod_{\iota \in J_1} A_\iota \times \prod_{\iota \in J_2} A_\iota$.

In general, if $(J_\lambda)_{\lambda \in L}$ is any *partition* of the set I, the mapping $z \to (\mathrm{pr}_{J_\lambda} z)_{\lambda \in L}$ is a bijection (called *canonical*) of $\prod_{\iota \in I} A_\iota$ onto the product $\prod_{\lambda \in L} \left(\prod_{\iota \in J_\lambda} A_\iota \right)$. This may also be expressed by saying that the product of a family of sets is *associative*.

12. The following propositions generalize those of §3, no. 3; (X_ι), (Y_ι) denote families of subsets of E such that $X_\iota \subset A_\iota$ and $Y_\iota \subset A_\iota$ for all $\iota \in I$; Z denotes any subset of $\prod_\iota A_\iota$.

(a) If $\prod_\iota X_\iota \neq \emptyset$, the relation "$\prod_\iota X_\iota \subset \prod_\iota Y_\iota$" is *equivalent* to "for all $\iota \in I$, $X_\iota \subset Y_\iota$".

(b) We have $\overset{-1}{\mathrm{pr}}_\varkappa(X_\varkappa) = \prod_\iota Y_\iota$, where $Y_\varkappa = X_\varkappa$ and $Y_\iota = A_\iota$ whenever $\iota \neq \varkappa$. Hence

(47) $$\prod_{\iota \in I} X_\iota = \bigcap_{\iota \in I} \overset{-1}{\mathrm{pr}}_\iota(X_\iota).$$

(c) If $\prod_\iota X_\iota \neq \emptyset$, then

(48) $$\mathrm{pr}_\varkappa\left(\prod_\iota X_\iota\right) = X_\varkappa.$$

(d) For all Z we have

(49) $$Z \subset \prod_\iota \mathrm{pr}_\iota(Z).$$

(e) Let (J_1, J_2) be a partition of I into two sets, let $(a_\iota)_{\iota \in J_1}$ be a family of elements of E, and let $(X_\iota)_{\iota \in J_2}$ be a family of subsets of E, such that $a_\iota \in A_\iota$ for all $\iota \in J_1$, and $X_\iota \subset A_\iota$ for all $\iota \in J_2$. Then the product $\prod_{\iota \in I} Y_\iota$, where $Y_\iota = \{a_\iota\}$ when $\iota \in J_1$, and $Y_\iota = X_\iota$ when $\iota \in J_2$, can be put in one-to-one correspondence with $\prod_{\iota \in J_2} X_\iota$ by projecting onto this latter set.

13. Let $(A_\iota)_{\iota \in I}$ be a family of subsets of a set F, and let f be a mapping of a set E into the product $\prod_{\iota \in I} A_\iota$. If we put $f_\iota(x) = \mathrm{pr}_\iota(f(x))$, then f_ι is a mapping of E into A_ι, and f is the mapping $x \to (f_\iota(x))$. Conversely, if for each index $\iota \in I$, f_ι is a mapping of E into A_ι, then

$$x \to (f_\iota(x))$$

is a mapping of E into $\prod_{\iota \in I} A_\iota$, which is denoted by (f_ι) (by abuse of language, because this notation already denotes the family of mappings f_ι). Thus we define a bijection (called canonical) of the set $\left(\prod_{\iota \in I} A_\iota\right)^E$ onto the set $\prod_{\iota \in I} (A_\iota^E)$.

14. Let E, F, G be three sets. For each mapping f of $F \times G$ into E and for each $y \in G$, let f_y denote the partial mapping $x \to f(x, y)$ of F into E. Then $y \to f_y$ is a mapping of G into E^F. Conversely, for each mapping g of G into E^F there exists a unique mapping f of $F \times G$ into E such that $f_y = g(y)$ for all $y \in G$. Thus we define a bijection (called *canonical*) of the set $E^{F \times G}$ onto the set $(E^F)^G$.

15. Let $(A_\iota)_{\iota \in I}$ be a family of non-empty subsets of a set E. For each $\iota \in I$ let f_ι be a mapping of A_ι into a set F such that, for each pair of indices $(\iota, \varkappa)$, f_ι and $f_\varkappa$ agree on $A_\iota \cap A_\varkappa$. Then, if

$$A = \bigcup_{\iota \in I} A_\iota,$$

there exists a unique mapping f of A into F such that the *restriction* of f to each A_ι is equal to f_ι. In particular, if $A_\iota \cap A_\varkappa = \emptyset$ whenever $\iota \neq \varkappa$, we see that sets F^A and $\prod_{\iota \in I} F^{A_\iota}$ are in one-to-one correspondence (called *canonical*).

5. EQUIVALENCE RELATIONS AND QUOTIENT SETS

1. Let $(A_\iota)_{\iota \in I}$ be a *partition* of a set E. The relation $R\{x, y\}$: "there exists $\iota \in I$ such that $x \in A_\iota$ and $y \in A_\iota$" between two generic elements x, y of E satisfies the following conditions :

(a) $R\{x, x\}$ is an *identity* (*reflexivity* of R).

(b) $R\{x, y\}$ and $R\{y, x\}$ are *equivalent* (*symmetry* of R).

(c) The relation "$R\{x, y\}$ and $R\{y, z\}$" *implies* $R\{x, z\}$ (*transitivity* of R).

If C denotes the subset of $E \times E$ defined by the relation R, the conditions (a), (b), (c) are respectively equivalent to the following conditions : (a′) $\Delta \subset C$; (b′) $\overset{-1}{C} = C$; (c′) $C \circ C \subset C$. From (a′) and (c′) it follows that $C \circ C = C$.

2. Conversely, let $R\{x, y\}$ be a *reflexive*, *symmetric*, and *transitive* relation, and let C be its graph in $E \times E$. Then the image $\mathfrak{F}$ of E under the mapping $x \to C(x)$ of E into $\mathfrak{P}(E)$ is a *partition* of E, and the relation

"there exists a subset $X \in \mathfrak{F}$ such that $x \in X$ and $y \in X$" is *equivalent* to $R\{x, y\}$.

Every relation which satisfies conditions (a), (b), and (c) is called an *equivalence relation* on E. The partition $\mathfrak{F}$ which it defines, considered as a subset of $\mathfrak{P}(E)$, is called the *quotient set of* E *by the relation* R, and is denoted by E/R; its elements are called *equivalence classes with respect to* R. The mapping $x \to C(x)$ of E *onto* E/R, which maps each element x of E to the equivalence class which contains x, is called the *canonical mapping* of E onto E/R.

The relation of *equality* $x = y$ is an equivalence relation. The canonical mapping of E onto the corresponding quotient set is just $x \to \{x\}$, and is *bijective.*

If R is an equivalence relation, the notation "$x \equiv y \pmod{R}$" is sometimes used as a synonym for $R\{x, y\}$; it is read "*x is equivalent to y modulo* R".

3. On a product set $E \times F$, the relation "$pr_1 z = pr_1 z'$" between z and z' is an equivalence relation R, and the quotient set $(E \times F)/R$ can be put in one-to-one correspondence with E (this is the origin of the name *quotient set*).

More generally, let f be a mapping of a set E into a set F. Then the relation "$f(x) = f(y)$" is an equivalence relation on E. If we denote this relation by R, the mapping $z \to \overset{-1}{f}(z)$ (where $\overset{-1}{f}(z)$ is considered as an element of E/R) is a *bijection* of $f(E)$ onto E/R.

It follows that f may be considered as the *composition* of the following three mappings, in the given order:

(1) the *canonical* mapping of the subset $f(E)$ of F into the set F;

(2) the *bijective* mapping of E/R onto $f(E)$, whose inverse has just been defined;

(3) the *canonical* mapping of E onto E/R.

This decomposition of a mapping is called its *canonical decomposition* or *canonical factorization.*

4. Every equivalence relation R on a set E may be defined by means of a mapping as in the previous subsection; for, if C is the graph of R, the relation "$C(x) = C(y)$" is equivalent to $R\{x, y\}$.

5. Let R be an equivalence relation on a set E, and let A be a subset of E. Then the relation $R\{x, y\}$ between two generic elements x, y of A is an equivalence relation *on* A, called the relation *induced* by R on A, and denoted by R_A. Let f be the canonical mapping of E onto E/R and let g be that of A onto A/R_A. By making an element of E/R and an element of A/R_A correspond if they are the images of the same element of E under f and g respectively, we have a *one-to-one* correspondence between the image $f(A)$ of A under f and the quotient set A/R_A. If φ denotes the canonical mapping of A into E, this corres-

pondence is realized by the mapping $z \to f(\varphi(\overset{-1}{g}(z)))$ and its inverse, both of which are also called *canonical.*

6. A subset A of E is said to be *saturated* with respect to the equivalence relation R if for each $x \in A$ the equivalence class of x with respect to R is contained in A. In other words, the saturated sets with respect to R are *unions of equivalence classes with respect to* R. If f is the canonical mapping of E onto E/R, then a set is saturated if it is of the form $\overset{-1}{f}(X)$, where $X \subset E/R$.

Let A be a subset of E. The intersection of the saturated sets which contain A is $\overset{-1}{f}(f(A))$. This set may also be defined as the union of the equivalence classes of the elements of A, and is called the *saturation* of A (with respect to R).

7. Let $P\{x, y, z\}$ be a relation in which there appears a generic element x of E. Then P is said to be *compatible* (in x) *with the equivalence relation* R if the relation "$P\{x, y, z\}$ and $x \equiv x' \pmod{R}$" *implies* $P\{x', y, z\}$.

Let f be the canonical mapping of E onto E/R, and let t be a generic element of E/R. The relation "there exists $x \in \overset{-1}{f}(t)$ such that $P\{x, y, z\}$" is then equivalent to "for all $x \in \overset{-1}{f}(t)$, $P\{x, y, z\}$"; the latter is a relation between t, y, z, said to be *induced* by P on *passing to the quotient* (with respect to x). If we denote it by $P'\{t, y, z\}$, then $P\{x, y, z\}$ is equivalent to $P'\{f(x), y, z\}$.

There are analogous definitions for a relation involving any number of arguments, and for the case in which the relation is compatible with R in *several* of its arguments. For example, if A is a subset of E, to say that the relation "$x \in A$" is compatible (in x) with R is equivalent to saying that A is saturated with respect to R. If φ is a mapping of E into a set F, to say that the functional relation "$y = \varphi(x)$" is compatible (in x) with R is to say that φ is constant on each equivalence class with respect to R. Passing to the quotient, R therefore induces a relation between y and a generic element t of E/R; this relation is functional in y and so determines a mapping φ' of E/R into F, satisfying the identity $\varphi(x) = \varphi'(f(x))$.

8. Let R be an equivalence relation on a set E, let S be an equivalence relation on a set F, and let f be a mapping of E into F. The mapping f is said to be *compatible with* R *and* S if the relation $x \equiv x' \pmod{R}$ *implies* $f(x) \equiv f(x') \pmod{S}$. If g is the canonical mapping of F onto F/S, then the composite function $g \circ f$ has the same value at all elements of an equivalence class z with respect to R; if we denote this common value by $h(z)$, then h is a mapping of E/R into F/S, and is said to be *induced by f on passing to the quotients.*

9. Let R be an equivalence relation on E, and let S be an equivalence relation on E/R. If f is the canonical mapping of E onto E/R, then

"$f(x) \equiv f(y) \pmod{S}$" is an equivalence relation T on E. An equivalence class with respect to T is therefore the union in E of equivalence classes with respect to R which are equivalent to each other with respect to S; and the relation "$x \equiv y \pmod{R}$" implies "$x \equiv y \pmod{T}$". If g and φ are the canonical mappings of E/R onto (E/R)/S and E onto E/T, respectively, we obtain a one-to-one correspondence (called *canonical*) between (E/R)/S and E/T by making an element of (E/R)/S and an element of E/T correspond to each other if they are the images of the same element of E under $g \circ f$ and φ, respectively.

Conversely, let R and T be two equivalence relations on E such that "$x \equiv y \pmod{R}$" implies "$x \equiv y \pmod{T}$". Then T is compatible (in the sense of no. 7) with R, both in x and in y; by passing to the quotient E/R (with respect to x and y), T induces an equivalence relation S on E/R. If f again denotes the canonical mapping of E onto E/R, the relation "$f(x) = f(y) \pmod{S}$" is equivalent to "$x \equiv y \pmod{T}$". The equivalence relation S is called the *quotient* of T by R, and is denoted by T/R. From the previous paragraph we see that there exists a one-to-one correspondence (the canonical correspondence) beween (E/R)/(T/R) and E/T.

10. Now let E, F be any two sets, which may or may not be distinct. Let $R\{x, y\}$ be an equivalence relation on E and let $S\{z, t\}$ be an equivalence relation on F. Then the relation "$R\{x, y\}$ and $S\{z, t\}$" between elements (x, z) and (y, t) of the product set $E \times F$ is an equivalence relation on $E \times F$, called the *product of* R *by* S, and denoted by $R \times S$. Every equivalence class with respect to $R \times S$ is the product of an equivalence class with respect to R and an equivalence class with respect to S. If u denotes a generic element of E/R, and v a generic element of F/S, then $(u, v) \to u \times v$ is a *bijection* (called *canonical*) of $(E/R) \times (F/S)$ onto $(E \times F)/(R \times S)$.

6. ORDERED SETS

1. A relation $\omega\{x, y\}$ between two generic elements x, y of a set E is said to be an *order relation* on E, if it satisfies the following two conditions:

(a) The relation "$\omega\{x, y\}$ and $\omega\{y, z\}$" *implies* $\omega\{x, z\}$ (transitivity).

(b) The relation "$\omega\{x, y\}$ and $\omega\{y, x\}$" is *equivalent* to "$x = y$".

Condition (b) implies that ω is reflexive.

Let C be the subset of $E \times E$ defined by the relation $\omega\{x, y\}$ as a property of the pair (x, y). Then conditions (a) and (b) are respectively equivalent to the following: (a′) $C \circ C \subset C$; (b′) $C \cap \overset{-1}{C} = \Delta$. These imply that $C \circ C = C$.

When we consider a particular order relation on a set E, we say that E is *ordered* by this relation, and that the relation defines an *order structure* (cf. §8) (or an *ordering*) on E.

If $\omega\{x, y\}$ is an order relation on E, then so is $\omega\{y, x\}$. These two order relations, and the orderings they define, are said to be *opposites* of each other.

Let $\omega\{x, y\}$ be an order relation on E, and let A be any subset of E. The relation $\omega\{x, y\}$ between two generic elements x, y of A is then an order relation *on* A, and the ordering it defines on A is said to be *induced* by the ordering defined by $\omega\{x, y\}$ on E. The given ordering on E is said to be an *extension* of the induced ordering on A.

A *reflexive* and *transitive* relation $\varpi\{x, y\}$ between two generic elements x, y of E is called a *preorder relation* on E. The relation "$\varpi\{x, y\}$ and $\varpi\{y, x\}$" is an *equivalence relation* R on E, and $\varpi\{x, y\}$ is compatible (in x and y) with this relation. Passing to the quotient (with respect to x and y), $\varpi\{x, y\}$ induces on the set E/R an *order relation*, said to be *associated* with $\varpi\{x, y\}$. A set endowed with a preorder relation is called a *preordered set.*

2. The inclusion relation "$Y \subset X$" is an order relation on the set of subsets $\mathfrak{P}(E)$ of any set E.

If E and F are two sets, which may or may not be distinct, the relation "g extends f" is an order relation on the set of all mappings of subsets of E into F.

The set **N** of natural integers (*) is ordered by the relation "$x \leqslant y$".

3. By analogy with this last example, when a set E is ordered by a relation $\omega\{x, y\}$, it is often convenient to denote the relation $\omega\{x, y\}$ by $x \leqslant y$, or $y \geqslant x$; these relations are read "x is *less* than y", or "y is *greater* than x" (†). The relations "$x < y$" and "$y > x$" (read "x is *strictly less* than y", or "y is *strictly greater* than x") are by definition equivalent to "$x \leqslant y$ and $x \neq y$".

The relation "$x \leqslant y$" is *equivalent* to "$x < y$ or $x = y$". The relation "$x \leqslant y$ and $y < z$" implies "$x < z$"; similarly, "$x < y$ and $y \leqslant z$" implies "$x < z$".

(*) In accordance with our point of view in this Summary of Results, we assume the theory of integers as known. But it should not be thought that this theory is necessary for building up the theory of sets; the reader will see, by referring to Chapter III that, on the contrary, the integers can be defined, and all their known properties proved, from the results of the theory of sets.

In our terminology, 0 belongs to **N**.

(†) Thus, in our terminology, "less than" and "greater than" do *not* exclude "equal to".

4. A subset X of a set E, ordered by a relation "$x \leqslant y$", is said to be *totally ordered* by this relation if, for all $x \in X$ and all $y \in X$, we have either $x \leqslant y$ or $y \leqslant x$ (or, equivalently, either $x < y$ or $x = y$ or $x > y$, these three relations being mutually exclusive).

The empty subset of an ordered set is always totally ordered. It may happen that the whole set E is totally ordered, as is the case for the set **N** and the relation $x \leqslant y$.

Every subset of a totally ordered set is totally ordered by the induced ordering.

In an ordered set E, if a and b are two elements such that $a \leqslant b$, the subset of E consisting of the elements x such that $a \leqslant x \leqslant b$ is called the *closed interval with left-hand endpoint a and right-hand endpoint b*, and is denoted by $[a, b]$. The set of all $x \in E$ such that $a \leqslant x < b$ (resp. $a < x \leqslant b$) is called the *interval half-open on the right* (resp. *on the left*) with endpoints a and b, and is denoted by $[a, b[$ (resp. $]a, b]$). The set of all $x \in E$ such that $a < x < b$ is called the *open interval* with endpoints a and b, and is denoted by $]a, b[$.

The set of all $x \in E$ such that $x \leqslant a$ (resp. $x < a$) is called the *closed* (resp. *open*) *interval unbounded on the left, with right-hand endpoint a*, and is denoted by $]\leftarrow, a]$ (resp. $]\leftarrow, a[$); likewise, the set of all $x \in E$ such that $x \geqslant a$ (resp. $x > a$) is called the closed (resp. *open*) *interval unbounded on the right, with left-hand endpoint a*, and is denoted by $[a, \rightarrow[$ (resp. $]a, \rightarrow[$). Finally, we consider E itself as an *open interval unbounded in both directions*, and as such we denote it by $]\leftarrow, \rightarrow[$.

5. If X is a subset of an ordered set E, there is at most one element a of X such that $a \leqslant x$ for all $x \in X$; when there exists an element a with this property, it is called *the least element of* X. Likewise, there exists at most one element b such that $x \leqslant b$ for all $x \in X$; b, if it exists, is called *the greatest element of* X.

In a *totally ordered* set E, every *finite* non-empty subset has a greatest element and a least element, called the *maximum* and *minimum* elements respectively of the subset.

An ordered set in which every non-empty subset has a least element is said to be *well-ordered*. The set **N** of natural integers is well-ordered; a subset of **N** has a greatest element if and only if it is finite and non-empty. It is shown, by use of the axiom of choice, that on every set there exists a well-ordering (*Zermelo's theorem*).

In the set $\mathfrak{P}(E)$ of subsets of a set E, ordered by inclusion, a subset $\mathfrak{F}$ of $\mathfrak{P}(E)$ has a least element if and only if the intersection of the sets of $\mathfrak{F}$ belongs to $\mathfrak{F}$, and this intersection is then the least element. Similarly, $\mathfrak{F}$ has a greatest element if and only if the union of the sets of $\mathfrak{F}$ belongs to $\mathfrak{F}$, and this union is then the greatest element.

A subset X of an ordered set E is said to be *cofinal* (resp. *coinitial*) in E if, for each $y \in E$, there exists $x \in X$ such that $y \leqslant x$ (resp. $y \geqslant x$). To say that an ordered set E has a greatest (resp. least) element means that there exists a cofinal (resp. coinitial) subset of E consisting of a single element.

6. Let X be a subset of an ordered set E. Every $x \in X$ such that there exists no element $z \in X$ for which $z < x$ is called a *minimal* element of X. Every $y \in X$ such that there exists no element $z \in X$ for which $z > y$ is called a *maximal* element of X. The set of maximal elements (or the set of minimal elements) may be empty; it may also be infinite. If X has a least element a, then a is the *only* minimal element of X; likewise, if X has a greatest element b, then b is the *only* maximal element of X.

7. Let X be a subset of an ordered set E. If an element $x \in E$ is such that $z \leqslant x$ for all $z \in X$, then z is called *an upper bound* of X. Similarly, an element $y \in E$ such that $z \geqslant y$ for all $z \in X$ is called *a lower bound* of X.

The set of upper bounds (or the set of lower bounds) of a subset X may be empty. A subset X whose set of upper (resp. lower) bounds is not empty is said to be *bounded above* (resp. *bounded below*). A set which is bounded both above and below is said simply to be *bounded*. Every element greater than an upper bound of X is an upper bound of X; every element less than a lower bound of X is a lower bound of X.

If the set of upper bounds of a subset X has a least element a, then a is called the *least upper bound* or *supremum* of X. If the set of lower bounds of X has a greatest element b, then b is called the *greatest lower bound* or *infimum* of X. If these bounds exist, they are unique by definition, and they are denoted respectively by $\sup_E X$ (or $\sup X$), $\inf_E X$ (or $\inf X$). If X has a greatest element, it is its least upper bound; if X has a least element, it is its greatest lower bound. Conversely, if the least upper bound (resp. greatest lower bound) of X exists and belongs to X, it is the greatest (resp. least) element of X.

Let f be a mapping of a set A into E. If $f(A)$ is bounded above (resp. bounded below, bounded) in E, then f is said to be *bounded above* (resp. *bounded below, bounded*). If $f(A)$ has a least upper bound (resp. greatest lower bound) in E, this bound is called the *least upper bound* (resp. *greatest lower bound*) of f and is denoted by $\sup_{x \in A} f(x)$ (resp. $\inf_{x \in A} f(x)$).

8. A preordered set E in which every *finite* non-empty subset of E is *bounded above* (resp. *bounded below*) is said to be *right directed* (or *directed*) (resp. *left directed*).

An ordered set E in which every *finite* non-empty subset of E has a least upper bound and a greatest lower bound is called a *lattice*.

The set of subsets of any set, ordered by inclusion, is a lattice. Every totally ordered set is a lattice.

9. An ordered set E is said to be *inductive* if it satisfies the following condition : *every totally ordered subset of* E *has an upper bound.*

The set $\mathfrak{P}(E)$, ordered by inclusion, is inductive. So is the set of mappings of subsets of a set E into a set F when ordered by the relation "g extends f" between f and g.

An arbitrary subset of an inductive set is not in general inductive. But if a is any element of an inductive set E, then the subset of E consisting of all elements $x \in E$ such that $x \geqslant a$ is inductive.

10. The following proposition is proved with the help of the axiom of choice, and is known as *Zorn's lemma :*

Every inductive ordered set has at least one maximal element.

11. In the set of subsets $\mathfrak{P}(E)$ of a set E, ordered by inclusion, the least upper bound of a set $\mathfrak{F}$ of subsets of E is the *union* of the sets of $\mathfrak{F}$. Application of Zorn's lemma gives the following result :

If $\mathfrak{F}$ *is a set of subsets of a set* E *such that, for each subset* $\mathfrak{G}$ *of* $\mathfrak{F}$ *which is totally ordered by the relation of inclusion, the union of the sets of* $\mathfrak{G}$ *belongs to* $\mathfrak{F}$, *then* $\mathfrak{F}$ *has at least one maximal element* (that is, a subset of E which belongs to $\mathfrak{F}$ but is not contained in any other subset of E belonging to $\mathfrak{F}$).

A set $\mathfrak{F}$ of subsets of a set E is said to be *of finite character* if the property "$X \in \mathfrak{F}$" is *equivalent* to the property "every *finite* subset of X belongs to $\mathfrak{F}$". With this definition, we have the following theorem :

Every set of subsets of E *of finite character has at least one maximal element.*

12. A mapping f of a subset A of an ordered set E into an ordered set F is said to be *increasing* (resp. *decreasing*) if the relation $x \leqslant y$ between generic elements of A implies $f(x) \leqslant f(y)$ (resp. $f(y) \leqslant f(x)$). Every constant function on A is thus both increasing and decreasing, and the converse is true if A is (right or left) directed.

A mapping f is said to be *strictly increasing* (resp. *strictly decreasing*) if the relation $x < y$ implies $f(x) < f(y)$ (resp. $f(x) < f(y)$).

If E is *totally ordered*, every *strictly increasing* (or *strictly decreasing*) mapping of a subset A of E into an ordered set F is *injective.*

If I is an *ordered* set of indices, a family $(X_\iota)_{\iota \in I}$ of subsets of a set E is said to be *increasing* (resp. *decreasing*) if the mapping $\iota \rightarrow X_\iota$ of I into $\mathfrak{P}(E)$, ordered by inclusion, is increasing (resp. decreasing).

13. Let I be a directed set, and let $(E_\alpha)_{\alpha \in I}$ be a family of sets indexed by I. For each pair (α, β) of indices in I such that $\alpha \leqslant \beta$, let $f_{\beta\alpha}$ be a mapping *of* E_α *into* E_β, and suppose that the relations $\alpha \leqslant \beta \leqslant \gamma$ imply $f_{\gamma\alpha} = f_{\gamma\beta} \circ f_{\beta\alpha}$.

Let F be the sum of the family of sets $(E_\alpha)_{\alpha \in I}$. By abuse of language, we shall identify the E_α with the corresponding subsets of F. Given two elements x and y in F, let $R\{x, y\}$ be the following relation (where α

and β denote the elements of I such that $x \in E_\alpha$ and $y \in E_\beta$) : "there exists $\gamma \in I$ such that $\gamma \geqslant \alpha$ and $\gamma \geqslant \beta$ and $f_{\gamma\alpha}(x) = f_{\gamma\beta}(y)$".

Then R is an equivalence relation on F. Let E be the quotient set F/R, and let f be the canonical mapping $F \to F/R$. The set E is called the *direct limit of the family* $(E_\alpha)_{\alpha \in I}$ *with respect to the family of mappings* $(f_{\beta\alpha})$, and the restriction f_α of f to E_α is called the *canonical mapping of* E_α *into* E. We have $f_\beta \circ f_{\beta\alpha} = f_\alpha$ whenever $\alpha \leqslant \beta$. We write

$$E = \varinjlim (E_\alpha, f_{\beta\alpha})$$

or simply $E = \varinjlim E_\alpha$ when there is no risk of confusion. By abuse of language, the pair $((E_\alpha), (f_{\beta\alpha}))$ is called a *direct system of sets relative to* I.

If the $f_{\beta\alpha}$ are injective, then the f_α are injective. In this case we shall usually identify E_α and $f_\alpha(E_\alpha)$, and thus consider E as the *union* of the E_α. Conversely, if a set E' is the union of a family $(E'_\alpha)_{\alpha \in I}$ of subsets such that the relation $\alpha \leqslant \beta$ implies $E'_\alpha \subset E'_\beta$, and if (for $\alpha \leqslant \beta$) $j_{\beta\alpha}$ denotes the canonical injection of E'_α into E'_β, then we may identify $\varinjlim (E'_\alpha, j_{\alpha\beta})$ with E', and the canonical mappings of the E'_α into $\varinjlim (E'_\alpha, j_{\beta\alpha})$ with the canonical injections of the E'_α into E.

More generally, let $(E_\alpha, f_{\beta\alpha})$ be a direct system of sets relative to I, and for each $\alpha \in I$ let g_α be a mapping of E_α into a set E' such that the relation $\alpha \leqslant \beta$ implies $g_\beta \circ f_{\beta\alpha} = g_\alpha$. Then there exists a unique mapping g of $E = \varinjlim E_\alpha$ into E' such that $g_\alpha = g \circ f_\alpha$ for all $\alpha \in I$. The mapping g is surjective if and only if E' is the union of the $g_\alpha(E_\alpha)$. The mapping g is injective if and only if, for each $\alpha \in I$, the relations $x \in E_\alpha$, $y \in E_\alpha$, $g_\alpha(x) = g_\alpha(y)$ imply that there exists $\beta \geqslant \alpha$ such that $f_{\beta\alpha}(x) = f_{\beta\alpha}(y)$. If g is bijective, E' is sometimes identified with the direct limit of the E_α.

Let $(A_\alpha, \varphi_{\beta\alpha})$ and $(B_\alpha, \psi_{\beta\alpha})$ be two direct systems of sets relative to the same index set I. Let $A = \varinjlim (A_\alpha, \varphi_{\beta\alpha})$, $B = \varinjlim (B_\alpha, \psi_{\beta\alpha})$, and for each $\alpha \in I$ let φ_α (resp. ψ_α) denote the canonical mapping of A_α into A (resp. of B_α into B). For each $\alpha \in I$ let u_α be a mapping of A_α into B_α such that $u_\beta \circ \varphi_{\beta\alpha} = \psi_{\beta\alpha} \circ u_\alpha$ whenever $\alpha \leqslant \beta$. The family (u_α) is called a *direct system of mappings of* $(A_\alpha, \varphi_{\beta\alpha})$ *into* $(B_\alpha, \psi_{\beta\alpha})$. Under these conditions there exists a unique mapping $u: A \to B$ such that $u \circ \varphi_\alpha = \psi_\alpha \circ u_\alpha$ for all $\alpha \in I$. The mapping u is called the *direct limit* of the u_α, and is written $u = \varinjlim u_\alpha$, provided that there is no risk of confusion. Let $(C_\alpha, \theta_{\beta\alpha})$ be another direct system of sets relative to I, let (v_α) be a direct system of mappings of $(B_\alpha, \psi_{\beta\alpha})$ into $(C_\alpha, \theta_{\beta\alpha})$, and let $v = \varinjlim v_\alpha$. Then $\varinjlim (v_\alpha \circ u_\alpha) = v \circ u$.

Keeping the above notation, let $D_\alpha = A_\alpha \times B_\alpha$ and $\omega_{\beta\alpha} = \varphi_{\beta\alpha} \times \psi_{\beta\alpha}$. Then the family $(D_\alpha, \omega_{\beta\alpha})$ is a direct system of sets. Let $D = \varinjlim (D_\alpha, \omega_{\beta\alpha})$, let ω_α be the canonical mapping of D_α into D, let $D' = A \times B$, and let $\omega'_\alpha = \varphi_\alpha \times \psi_\alpha$. Then there exists a unique bijection (called *canonical*) $f: D \to D'$ such that $f \circ \omega_\alpha = \omega'_\alpha$ for all $\alpha \in I$. We shall usually identify the product D' of the direct limits with the direct limit D of the products D_α.

Let J be a cofinal subset of I; then J is also directed. If $((E_\alpha)_{\alpha \in I}, (f_{\beta\alpha})_{\alpha \in I, \beta \in I})$ is a direct system of sets relative to I, then $((E_\alpha)_{\alpha \in J}, (f_{\beta\alpha})_{\alpha \in J, \beta \in J})$ is a direct system of sets relative to J. Let E' be its direct limit, and let f'_α be the canonical mapping of E_α into E' for all $\alpha \in J$. Then there exists a unique mapping $g: E \to E'$ such that $g(f'_\alpha(x)) = f_\alpha(x)$ for all $\alpha \in J$ and all $x \in E_\alpha$ (where f_α denotes the canonical mapping of E_α into E). This mapping is a *bijection*, by means of which E' is usually identified with E.

14. Let I be a preordered set and let $(E_\alpha)_{\alpha \in I}$ be a family of sets indexed by I. For each pair (α, β) of indices of I such that $\alpha \leqslant \beta$, let $f_{\alpha\beta}$ be a mapping *of* E_β *into* E_α, and suppose that the relations $\alpha \leqslant \beta \leqslant \gamma$ imply $f_{\alpha\gamma} = f_{\alpha\beta} \circ f_{\beta\gamma}$.

Let G be the product of the family of sets $(E_\alpha)_{\alpha \in I}$. Let E be the subset of G consisting of all elements x which satisfy *all* the relations $\mathrm{pr}_\alpha x = f_{\alpha\beta}(\mathrm{pr}_\beta x)$, for every pair of indices α, β such that $\alpha \leqslant \beta$. The set E is called the *inverse limit of the family* $(E_\alpha)_{\alpha \in I}$ *with respect to the family of mappings* $(f_{\alpha\beta})$, and the restriction f_α of pr_α to E is called the *canonical mapping of* E *into* E_α. We have $f_\alpha = f_{\alpha\beta} \circ f_\beta$ whenever $\alpha \leqslant \beta$. We write $E = \varprojlim (E_\alpha, f_{\alpha\beta})$, or simply $E = \varprojlim E_\alpha$ when there is no risk of confusion. By abuse of language, the pair $((E_\alpha), (f_{\alpha\beta}))$ is called an *inverse system of sets relative to* I.

It should be mentioned that E can be empty, even when all the E_α are non-empty and all the mappings $f_{\alpha\beta}$ are surjective.

For each $\alpha \in I$ let g_α be a mapping of a set E' into E_α such that the relation $\alpha \leqslant \beta$ implies $f_{\alpha\beta} \circ g_\beta = g_\alpha$. Then there exists a unique mapping g of E' into E such that $g_\alpha = f_\alpha \circ g$ for all $\alpha \in I$. For g to be injective it is necessary and sufficient that for each pair of distinct elements x', y' of E', there should exist $\alpha \in I$ such that $g_\alpha(x') \neq g_\alpha(y')$.

Let $(A_\alpha, \varphi_{\alpha\beta})$ and $(B_\alpha, \psi_{\alpha\beta})$ be two inverse systems of sets relative to the same index set I. Let $A = \varprojlim (A_\alpha, \varphi_{\alpha\beta})$, $B = \varprojlim (B_\alpha, \psi_{\alpha\beta})$, and for each $\alpha \in I$ let φ_α (resp. ψ_α) be the canonical mapping of A into A_α (resp. of B into B_α). For each $\alpha \in I$ let u_α be a mapping of A_α into B_α such that $\psi_{\alpha\beta} \circ u_\beta = u_\alpha \circ \varphi_{\alpha\beta}$ whenever $\alpha \leqslant \beta$. The family (u_α) is called an *inverse system of mappings of* $(A_\alpha, \varphi_{\alpha\beta})$ *into* $(B_\alpha, \psi_{\alpha\beta})$. Under these conditions there exists a unique mapping $u: A \to B$ such that $\psi_\alpha \circ u = u_\alpha \circ \varphi_\alpha$ for all $\alpha \in I$. The mapping u is called the *inverse limit* of the u_α, and is written $u = \varprojlim u_\alpha$ when there is no risk of confusion. Let $(C_\alpha, \theta_{\alpha\beta})$ be another inverse system of sets relative to I, let v_α be an inverse system of mappings of $(B_\alpha, \psi_{\alpha\beta})$ into $(C_\alpha, \theta_{\alpha\beta})$, and let $v = \varprojlim v_\alpha$. Then we have $\varprojlim (v_\alpha \circ u_\alpha) = v \circ u$.

Let J be a cofinal subset of I, and suppose that J is *directed* (with respect to the ordering induced from I). If $((E_\alpha)_{\alpha \in I}, (f_{\alpha\beta})_{\alpha \in I, \beta \in I})$ is an inverse system of sets relative to I, with inverse limit E, then $((E_\alpha)_{\alpha \in J}, (f_{\alpha\beta})_{\alpha \in J, \beta \in J})$ is also an inverse system of sets, relative to J. Let E′ be its inverse limit and let f'_α be the canonical mapping of E′ into E_α, where $\alpha \in J$. For each $x \in E$ let $g(x) = (f_\alpha(x))_{\alpha \in J} \in E'$ (where f_α denotes the canonical mapping of E into E_α). Then g is a *bijection* of E onto E′, by means of which E′ is usually identified with E.

7. POWERS. COUNTABLE SETS

1. Two sets E, F are said to be *equipotent* if they can be put in one-to-one correspondence.

Two sets, each equipotent to a third, are themselves equipotent.

If E and F are equipotent, then $\mathfrak{P}(E)$ and $\mathfrak{P}(F)$ are equipotent.

If E and F, E′ and F′, E″ and F″ are respectively equipotent, then $E \times E' \times E''$ and $F \times F' \times F''$ are equipotent. This proposition extends to the product of any number of sets.

2. Let X and Y be two generic subsets of a set E. The relation "X and Y are equipotent" is an *equivalence relation* on $\mathfrak{P}(E)$. The equivalence class (with respect to this relation) to which X belongs is called the *power* (*) of X, and the set of these classes (i.e., the quotient of $\mathfrak{P}(E)$ by the above relation) is called the *set of powers* of subsets of E.

If E and F are two distinct sets, the relation "X and Y are equipotent" between a subset X of E and a subset Y of F is expressed by saying that the power of X and the power of Y are *equivalent*. In this way we have a one-to-one correspondence between a subset of the set of powers of subsets of E and a subset of the set of powers of subsets of F.

(*) In formalized set theory (cf. Chapter III, § 3) we define the notion of the *cardinal* of a set, which we also call (by abuse of language) the *power* of the set. This abuse of language does not, however, cause any confusion, because two subsets of a set have the same power (in the sense defined above) if and only if they have the same cardinal; likewise, the power of a subset A of a set E is less than the power of a subset B of a set F (no. 3) if and only if the cardinal of A is less than that of B; and finally, if the power of A is the sum (no. 5) of the powers of a family (A_ι) of subsets of E, then the cardinal of A is the sum of the cardinals of the A_ι.

3. Let E, F be any two sets, which may or may not be distinct, and let $\mathfrak{a}$ (resp. $\mathfrak{b}$) be an element of the set of powers of subsets of E (resp. F). Then $\mathfrak{a}$ is said to be *less than* $\mathfrak{b}$, or $\mathfrak{b}$ *greater than* $\mathfrak{a}$, if there exists a one-to-one mapping of a subset $X \subset E$ with power $\mathfrak{a}$ *into* a subset $Y \subset F$ with power $\mathfrak{b}$. If also $\mathfrak{a}$ and $\mathfrak{b}$ are not equivalent powers, than $\mathfrak{a}$ is said to be *strictly less than* $\mathfrak{b}$, or $\mathfrak{b}$ *strictly greater than* $\mathfrak{a}$.

If $\mathfrak{a}$ and $\mathfrak{b}$ are equivalent, then $\mathfrak{a}$ is both greater and less than $\mathfrak{b}$. Conversely, it is a theorem that *if* $\mathfrak{a}$ *is both greater and less than* $\mathfrak{b}$, *then* $\mathfrak{a}$ *and* $\mathfrak{b}$ *are equivalent.* It follows in particular that the set of powers of subsets of a set E is *ordered* by the relation "$\mathfrak{a}$ is less than $\mathfrak{b}$"; whenever we speak of this set as an ordered set, it is always the ordering defined by this relation that we mean.

Furthermore, using Zorn's lemma (and hence the axiom of choice), it is shown that *the set of powers of subsets of a set* E *is well-ordered.*

4. The power of a set E is *strictly less than* that of the set $\mathfrak{P}(E)$.

If f is a mapping of a set E into a set F, then the power of the image $f(X)$ of any subset X of E is *less than* the power of X.

5. Let $(X_\iota)_{\iota \in I}$ be a family of subsets of a set E such that $X_\iota \cap X_\varkappa = \emptyset$ whenever $\iota \neq \varkappa$. Let $(Y_\iota)_{\iota \in I}$ be a family of subsets of a set F, indexed by the same set I and such that the power of Y_ι is *less than* that of X_ι for all $\iota \in I$. Then the power of the union $\bigcup_{\iota \in J} Y_\iota$ is *less than* that of $\bigcup_{\iota \in J} X_\iota$ for all subsets J of I.

If also $Y_\iota \cap Y_\varkappa = \emptyset$ whenever $\iota \neq \varkappa$, and if X_ι and Y_ι are *equipotent* for all $\iota \in I$, then $\bigcup_{\iota \in J} X_\iota$ is *equipotent* to $\bigcup_{\iota \in J} Y_\iota$.

In particular, if F and E are the same set, we see that the power of the union of a set of *pairwise disjoint* subsets of E depends only on the powers of these subsets; it is called the *sum* of these powers (thus this function is defined for a family $(\mathfrak{a}_\iota)$ of elements of the set of powers only if there exists a family (X_ι) of pairwise disjoint subsets of E such that X_ι has power $\mathfrak{a}_\iota$ for each index ι).

If $(X_\iota)_{\iota \in I}$ and $(Y_\iota)_{\iota \in I}$ are families of subsets of E and F, respectively, indexed by the same set I such that X_ι and Y_ι are *equipotent* for all ι, then the products $\prod_\iota X_\iota$ and $\prod_\iota Y_\iota$ are *equipotent.*

6. The set **N** of natural integers may be considered as the set of powers of *finite subsets* of an *infinite* set. The order relation "$x \leqslant y$" on **N** is just the relation ordering this set of powers, and the *sum* of two natural integers is a function identical with the sum of two powers as defined above.

7. A set is said to be *countable* if it is equivalent to a subset of the set **N** of natural integers. Every *finite* set is therefore countable; if it has n elements, it is equipotent to the interval $[0, n-1]$ of the set **N**. Every

countable infinite set is equipotent to **N**; in particular, every infinite subset of **N** has the same power as **N**.

If E is an *infinite* set, there exists a *partition* of E into *countable infinite* sets; in particular, every infinite set has a power *greater than* that of **N**.

If E is an *infinite* set, the sets $E \times E$ and $E \times \mathbf{N}$ are both *equipotent* to E, and the set of *finite subsets* of E is *equipotent* to E. In particular, $\mathbf{N} \times \mathbf{N}$ is a *countable infinite* set.

8. A *sequence of elements* of a set E is by definition a family of elements of E, indexed by the set **N** or a subset of **N**. A sequence whose index set is **N** is therefore written $(x_n)_{n \in \mathbf{N}}$, or more simply (x_n) when there is no likelihood of confusion. If n denotes a generic integer, x_n is said to be the *general term* of the sequence, or the *nth term*. The latter terminology is also used when n is replaced by a particular integer. The set of elements of a sequence is countable.

A sequence is said to be *infinite* or *finite* according as the index set is an infinite or finite subset of **N**. The set of elements of a finite sequence is finite.

Every subfamily of a sequence is again a sequence, called a *subsequence* of the given sequence. Every subsequence of a finite sequence is a finite sequence.

A family of elements whose index set is $\mathbf{N} \times \mathbf{N}$, or a subset of $\mathbf{N} \times \mathbf{N}$, is called a *double sequence*. A double sequence indexed by $\mathbf{N} \times \mathbf{N}$ is written $(x_{m,n})$, or more simply (x_{mn}) if there is no risk of confusion. Similarly for sequences with more than two indices.

Two sequences (x_n), (y_n) are said *to differ only in the order of their terms* if there exists a permutation f of the index set such that $y_n = x_{f(n)}$ for all n.

With any family of elements $(x_\iota)_{\iota \in I}$ whose index set I is countably infinite we may associate an infinite sequence, as follows: there exists a bijection $n \to f(n)$ of **N** onto I; putting $y_n = x_{f(n)}$, the sequence (y_n) is said to be obtained by *ranging the family* (x_ι) *in the order defined by* f. Thus the sequences corresponding to two distinct bijections of **N** onto I differ only in the order of their terms.

Operating in the same way when I is *finite*, we obtain a *finite sequence* associated with the family (x_ι).

9. The union, intersection, and product of a family $(X_\iota)_{\iota \in I}$ of subsets of a set E are said to be *countable* if I is a countable set, *finite* if I is finite.

If I is *countable*, and if the power of X_ι is *less than* a given infinite power $\mathfrak{a}$ for all $\iota \in I$, then the power of the union $\bigcup_\iota X_\iota$ is *less than* $\mathfrak{a}$. If also at least one of the X_ι has power $\mathfrak{a}$, then $\bigcup_\iota X_\iota$ has power $\mathfrak{a}$. In particular, every countable union of sets of power $\mathfrak{a}$ also has power $\mathfrak{a}$; every countable union of countable sets is a countable set.

8. SCALES OF SETS. STRUCTURES

1. Given, for example, three *distinct* sets E, F, G, we may form other sets from them by taking their sets of subsets, or by forming the product of one of them by itself, or again by forming the product of two of them taken in a certain order. In this way we obtain *twelve* new sets. If we add these to the three original sets E, F, G, we may repeat the same operations on these fifteen sets, omitting those which give us sets already obtained; and so on. In general, any one of the sets obtained by this procedure (according to an explicit scheme) is said to belong to the *scale of sets on* E, F, G *as base*.

For example, let M, N, P be three sets of this scale, and let $R\{x, y, z\}$ be a relation between generic elements x, y, z of M, N, P, respectively. Then R defines a subset of $M \times N \times P$, hence (via a canonical correspondence) a subset of $(M \times N) \times P$, i.e., an element of $\mathfrak{P}((M \times N) \times P)$. Thus to give a *relation* between elements of several sets in the same scale is the same as to give an *element* of another set in the scale. Likewise, to give a mapping of M into N, for example, amounts (by considering the graph of this mapping) to giving a subset of $M \times N$, i.e., an element of $\mathfrak{P}(M \times N)$, which is again a set in the scale. Finally, to give two elements (for example) of M amounts to giving a single element in the product set $M \times M$.

Thus being given a certain number of elements of sets in a scale, relations between generic elements of these sets, and mappings of subsets of certain of these sets into others, all comes down in the final analysis to being given a *single element* of one of the sets in the scale.

2. We said earlier (§6) that an element C of the set $\mathfrak{P}(E \times E)$ defines an order structure on E if it has the properties (a) $C \circ C \subset C$ and (b) $C \cap \overset{-1}{C} = \Delta$.

In general, consider a set M in a scale of sets whose base consists, for the sake of example, of three sets E, F, G. Let us give ourselves a certain number of explicitly stated properties of a generic element of M, and let T be the intersection of the subsets of M defined by these properties. An element σ of T is said to define a *structure* of the *species* T on E, F, G. The structures of species T are therefore characterized by the scheme of formation of M from E, F, G, and by the properties defining T, which are called the *axioms* of these structures. We give a specific name to all the structures of the same species. Every proposition which is a consequence of the proposition "$\sigma \in T$" (i.e., of the axioms defining T) is said to belong to the *theory* of the structures of species T; for example, the propositions stated in § 6 belong to the theory of structures of ordered sets.

In the last example, the axioms may be stated for a completely arbitrary base set E. Hence we give the same name to the structures which satisfy these axioms, independently of the set on which they are defined; and the propositions deduced from these axioms are valid in any set, because their formulation does not involve any special properties of the set E. Such remarks apply whenever the axioms are of this nature (*).

Most often when a scale is used with a base consisting of several sets E, F, G, one of these sets, say E, plays a preponderant role in the structures under considerations. Therefore, by abuse of language, these structures are said to be defined in the set E, with F and G considered as auxiliary sets.

Finally, to simplify the language, a particular name is often given to a set which has been endowed with a structure of a definite species. Thus we speak of an *ordered set*, and in later parts of this series we shall define the notions of *group*, *ring*, *field*, *topological space*, *uniform space*, etc., all of which are words denoting sets endowed with certain structures.

3. Consider the structures of the same species T, where T is a subset of a set M in a scale of sets. If we adjoin new "axioms" to those which define T, the system of axioms thus obtained defines a subset U of M, contained in T. The structures of the species U are said to be *richer* than the structures of the species T. For example, the structures of *totally ordered* sets are richer than the structures of ordered sets, because the element C of $\mathfrak{P}(E \times E)$ which defines such a structure has to satisfy the additional axiom $C \cup \overset{-1}{C} = E \times E$.

4. Let M, M′ be two sets in the same scale, say with E, F, G as base. Let T be a subset of M and T′ a subset of M′, each defined by certain explicitly stated axioms. Whenever we can define explicitly a *one-to-one mapping of* T *onto* T′, we consider two elements $\sigma \in T$, $\sigma' \in T'$, which correspond to each other under this mapping, as defining the *same* structure on E, F, G; and the systems of axioms which define T and T′ are said to be *equivalent*.

The *topological* structures provide an example of this situation; they can be defined by means of several equivalent systems of axioms, two of which systems are particularly useful (see *General Topology*, Chapter I, § 1).

5. Let E, F, G be three sets, and suppose we are given *bijective* mappings of E, F, G onto three other sets E′, F′, G′, respectively. Since we know how to define the *canonical extensions* of bijective mappings to sets of subsets (§ 2, no. 9) and to product sets (§ 3, no. 14), we can define, step by step,

(*) The reader may have observed that the indications given here are left rather vague; they are not intended to be other than heuristic, and indeed it seems scarcely possible to state general and precise definitions for structures outside of the framework of formal mathematics (see Chapter IV).

the *extension* of the given bijections to two sets M, M′ constructed respectively according to the *same* scheme in the scale of sets based on E, F, G, and that based on E′, F′, G′. Let f be the bijection of M onto M′ so obtained. If σ is a structure on E, F, G which is an element of a subset T of M, we say that $f(\sigma)$ is the structure obtained by *transporting* the structure σ onto E′, F′, G′ by means of the given bijections of E onto E′, F onto F′, G onto G′. Every proposition relating to the structure σ on E, F, G gives rise (by use of appropriate extensions) to a proposition relating to the structure $f(\sigma)$ on E′, F′, G′.

Conversely, a structure σ on E, F, G and a structure σ' on E′, F′, G′ are said to be *isomorphic* if σ' can be obtained by *transporting* σ by means of bijections of E, F, G onto E′, F′, G′, respectively; these mappings are then said to constitute an *isomorphism* of σ onto σ'.

When we are concerned with structures on a single set E, the bijection of E onto E′ which transports σ into σ' is also called an *isomorphism of the set* E, *endowed with the structure* σ, *onto the set* E′, *endowed with the structure* σ'.

This mapping is also called an isomorphism when F and G are two auxiliary sets, and the bijections for these two sets are the *identity* mappings of F and G onto themselves.

An isomorphism of a set E, endowed with a structure σ, onto itself is called an *automorphism*.

When there exists an isomorphism of a set E, endowed with a structure σ, onto a set E′, endowed with a structure σ', it is often convenient to *identify* E with E′, i.e., to give *the same name* to an element of a set M in the scale based on E and to the element which is its image under the appropriate extension of f to the set M.

6. Given a system of axioms defining a subset T of a set M in a scale of sets, we should make sure, before speaking of the structures which satisfy these axioms, that the set T *is not necessarily empty;* otherwise the axioms would be said to be *contradictory* or *inconsistent*.

7. It may happen that a system of axioms defining a structure on a set can be stated for an arbitrary set, but that when we consider two structures satisfying these axioms and defined on two distinct sets E, F, we find from the axioms that these structures (if they exist) are necessarily *isomorphic* (which implies in particular that E and F are *equipotent*). Then the theory of the structures satisfying these axioms is said to be *univalent*; otherwise they are said to be *multivalent*.

The theory of integers, the theory of real numbers, and classical Euclidean geometry are univalent theories; the theory of ordered sets, group theory, and topology are multivalent theories. The study of multivalent theories is the most striking feature which distinguishes modern mathematics from classical mathematics.

INDEX OF NOTATION

The reference numbers indicate the chapter, section, and subsection, in that order. (R refers to the Summary of Results.)

INDEX OF TERMINOLOGY

The reference numbers indicate the chapter, section, and sub-section (or exercise) in that order. (R refers to Summary of Results.)

INDEX OF TERMINOLOGY

AXIOMS AND SCHEMES OF THE THEORY OF SETS

S1. If A is a relation, (A or A) $\Longrightarrow A$ is an axiom.
S2. If A and B are relations, the relation $A \Longrightarrow$ (A or B) is an axiom.
S3. If A and B are relations, the relation (A or B) $\Longrightarrow$ (B or A) is an axiom.
S4. If A, B, and C are relations, the relation

$$(A \Longrightarrow B) \Longrightarrow ((C \text{ or } A) \Longrightarrow (C \text{ or } B))$$

is an axiom.
S5. If R is a relation, T a term, and x a letter, the relation $(T|x)R \Longrightarrow (\exists x)R$ is an axiom.
S6. Let x be a letter, T and U terms, and $R\{x\}$ a relation. Then the relation

$$(T = U) \Longrightarrow (R\{T\} \Longleftrightarrow R\{U\})$$

is an axiom.
S7. Let R and S be relations and x a letter. Then the relation

$$((\forall x)(R \Longleftrightarrow S)) \Longrightarrow (\tau_x(R) = \tau_x(S))$$

is an axiom.
S8. Let R be a relation, x and y distinct letters, X and Y letters distinct from x and y which do not appear in R. Then the relation

$$(\forall y)(\exists X)(\forall x)(R \Longrightarrow (x \in X)) \Longrightarrow (\forall Y) \operatorname{Coll}_x ((\exists y)((y \in Y) \text{ and } R))$$

is an axiom.
A1. $(\forall x)(\forall y)((x \subset y \text{ and } y \subset x) \Longrightarrow (x = y))$.
A2. $(\forall x)(\forall y) \operatorname{Coll}_z (z = x \text{ or } z = y)$.
A3. $(\forall x)(\forall x')(\forall y)(\forall y')(((x, y) = (x', y')) \Longrightarrow (x = x' \text{ and } y = y'))$.
A4. $(\forall X) \operatorname{Coll}_Y (Y \subset X)$.
A5. There exists an infinite set.

Printing and Binding: Strauss GmbH, Mörlenbach